CONCENTRATION INEQUALITIES

Concentration Inequalities

A Nonasymptotic Theory of Independence

STÉPHANE BOUCHERON

GÁBOR LUGOSI

PASCAL MASSART

Great Clarendon Street, Oxford, OX2 6DP,
United Kingdom

Oxford University Press is a department of the University of Oxford.
It furthers the University's objective of excellence in research, scholarship,
and education by publishing worldwide. Oxford is a registered trade mark of
Oxford University Press in the UK and in certain other countries

First published 2013
First published in paperback 2016

Published in the United States of America by Oxford University Press
198 Madison Avenue, New York, NY 10016, United States of America

British Library Cataloguing in Publication Data
Data available

Library of Congress Cataloging in Publication Data
Data available

ISBN 978–0–19–953525–5 (Hbk.)
ISBN 978–0–19–876765–7 (Pbk.)

FOREWORD

Measure concentration ideas developed during the last century in various parts of mathematics including functional analysis, probability theory, and statistical mechanics, areas typically dealing with models involving an infinite number of variables. After early observations, and in particular a geometric interpretation of the law of large numbers by E. Borel, the real birth of measure concentration took place in the early 1970s with the new proof by V. Milman, relying on Lévy's inequality (of isoperimetric nature), of Dvoretzky's theorem on spherical sections of convex bodies in high dimension. The inherent concept of measure concentration emphasized by V. Milman through this proof turned out to be one of the main achievements of analysis of the second part of the last century. It opened a posteriori completely new perspectives and developments with applications to various fields of mathematics. In particular, prompted by the concept and results, in the 1980s and 1990s M. Talagrand undertook a deep investigation of concentration inequalities for product measures, emphasizing a revolutionary new look at independence. Viewing namely random variables depending (in a smooth way) on the influence of many independent random variables (but not too much on any of them) as essentially constant led him to groundbreaking achievements and striking applications. Particularly with the tool of celebrated convex distance inequality, M. Talagrand developed applications to combinatorial probability, statistical mechanics, and empirical processes. Simultaneously, the entropic method, relying on an early observation by I. Herbst in the context of logarithmic Sobolev inequalities and developing information theoretic ideas, became a powerful additional and flexible method in the investigation of new concentration properties.

Since then, the concentration-of-measure phenomenon has spread out to an impressively wide range of illustrations and applications, and became a central tool and viewpoint in the quantitative analysis of a number of asymptotic properties in numerous topics of interest including geometric analysis, probability theory, statistical mechanics, mathematical statistics and learning theory, random matrix theory or quantum information theory, stochastic dynamics, randomized algorithms, complexity, and so on.

This book by S. Boucheron, G. Lugosi, and P. Massart is a most welcome and complete account of the modern developments of concentration inequalities in the context of the probabilistic method. The monograph covers most of the important and recent developments, with constant attention to illustrations and applications which make the theory so fruitful and attractive. The emphasis put on information theoretic methods is one main feature of the exposition and there is considerable benefit in this approach for a number of fundamental results and tools, for example the convex distance inequality or sharp bounds on empirical processes of fundamental importance in statistical applications. The monograph covers further basic and most illustrative examples of the current research, including dimension reduction, random matrices, Boolean analysis, transportation inequalities, and isoperimetric-type bounds. The style adopted by the authors is a perfect balance from the basic and classical material up to the most sophisticated and powerful results, always accessible and clearly reachable. Young and confirmed scientists, independently of their background, will find with this book the ideal path toward the powerful ideas and tools of concentration inequalities, suggested and illustrated with the most relevant applications and developments.

It is an honour and a pleasure to write this preface to this wonderful book, which is sure to be a huge success.

Michel Ledoux
Université de Toulouse

CONTENTS

1

Introduction

The topic of this book is the study of random fluctuations of functions of independent random variables. *Concentration inequalities* quantify such statements, typically by bounding the probability that such a function differs from its expected value (or from its median) by more than a certain amount.

The search for concentration inequalities has been a topic of intensive research in the last decades in a variety of areas because of their importance in numerous applications. Among the areas of applications, without trying to be exhaustive, we mention statistics, learning theory, discrete mathematics, statistical mechanics, random matrix theory, information theory, and high-dimensional geometry.

While concentration properties for sums of independent random variables were thoroughly studied and fairly well understood in classical probability theory, powerful tools to handle more general functions of independent random variables were not introduced until the appearance of martingale methods in the 1970s; see Yurinskii (1976), Maurey (1979), Milman and Schechtman (1986), Shamir and Spencer (1987), and McDiarmid (1989).

A remarkable series of papers in the mid-1990s by Michel Talagrand provided major new insight into the problem and opened many exciting new research directions. The main principle, as summarized by Talagrand (1995), is that "a random variable that smoothly depends on the influence of many independent random variables satisfies Chernoff type bounds." This book provides answers to the natural question hidden behind this citation: What kind of smoothness conditions should we put on a function f of independent random variables $X_1, \ldots, X_n$ in order to get concentration bounds for $Z = f(X_1, \ldots, X_n)$ around its mean or its median?

In this introductory chapter we briefly review the history of the subject and outline the contents, as an appetizer for the rest of the book.

Before getting started, we emphasize that one of the main driving forces behind the development of the theory was the need to understand random fluctuations of suprema of empirical processes defined as follows. Let $\mathcal{T}$ be a set that for now we assume to be finite and let $X_1, \ldots, X_n$ be independent random vectors taking values in $\mathbb{R}^{\mathcal{T}}$. We are interested in concentration properties of $\sup_{s\in\mathcal{T}} \sum_{i=1}^n X_{i,s}$ (where $X_i = (X_{i,s})_{s\in\mathcal{T}}$). Throughout the book we will regularly return to this example and discuss implications of the general theory.

1.1 Sums of Independent Random Variables and the Martingale Method

The simplest and most thoroughly studied example is the sum of independent real-valued random variables. The key to the study of this case is summarized by the trivial but fundamental additive formulas

$$\mathrm{Var}\left(\sum_{i=1}^n X_i\right) = \sum_{i=1}^n \mathrm{Var}\,(X_i)$$

and

$$\psi_{\sum_{i=1}^n X_i}(\lambda) = \sum_{i=1}^n \psi_{X_i}(\lambda) \tag{1.1}$$

where $\psi_Y(\lambda) = \log \mathbf{E}e^{\lambda Y}$ denotes the logarithm of the moment generating function of the random variable Y. These formulae allow one to derive concentration inequalities for $Z = X_1 + \cdots + X_n$ around its expectation via Markov's inequality, as shown in Chapter 2.

Hoeffding's inequality

One of the basic benchmark inequalities for sums of independent bounded random variables is Hoeffding's inequality (Theorem 2.8). It may be proved by noting that for a random variable Y taking values in an interval $[a, b]$,

$$\mathrm{Var}\,(Y) \leq \frac{(b-a)^2}{4}$$

which, through an exponential change of the underlying probability measure detailed in Lemma 2.2, leads to the following bound for the log-moment generating function of $Y - \mathbf{E}Y$:

$$\psi_{Y-\mathbf{E}Y}(\lambda) \leq \frac{\lambda^2(b-a)^2}{8}.$$

If $X_1, \ldots, X_n$ are independent random variables taking values in $[a_1, b_1], \ldots, [a_n, b_n]$ the additivity formula (1.1) implies that

$$\psi_{Z-\mathbf{E}Z}(\lambda) \leq \frac{\lambda^2 v}{2} \text{ for every } \lambda \in \mathbb{R}$$

where $v = \sum_i (b_i - a_i)^2/4$. Since the right-hand side corresponds to the log-moment generating function of a centered normal random variable with variance v, $Z - \mathbf{E}Z$ is said to be *sub-Gaussian* with variance factor v. The sub-Gaussian property implies that $Z - \mathbf{E}Z$ has a sub-Gaussian-like tail. More precisely, as it is proved in Section 2.6, for all $t > 0$,

$$\mathbf{P}\left\{|Z - \mathbf{E}Z| \geq t\right\} \leq 2e^{-t^2/(2v)}.$$

In his influential paper Hoeffding (1963) points out that the same result holds true under the weaker assumption that Z is a martingale with bounded increments. This simple observation is the basis of the *martingale method* for proving concentration inequalities, a powerful methodology that is still actively investigated. Hoeffding's inequality for martingales was more explicitly stated in the subsequent work of Azuma (1967) and Hoeffding's inequality for martingales with bounded increments is often referred to as Azuma's inequality or the Azuma–Hoeffding inequality. However, it took some time before the power of the martingale approach for the study of functions of independent variables was realized, see McDiarmid (1989, 1998), Chung and Lu (2006*a*, 2006*b*), and Dubhashi and Panconesi (2009) for surveys.

The bounded differences condition

One of the simplest and more natural smoothness assumptions that one may consider is the so-called *bounded differences condition*. A function $f : \mathcal{X}^n \to \mathbb{R}$ of n variables (all taking values in some measurable set $\mathcal{X}$) is said to satisfy the bounded differences condition if there exists constants $c_1, \ldots, c_n > 0$ such that for every $x_1, \ldots, x_n, y_1, \ldots, y_n \in \mathcal{X}^n$ and for all $i = 1, \ldots, n$,

$$\left|f(x_1, \ldots, x_i, \ldots, x_n) - f(x_1, \ldots, x_{i-1}, y_i, x_{i+1} \ldots, x_n)\right| \leq c_i.$$

In other words, changing any of the n variables, while keeping the rest fixed, cannot cause a big change in the value of the function. Equivalently, one may interpret this as a Lipschitz condition. Indeed, defining the weighted Hamming distance d_c on the product space $\mathcal{X}^n$ as

$$d_c(x, y) = \sum_{i=1}^{n} c_i \mathbb{1}_{\{x_i \neq y_i\}},$$

the bounded differences condition means that f is 1-Lipschitz with respect to the metric d_c.

The sum of bounded variables is the simplest example of a function of bounded differences. Indeed, if $X_1, \ldots, X_n$ are real-valued independent random variables such that X_i takes its values in the interval $[a_i, b_i]$, then $f(X_1, \ldots, X_n) = \sum_{i=1}^{n} X_i$ satisfies the bounded differences condition with $c_i = b_i - a_i$. The basic argument behind the martingale-based approach is that once the function f satisfies the bounded differences condition, $Z = f(X_1, \ldots, X_n)$ may be interpreted as a martingale with bounded increments with respect to Doob's filtration. In other words, one may write

$$Z - \mathbf{E}Z = \sum_{i=1}^{n} \Delta_i \tag{1.2}$$

where $\Delta_i = \boldsymbol{E}[Z|X_1,\ldots,X_i] - \boldsymbol{E}[Z|X_1,\ldots,X_{i-1}]$ for $i = 2,\ldots,n$ and $\Delta_1 = \boldsymbol{E}[Z|X_1] - \boldsymbol{E}Z$, and notice that the bounded differences condition implies that, conditionally on $X_1,\ldots,X_{i-1}$, the martingale increment Δ_i takes its values in an interval of length at most c_i. Hence Hoeffding's inequality remains valid for Z with $v = (1/4)\sum_{i=1}^n c_i^2$. This result is known as the *bounded differences inequality*, also often referred to as *McDiarmid's inequality*. In this book we offer various alternative proofs and variants of this fundamental inequality (see Sections 6.1 and 8.1).

Another approach to understanding the concentration properties of Lipschitz functions of independent variables is based on investigating how product measures concentrate in high-dimensional spaces. The main ideas behind this approach, dominant in Talagrand's work, are briefly explained next.

1.2 The Concentration-of-Measure Phenomenon

Isoperimetric inequalities and concentration

The classical isoperimetric theorem (see Section 7.2) states that among all compact sets $A \subset \mathbb{R}^n$ with smooth boundary and a fixed volume, Euclidean balls are the ones with smallest surface area. This result has the following equivalent formulation that allows one to ask and investigate the same question in general metric spaces. Writing $d(x,A) = \inf_{y\in A} d(x,y)$ and

$$A_t = \left\{x \in \mathbb{R}^n : d(x,A) < t\right\}$$

for the *t-blowup* of A (with respect to the Euclidean distance d), the isoperimetric theorem states that for any compact set A and a Euclidean ball B with the same volume, $\lambda(A_t) \geq \lambda(B_t)$ for all $t > 0$. Here the Lebesgue measure λ and the Euclidean distance d play a fundamental role but the same question may be asked for more general measures and distance functions. For our purposes, probability measures are closer to the heart of the matter. An equally interesting, though somewhat less known, case is the isoperimetric problem on the sphere. The corresponding isoperimetric theorem is usually referred to as *Lévy's isoperimetric theorem* – proved independently by Lévy (1951) and Schmidt (1948). Again this theorem can be stated in two equivalent ways but the one that is more important for our goals is as follows: Let $S^{n-1} = \{x \in \mathbb{R}^n : \|x\| = 1\}$ denote the unit sphere in $\mathbb{R}^n$ and let μ denote the uniform (i.e. rotation invariant) probability measure on S^{n-1}. For any measurable set $A \subset S^{n-1}$, if B is a geodesic ball (i.e. a spherical cap) with $\mu(B) = \mu(A)$, then, for all $t > 0$,

$$\mu(A_t) \geq \mu(B_t),$$

where the t-blowups A_t and B_t are understood with respect to the geodesic distance on the sphere. The first appearance of the concentration-of-measure principle may be deduced from this statement. Indeed, by considering a half-sphere B, one may explicitly compute the measure of the spherical cap B_t^c and Lévy's isoperimetric theorem implies that for any set $A \subset S^{n-1}$ with $\mu(A) \geq 1/2$, the complement A_t^c of the t-blowup of A satisfies

$$\mu\left(A_t^c\right) \leq e^{-(n-1)t^2/2}.$$

In other words, as soon as $\mu(A) \geq 1/2$, the measure of A_t^c decreases very fast as a function of t. This is the essence of the *concentration-of-measure phenomenon* whose importance was perhaps first fully recognized by Vitali Milman in his proof of Dvoretzky's theorem. Unlike the original formulation of the isoperimetric theorem, the inequality above may be generalized to measures on abstract metric spaces without any reference to geometry.

Lipschitz functions

Consider a metric space $(\mathcal{X}, d)$ and a continuous functional $f : \mathcal{X} \to \mathbb{R}$. Given a probability measure $\mathbf{P}$ on $\mathcal{X}$, one is interested in bounding the deviation probabilities

$$\mathbf{P}\left\{f(X) \geq \mathbf{M}f(X) + t\right\} \quad \text{and} \quad \mathbf{P}\left\{|f(X) - \mathbf{M}f(X)| \geq t\right\}$$

where X is a random variable taking values in $\mathcal{X}$ with distribution $\mathbf{P}$ and $\mathbf{M}f(X)$ is a median of $f(X)$. Given a Borel set $A \subset \mathcal{X}$, let

$$A_t = \left\{x \in \mathcal{X} : d(x, A) < t\right\}$$

denote the t-blowup of A where $t > 0$. Now observe that if f is 1-Lipschitz (i.e. $f(x) - f(y) \leq d(x, y)$ for all $x, y \in \mathcal{X}$), then taking $A = \left\{x \in \mathcal{X} : f(x) \leq \mathbf{M}f(X)\right\}$, for all $x \in A_t$,

$$f(x) < \mathbf{M}f(X) + t,$$

and therefore

$$\mathbf{P}\left\{f(X) \geq \mathbf{M}f(X) + t\right\} \leq \mathbf{P}\left\{A_t^c\right\} = \mathbf{P}\{d(X, A) \geq t\}.$$

We can now forget what exactly the set A is and just use the fact that $\mathbf{P}\{A \geq 1/2\}$. Indeed, defining the *concentration function*

$$\alpha(t) = \sup_{A \subset \mathcal{X} : \mathbf{P}\{A\} \geq \frac{1}{2}} \mathbf{P}\{d(X, A) \geq t\},$$

we obtain

$$\mathbf{P}\left\{f(X) \geq \mathbf{M}f(X) + t\right\} \leq \alpha(t).$$

Changing f into $-f$, one also gets

$$\mathbf{P}\left\{f(X) \leq \mathbf{M}f(X) - t\right\} \leq \alpha(t).$$

Combining these inequalities of course implies the concentration inequality

$$P\left\{|f(X) - Mf(X)| \geq t\right\} \leq 2\alpha(t).$$

The conclusion is that if one can control the concentration function α, as in the case of the uniform probability measure on the sphere, then one immediately gets a concentration inequality for any Lipschitz function.

What makes this general principle attractive is that the concentration function α may be controlled without determining the extremal sets of the isoperimetric problem and any upper bound for the function α yields concentration inequalities for all Lipschitz functions.

The Gaussian case

The principle described above is nicely illustrated in the case when $(\mathcal{X}, d)$ is the n-dimensional Euclidean space $\mathbb{R}^n$ and P is the standard Gaussian probability measure in $\mathbb{R}^n$. Indeed, in this case the isoperimetric problem is connected to that of the sphere via Poincaré's limit procedure. The Gaussian isoperimetric problem was completely solved independently by Borell (1975) and Tsirelson, Ibragimov, and Sudakov (1976). The Gaussian isoperimetric theorem, stated and proved in Section 10.4, states that for any Borel set $A \subset \mathbb{R}^n$, if $H \subset \mathbb{R}^n$ is a half-space with $P(H) = P(A)$, then $P(A_t^c) \leq P(H_t^c)$ for all $t > 0$.

The Gaussian isoperimetric theorem reveals the exact form of the concentration function. Indeed, define the standard normal tail function by

$$\overline{\Phi}(t) = \frac{1}{\sqrt{2\pi}} \int_t^{\infty} e^{-u^2/2} du,$$

and for any Borel set $A \subset \mathbb{R}^n$, let $t_A \in \mathbb{R}$ be such that $1 - \overline{\Phi}(t_A) = P(A)$. Then, taking H to be the half-space $(-\infty, t_A) \times \mathbb{R}^{n-1}$, we see that

$$P(A) = P(H) \quad \text{and} \quad P(H_t^c) = \overline{\Phi}(t_A + t).$$

Now, if $P(A) \geq 1/2$, then $t_A \geq 0$, and therefore $P(H_t^c) \leq \overline{\Phi}(t)$. Hence, the Gaussian isoperimetric theorem implies that the concentration function α of the standard Gaussian measure P is exactly equal to the standard Gaussian tail function $\overline{\Phi}$.

Putting things together, we see that if X is a standard Gaussian vector in $\mathbb{R}^n$ and $f : \mathbb{R}^n \to \mathbb{R}$ is a 1-Lipschitz function, then, for all $t > 0$,

$$P\left\{f(X) - Mf(X) \geq t\right\} \leq \overline{\Phi}(t) \leq e^{-t^2/2}.$$

Concentration of product measures

The Gaussian isoperimetric inequality implies sharp concentration inequalities for smooth functions of independent normal random variables. However, if we wish to understand random fluctuations of functions of more general independent random variables, then we need to study the concentration of general product measures. In order to do this, the first step is to define an appropriate distance on a product space $\mathcal{X}^n$. A natural candidate is the *Hamming distance,* or more generally, a weighted Hamming distance which offers more flexibility. For any vector $\alpha = (\alpha_1, \ldots, \alpha_n)$ of non-negative real numbers and for any $x = (x_1, \ldots, x_n), y = (y_1, \ldots, y_n) \in \mathcal{X}^n$, define

$$d_\alpha(x,y) = \sum_{i=1}^{n} \alpha_i \mathbb{1}_{\{x_i \neq y_i\}}.$$

Let $X = (X_1, \ldots, X_n)$ be a vector of independent random variables, each taking values in $\mathcal{X}$ and denote by $\mathbf{P}$ the distribution of X. Then by a simple consequence of the bounded differences inequality, we have the following concentration property of the product probability measure $\mathbf{P}$ with respect to the weighted Hamming distance d_α: for every $A \subset \mathcal{X}^n$ with $\mathbf{P}\{X \in A\} \geq 1/2$,

$$\mathbf{P}\left\{d_\alpha(X,A) \geq t\right\} \leq e^{-t^2/(2\|\alpha\|^2)} \tag{1.3}$$

where $\|\alpha\|$ denotes the Euclidean norm of the vector α. (See Section 7.4 for the proof.)

This implies that if $f : \mathcal{X}^n \to \mathbb{R}$ is 1-Lipschitz with respect to the distance d_α, it satisfies the sub-Gaussian tail bound

$$\mathbf{P}\left\{f(X) \geq \mathbf{M}f(X) + t\right\} \leq e^{-t^2/(2\|\alpha\|^2)}.$$

We illustrate this inequality by considering the special case of the supremum of a *Rademacher process.* Let $\mathcal{X}^n = \{-1, 1\}^n$ and

$$f(x) = \max_{t \in \mathcal{T}} \sum_{i=1}^{n} \alpha_{i,t} x_i = \sum_{i=1}^{n} \alpha_{i,t^*(x)} x_i,$$

where $\mathcal{T}$ is a finite set and $(\alpha_{i,t})$ is a collection of real numbers indexed by $i = 1, \ldots, n$ and $t \in \mathcal{T}$, and $t^*(x) \in \mathcal{T}$ denotes an index for which the maximum is achieved. For all $x, y \in \{-1, 1\}^n$,

$$f(x) - f(y) \leq \sum_{i=1}^{n} \alpha_{i,t^*(x)}(x_i - y_i) \leq 2 \sum_{i=1}^{n} \max_{t \in \mathcal{T}} \left|\alpha_{i,t}\right| \mathbb{1}_{\{x_i \neq y_i\}}.$$

Thus, f is 1-Lipschitz with respect to the weighted Hamming distance d_α where $\alpha_i = 2 \max_{t \in \mathcal{T}} \left|\alpha_{i,t}\right|$ for all i. As a consequence, if X is uniformly distributed on the hypercube $\{-1, 1\}^n$, the random variable

$$f(X) = \max_{t \in \mathcal{T}} \sum_{i=1}^{n} \alpha_{i,t} X_i$$

satisfies

$$\mathbf{P}\left\{f(X) \geq \mathbf{M}f(X) + t\right\} \leq e^{-t^2/(2v)}$$

where the "variance factor" v is defined by $v = 4\sum_{i=1}^n \max_{t\in\mathcal{T}} \alpha_{i,t}^2$. This result is not completely satisfactory as v can be much larger than the largest variance of the individual random variables $\sum_{i=1}^n \alpha_{i,t}X_i$. One would ideally expect to be able to exchange the order of the sum and the maximum in the above definition of v.

Indeed, such a result is possible (by paying the modest price of losing some absolute multiplicative constant in the exponent), thanks to the celebrated *convex distance inequality* of Talagrand (proved in Section 7.4) which is one of the major milestones of the theory.

To see how this works, note first that setting $\alpha_i(x) = 2\left|\alpha_{i,t^*(x)}\right|$, the supremum of the Rademacher process f defined above satisfies

$$f(x) - f(y) \le \sum_{i=1}^n \alpha_i(x)\mathbb{1}_{\{x_i\neq y_i\}}, \tag{1.4}$$

a relaxed regularity condition as compared to the Lipschitz property with respect to some given weighted Hamming distance d_α. The beauty of Talagrand's convex distance inequality is that it guarantees that the following uniform version of (1.3) holds for all $v > 0$ and for every set $A \subset \mathcal{X}^n$ with $\mathbf{P}\{X \in A\} \ge 1/2$:

$$\mathbf{P}\left\{\sup_{\alpha\in[0,\infty)^n:\|\alpha\|^2\le v} d_\alpha(X,A) \ge t\right\} \le 2e^{-t^2/(4v)}.$$

Now one can play a similar game as for the case of Lipschitz functions before. Choosing $A = \{x \in \mathcal{X}^n : f(x) \le \mathbf{M}f(X)\}$, for every $x \in \mathcal{X}^n$ such that $d_{\alpha(x)}(x,A) < t$, the regularity condition (1.4) implies that $f(x) < \mathbf{M}f(X) + t$. Hence, taking $v = \sup_{x\in\mathcal{X}^n}\sum_{i=1}^n \alpha_i^2(x)$, we have

$$\begin{aligned}\{x \in \mathcal{X}^n : f(x) \ge \mathbf{M}f(X) + t\} &\subset \{x \in \mathcal{X}^n : d_{\alpha(x)}(x,A) \ge t\} \\ &\subset \left\{x \in \mathcal{X}^n : \sup_{\|\alpha\|^2\le v} d_\alpha(x,A) \ge t\right\},\end{aligned}$$

and therefore,

$$\mathbf{P}\{f(X) \ge \mathbf{M}f(X) + t\} \le 2e^{-t^2/(4v)}.$$

If we consider again the example of the maximum of a Rademacher process, we see that $v \le 4\sup_{t\in\mathcal{T}}\sum_{i=1}^n \alpha_{i,t}^2$ and we obtain a concentration inequality of the desired form. This example highlights the power of Talagrand's convex distance inequality and the interest in considering the relaxed regularity condition (1.4) as opposed to Lipschitz regularity with respect to some given weighted Hamming distance. Indeed, the convex distance inequality became the key tool for obtaining improved concentration inequalities in countless applications, some of them shown in detail in this book.

Nevertheless, this regularity condition may be too restrictive in some cases. To understand the shortcomings of this condition, consider the fundamental example of the supremum of an empirical process defined as follows. Let $\mathcal{T}$ be a finite set and for

$i = 1, \ldots, n$, let $x_i = (x_{i,t})_{t \in \mathcal{T}}$ be a vector whose components are indexed by $\mathcal{T}$. Writing $x = (x_1, \ldots, x_n)$, we may define $f(x) = \max_{t \in \mathcal{T}} \sum_{i=1}^{n} x_{i,t}$. Note that the maximum of a Rademacher process is a special case. However, the study of suprema of general empirical processes is more involved. Indeed, if we try to use the approach that turned out to be successful for Rademacher processes, the increments of f are controlled by

$$f(x) - f(y) \leq \sum_{i=1}^{n} x_{i,t^*(x)} - y_{i,t^*(x)}$$

where $t^*(x) \in \mathcal{T}$ is a point at which the maximum of $\sum_{i=1}^{n} x_{i,t}$ is achieved. At this point we see how lucky we were in the case of Rademacher processes by simultaneously benefiting from the special structure of $x_{i,j} = \alpha_{i,j} x_i$ and the boundedness of the x_i's to end up satisfying (1.4). Dealing with general empirical processes is a significantly more intricate issue. By a substantial deepening of the approach that led to the convex distance inequality, Talagrand (1996*b*) was able to derive a Bennett-type concentration inequality for the suprema of empirical processes (see Theorem 12.5 for a somewhat sharper version). The authors of this book were awestruck by this achievement of Talagrand but collectively confess that they were unable to go further than a line-by-line reading of the proof. However, Talagrand's work stimulated intensive research partly in the search for more transparent proofs. Today, following the path opened by Ledoux (1997), a more accessible proof is available by what we call the *entropy method*. This method, briefly sketched in the next section, is one of the central topics of this book. We feel that many of the most important concentration inequalities can be obtained in a principled and transparent way by the entropy method. In particular, the reader will find in this book a complete proof of Talagrand's inequality for empirical processes.

We would like to emphasize that, apart from an exciting mathematical challenge, the study of concentration properties of the supremum of an empirical process is strongly motivated by applications in mathematical statistics, machine learning, and other areas. This is why we keep this example as one of the recurring themes of this book.

1.3 The Entropy Method

The entropy method replaces Talagrand's subtle induction arguments by *sub-additive inequalities* (often called "tensorization" inequalities in the literature) that follow naturally from the convexity of entropy and related quantities like the variance.

The Efron–Stein inequality

Perhaps the simplest inequality of this type is the *Efron–Stein inequality* that, in spite of its simplicity, turns out to be a surprisingly powerful tool for bounding the variance of general functions of independent random variables. This inequality, studied in depth in Chapter 3, can be stated as follows. Let $X = (X_1, \ldots, X_n)$ be a vector of independent random variables and denote by $X^{(i)} = (X_1, \ldots, X_{i-1}, X_{i+1}, \ldots, X_n)$ the $(n-1)$-vector obtained by dropping

X_i. Let $\mathbf{E}^{(i)}$ and $\mathrm{Var}^{(i)}$ denote the conditional expectation and variance operators given $X^{(i)}$. Then $Z = f(X_1, \ldots, X_n)$ satisfies

$$\mathrm{Var}(Z) \leq \mathbf{E} \sum_{i=1}^{n} \mathrm{Var}^{(i)}(Z).$$

This inequality was proved by Efron and Stein (1981) under the additional assumption that f is symmetric and by Steele (1986) in the general case. As pointed out by Rhee and Talagrand (1986), the Efron–Stein inequality may be viewed as a martingale inequality. The argument, detailed in Section 3.1, may be summarized as follows. Since the martingale increments are orthogonal in $\mathbb{L}_2$, the decomposition (1.2) implies that

$$\mathrm{Var}(Z) = \sum_{i=1}^{n} \mathbf{E}\Delta_i^2.$$

Now using the independence of the X_i, the martingale increments may be rewritten as $\Delta_i = \mathbf{E}\left[Z - \mathbf{E}^{(i)}Z | X_1, \ldots, X_i\right]$ and the Efron–Stein inequality is obtained by a simple use of Jensen's inequality. This proof emphasizes the role of the Efron–Stein inequality as a substitute for the additivity of the variance for independent random variables.

Sub-additivity of entropy

The Efron–Stein inequality has an interpretation that gives rise to far-reaching generalizations. In particular, it paves the way to appropriate generalizations of (1.1) which was the key to exponential inequalities for sums of independent random variables. Indeed the variance may be viewed as a special case of a Φ-*entropy* defined as follows. If Φ denotes a convex function defined on an interval I and Y is an integrable random variable taking its values in I, then the Φ-entropy of Y is defined by

$$H_\Phi(Y) = \mathbf{E}\Phi(Y) - \Phi(\mathbf{E}Y).$$

By Jensen's inequality, the Φ-entropy is non-negative and it is finite if and only if $\Phi(Y)$ is integrable. The variance corresponds to the choice $\Phi(x) = x^2$, while taking $\Phi(x) = x \log x$ leads to the definition of the "usual" notion of entropy $\mathrm{Ent}(Y)$ of a nonnegative random variable Y.

As it turns out, the sub-additive property of the variance expressed by the Efron–Stein inequality remains true for a large class of Φ-entropies (characterized in Chapter 14) that includes the ordinary entropy. More precisely, if Y is a nonnegative function of the independent random variables $X_1, \ldots, X_n$, then

$$\mathrm{Ent}(Y) \leq \mathbf{E} \sum_{i=1}^{n} \mathrm{Ent}^{(i)}(Y)$$

where $\mathrm{Ent}^{(i)}(Y) = \mathbf{E}^{(i)}\Phi(Y) - \Phi\left(\mathbf{E}^{(i)}(Y)\right)$ with $\Phi(x) = x \log x$. Applying this sub-additive inequality to the random variable $Y = e^{\lambda Z}$ is the basis of the entropy method.

Herbst's argument

The sub-additivity property of entropy seems to have appeared first in the proof of the *Gaussian logarithmic Sobolev inequality* of Gross (1975). In fact, the Gaussian logarithmic Sobolev inequality, combined with an elegant argument attributed to Herbst, leads smoothly to the Gaussian concentration inequality. We sketch the argument here and refer to Chapter 5 for the details. To handle distributions other than Gaussian, one needs to modify the argument as the logarithmic Sobolev inequality does not hold in general. This is done in Chapter 6.

The Gaussian logarithmic Sobolev inequality states that if X is a standard Gaussian vector in $\mathbb{R}^n$ and $g : \mathbb{R}^n \to \mathbb{R}$ is a continuously differentiable function, then

$$\mathrm{Ent}\left(g^2(X)\right) \leq 2E\left[\left\|\nabla g(X)\right\|^2\right].$$

The proof of this inequality relies on the sub-additivity of entropy. The connection between the Gaussian logarithmic Sobolev inequality and concentration is established by Herbst's argument that we will face in various contexts. In the Gaussian framework it is especially simple to explain.

Indeed, if $f : \mathbb{R}^n \to \mathbb{R}$ is a continuously differentiable 1-Lipschitz function, then for all $x \in \mathbb{R}^n$, $\left\|\nabla f(x)\right\| \leq 1$, and for any $\lambda > 0$, we may apply the Gaussian logarithmic Sobolev inequality to the function $g = e^{\lambda f/2}$. Since for all $x \in \mathbb{R}$,

$$\left\|\nabla g(x)\right\|^2 = \frac{\lambda^2}{4}\left\|\nabla f(x)\right\|^2 e^{\lambda f(x)} \leq \frac{\lambda^2}{4} e^{\lambda f(x)},$$

we derive from the Gaussian logarithmic Sobolev inequality that

$$\frac{\mathrm{Ent}\left(e^{\lambda f(X)}\right)}{Ee^{\lambda f(X)}} \leq \frac{\lambda^2}{2}.$$

Now the next crucial observation is that, defining $F(\lambda) = \log Ee^{\lambda(f(X)-Ef(X))}$,

$$\frac{\mathrm{Ent}\left(e^{\lambda f(X)}\right)}{Ee^{\lambda f(X)}} = \lambda F'(\lambda) - F(\lambda).$$

This way we obtain the following differential inequality for the logarithm of the moment generating function

$$\frac{d}{d\lambda}\left(\frac{F(\lambda)}{\lambda}\right) = \frac{F'(\lambda)}{\lambda} - \frac{F(\lambda)}{\lambda^2} \leq \frac{1}{2}$$

which one can integrate and obtain that for all $\lambda > 0$,

$$F(\lambda) \leq \frac{\lambda^2}{2}.$$

This leads to the Gaussian concentration bound

$$\mathbf{P}\{f(X) - \mathbf{E}f(X) \geq t\} \leq e^{-t^2/2}.$$

This bound has the same flavor as that which we obtained from the Gaussian isoperimetric theorem, except that the median is replaced by the mean. This is typically what happens when one uses the entropy method rather than the isoperimetric method. If one does not care too much about absolute constants in the exponential bounds, this difference is negligible since starting from a concentration inequality around the median one can obtain a concentration inequality around the mean and vice versa, simply because the difference between the median and the mean is under control.

1.4 The Transportation Method

We also discuss an alternative way of proving concentration inequalities, the so-called *transportation method*. The method was initiated by Marton (1986) who built on ideas from information theory due to Ahlswede, Gács, and Körner (1976) and Csiszár and Körner (1981). The method is based on a beautiful coupling idea. Given some cost function d, the *transportation cost* between two probability measures P and Q is defined by

$$\min_{\mathbf{P} \in \mathcal{P}(P,Q)} \mathbf{E}_{\mathbf{P}} d(X, Y),$$

where $\mathcal{P}(P, Q)$ denotes the class of joint distributions of the random variables X and Y such that the marginal distribution of X is P and that of Y is Q. The transportation cost measures the amount of effort required to "transport" a mass distributed according to P into a mass distributed according to Q, relative to the cost function d. The transportation problem consists of constructing an optimal *coupling* $\mathbf{P} \in \mathcal{P}(P, Q)$, that is, a minimizer of $\mathbf{E}_{\mathbf{P}} d(X, Y)$. In order to explain the link between the transportation cost problem and concentration, we describe the main ideas within the Gaussian framework.

The core of this connection lies in bounding the transportation cost by some function of the Kullback–Leibler divergence $D(Q\|P)$ where we recall that whenever Q is absolutely continuous with respect to P, $D(Q\|P) = \operatorname{Ent}(dQ/dP)$. In the Gaussian case such a transportation inequality is available for the quadratic cost. In particular, the following inequality, due to Talagrand (1996*d*), is proved in Section 8.5: Let P be the standard Gaussian probability measure on $\mathbb{R}^n$ and let Q be any probability measure which is absolutely continuous with respect to P. Then

$$\min_{\mathbf{P} \in \mathcal{P}(P,Q)} \sum_{i=1}^{n} \mathbf{E}_{\mathbf{P}} (X_i - Y_i)^2 \leq 2D(Q\|P).$$

The Gaussian concentration inequality may now be derived from this transportation inequality by an argument due to Bobkov and Götze (1999). The sketch of the argument is as follows: Assume that $f : \mathbb{R}^n \to \mathbb{R}$ is a 1-Lipschitz function, that is,

$$f(y) - f(x) \leq \left(\sum_{i=1}^{n} (x_i - y_i)^2 \right)^{1/2} \quad \text{for all} \quad x, y \in \mathbb{R}^n.$$

Then Jensen's inequality implies that for any probability distribution $\mathbf{P}$ coupling P to $Q \ll P$, one has

$$E_Q f - E_P f = \mathbf{E}_{\mathbf{P}}[f(Y) - f(X)] \leq \left(\sum_{i=1}^{n} \mathbf{E}_{\mathbf{P}} (X_i - Y_i)^2 \right)^{1/2}.$$

Hence, the transportation inequality implies that

$$E_Q f - E_P f \leq \sqrt{2D(Q\|P)}.$$

Now the concentration of the random variable $Z = f(X)$ (where X is a standard Gaussian random vector) may be obtained by the following classical duality formula for entropy that we prove in Section 4.9:

$$\psi_{Z-\mathbf{E}Z}(\lambda) = \sup_{Q \ll P} \left[\lambda \left(E_Q f - E_P f \right) - D(Q\|P) \right].$$

Combining the last two inequalities, we get that for any $\lambda > 0$,

$$\psi_{Z-\mathbf{E}Z}(\lambda) \leq \sup_{Q \ll P} \left[\lambda \sqrt{2D(Q\|P)} - D(Q\|P) \right] \leq \frac{\lambda^2}{2},$$

simply because $2ab - a^2 \leq b^2$. This implies the same Gaussian concentration inequality as that derived from the Gaussian logarithmic Sobolev inequality and Herbst's argument.

In Chapter 8 we offer a detailed account of the transportation method for proving concentration inequalities, pioneered by Marton (1996*a*, 1996*b*). In particular, we show how this method allows one to prove not only the bounded differences inequality but also Talagrand's convex distance inequality.

As far as we know, there is no clear hierarchy between the entropy method and the transportation method. As we will see, there are various results that one can prove by using one method or the other and there are also results that one can get by one method but not the other. The entropy method is quite versatile, easy to use, and performs especially well when dealing with suprema of empirical processes. However, the entropy method often faces difficulties when one tries to use it to prove inequalities for the left tail (i.e. upper bounds for $\mathbf{P}\{Z < \mathbf{E}Z - t\}$ for $t > 0$). On the other hand, the transportation method is often more efficient for left tails but turns out to be less flexible than the entropy method, especially when used for empirical processes.

1.5 Reading Guide

We were guided by two principles while organizing the material of this book (see Fig. 1.1). First, we tried to keep the exposition as elementary as possible and illustrate the theory with numerous examples and applications. Our intention was to make most of the material accessible to researchers and mathematically mature graduate students and to introduce the reader to the main ideas of the theory while keeping technicalities at a minimal level, at least in the first half of the book. This led us to a somewhat nonlinear structure in which the same topic is revisited several times throughout the book, but with different degrees of depth. Very roughly, the material may be split into two parts going from Chapter 1 to 9 and from Chapter 10 to 15. The first chapters expose the general tools required to prove concentration inequalities together with applications of the theory to many examples. Chapters 2, 3, and 4 include inequalities for sums of independent random variables (such as Hoeffding's inequality that we introduced above), variance bounds for functions of independent variables related to the Efron–Stein inequality, and the basic information-theoretic tools needed to develop the entropy method, such as the sub-additive inequality for entropy. In Chapters 5 and 6 we present the essence of the entropy method building upon logarithmic Sobolev inequalities (or their modifications) and Herbst's argument. In Chapter 7 we investigate the connection between isoperimetry and concentration while Chapter 8 is devoted to the transportation method. Chapter 9 is entirely dedicated to the intricate concentration and isoperimetric properties of the simplest product space, the binary hypercube. We describe some fascinating applications to the study of threshold phenomena. More precisely, we consider general monotone functions $f : \{-1, 1\}^n \to \{0, 1\}$ of several binary random variables and consider independent binary random variables $X_1, \ldots, X_n$ with distribution $\mathbf{P}\{X_i = 1\} = 1 - \mathbf{P}\{X_i = -1\} = p$. We are interested in the behavior of $\mathbf{P}\{f(X_1, \ldots, X_n) = 1\}$ as a function of the parameter $p \in [0, 1]$. If f is increasing in each variable (and not constant), this probability grows monotonically from 0 to 1. Using the technology based on logarithmic Sobolev and isoperimetric inequalities, we establish surprisingly general sufficient conditions under which $\mathbf{P}\{f(X_1, \ldots, X_n) = 1\}$ "jumps" from near 0 to near 1 in a very short interval of the value of the parameter p.

The second half of the book contains some more advanced material. It includes a deepening of the general tools and their applications. In Chapter 10 we further investigate isoperimetric problems in the binary hypercube and Gaussian spaces. In particular, we reproduce Bobkov's elegant proof of the Gaussian isoperimetric theorem.

Chapters 11–13 are devoted to our canonical example; the supremum of an empirical process. Chapter 11 covers inequalities for the variance of the maximum, mostly building on the Efron–Stein inequality, while in Chapter 12 we derive various exponential inequalities. In Chapter 13 we present some tools to control the expectation of the supremum of an empirical process and combine them with the concentration inequalities established in the previous chapters.

Finally, in Chapters 14 and 15 we describe a method for proving moment inequalities for functions of independent random variables. The method is based on a natural extension of the entropy method that leads to moment inequalities interpolating between the Efron–Stein inequality and exponential concentration inequalities.

Each chapter is supplemented by a list of exercises. These exercise sections have several roles. Some ask the reader to complete arguments that have only been sketched in the text. Our intention was to make the main text as self-contained as possible but the proof of a few results that are somewhat technical and not crucial for the main stream of the arguments are relegated to the exercise sections. Most of these exercises come with detailed hints and

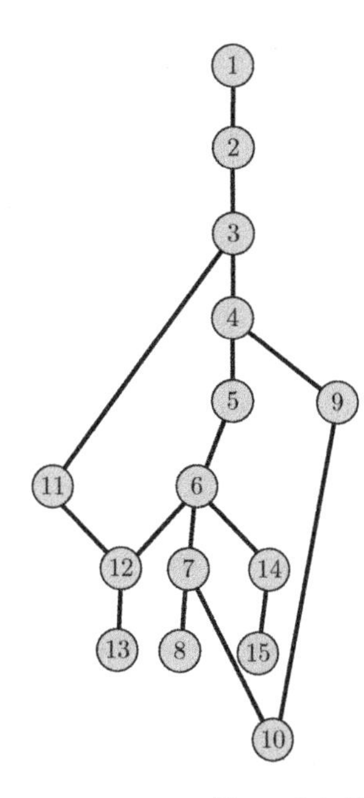

1. Introduction
2. Basic inequalities
3. Bounding the variance
4. Basic information inequalities
5. Logarithmic Sobolev inequalities
6. The entropy method
7. Concentration and isoperimetry
8. The transportation method
9. Influences and threshold phenomena
10. Isoperimetry on the hypercube and Gaussian spaces
11. The variance of suprema of empirical processes
12. Suprema of empirical processes: exponential inequalities
13. The expected value of suprema of empirical processes
14. Φ-entropies
15. Moment inequalities

Figure 1.1 Dependence structure of the chapters

the reader should not have major difficulties in filling in the details. Many other exercises describe related results from the literature whose proof may be more difficult. In all cases, the exercises provide important supplementary information and we encourage the reader to look at them.

In order to avoid interrupting the flow of the arguments with references to the literature, we postpone all bibliographical remarks to the end of each chapter where the reader may find the source of the material described in the chapter and pointers for further reading and related material.

We emphasize at this point that the reader looking for a comprehensive account of concentration inequalities will be disappointed as there are large chunks of the literature that we do not cover. For example, we only superficially touch upon the martingale method, the "classical" approach to concentration inequalities. Martingales are still the most adequate tool for some problems and the interested reader is referred to the surveys of McDiarmid (1989, 1998), Schechtman (2003), Chung and Lu (2006*a*, 2006*b*), and Dubhashi and Panconesi (2009).

An important extension that we entirely avoid in this book concerns concentration inequalities for functions of dependent random variables. For concentration inequalities for functions of mixing processes, Markov chains, and Markov random fields, we refer the reader to Marton (1996*b*, 2003, 2004), Rio (2000), Samson (2000), Catoni (2003), Külske (2003), Collet (2006), Chazottes *et al.* (2007), and Kontorovich and Ramanan (2008), just to name a few important papers from a continuously growing body of research. The methods used in the above-mentioned papers range from combinations of martingale methods with coupling techniques to refinements of the transportation method. Chatterjee (2007) developed a general, elegant, and powerful method for proving concentration bounds for dependent random variables, based on an adaptation of Stein's method; see also Chatterjee and Dey (2010).

Mostly motivated by the need to understand the behavior of the number of copies of small subgraphs (such as triangles) in a random graph, an important body of research that we do not cover in this book, is devoted to finding sharp concentration inequalities for low-degree polynomials of independent Bernoulli random variables. The interested reader may find a long and fascinating story that unfolds in the series of papers of Kim and Vu (2000, 2004), Vu (2000, 2001), Janson and Ruciński (2004, 2002), Janson, Oleszkiewicz, and Ruciński (2004), Bolthausen, Comets, and Dembo (2009), Döring and Eichelsbacher (2009), Chatterjee (2010), DeMarco and Kahn (2010), and Schudy and Sviridenko (2012).

A related important subject that we only tangentially touch upon is the theory of U-statistics and U-processes. Introduced by Hoeffding (1948), this special class of functions of independent random variables has attracted considerable attention. We only discuss briefly some special cases such as a Gaussian chaos of order two (see Example 2.12). For general moment and exponential inequalities for U-statistics, we refer the interested reader to the book of de la Peña and Giné (1999). For a sample of concentration inequalities for U-statistics and U-processes, some of which are established with the help of the general techniques described in this book, see Adamczak (2006), Clémençon, Lugosi, and Vayatis (2008), Giné, Latała, and Zinn (2000), Houdré and Reynaud-Bouret (2003), Major (2005, 2006, 2007), and Verzelen (2010).

Many important geometrical aspects of the concentration-of-measure phenomenon omitted from this book are treated in Ledoux's outstanding monograph (2001). Ledoux's book describes the concentration-of-measure phenomenon from the perspective of geometry and functional analysis. A decade earlier, the influential book by Ledoux and Talagrand (1991) emphasized the use of concentration arguments in the analysis of sums of independent random vectors. During the 1990s, it became clear that functional inequalities may lead to powerful concentration inequalities and even to sharp isoperimetric estimates (see, e.g., Chapter 10).

1.6 Acknowledgments

The idea for this book was born several years ago in conversations with Olivier Bousquet. The original plan was to have a joint project with four authors, including Olivier, but unfortunately (for us) he had to quit at an early stage. Still, this book enormously benefited from our early discussions with him and we are infinitely grateful for his early push, without which this book would never have been started. Had he stayed on board, the reader could enjoy a much better book now.

We are indebted to several colleagues and friends who read and commented on early versions of the manuscript. We especially thank Gérard Biau, Djalil Chafaï, Sjoerd Dirksen,

Guillaume Lecué, Michel Ledoux, Mylène Maïda, Andrew Nobel, and Raphaël Rossignol for their thorough reading and many very good comments on different parts of the book.

Our view of the subject formed, shaped, and matured thanks to conversations and discussions with our colleagues and collaborators, such as Louigi Addario-Berry, Sylvain Arlot, Peter Bartlett, Lucien Birgé, Charles Bordenave, Nicolas Broutin, Sébastien Bubeck, Ismael Castillo, Nicolò Cesa-Bianchi, Stéphan Clémençon, Pedro Delicado, Luc Devroye, Maxime Fevrier, Evarist Giné, András György, Laci Györfi, Gérard Kérkyacharian, Vladimir Koltchinskii, Tamás Linder, Edouard Maurel-Segala, Andreas Maurer, Colin McDiarmid, Shahar Mendelson, Richard Nickl, Omiros Papaspiliopoulos, Sasha Rakhlin, Bruce Reed, Adrien Saumard, Gilles Stoltz, Sasha Tsybakov, Frederic Udina, Ramon van Handel, Nicolas Vayatis, and Marten Wegkamp. We have been extremely lucky to be able to work with and learn from such a wonderful group of people.

We are grateful to Sohail Bahmani, Fabian Gieringer, Laci Györfi, Kevin Jamieson, Kengo Kato, Santosh Kumar, Andrew Nobel, Yuval Peres, Lam Pham, Sasha Rakhlin, Sergio Verdú, and Songfeng Zheng for pointing out mistakes and misprints in the first printing. Needless to say, all remaining errors are solely the authors' responsibility.

2

Basic Inequalities

Our main concern in this book is to understand under what conditions random variables are concentrated around their expected values. The random variables on which we focus are functions of several independent random variables. In a certain sense, this book is a study of independence, possibly the most important notion of probability theory.

The most basic concentration results are the laws of large numbers that state that averages of independent random variables are, under mild integrability conditions, close to their expectations with high probability. Of course, laws of large numbers have been thoroughly studied in classical probability theory. More recent results reveal that such concentration behavior is shared by a large class of general functions of independent variables, and this is precisely the subject of our book.

While laws of large numbers are asymptotic in nature, we are interested in more quantitative results. Throughout the book we focus on concentration inequalities that hold for a fixed sample size. In this chapter we recall some useful inequalities for sums (or averages) of independent random variables. This exercise is useful not only because the results will serve as a reference for comparison with other, more general, concentration inequalities, but also because some of the basic proof techniques appear in more general contexts.

Concentration properties of sums of independent variables are sensitive to the integrability of the individual terms. In the most favorable situations one can derive exponential tail bounds. We pay special attention to the cases when the sums exhibit a certain sub-Gaussian behavior, though often tail probabilities decrease significantly slower than those of a Gaussian random variable. In such problems inequalities for moments of the random variable in focus may prove to be useful.

We start this chapter by reviewing some elementary facts about tail probabilities. Then, in Section 2.2, we describe the so-called Cramér–Chernoff method, the basic technique for deriving exponential upper bounds for tail probabilities. In Sections 2.3 and 2.4 we single out two types of tail behaviors that we often face. We call these sub-Gaussian and sub-gamma random variables and we characterize them in terms of the behavior of their moments.

In Section 2.5 a simple useful inequality is presented for bounding the expected maximum of random variables.

Hoeffding's inequality, Bennett's inequality, and Bernstein's inequality are three classical benchmark inequalities for sums of independent random variables that are shown and proved in Sections 2.6, 2.7, and 2.8.

In Section 2.9 we describe the Johnson–Lindenstrauss lemma as an interesting application of concentration of sums of independent random variables. Later in the book we return to this example and its modifications to illustrate some of the results.

Finally, some simple association inequalities are presented in Section 2.10, while Minkowski's inequality is the subject of Section 2.11.

2.1 From Moments to Tails

In this book, by a concentration inequality we usually mean an upper bound for the probability that a real-valued random variable Z differs from its expected value by more than a certain amount. In other words, we seek upper bounds for tail probabilities of the form

$$\mathbf{P}\{Z - \mathbf{E}Z \geq t\} \quad \text{and} \quad \mathbf{P}\{Z - \mathbf{E}Z \leq -t\}$$

where $t > 0$. Of course, here we assume implicitly that the expected value $\mathbf{E}Z$ exists.

An elementary, yet powerful device to bound such tail probabilities is based on *Markov's inequality*. To derive Markov's inequality, simply note that, given a nonnegative random variable Y, for all $t > 0$, $Y\mathbb{1}_{\{Y \geq t\}} \geq t\mathbb{1}_{\{Y \geq t\}}$. Taking expectations of both sides of this inequality, we get Markov's inequality:

$$\mathbf{P}\{Y \geq t\} \leq \frac{\mathbf{E}\left[Y\mathbb{1}_{\{Y \geq t\}}\right]}{t} \leq \frac{\mathbf{E}Y}{t}.$$

Of course, this inequality is interesting only if $\mathbf{E}Y < \infty$, that is, if Y is integrable. An obvious way of using Markov's inequality to obtain concentration inequalities is to apply it to $Y = |Z - \mathbf{E}Z|$. However, with a simple trick Markov's inequality can be boosted, leading to much sharper estimates. Such an improvement is possible whenever Z satisfies stronger integrability conditions. The idea is to apply Markov's inequality to a convenient transformation of $Z - \mathbf{E}Z$ rather than to just $|Z - \mathbf{E}Z|$. If ϕ denotes a nondecreasing and nonnegative function defined on a (possibly infinite) interval $I \subset \mathbb{R}$, and if Y denotes a random variable taking values in I, then Markov's inequality implies that for every $t \in I$ with $\phi(t) > 0$,

$$\mathbf{P}\{Y \geq t\} \leq \mathbf{P}\{\phi(Y) \geq \phi(t)\} \leq \frac{\mathbf{E}\phi(Y)}{\phi(t)}. \tag{2.1}$$

The most common application of this is in *Chebyshev's inequality*, obtained by taking $\phi(t) = t^2$ over $I = (0, \infty)$ and $Y = |Z - \mathbf{E}Z|$. In this case we get

$$\mathbf{P}\left\{|Z - \mathbf{E}Z| \geq t\right\} \leq \frac{\operatorname{Var}(Z)}{t^2}.$$

More generally, we may take $\phi(t) = t^q$ for some $q > 0$. Then for all $t > 0$ we have

$$P\left\{|Z - EZ| \geq t\right\} \leq \frac{E\left[|Z - EZ|^q\right]}{t^q}.$$

If the random variable Z is such that $E|Z|^q < \infty$ for all $q > 0$ then one may choose the value of q to optimize the obtained upper bound.

The prominent role of Chebyshev's inequality is not only explained historically, but also because among all the absolute moments of the form $E\left[|Z - EZ|^q\right]$, the variance is typically the easiest to handle. This is certainly the case when Z is a sum of independent random variables $Z = X_1 + \cdots + X_n$. In this case, since the expected value of a product of independent random variables equals the product of their expectations, we have

$$\operatorname{Var}(Z) = \sum_{i=1}^{n} \operatorname{Var}(X_i)$$

and Chebyshev's inequality becomes

$$P\left\{\frac{1}{n}\left|\sum_{i=1}^{n}(X_i - EX_i)\right| \geq t\right\} \leq \frac{\sigma^2}{nt^2},$$

where $\sigma^2 = n^{-1}\sum_{i=1}^{n}\operatorname{Var}(X_i)$.

There is a whole family of choices of the function ϕ for which the upper tail bound obtained by Markov's inequality can be conveniently handled for sums of independent random variables. These are exponential functions of the form $\phi(t) = e^{\lambda t}$ where λ is a positive number. In this case Markov's inequality implies

$$P\{Z \geq t\} \leq \frac{Ee^{\lambda Z}}{e^{\lambda t}},$$

that is, the *moment generating function* $F(\lambda) = Ee^{\lambda Z}$, defined for all $\lambda \in \mathbb{R}$, appears in the upper bound. If $Z = X_1 + \cdots + X_n$ is a sum of independent random variables, then by independence,

$$Ee^{\lambda\sum_{i=1}^{n}(X_i - EX_i)} = \prod_{i=1}^{n} Ee^{\lambda(X_i - EX_i)}.$$

This simple observation forms the basis of the *Cramér–Chernoff method* that we study in the next section. The main idea is to control the moment-generating function of a random variable and then to optimize, in λ, the tail bound obtained by Markov's inequality. Even though moment bounds are sharper than the ones obtained by the Cramér–Chernoff method (see Exercise 2.5), the advantages offered by the equation above make the Cramér–Chernoff method an attractive and convenient tool for bounding tail probabilities of sums of independent random variables. When the moment-generating function exists for non-zero values of λ, this technique leads to exponential bounds for the tail

$$P\left\{|Z - EZ| \geq t\right\}.$$

Since this probability is bounded by

$$P\{Z - EZ \geq t\} + P\{EZ - Z \geq t\},$$

considering either $\widetilde{Z} = Z - EZ$ or $\widetilde{Z} = EZ - Z$, we can focus on exponential bounds for $P\{Z \geq t\}$ where Z is a centered random variable.

2.2 The Cramér–Chernoff Method

In this section we describe and formalize the Cramér–Chernoff bounding method. This method determines the best possible bound for a tail probability that one can possibly obtain using Markov's inequality with an exponential function $\phi(t) = e^{\lambda t}$ in (2.1). This simple technique leads to surprisingly sharp bounds in many cases. We work out some simple examples.

Let Z be a real-valued random variable. For $\lambda \geq 0$, Markov's inequality (2.1) implies

$$P\{Z \geq t\} \leq e^{-\lambda t} E e^{\lambda Z}.$$

Since this inequality holds for all values of $\lambda \geq 0$, one may choose λ to minimize the upper bound. Defining the logarithm of the moment-generating function as

$$\psi_Z(\lambda) = \log E e^{\lambda Z} \quad \text{for all} \quad \lambda \geq 0,$$

and introducing

$$\psi_Z^*(t) = \sup_{\lambda \geq 0} (\lambda t - \psi_Z(\lambda)),$$

we obtain *Chernoff's inequality*:

$$P\{Z \geq t\} \leq \exp(-\psi_Z^*(t)).$$

The function ψ_Z^* is called the *Cramér transform* of Z. Since $\psi_Z(0) = 0$, ψ_Z^* is a nonnegative function. If EZ exists, then the convexity of the exponential function and Jensen's inequality imply that $\psi_Z(\lambda) \geq \lambda EZ$ and therefore, for all negative values of λ, $\lambda t - \psi_Z(\lambda) \leq 0$ whenever $t \geq EZ$. This means that we may formally extend the supremum over all $\lambda \in \mathbb{R}$ in the definition of the Cramér transform:

$$\psi_Z^*(t) = \sup_{\lambda \in \mathbb{R}} (\lambda t - \psi_Z(\lambda)).$$

The expression of the right-hand side is known as the *Fenchel–Legendre dual function* of ψ_Z. Thus, at every $t \geq EZ$, the Cramér transform $\psi_Z^*(t)$ coincides with the Fenchel–Legendre dual.

Of course Chernoff's inequality is trivial whenever $\psi_Z^*(t) = 0$. This is the case if $\psi_Z(\lambda) = \infty$ for all positive λ or if $t \leq \boldsymbol{E}Z$ (using again the lower bound $\psi_Z(\lambda) \geq \lambda \boldsymbol{E}Z$). To avoid such trivialities, we assume that there exists a $\lambda > 0$ such that $\boldsymbol{E}e^{\lambda Z} < \infty$. It is easy to see (e.g. by applying Hölder's inequality) that the set of all such positive values of λ is an interval whose left end point equals 0. Denote by b the supremum of this interval so that $0 < b \leq \infty$. Then ψ_Z is convex (strictly convex whenever Z is not almost surely constant) and infinitely many times differentiable on $I = (0, b)$.

The case when Z is centered (i.e. $\boldsymbol{E}Z = 0$) is of special interest. In such a case ψ_Z is continuously differentiable on $[0, b)$ with $\psi_Z'(0) = \psi_Z(0) = 0$. We can also write the Cramér transform as $\psi_Z^*(t) = \sup_{\lambda \in I} (\lambda t - \psi_Z(\lambda))$. We leave the proof of these basic properties of ψ_Z to the reader (see Exercise 2.6).

Differentiability of ψ_Z implies that the Cramér transform can be computed by differentiating $\lambda t - \psi_Z(\lambda)$ with respect to λ. The optimizing value of λ is found by setting the derivative to zero, that is,

$$\psi_Z^*(t) = \lambda_t t - \psi_Z(\lambda_t)$$

where λ_t is such that $\psi_Z'(\lambda_t) = t$. The strict convexity of ψ_Z implies that ψ_Z' has an increasing inverse $(\psi_Z')^{-1}$ on the interval $\psi_Z'(I) \stackrel{\text{def}}{=} (0, B)$ and therefore, for any $t \in (0, B)$,

$$\lambda_t = (\psi_Z')^{-1}(t).$$

In the rest of this section we use this simple formula to compute the Cramér transform explicitly in three illustrative cases.

Normal random variables Let Z be a centered normal random variable with variance σ^2. Then

$$\psi_Z(\lambda) = \frac{\lambda^2 \sigma^2}{2} \quad \text{and} \quad \lambda_t = \frac{t}{\sigma^2}$$

and therefore, for every $t > 0$,

$$\psi_Z^*(t) = \frac{t^2}{2\sigma^2}.$$

Hence, Chernoff's inequality implies, for all $t > 0$,

$$\boldsymbol{P}\{Z \geq t\} \leq e^{-t^2/(2\sigma^2)}.$$

Chernoff's inequality appears to be quite sharp in this case. In fact, one can show that it cannot be improved uniformly by more than a factor of $1/2$ (see Exercise 2.7).

Poisson random variables Let Y be a Poisson random variable with parameter v, that is, $\mathbb{P}\{Y = k\} = e^{-v}v^k/k!$ for all $k = 0, 1, 2, \ldots$. Let $Z = Y - v$ be the corresponding centered variable. Then by direct calculation,

$$\mathbb{E}e^{\lambda Z} = e^{-\lambda v} \sum_{k=0}^{\infty} e^{\lambda k} e^{-v} \frac{v^k}{k!} = e^{-\lambda v - v} \sum_{k=0}^{\infty} \frac{\left(ve^{\lambda}\right)^k}{k!} = e^{-\lambda v - v} e^{ve^{\lambda}},$$

and consequently,

$$\psi_Z(\lambda) = v\left(e^{\lambda} - \lambda - 1\right) \quad \text{and} \quad \lambda_t = \log\left(1 + \frac{t}{v}\right).$$

Therefore the Cramér transform equals, for every $t > 0$,

$$\psi_Z^*(t) = vh\left(\frac{t}{v}\right)$$

where the function h is defined, for all $x \geq -1$, by $h(x) = (1 + x)\log(1 + x) - x$. Similarly, for every $t \leq v$,

$$\psi_{-Z}^*(t) = vh\left(-\frac{t}{v}\right).$$

Bernoulli random variables In our third principal example, let Y be a Bernoulli random variable with probability of success p, that is, $\mathbb{P}\{Y = 1\} = 1 - \mathbb{P}\{Y = 0\} = p$. Denote by $Z = Y - p$ the centered version of Y. If $0 < t < 1 - p$, we have

$$\psi_Z(\lambda) = \log\left(pe^{\lambda} + 1 - p\right) - \lambda p \quad \text{and} \quad \lambda_t = \log\frac{(1-p)(p+t)}{p(1-p-t)}$$

and therefore, for every $t \in (0, 1-p)$,

$$\psi_Z^*(t) = (1 - p - t)\log\frac{1-p-t}{1-p} + (p+t)\log\frac{p+t}{p}.$$

Equivalently, setting $a = t + p$ for every $a \in (p, 1)$,

$$\psi_Z^*(t) = h_p(a) \stackrel{\text{def}}{=} (1-a)\log\frac{1-a}{1-p} + a\log\frac{a}{p}.$$

We note here that $h_p(a)$ is just the *Kullback–Leibler divergence* $D(P_a \| P_p)$ between a Bernoulli distribution P_a of parameter a and a Bernoulli distribution P_p of parameter p (see Chapter 4 for the definition).

Sums of independent random variables The reason why Chernoff's inequality became popular is that it is very simple to use when applied to a sum of independent random variables. As an illustration, assume that $Z = X_1 + \cdots + X_n$ where $X_1, \ldots, X_n$ are independent and identically distributed real-valued random variables. Denote the logarithm of the moment-generating function of the X_i by $\psi_X(\lambda) = \log \boldsymbol{E}e^{\lambda X_i}$, and the corresponding Cramér transform by $\psi_X^*(t)$. Then, by independence, for all λ for which $\psi_X(\lambda) < \infty$,

$$\psi_Z(\lambda) = \log \boldsymbol{E}e^{\lambda \sum_{i=1}^n X_i} = \log \prod_{i=1}^n \boldsymbol{E}e^{\lambda X_i} = n\psi_X(\lambda)$$

and consequently,

$$\psi_Z^*(t) = n\psi_X^*\left(\frac{t}{n}\right).$$

As an example, consider a random variable Y with binomial distribution with parameters n and p. In other words, Y is the sum of n independent and identically distributed Bernoulli (p) random variables. Then, for all $0 < t < n(1-p)$, the Cramér transform of $Z = Y - np$ equals

$$\psi_Z^*(t) = nh_p(t/n + p)$$

and therefore, by Chernoff's inequality,

$$\boldsymbol{P}\{Z \geq t\} \leq \exp\left(-nh_p(t/n + p)\right).$$

We refer to the exercises for several simple versions of Chernoff's inequality for binomial random variables.

2.3 Sub-Gaussian Random Variables

Many important classes of random variables have tail probabilities decreasing at least as rapidly as normally distributed random variables. In order to facilitate the exploration of this phenomenon, we find it useful to formalize the notion of a sub-Gaussian random variable. There are several ways to do this and we propose the following definition, based on the logarithmic moment-generating function $\psi_X(\lambda) = \log \boldsymbol{E}e^{\lambda X}$ of a random variable X: A centered random variable X is said to be *sub-Gaussian* with *variance factor* v if

$$\psi_X(\lambda) \leq \frac{\lambda^2 v}{2} \quad \text{for every} \quad \lambda \in \mathbb{R}.$$

We denote the collection of such random variables by $\mathcal{G}(v)$.

Note that this definition does not require the variance of X to be equal to v, just that it is bounded by v (see Exercise 2.16). This definition is natural as we know from the previous section that $\exp(\lambda^2 v/2)$ is the moment-generating function of a centered normal random variable Y with variance v. Hence, the above definition says that a centered random variable X belongs to $\mathcal{G}(v)$ if the moment-generating function of X is dominated by the moment-generating function of Y. This notion is also convenient because it is naturally stable under convolution in the sense that if $X_1, \ldots, X_n$ are independent such that for every i, $X_i \in \mathcal{G}(v_i)$, then $\sum_{i=1}^n X_i \in \mathcal{G}\left(\sum_{i=1}^n v_i\right)$.

Characterization Next we connect the notion of a sub-Gaussian random variable with some other standard ways of defining sub-Gaussian distributions. First observe that Chernoff's inequality implies that the tail probabilities of a sub-Gaussian random variable are dominated by the corresponding Gaussian tail probabilities. More precisely, if X belongs to $\mathcal{G}(v)$, then for every $t > 0$,

$$\mathbf{P}\{X > t\} \vee \mathbf{P}\{-X > t\} \leq e^{-t^2/(2v)}$$

where $a \vee b$ denotes the maximum of a and b. In fact, one can characterize sub-Gaussian variables in terms of their tail probabilities and also in terms of the growth of their moments, as summarized in the following theorem.

Theorem 2.1 *Let X be a random variable with $\mathbf{E}X = 0$. If for some $v > 0$*

$$\mathbf{P}\{X > x\} \vee \mathbf{P}\{-X > x\} \leq e^{-x^2/(2v)} \quad \textit{for all } x > 0 \tag{2.2}$$

then for every integer $q \geq 1$,

$$\mathbf{E}\left[X^{2q}\right] \leq 2q!(2v)^q \leq q!(4v)^q. \tag{2.3}$$

Conversely, if for some positive constant C

$$\mathbf{E}\left[X^{2q}\right] \leq q!C^q,$$

then $X \in \mathcal{G}(4C)$ (and therefore (2.2) holds with $v = 4C$).

Proof Assume first (2.2). We may assume that $v = 1$ since otherwise one can apply the result for the random variable $X/\sqrt{v}$. We have

$$\begin{aligned}
\mathbf{E}\left[X^{2q}\right] &= \int_0^\infty \mathbf{P}\left\{|X|^{2q} > x\right\} dx \\
&= 2q \int_0^\infty x^{2q-1}\mathbf{P}\left\{|X| > x\right\} dx \leq 4q \int_0^\infty x^{2q-1}e^{-x^2/2}dx.
\end{aligned}$$

By setting $x = \sqrt{2t}$, the previous inequality becomes

$$\mathbf{E}\left[X^{2q}\right] \leq 4q \int_0^\infty (2t)^{q-1}e^{-t}dt = 2^{q+1}q!,$$

which implies (2.3). Conversely, assume $\mathbf{E}\left[X^{2q}\right] \leq q!C^q$ and introduce an independent copy X' of X. Then by symmetry of $X - X'$ we have

$$Ee^{\lambda X}Ee^{-\lambda X} = Ee^{\lambda(X-X')} = \sum_{q=0}^{\infty} \frac{\lambda^{2q}E\left[(X-X')^{2q}\right]}{(2q)!}$$

for every $\lambda \in \mathbb{R}$. Now, by convexity of $x \to x^{2q}$,

$$E\left[(X-X')^{2q}\right] \leq 2^{2q-1}\left(E\left[X^{2q}\right]+E\left[X'^{2q}\right]\right) = 2^{2q}E\left[X^{2q}\right]$$

and therefore, using our assumption for the moments of X, we have

$$Ee^{\lambda X}Ee^{-\lambda X} = \sum_{q=0}^{\infty} \frac{\lambda^{2q}E\left[(X-X')^{2q}\right]}{(2q)!} \leq \sum_{q=0}^{\infty} \frac{\lambda^{2q}2^{2q}C^q q!}{(2q)!}.$$

Observe that since X is centered, $Ee^{-\lambda X} \geq 1$ and that for every integer $q \geq 1$,

$$\frac{(2q)!}{q!} = \prod_{j=1}^{q}(q+j) \geq \prod_{j=1}^{q}(2j) = 2^q q!.$$

Using these observations, we conclude that

$$Ee^{\lambda X} \leq \sum_{q=0}^{\infty} \frac{\lambda^{2q}2^q C^q}{q!} = e^{2\lambda^2 C},$$

that is, $X \in \mathcal{G}(4C)$. □

Finally, we mention that the growth condition for the moments of X given in Theorem 2.1 is equivalent to another condition that is often used as an alternative definition of sub-Gaussian variables. As this condition states that for some $\alpha > 0$,

$$E\exp\left(\alpha X^2\right) \leq 2 \tag{2.4}$$

then clearly

$$\sum_{q=1}^{\infty} \frac{\alpha^q E\left[X^{2q}\right]}{q!} \leq 1,$$

which implies that $E\left[X^{2q}\right] \leq \alpha^{-q}q!$ (and therefore that $X \in \mathcal{G}(4/\alpha)$). Conversely, if $E\left[X^{2q}\right] \leq C^q q!$ for every integer q (which holds with $C = 4v$ whenever $X \in \mathcal{G}(v)$), then, setting $\alpha = 1/(2C)$,

$$E\exp\left(\alpha X^2\right) = \sum_{q=0}^{\infty} \frac{\alpha^q E\left[X^{2q}\right]}{q!} \leq \sum_{q=0}^{\infty} 2^{-q} = 2.$$

Therefore, for a centered random variable X, condition (2.4) holds for some positive α if, and only if, X is sub-Gaussian with variance factor v, for some $v \in [2/\alpha, 4/\alpha]$.

Bounded variables Bounded variables are an important class of sub-Gaussian random variables. The sub-Gaussian property of bounded random variables is established by the following lemma:

Lemma 2.2 (HOEFFDING'S LEMMA) *Let Y be a random variable with $\mathbf{E}Y = 0$, taking values in a bounded interval $[a, b]$ and let $\psi_Y(\lambda) = \log \mathbf{E}e^{\lambda Y}$. Then $\psi_Y''(\lambda) \le (b-a)^2/4$ and $Y \in \mathcal{G}((b-a)^2/4)$.*

Proof Observe first that

$$\left| Y - \frac{(b+a)}{2} \right| \le \frac{(b-a)}{2}$$

and therefore

$$\mathrm{Var}(Y) = \mathrm{Var}(Y - (b+a)/2) \le \frac{(b-a)^2}{4}.$$

Now, let P denote the distribution of Y and let P_λ be the probability distribution with density

$$x \to e^{-\psi_Y(\lambda)} e^{\lambda x}$$

with respect to P. Since P_λ is concentrated on $[a, b]$, the variance of a random variable Z with distribution P_λ is bounded by $(b-a)^2/4$. Hence, by an elementary computation,

$$\begin{aligned}\psi_Y''(\lambda) &= e^{-\psi_Y(\lambda)} \mathbf{E}\left[Y^2 e^{\lambda Y}\right] - e^{-2\psi_Y(\lambda)} \left(\mathbf{E}\left[Y e^{\lambda Y}\right]\right)^2 \\ &= \mathrm{Var}(Z) \le \frac{(b-a)^2}{4}.\end{aligned}$$

The sub-Gaussian property follows by noting that $\psi_Y(0) = \psi_Y'(0) = 0$, and by Taylor's theorem that implies that, for some $\theta \in [0, \lambda]$,

$$\psi_Y(\lambda) = \psi_Y(0) + \lambda\psi_Y'(0) + \frac{\lambda^2}{2}\psi_Y''(\theta) \le \frac{\lambda^2(b-a)^2}{8}.$$ □

The upper bound on the variance factor is sharp in the special case of a *Rademacher* random variable X whose distribution is defined by $\mathbf{P}\{X = -1\} = \mathbf{P}\{X = 1\} = 1/2$. Then one may take $a = -b = 1$ and $\mathrm{Var}(X) = 1 = (b-a)^2/4$.

2.4 Sub-Gamma Random Variables

Apart from sub-Gaussian random variables, we will often encounter random variables that are not quite sub-Gaussian, but nearly. In order to understand these variables, we consider here a somewhat less stringent condition on the moment-generating function. A real-valued centered random variable X is said to be *sub-gamma on the right tail with variance factor v and scale parameter c* if

$$\psi_X(\lambda) \le \frac{\lambda^2 v}{2(1-c\lambda)} \text{ for every } \lambda \quad \text{such that} \quad 0 < \lambda < 1/c.$$

We denote the collection of such random variables by $\Gamma_+(v, c)$. Similarly, X is said to be *sub-gamma on the left tail with variance factor v and scale parameter c* if $-X$ is sub-gamma on the right tail with variance factor v and tail parameter c. We denote the collection of such random variables by $\Gamma_-(v, c)$. Finally, X is simply said to be *sub-gamma with variance factor v and scale parameter c* if X is sub-gamma both on the right and left tails with the same variance factor v and scale parameter c. The collection of such random variables is denoted by $\Gamma(v, c)$. Observe that $\Gamma(v, 0) = \mathcal{G}(v)$. To explain the terminology, consider a random variable Y with *gamma distribution* with parameters $a, b \ge 0$. Then its centered version $X = Y - \mathbf{E}Y$ is a typical example of a sub-gamma variable. To see this, recall first that Y has density

$$f(x) = \frac{x^{a-1}e^{-x/b}}{\Gamma(a)b^a}, \quad x \ge 0$$

where $\Gamma(a) = \int_0^\infty x^{a-1}e^{-x}dx$ is Euler's Gamma function. It is easy to see that $\mathbf{E}Y = ab$ and $\text{Var}(Y) = ab^2$. Then, for all $\lambda < 1/b$,

$$\mathbf{E}e^{\lambda X} = \int_0^\infty e^{\lambda(x-ab)}f(x)dx = \exp\left(-\lambda ab - a\log(1-\lambda b)\right).$$

It is also easy to see that for all $u \in (0, 1)$,

$$-\log(1-u) - u \le \frac{u^2}{2(1-u)}$$

(see Exercise 2.8), so the logarithmic moment-generating function of X may be bounded, for all $\lambda \in (0, 1/b)$, as

$$\psi_X(\lambda) = a\left(-\log(1-\lambda b) - \lambda b\right) \le \frac{\lambda^2 v}{2(1-c\lambda)}$$

where $v = ab^2$ and $c = b$. This shows that X is a sub-gamma random variable on the right tail, with variance factor ab^2 and scale parameter b, that is, X belongs to $\Gamma_+(ab^2, b)$. Since the distribution of X is not symmetric around 0, the behavior of X on the left tail is slightly different. Indeed, for all $u < 0$,

$$-\log(1-u) - u \le \frac{u^2}{2},$$

and therefore, for all $\lambda < 0$,

$$\psi_X(\lambda) = a\left(-\log(1-\lambda b) - \lambda b\right) \le \frac{\lambda^2 v}{2}$$

where $v = ab^2$. This shows that X is more concentrated on the left tail than on the right tail. In fact, the left tail of X has a sub-Gaussian behavior. X belongs to $\Gamma_-(ab^2, 0)$ and to $\Gamma_+(ab^2, b)$ and therefore, to $\Gamma(ab^2, b)$.

Characterization Similarly to the sub-Gaussian property, the sub-gamma property can be characterized in terms of tail or moment conditions. We start by computing the Fenchel–Legendre dual function of

$$\psi(\lambda) = \frac{v\lambda^2}{2(1 - c\lambda)}.$$

Setting

$$h_1(u) = 1 + u - \sqrt{1 + 2u} \text{ for } u > 0,$$

it follows by elementary calculation that for every $t > 0$,

$$\psi^*(t) = \sup_{\lambda \in (0,1/c)} \left(t\lambda - \frac{\lambda^2 v}{2(1 - c\lambda)} \right) = \frac{v}{c^2} h_1\left(\frac{ct}{v}\right). \tag{2.5}$$

Since h_1 is an increasing function from $(0, \infty)$ onto $(0, \infty)$ with inverse function $h_1^{-1}(u) = u + \sqrt{2u}$ for $u > 0$, we finally get

$$\psi^{*-1}(u) = \sqrt{2vu} + cu.$$

Hence, Chernoff's inequality implies that whenever X is a sub-gamma random variable on the right tail with variance factor v and scale parameter c, for every $t > 0$, we have

$$\mathbf{P}\{X > t\} \leq \exp\left(-\frac{v}{c^2} h_1\left(\frac{ct}{v}\right)\right),$$

or equivalently, for every $t > 0$,

$$\mathbf{P}\left\{X > \sqrt{2vt} + ct\right\} \leq e^{-t}.$$

Therefore, if X belongs to $\Gamma(v, c)$, then for every $t > 0$,

$$\mathbf{P}\left\{X > \sqrt{2vt} + ct\right\} \vee \mathbf{P}\left\{-X > \sqrt{2vt} + ct\right\} \leq e^{-t}.$$

Such a behavior for the tails essentially characterizes sub-gamma random variables. More precisely, we have the following.

Theorem 2.3 *Let X be a centered random variable. If for some $v > 0$*

$$\mathbf{P}\left\{X > \sqrt{2vt} + ct\right\} \vee \mathbf{P}\left\{-X > \sqrt{2vt} + ct\right\} \leq e^{-t} \text{ for every } t > 0, \tag{2.6}$$

then for every integer $q \geq 1$

$$\mathbf{E}\left[X^{2q}\right] \leq q!(8v)^q + (2q)!(4c)^{2q}. \tag{2.7}$$

Conversely, if for some positive constants A and B,

$$\mathbf{E}\left[X^{2q}\right] \leq q!A^q + (2q)!B^{2q}, \tag{2.8}$$

then $X \in \Gamma\left(4\left(A+B^2\right), 2B\right)$ (and therefore (2.6) also holds with $v = 4\left(A+B^2\right)$ and $c = 2B$).

Proof Assume first that (2.6) holds. Using integration by parts,

$$\mathbf{E}\left[X^{2q}\right] = 2q \int_0^\infty x^{2q-1}\mathbf{P}\left\{|X| > x\right\} dx.$$

Setting $x = \sqrt{2vt} + ct$ and using (2.6), this implies

$$\begin{aligned}
\mathbf{E}\left[X^{2q}\right] &\leq 4q \int_0^\infty \left(\sqrt{2vt} + ct\right)^{2q-1} \left(\frac{\sqrt{2vt} + 2ct}{2t}\right) e^{-t} dt \\
&\leq 2q \int_0^\infty \left(\sqrt{2vt} + 2ct\right)^{2q} \frac{e^{-t}}{t} dt.
\end{aligned}$$

By convexity of x^{2q}, $(a+b)^{2q} \leq 2^{2q-1}\left(a^{2q} + b^{2q}\right)$, and therefore

$$\begin{aligned}
\mathbf{E}\left[X^{2q}\right] &\leq q2^{2q} \int_0^\infty \left((2tv)^q + (2ct)^{2q}\right) \frac{e^{-t}}{t} dt \\
&\leq 2^{2q}\left(q!2^q v^q + \frac{(2q)!}{2}(2c)^{2q}\right),
\end{aligned}$$

(2.7) holds. Conversely, assuming (2.8), we may use the same symmetrization trick as for the characterization of the sub-Gaussian property in terms of moments. Considering an independent copy X' of X, we have

$$\mathbf{E}e^{\lambda X}\mathbf{E}e^{-\lambda X} = \mathbf{E}e^{\lambda(X-X')} = \sum_{q=0}^\infty \frac{\lambda^{2q}\mathbf{E}\left[(X-X')^{2q}\right]}{(2q)!}.$$

By convexity, again we note that

$$\mathbf{E}\left[(X-X')^{2q}\right] \leq 2^{2q-1}\left(\mathbf{E}\left[X^{2q}\right] + \mathbf{E}\left[X'^{2q}\right]\right) = 2^{2q}\mathbf{E}\left[X^{2q}\right]$$

and plugging this inequality together with (2.8) into the previous equation leads to

$$\mathbf{E}e^{\lambda X}\mathbf{E}e^{-\lambda X} \leq \sum_{q=0}^\infty \frac{\lambda^{2q}2^{2q}\left(A^q q! + B^{2q}(2q)!\right)}{(2q)!}.$$

Using again $q!/(2q)! \leq 2^{-q}/q!$ and that $EX = 0$ implies $Ee^{-\lambda X} \geq 1$, we obtain, for every λ with $2B|\lambda| < 1$,

$$Ee^{\lambda X} \leq e^{2A\lambda^2} + \frac{4B^2\lambda^2}{1 - 4B^2\lambda^2} \leq e^{2A\lambda^2} + \frac{4B^2\lambda^2}{1 - 2B|\lambda|}.$$

The final result follows from the elementary inequality

$$e^x + y \leq e^{x+y}$$

which holds for all $x, y > 0$. □

2.5 A Maximal Inequality

The purpose of this section is to show how information about the Cramér transform of random variables in a finite collection can be used to bound the expected maximum of these random variables.

The main idea is perhaps most transparent if we consider sub-Gaussian random variables. Let $Z_1, \ldots, Z_N$ be real-valued random variables where a $v > 0$ exists such that for every $i = 1, \ldots, N$, the logarithm of the moment-generating function of Z_i satisfies $\psi_{Z_i}(\lambda) \leq \lambda^2 v/2$ for all $\lambda > 0$. Then, by Jensen's inequality,

$$\begin{aligned} \exp\left(\lambda E \max_{i=1,\ldots,N} Z_i\right) &\leq E \exp\left(\lambda \max_{i=1,\ldots,N} Z_i\right) = E \max_{i=1,\ldots,N} e^{\lambda Z_i} \\ &\leq \sum_{i=1}^{N} Ee^{\lambda Z_i} \leq Ne^{\lambda^2 v/2}. \end{aligned}$$

Taking logarithms on both sides, we have

$$E \max_{i=1,\ldots,N} Z_i \leq \frac{\log N}{\lambda} + \frac{\lambda v}{2}.$$

The upper bound is minimized for $\lambda = \sqrt{2\log N/v}$, which yields

$$E \max_{i=1,\ldots,N} Z_i \leq \sqrt{2v \log N}.$$

This simple bound is asymptotically sharp if the Z_i are i.i.d. normal random variables (see Exercise 2.17).

Of course, the argument above may be generalized beyond sub-Gaussian variables. Next we formalize such a general inequality but first we start with a technical result that establishes a useful formula for the inverse of the Fenchel–Legendre dual of a smooth convex function.

Lemma 2.4 *Let ψ be a convex and continuously differentiable function defined on the interval $[0, b)$ where $0 < b \leq \infty$. Assume that $\psi(0) = \psi'(0) = 0$ and set, for every $t \geq 0$,*

$$\psi^*(t) = \sup_{\lambda \in (0,b)} (\lambda t - \psi(\lambda)).$$

Then ψ^* is a nonnegative convex and nondecreasing function on $[0, \infty)$. Moreover, for every $y \geq 0$, the set $\{t \geq 0 : \psi^*(t) > y\}$ is non-empty and the generalized inverse of ψ^*, defined by

$$\psi^{*-1}(y) = \inf\{t \geq 0 : \psi^*(t) > y\},$$

can also be written as

$$\psi^{*-1}(y) = \inf_{\lambda \in (0,b)} \left[\frac{y + \psi(\lambda)}{\lambda}\right].$$

Proof By definition, ψ^* is the supremum of convex and nondecreasing functions on $[0, \infty)$ and $\psi^*(0) = 0$ and therefore, ψ^* is a nonnegative, convex, and nondecreasing function on $[0, \infty)$. Moreover, given $\lambda \in (0, b)$, since $\psi^*(t) \geq \lambda t - \psi(\lambda)$, ψ^* is unbounded, which shows that for every $y \geq 0$, the set $\{t \geq 0 : \psi^*(t) > y\}$ is non-empty. Defining

$$u = \inf_{\lambda \in (0,b)} \left[\frac{y + \psi(\lambda)}{\lambda}\right],$$

for every $t \geq 0$, we have $u \geq t$ if, and only if, for every $\lambda \in (0, b)$

$$\frac{y + \psi(\lambda)}{\lambda} \geq t.$$

Since this inequality implies $y \geq \psi^*(t)$, we have $\{t \geq 0 : \psi^*(t) > y\} = (u, \infty)$. This proves that $u = \psi^{*-1}(y)$ by definition of ψ^{*-1}. □

The next result offers a convenient bound for the expected value of the maximum of finitely many exponentially integrable random variables. This type of bound has been used in so-called chaining arguments for bounding suprema of Gaussian or empirical processes (see Chapter 13).

Theorem 2.5 *Let $Z_1, \ldots, Z_N$ be real-valued random variables such that for every $\lambda \in (0, b)$ and $i = 1, \ldots, N$, the logarithm of the moment-generating function of Z_i satisfies $\psi_{Z_i}(\lambda) \leq \psi(\lambda)$ where ψ is a convex and continuously differentiable function on $[0, b)$ with $0 < b \leq \infty$ such that $\psi(0) = \psi'(0) = 0$. Then*

$$\mathbf{E} \max_{i=1,\ldots,N} Z_i \leq \psi^{*-1}(\log N).$$

In particular, if the Z_i are sub-Gaussian with variance factor v, that is, $\psi(\lambda) = \lambda^2 v/2$ for every $\lambda \in (0, \infty)$, then

$$\mathbf{E} \max_{i=1,\dots,N} Z_i \leq \sqrt{2v \log N}.$$

Proof By Jensen's inequality,

$$\exp\left(\lambda \mathbf{E} \max_{i=1,\dots,N} Z_i\right) \leq \mathbf{E} \exp\left(\lambda \max_{i=1,\dots,N} Z_i\right) = \mathbf{E} \max_{i=1,\dots,N} \exp\left(\lambda Z_i\right)$$

for any $\lambda \in (0, b)$. Thus, recalling that $\psi_{Z_i}(\lambda) = \log \mathbf{E} \exp(\lambda Z_i)$,

$$\exp\left(\lambda \mathbf{E} \max_{i=1,\dots,N} Z_i\right) \leq \sum_{i=1}^{N} \mathbf{E} \exp\left(\lambda Z_i\right) \leq N \exp\left(\psi(\lambda)\right).$$

Therefore, for any $\lambda \in (0, b)$,

$$\lambda \mathbf{E} \max_{i=1,\dots,N} Z_i - \psi(\lambda) \leq \log N,$$

which means that

$$\mathbf{E} \max_{i=1,\dots,N} Z_i \leq \inf_{\lambda \in (0,b)} \left(\frac{\log N + \psi(\lambda)}{\lambda}\right)$$

and the result follows from Lemma 2.4. □

We may also apply Theorem 2.5 to establish a bound for the expected maximum of sub-gamma random variables.

Corollary 2.6 *Let $Z_1, \dots, Z_N$ be real-valued random variables belonging to $\Gamma_+(v, c)$ (see Section 2.4 for the definition). Then*

$$\mathbf{E} \max_{i=1,\dots,N} Z_i \leq \sqrt{2v \log N} + c \log N.$$

Example 2.7 (CHI-SQUARED DISTRIBUTION) An important example of gamma-distributed random variables is a chi-square random variable. If p is a positive integer, a gamma random variable with parameters $a = p/2$ and $b = 2$ is said to have *chi-square distribution* with p *degrees of freedom*. (Note that if $Y_1, \dots, Y_p$ are independent standard normal random variables then $\sum_{i=1}^{p} Y_i^2$ has *chi-square distribution* with p *degrees of freedom*.) Corollary 2.6 implies that if $X_1, \dots, X_N$ have chi-square distribution with p degrees of freedom, then

$$\mathbf{E}\left[\max_{i=1,\dots,N} X_i - p\right] \leq 2\sqrt{p \log N} + 2 \log N.$$

2.6 Hoeffding's Inequality

In the next few sections we establish some of the classical inequalities for tail probabilities of sums of independent real-valued random variables. The Cramér–Chernoff method is especially relevant in this case. In fact, it was invented for the study of sums of independent random variables. The key to success is that the exponential moment-generating function converts sums into products and the expected value of a product of independent random variables is just the product of their expected values. Indeed if $X_1, \ldots, X_n$ are independent random variables with a finite mean such that for some non-empty interval I, $\boldsymbol{E}e^{\lambda X_i}$ is finite for all $i \leq n$ and all $\lambda \in I$, then defining

$$S = \sum_{i=1}^{n} (X_i - \boldsymbol{E}X_i),$$

by independence, for all $\lambda \in I$,

$$\psi_S(\lambda) = \sum_{i=1}^{n} \log \boldsymbol{E}e^{\lambda(X_i - \boldsymbol{E}X_i)}.$$

This expression may now be bounded under various assumptions on the X_i and Chernoff's inequality may be used. We start with the perhaps simplest version for sums of bounded random variables. Recall that Hoeffding's lemma (Lemma 2.2) establishes a sub-Gaussian property of bounded random variables. Hoeffding's inequality is a straightforward consequence of Hoeffding's lemma and Chernoff's inequality.

Indeed, if X_i takes its values in a bounded interval $[a_i, b_i]$, for all $i \leq n$, then by Lemma 2.2,

$$\psi_S(\lambda) \leq \frac{\lambda^2}{8} \sum_{i=1}^{n} (b_i - a_i)^2.$$

The obtained tail inequality is the following.

Theorem 2.8 (HOEFFDING'S INEQUALITY) *Let $X_1, \ldots, X_n$ be independent random variables such that X_i takes its values in $[a_i, b_i]$ almost surely for all $i \leq n$. Let*

$$S = \sum_{i=1}^{n} (X_i - \boldsymbol{E}X_i).$$

Then for every $t > 0$,

$$\boldsymbol{P}\{S \geq t\} \leq \exp\left(-\frac{2t^2}{\sum_{i=1}^{n} (b_i - a_i)^2}\right).$$

We may apply Hoeffding's inequality to sums of random variables of the form $X_i = \varepsilon_i \alpha_i$ where $\varepsilon_1, \ldots, \varepsilon_n$ are independent Rademacher random variables (i.e. symmetric sign variables with $\boldsymbol{P}\{\varepsilon_i = 1\} = \boldsymbol{P}\{\varepsilon_i = -1\} = 1/2$) and $\alpha_1, \ldots, \alpha_n$ are real numbers. We get

$$\boldsymbol{P}\{S \geq t\} \leq \exp\left(-\frac{t^2}{2\sum_{i=1}^{n} \alpha_i^2}\right).$$

Since in this case $\mathrm{Var}(S) = \sum_{i=1}^n \alpha_i^2$, Hoeffding's inequality implies a bona fide sub-Gaussian tail inequality. In general, however, the variance of S may be much smaller than $\sum_{i=1}^n (b_i - a_i)^2$. In such cases sharper bounds are called for. Bennett's and Bernstein's inequalities discussed in the next sections provide such improvements.

2.7 Bennett's Inequality

As in the proof of Hoeffding's inequality, our starting point is the fact that the logarithmic moment-generating function of an independent sum equals the sum of the logarithmic moment-generating functions of the centered summands, that is,

$$\psi_S(\lambda) = \sum_{i=1}^n \left(\log \mathbf{E} e^{\lambda X_i} - \lambda \mathbf{E} X_i\right).$$

Using $\log u \leq u - 1$ for $u > 0$,

$$\psi_S(\lambda) \leq \sum_{i=1}^n \mathbf{E}\left[e^{\lambda X_i} - \lambda X_i - 1\right]. \tag{2.9}$$

Both Bennett's and Bernstein's inequalities may be derived from this bound, under different integrability conditions for the X_i.

Theorem 2.9 (BENNETT'S INEQUALITY) *Let $X_1, \ldots, X_n$ be independent random variables with finite variance such that $X_i \leq b$ for some $b > 0$ almost surely for all $i \leq n$. Let*

$$S = \sum_{i=1}^n (X_i - \mathbf{E} X_i)$$

and $v = \sum_{i=1}^n \mathbf{E}\left[X_i^2\right]$. If we write $\phi(u) = e^u - u - 1$ for $u \in \mathbb{R}$, then, for all $\lambda > 0$,

$$\log \mathbf{E} e^{\lambda S} \leq n \log\left(1 + \frac{v}{nb^2}\phi(b\lambda)\right) \leq \frac{v}{b^2}\phi(b\lambda),$$

and for any $t > 0$,

$$\mathbf{P}\{S \geq t\} \leq \exp\left(-\frac{v}{b^2} h\left(\frac{bt}{v}\right)\right)$$

where $h(u) = (1 + u)\log(1 + u) - u$ for $u > 0$.

Proof By homogeneity we may assume that $b = 1$. Note that $u^{-2}\phi(u)$ is a nondecreasing function of $u \in \mathbb{R}$ (where at 0 we continuously extend the function). Hence, for all $i \leq n$ and $\lambda > 0$,

$$e^{\lambda X_i} - \lambda X_i - 1 \leq X_i^2\left(e^\lambda - \lambda - 1\right)$$

which, following expectations, yields

$$\mathbf{E} e^{\lambda X_i} - \lambda \mathbf{E} X_i - 1 \leq \mathbf{E}\left[X_i^2\right]\phi(\lambda).$$

Here, we refrain from invoking $\log u \leq u - 1$, and sum these inequalities for $i = 1, \ldots, n$ so as to get,

$$\psi_S(\lambda) \leq \sum_{i=1}^{n} \left(\log\left(1 + \lambda EX_i + E\left[X_i^2\right]\phi(\lambda)\right) - \lambda EX_i\right).$$

Now, using the concavity of the logarithm,

$$\psi_S(\lambda) \leq n\left(\log\left(1 + \lambda \frac{\sum_{i=1}^n EX_i}{n} + \frac{v}{n}\phi(\lambda)\right) - \lambda \frac{\sum_{i=1}^n EX_i}{n}\right).$$

Finally, using $\log(1+u) \leq u$, the latter inequality entails

$$\psi_S(\lambda) \leq v\phi(\lambda).$$

Recall from Section 2.2 that the upper bound is just the logarithm of the moment-generating function of a centered Poisson random variable with parameter v. Therefore, the Cramér transform of S is also bounded by that of a corresponding Poisson random variable, that is,

$$\psi_S^*(t) \geq vh\left(\frac{t}{v}\right)$$

which proves the theorem via Chernoff's inequality. □

The easy-to-prove inequality

$$h(u) \geq \frac{u^2}{2(1+u/3)}$$

(see Exercise 2.8) may help interpret Bennett's inequality. This inequality implies that, under the conditions of Theorem 2.9,

$$P\{S \geq t\} \leq \exp\left(-\frac{t^2}{2(v+bt/3)}\right). \tag{2.10}$$

This is known as *Bernstein's Inequality*. For $t \gg v/b$, it loses a logarithmic factor in the exponent with respect to Bennett's inequality. On the other hand, if v is the dominant term in the denominator of the exponent, Bennett's and Bernstein's inequalities are almost equivalent and both provide a sub-Gaussian type inequality.

In the next section we show that Bernstein's inequality holds under weaker assumptions than boundedness.

2.8 Bernstein's Inequality

The next inequality is somewhat more general than the classical form of Bernstein's inequality shown in the previous section. Here, instead of boundedness, we only require an appropriate control of moments.

Theorem 2.10 (BERNSTEIN'S INEQUALITY) *Let $X_1, \ldots, X_n$ be independent real-valued random variables. Assume that there exist positive numbers v and c such that $\sum_{i=1}^n \mathbf{E}\left[X_i^2\right] \le v$ and*

$$\sum_{i=1}^n \mathbf{E}\left[(X_i)_+^q\right] \le \frac{q!}{2} v c^{q-2} \quad \text{for all integers } q \ge 3,$$

where $x_+ = \max(x, 0)$.
If $S = \sum_{i=1}^n (X_i - \mathbf{E}X_i)$, then for all $\lambda \in (0, 1/c)$ and $t > 0$,

$$\psi_S(\lambda) \le \frac{v\lambda^2}{2(1 - c\lambda)}$$

and

$$\psi_S^*(t) \ge \frac{v}{c^2} h_1\left(\frac{ct}{v}\right),$$

where $h_1(u) = 1 + u - \sqrt{1 + 2u}$ for $u > 0$. In particular, for all $t > 0$,

$$\mathbf{P}\left\{S \ge \sqrt{2vt} + ct\right\} \le e^{-t}.$$

Proof Recall the notation $\phi(u) = e^u - u - 1$ and observe that for $u \le 0$,

$$\phi(u) \le \frac{u^2}{2}.$$

Hence, for $\lambda > 0$, we have, for all $i \le n$,

$$\phi(\lambda X_i) \le \frac{\lambda^2 X_i^2}{2} + \sum_{q=3}^{\infty} \frac{\lambda^q (X_i)_+^q}{q!}$$

which implies, by the monotone convergence theorem,

$$\mathbf{E}\phi(\lambda X_i) \le \frac{\lambda^2 \mathbf{E}\left[X_i^2\right]}{2} + \sum_{q=3}^{\infty} \frac{\lambda^q \mathbf{E}\left[(X_i)_+^q\right]}{q!},$$

and therefore, by the assumptions of the theorem,

$$\sum_{i=1}^n \mathbf{E}\phi(\lambda X_i) \le \frac{v}{2} \sum_{q=2}^{\infty} \lambda^q c^{q-2}.$$

This proves, on the one hand, that for any $\lambda \in (0, 1/c)$, $e^{\lambda X_i}$ is integrable for all $i \le n$, and on the other hand, using inequality (2.9), that for $\lambda \in (0, 1/c)$,

$$\psi_S(\lambda) \le \sum_{i=1}^n \mathbf{E}\phi(\lambda X_i) \le \frac{v\lambda^2}{2(1 - c\lambda)}.$$

Therefore,

$$\psi_S^*(t) \geq \sup_{\lambda \in (0,1/c)} \left(t\lambda - \frac{\lambda^2 v}{2(1 - c\lambda)} \right)$$

and the stated bound for $\psi_S^*(t)$ follows from (2.5). The tail inequality of the theorem follows easily from Chernoff's inequality and the calculations shown at the beginning of Section 2.4. □

In some cases the following form of Bernstein's inequality is more convenient.

Corollary 2.11 *Let $X_1, \ldots, X_n$ be independent real-valued random variables satisfying the conditions of Theorem 2.10 and let $S = \sum_{i=1}^n (X_i - \mathbf{E}X_i)$. Then for all $t > 0$,*

$$\mathbf{P}\{S \geq t\} \leq \exp\left(-\frac{t^2}{2(v + ct)} \right).$$

Proof The corollary follows from the elementary inequality

$$h_1(u) \geq \frac{u^2}{2(1 + u)}, \qquad u > 0$$

(see Exercise 2.8). Thus, Theorem 2.10 implies that

$$\psi_S^*(t) \geq \frac{t^2}{2(v + ct)}$$

and the statement follows from Chernoff's inequality. □

Finally, note that one may recover (2.10) from Corollary 2.11. Indeed if $X_1, \ldots, X_n$ are independent such that $X_i \leq b$ almost surely for all $i \leq n$, then the conditions of Theorem 2.10 hold with

$$v = \sum_{i=1}^n \mathbf{E}\left[X_i^2\right] \quad \text{and} \quad c = b/3.$$

Example 2.12 (GAUSSIAN CHAOS OF ORDER TWO) As an example, we derive tail bounds for a special second-order Gaussian U-statistics, known as *Gaussian chaos.* Let $X = (X_1, \ldots, X_n)$ be a vector of independent standard normal random variables and let $A = (a_{i,j})_{n \times n}$ be a symmetric matrix with zeroes in its diagonal, that is, $a_{i,i} = 0$ for $i = 1, \ldots, n$. Then the quadratic form

$$Z = X^T A X = \sum_{i=1}^n \sum_{j=1}^n a_{i,j} X_i X_j$$

is a zero-mean random variable. To derive a concentration inequality for Z, we use the fact that a symmetric matrix can be diagonalized, that is, decomposed as $A = B^T \Lambda B$

where B is an $n \times n$ orthogonal matrix (i.e. the columns of B are orthogonal vectors of norm 1) such that $B^{-1} = B^T$ and Λ is a diagonal matrix with the eigenvalues $\mu_1 \dots, \mu_n$ of A in the diagonal entries. Denoting by $b_{i,j}$ the entries of the matrix B, we have

$$Z = \sum_{i=1}^{n} \mu_i Y_i^2 \quad \text{where} \quad Y_i = \sum_{j=1}^{n} b_{i,j} X_j, \quad i = 1, \dots, n.$$

By the rotational invariance of the standard multivariate normal distribution, we see that the distribution of $Y = (Y_1, \dots, Y_n)$ is the same as that of X, that is, $Y_1, \dots, Y_n$ are independent standard normal random variables. This implies that Z has the same distribution as

$$\sum_{i=1}^{n} \mu_i X_i^2 = \sum_{i=1}^{n} \mu_i (X_i^2 - 1)$$

where we used the fact that $\sum_{i=1}^{n} \mu_i$ equals the trace of the matrix A which is zero since we assumed that A has zeros in its diagonal. As we have seen it in Section 2.5, the logarithmic moment-generating function of $X_i^2 - 1$ equals, for all $\lambda < 1/2$,

$$\log \mathbf{E} e^{\lambda(X_i^2 - 1)} = \frac{1}{2}\left(-\log(1 - 2\lambda) - 2\lambda\right) \leq \frac{\lambda^2}{1 - 2\lambda}.$$

Therefore, the logarithmic moment-generating function of the Gaussian chaos becomes, for all $\lambda \in (0, 1/(2 \max_i \mu_i))$,

$$\psi_Z(\lambda) = \sum_{i=1}^{n} \frac{1}{2}\left(-\log(1 - 2\mu_i\lambda) - 2\mu_i\lambda\right) \leq \sum_{i=1}^{n} \frac{\mu_i^2 \lambda^2}{1 - 2(\mu_i)_+ \lambda} \leq \frac{\lambda^2 \|A\|_{\mathrm{HS}}^2}{1 - 2\lambda \|A\|}$$

where $\|A\|_{\mathrm{HS}} = \left(\sum_{i=1}^{n} \mu_i^2\right)^{1/2}$ is the so-called *Hilbert–Schmidt norm* (or *Frobenius norm*) of the matrix A and $\|A\| = \max_i |\mu_i|$ is the *operator norm*. Now we may use (2.5) to obtain the following bound for the upper tail: for all $t > 0$,

$$\mathbf{P}\left\{Z > 2\|A\|_{\mathrm{HS}}\sqrt{t} + 2\|A\|t\right\} \leq e^{-t},$$

or, by Exercise 2.8,

$$P\{Z > t\} \leq \exp\left(\frac{-t^2}{4(\|A\|_{\mathrm{HS}}^2 + \|A\|t)}\right).$$

2.9 Random Projections and the Johnson–Lindenstrauss Lemma

Next we describe an application in which Chernoff's inequality for sums of independent sub-Gaussian random variables plays a crucial role, in a perhaps unexpected situation. This application is an example of the power and elegance of the *probabilistic method* that has played such an important role in a large variety of applications ranging from combinatorics to the asymptotic geometry of Banach spaces.

The celebrated Johnson–Lindenstrauss lemma states roughly that, given an arbitrary set of n points in a (high-dimensional) Euclidean space, there exists a linear embedding of these points in a d-dimensional Euclidean space such that all pairwise distances are preserved within a factor of $1 \pm \varepsilon$ if d is proportional to $(\log n)/\varepsilon^2$. It is remarkable that this result does not involve the dimension of the space to which the n points belong. In fact, the dimension of this space may even be infinite.

To describe the problem more rigorously, consider an arbitrary set $A \subset \mathbb{R}^D$ where typically, D is a large positive integer. We note here that, in fact, $\mathbb{R}^D$ can be replaced by any (separable) Hilbert space by a straightforward generalization of the argument. For simplicity, we stick to the finite-dimensional framework. In this section we consider the special case where $A = \{a_1, \ldots, a_n\}$ is a finite set of n elements, but in Sections 5.6 and 13.6 we return to the case of general subsets. Given $\varepsilon \in (0,1)$, a map $f : \mathbb{R}^D \to \mathbb{R}^d$ is called an *ε-isometry* on A if for every pair $a, a' \in A$, we have

$$(1-\varepsilon)\,\|a-a'\|^2 \le \|f(a)-f(a')\|^2 \le (1+\varepsilon)\,\|a-a'\|^2 .$$

Now a natural question is to find the smallest possible value of d for which a *linear* ε-isometry exists on A. The Johnson–Lindenstrauss lemma, stated and proved below, ensures that when A is a finite set with cardinality n, a linear ε-isometry exists whenever $d \ge \kappa\varepsilon^{-2}\log n$, where κ is an absolute constant. We emphasize again the remarkable fact that this value does not depend on the dimension D of the space.

The idea of the proof is simple: just try a random linear function to determine whether it is an ε-isometry. While one might think that this is like looking for a needle in a haystack, it may come as a surprise that, if the distribution of the random choice is chosen properly, most random tries will work. This phenomenon is not uncommon in applications of the probabilistic method.

In other words, we prove below that a randomly chosen projection of $\mathbb{R}^D$ to $\mathbb{R}^d$ is, with large probability, an ε-isometry on the finite set A if d is at least a constant times $\varepsilon^{-2}\log n$.

The basic idea is to construct a random projection $W : \mathbb{R}^D \to \mathbb{R}^d$ (i.e. a linear mapping) that is an exact isometry "in expectation," that is, for every $\alpha \in \mathbb{R}^D$,

$$\boldsymbol{E}\left[\|W(\alpha)\|^2\right] = \|\alpha\|^2.$$

In other words, denoting by $L_{2,d}$ the space of square-integrable $\mathbb{R}^d$-valued random vectors, W is an isometry from $\mathbb{R}^D$ into $L_{2,d}$.

To construct W, let $X_{i,j}$, $i = 1, \ldots, d$, $j = 1, \ldots, D$ be independent and identically distributed real-valued random variables such that $\boldsymbol{E}X_{i,j} = 0$ and $\mathrm{Var}(X_{i,j}) = 1$. For every $\alpha = (\alpha_1, \ldots, \alpha_D) \in \mathbb{R}^D$ and $i \in \{1, \ldots, d\}$, define

$$W_i(\alpha) = \sum_{j=1}^{D} \alpha_j X_{i,j}.$$

$W_i(\alpha)/\sqrt{d}$ is the i-th component of the random vector $W(\alpha)$, that is, W is defined by

$$W(\alpha) = \left(\frac{1}{\sqrt{d}} W_i(\alpha)\right)_{i=1}^{d}.$$

Observe that by independence of the $X_{i,j}$, for every $i = 1, \ldots, d$,

$$\boldsymbol{E}\left[W_i(\alpha)^2\right] = \sum_{j=1}^{D} \alpha_j^2 \boldsymbol{E}\left[X_{i,j}^2\right] = \|\alpha\|^2.$$

Therefore, for every $\alpha \in \mathbb{R}^D$,

$$\boldsymbol{E}\left[\|W(\alpha)\|^2\right] = \frac{1}{d}\sum_{i=1}^{d} \boldsymbol{E}\left[W_i(\alpha)^2\right] = \|\alpha\|^2,$$

and indeed, in expectation, W is an isoperimetry from $\mathbb{R}^D$ into $L_{2,d}$.

It remains to show that on a sufficiently small subset $A \subset \mathbb{R}^D$, the random projection W defines an approximate isometry with large probability. To this end, we need convenient exponential integrability conditions on the distribution of the $X_{i,j}$. Traditionally the $X_{i,j}$ are taken to be standard normal variables. Here we show that it suffices if they are sub-Gaussian.

Theorem 2.13 (JOHNSON–LINDENSTRAUSS LEMMA) *Let A be a finite subset of $\mathbb{R}^D$ with cardinality n. Assume that for some $v \geq 1$, $X_{i,j} \in \mathcal{G}(v)$ and let $\varepsilon, \delta \in (0,1)$. If $d \geq 100v^2\varepsilon^{-2}\log\left(n/\sqrt{\delta}\right)$, then with probability at least $1-\delta$, W is an ε-isometry on A.*

Proof Denote by S the unit sphere of $\mathbb{R}^D$ and let T be the subset of S defined by

$$T = \left\{\frac{a-a'}{\|a-a'\|} : a, a' \in A, a \neq a'\right\}.$$

Then T has cardinality $N \leq n(n-1)/2$. We need to show that, under the stated condition for d,

$$\sup_{\alpha \in T} \left|\|W(\alpha)\|^2 - 1\right| \leq \varepsilon.$$

First note that for all $\alpha \in S$ and $i \leq d$, using the fact that the $X_{i,j}$ are sub-Gaussian,

$$\begin{aligned}
\boldsymbol{E}\exp\left(\lambda W_i(\alpha)\right) &= \boldsymbol{E}\exp\left(\lambda \sum_{j=1}^{D} \alpha_j X_{i,j}\right) \\
&= \prod_{j=1}^{D} \boldsymbol{E}\exp\left(\lambda \alpha_j X_{i,j}\right) \\
&\leq \exp\left(\lambda^2 \sum_{j=1}^{D} \alpha_j^2 v/2\right) \\
&= \exp\left(\lambda^2 v/2\right)
\end{aligned}$$

and therefore $W_i(\alpha) \in \mathcal{G}(v)$. Thus, by Theorem 2.1, for every integer $q \geq 2$,

$$\boldsymbol{E}\left[W_i(\alpha)^{2q}\right] \leq \frac{q!}{2} \times 4(2v)^q \leq \frac{q!}{2}(4v)^q.$$

Hence, since for each α the random variables $W_i(\alpha)$, $i = 1, \ldots, d$ are independent, we may use Bernstein's inequality (Theorem 2.10) for $\sum_{i=1}^d W_i(\alpha)^2$ with $v \leftarrow d(4v)^2$ and $c \leftarrow 4v$ to obtain, for every $\alpha \in T$ and $t > 0$,

$$\boldsymbol{P}\left\{\left|\sum_{i=1}^{d}\left(W_i(\alpha)^2 - 1\right)\right| \geq 4v\sqrt{2dt} + 4vt\right\} \leq 2e^{-t}.$$

This implies, by the union bound,

$$\mathbb{P}\left\{\sup_{\alpha\in T}\left|\sum_{i=1}^{d}\left(W_i(\alpha)^2 - 1\right)\right| \geq \sqrt{8vdt} + 4vt\right\} \leq 2Ne^{-t} \leq n^2e^{-t}.$$

Setting $t = \log(n^2/\delta)$, we have

$$\boldsymbol{P}\left\{\sup_{\alpha\in T}\left|\sum_{i=1}^{d}\left(W_i(\alpha)^2 - 1\right)\right| \geq 8v\left(\sqrt{d\log\frac{n}{\sqrt{\delta}}}\right) + 8v\log\frac{n}{\sqrt{\delta}}\right\} \leq \delta$$

or, equivalently,

$$\boldsymbol{P}\left\{\sup_{\alpha\in T}\left|\|W(\alpha)\|^2 - 1\right| \geq 8v\sqrt{\frac{\log\left(n/\sqrt{\delta}\right)}{d} + \frac{8v\log\left(n/\sqrt{\delta}\right)}{d}}\right\} \leq \delta.$$

Finally, we see that $d \geq 100v^2\varepsilon^{-2}\log\left(n/\sqrt{\delta}\right)$ implies that

$$8v\sqrt{\frac{\log\left(n/\sqrt{\delta}\right)}{d} + \frac{8v\log\left(n/\sqrt{\delta}\right)}{d}} \leq \frac{4\varepsilon}{5} + \frac{2\varepsilon^2}{25v} \leq \varepsilon$$

and therefore, with probability at least $1 - \delta$,

$$\sup_{\alpha\in T}\left|\|W(\alpha)\|^2 - 1\right| \leq \varepsilon$$

which is exactly what we wanted to prove. □

Remark 2.11 Note that by working with the general assumption of sub-Gaussian random variables, one loses a constant factor in the bound of the Johnson–Lindenstrauss lemma. Indeed, if we assume that the $X_{i,j}$ are standard normal, then $\sum_{i=1}^{d} W_i(\alpha)^2$ is a chi-squared random variable with d degrees of freedom and the inequality shown in Example 2.7 for gamma random variables implies

$$P\left\{\left|\sum_{i=1}^{d}\left(W_i(\alpha)^2-1\right)\right| \geq 2\sqrt{dt}+2t\right\} \leq 2e^{-t}.$$

This implies that with probability at least $1-\delta$, W is an ε-isometry on A whenever $d \geq 8\varepsilon^{-2}\log\left(n/\sqrt{\delta}\right)$. In Section 5.6 we take a closer look at random projections based on standard normal variables.

2.10 Association Inequalities

Next we recall some simple association inequalities. The first result states that if f and g are both increasing functions of a real variable, then for any random variable X, the correlation of $f(X)$ and $g(X)$ is positive.

Theorem 2.14 (CHEBYSHEV'S ASSOCIATION INEQUALITY) *Let f and g be nondecreasing real-valued functions defined on the real line. If X is a real-valued random variable and Y is a nonnegative random variable, then*

$$\mathbf{E}[Y]\mathbf{E}[Yf(X)g(X)] \geq \mathbf{E}[Yf(X)]\mathbf{E}[Yg(X)].$$

If f is nonincreasing and g is nondecreasing then

$$\mathbf{E}[Y]\mathbf{E}[Yf(X)g(X)] \leq \mathbf{E}[Yf(X)]\mathbf{E}[Yg(X)].$$

Remark 2.12 This is a slight generalization of what is usually referred to as Chebyshev's association inequality which can be recovered by taking $Y \equiv 1$.

Proof Let the pair of random variables (X', Y') be distributed as the pair (X, Y) and independent of it. If f and g are nondecreasing, $YY'(f(X)-f(X'))(g(X)-g(X')) \geq 0$, so obviously

$$\mathbf{E}[YY'(f(X)-f(X'))(g(X)-g(X'))] \geq 0.$$

Expand this expectation to obtain the first inequality. The proof of the second is similar. □

An important generalization of Chebyshev's association inequality is described as follows. A real-valued function f defined on $\mathbb{R}^n$ is said to be nondecreasing (nonincreasing) if it is nondecreasing (nonincreasing) in each variable while keeping all other variables fixed at any value.

Theorem 2.15 (HARRIS'S INEQUALITY) *Let $f, g : \mathbb{R}^n \to \mathbb{R}$ be nondecreasing functions. Let $X_1, \ldots, X_n$ be independent real-valued random variables and define the random vector $X = (X_1, \ldots, X_n)$ taking values in $\mathbb{R}^n$. Then*

$$E[f(X)g(X)] \geq E[f(X)]E[g(X)].$$

Similarly, if f is nonincreasing and g is nondecreasing then

$$E[f(X)g(X)] \leq E[f(X)]E[g(X)].$$

Proof Again, it suffices to prove the first inequality. We proceed by induction. For $n = 1$ the statement is just Chebyshev's association inequality. Now suppose the statement is true for $m < n$. Then

$$\begin{aligned} E[f(X)g(X)] &= EE[f(X)g(X)|X_1, \ldots, X_{n-1}] \\ &\geq E\left[E[f(X)|X_1, \ldots, X_{n-1}]E[g(X)|X_1, \ldots, X_{n-1}]\right] \end{aligned}$$

because given $X_1, \ldots, X_{n-1}$, both f and g are nondecreasing functions of the n-th variable. Now it follows by independence that the functions $f', g' : \mathbb{R}^{n-1} \to \mathbb{R}$ defined by $f'(x_1, \ldots x_{n-1}) = E[f(X)|X_1 = x_1, \ldots, X_{n-1} = x_{n-1}]$ and $g'(x_1, \ldots x_{n-1}) = E[g(X)|X_1 = x_1, \ldots, X_{n-1} = x_{n-1}]$ are nondecreasing functions, so by the induction hypothesis

$$\begin{aligned} & E[f'(X_1, \ldots, X_{n-1})g'(X_1, \ldots, X_{n-1})] \\ & \geq E[f'(X_1, \ldots, X_{n-1})]E[g'(X_1, \ldots, X_{n-1})] \\ & = E[f(X)]E[g(X)] \end{aligned}$$

as desired. □

2.11 Minkowski's Inequality

We close this chapter by proving a general version of Minkowski's inequality. The best known versions of this inequality may be considered as triangle inequalities for L_q norms of vectors or random variables. For example, one version states that if X_1 and X_2 are two real-valued random variables, then for $q \geq 1$,

$$E\left[|X_1 + X_2|^q\right]^{1/q} \leq E\left[|X_1|^q\right]^{1/q} + E\left[|X_2|^q\right]^{1/q}.$$

In this book (see Chapters 5 and 10), we will need the following, more general, formulation of Minkowski's inequality.

Theorem 2.16 (MINKOWSKI'S INEQUALITY) *Let X and Y be independent random variables taking their values in the sets $\mathcal{X}$ and $\mathcal{Y}$, respectively. Let $f : \mathcal{X} \times \mathcal{Y} \to \mathbb{R}$ be a real-valued measurable function and define the random variable $Z = f(X, Y)$. If $q \geq 1$, then*

$$\left(E_X\left[|E_Y Z|^q\right]\right)^{1/q} \leq E_Y\left[\left(E_X|Z|^q\right)^{1/q}\right]$$

where $\mathbf{E}_X$ *and* $\mathbf{E}_Y$ *denote expectations taken with respect to the distributions of* X *and* Y, *respectively (i.e.* $\mathbf{E}_X Z = \mathbf{E}[Z|Y]$ *and* $\mathbf{E}_Y Z = \mathbf{E}[Z|X]$*).*

Before proving the theorem, note that the classical version of Minkowski's inequality cited above may be recovered by letting Y be uniformly distributed on the set $\mathcal{Y} = \{1, 2\}$ and defining $X = (X_1, X_2), f(X, 1) = X_1$, and $f(X, 2) = X_2$.

Proof The inequality is obvious for $q \in \{1, \infty\}$ so we may assume that $1 < q < \infty$. Without loss of generality, we may assume that Z is nonnegative. Let Y' be an independent copy of Y, independent of X as well. Then

$$\begin{aligned}
\mathbf{E}_X\left[(\mathbf{E}_Y Z)^q\right] &= \mathbf{E}_X\left[(\mathbf{E}_{Y'} f(X,Y'))^{q-1}\,\mathbf{E}_Y f(X,Y)\right] \\
&= \mathbf{E}_X\left[\mathbf{E}_Y\left[(\mathbf{E}_{Y'} f(X,Y'))^{q-1} f(X,Y)\right]\right] \\
&= \mathbf{E}_Y\left[\mathbf{E}_X\left[(\mathbf{E}_{Y'} f(X,Y'))^{q-1} f(X,Y)\right]\right] \\
&\quad \text{(by Fubini's theorem)} \\
&\le \mathbf{E}_Y\left[\left(\mathbf{E}_X\left[(\mathbf{E}_{Y'} f(X,Y'))^q\right]\right)^{(q-1)/q}\left(\mathbf{E}_X\left[f^q(X,Y)\right]\right)^{1/q}\right] \\
&\quad \text{(by Hölder's inequality)} \\
&= \left(\mathbf{E}_X\left[(\mathbf{E}_{Y'} f(X,Y'))^q\right]\right)^{1-1/q}\mathbf{E}_Y\left[\left(\mathbf{E}_X f^q(X,Y)\right)^{1/q}\right] \\
&= \left(\mathbf{E}_X\left[(\mathbf{E}_Y Z)^q\right]\right)^{1-1/q}\mathbf{E}_Y\left[\left(\mathbf{E}_X Z^q\right)^{1/q}\right].
\end{aligned}$$

Dividing both sides by $\left(\mathbf{E}_X\left[(\mathbf{E}_Y Z)^q\right]\right)^{1-1/q}$, we obtain the desired inequality. □

2.12 Bibliographical Remarks

Exponential tail inequalities for sums of independent random variables have been proved from the early days of mathematical probability theory. Among the pioneers we mention Bernstein (1946), Craig (1933), Uspensky (1937), Chernoff (1952), Okamoto (1958), Bennett (1962), and Hoeffding (1963).

The proof of the maximal inequality of Theorem 2.5 is based on an argument used by Pisier (1983) to control the expectation of the supremum of variables belonging to some Orlicz space. For exponentially integrable variables it is possible to optimize Pisier's argument with respect to the parameter involved in the definition of the moment-generating function. This is exactly what is done in Theorem 2.5.

Hoeffding's lemma (Lemma 2.2) and Hoeffding's inequality (Theorem 2.8) are due to Hoeffding (1963). Bennett's inequality is taken from Bennett (1962), while Bernstein's inequality in its original form is in Bernstein (1946). Bernstein's inequality for unbounded variables can be found in Uspensky (1937). Theorem 2.10 appears in Birgé and Massart (1998). Note that the usual assumption in Bernstein's inequality involves conditions for the absolute moments of the X_i, instead of their positive part as in Theorem 2.10. This refinement was suggested to us by Emmanuel Rio.

The inequality derived in Example 2.12 for a Gaussian chaos of order 2 is described by Hanson and Wright (1971). The fact that a quadratic form of standard normal random variables has the same distribution as a weighted sum of independent random variables with chi-squared distribution is usually referred to as *Cochran's theorem*. For extensions to higher-order chaoses, we refer to Arcones and Giné (1993) and Latała (2006).

The Johnson–Lindenstrauss lemma first appears in Johnson and Lindenstrauss (1984), though its original proof is not probabilistic. The idea of random projections was introduced by Frankl and Maehara (1988, 1990), but see also Gupta and Dasgupta (2002) for a particularly simple proof. Achlioptas (2003) considered projections based on Rademacher random variables. The proof of Theorem 2.13 is adapted from the arguments of Achlioptas. Random projections have been used successfully in a variety of applications, for example, Linial, London, and Rabinovich (1995), Kleinberg (1997), and Indyk and Motwani (1998). For a survey we refer the reader to the book of Vempala (2004).

Theorem 2.14 is attributed to Chebyshev (see, e.g. Hall, Littlewood and Pólya (1952)). We note here that association properties may often be used to derive concentration properties. We refer the reader to the survey of Dubhashi and Ranjan (1998). Theorem 2.15 is due to Harris (1960), though sometimes it is referred to as the FKG inequality because of a generalization established by Fortuin, Kasteleyn, and Ginibre (1971).

The proof of the generalized Minkowski inequality (Theorem 2.16) presented here was first found by F. Riesz (see Steele (2004), Zygmund (1959)).

2.13 EXERCISES

2.1. Let MZ be a median of the square-integrable random variable Z (i.e. $P\{Z \geq MZ\} \geq 1/2$ and $P\{Z \leq MZ\} \geq 1/2$). Show that

$$|MZ - EZ| \leq \sqrt{\mathrm{Var}(Z)}.$$

2.2. Let X be a random variable with median MX such that positive constants a and b exist so that for all $t > 0$,

$$P\left\{|X - MX| > t\right\} \leq ae^{-t^2/b}.$$

Show that $|MX - EX| \leq \min\left(\sqrt{ab}, a\sqrt{b\pi}/2\right)$.

2.3. (CHEBYSHEV–CANTELLI INEQUALITY) Prove the following one-sided improvement of Chebyshev's inequality: for any real-valued random variable Y and $t > 0$,

$$P\{Y - EY \geq t\} \leq \frac{\mathrm{Var}(Y)}{\mathrm{Var}(Y) + t^2}.$$

2.4. (PALEY–ZYGMUND INEQUALITY) Show that if Y is a nonnegative random variable, then for any $a \in (0, 1)$,

$$P\{Y \geq aEY\} \geq (1-a)^2 \frac{(EY)^2}{E[Y^2]}.$$

2.5. (MOMENTS VS. CHERNOFF BOUNDS) Show that moment bounds for tail probabilities are always better than Cramér–Chernoff bounds. More precisely, let Y be a nonnegative random variable and let $t > 0$. The best moment bound for the tail probability $P\{Y \geq t\}$ is $\min_q E[Y^q]t^{-q}$ where the minimum is taken over all positive integers. The best Cramér–Chernoff bound is $\inf_{\lambda>0} Ee^{\lambda(Y-t)}$. Prove that

$$\min_q E[Y^q]t^{-q} \leq \inf_{\lambda>0} Ee^{\lambda(Y-t)}.$$

(See Philips and Nelson (1995).)

2.6. Let Z be a real-valued random variable. Show that the set of positive numbers $S = \{\lambda > 0 : Ee^{\lambda Z} < \infty\}$ is either empty, or an interval with left end point equal to 0. Let $b = \sup S$. Show that $\psi_Z(\lambda) = \log Ee^{\lambda Z}$ is convex and infinitely many times differentiable on $I = (0, b)$. Show that if $EZ = 0$, ψ_Z is continuously differentiable on $[0, b)$ with $\psi'_Z(0) = \psi_Z(0) = 0$ and the Cramér transform of Z equals $\psi^*_Z(t) = \sup_{\lambda \in I} (\lambda t - \psi_Z(\lambda))$.

2.7. Prove that if Z is a centered normal random variable with variance σ^2 then

$$\sup_{t>0} \left(P\{Z \geq t\} \exp\left(\frac{t^2}{2\sigma^2}\right)\right) = \frac{1}{2}.$$

2.8. (ELEMENTARY INEQUALITIES) Prove the following inequalities appearing in the text:

$$-\log(1-u) - u \leq \frac{u^2}{2(1-u)} \quad \text{for } u \in (0,1);$$

$$h(u) = (1+u)\log(1+u) - u \geq \frac{u^2}{2(1+u/3)} \quad \text{for } u > 0;$$

$$h_1(u) = 1 + u - \sqrt{1+2u} \geq \frac{u^2}{2(1+u)}, \quad \text{for } u > 0.$$

2.9. (SUB-GAUSSIAN LOWER TAIL FOR NONNEGATIVE RANDOM VARIABLES) Let X be a nonnegative random variable with finite second moment. Show that for any $\lambda > 0$, $Ee^{-\lambda(X-EX)} \leq e^{\lambda^2 E[X^2]/2}$. In particular, if $X_1, \ldots, X_n$ are independent nonnegative random variables, then for any $t > 0$,

$$P\{S \leq -t\} \leq \exp\left(\frac{-t^2}{2v}\right)$$

where $S = \sum_{i=1}^n (X_i - EX_i)$ and $v = \sum_{i=1}^n E\left[X_i^2\right]$.

2.10. Let $X_1, \ldots, X_n$ be independent Bernoulli random variables with parameters $p_1, \ldots, p_n$, respectively. Let $p = (1/n)\sum_{i=1}^n p_i$ and $S_n = \sum_{i=1}^n X_i$. Prove that

$$\mathbf{P}\{S_n - np \geq n\varepsilon\} \leq e^{-np\varepsilon^2/3} \quad \text{and} \quad \mathbf{P}\{S_n - np \leq -n\varepsilon\} \leq e^{-np\varepsilon^2/2}$$

(Angluin and Valiant (1979), see also Hagerup and Rüb (1990)).

2.11. Let B be binomially distributed with parameters (n, p). Show that for $p \leq a < 1$,

$$\mathbf{P}\{B > an\} \leq \left(\left(\frac{p}{a}\right)^a \left(\frac{1-p}{1-a}\right)^{1-a}\right)^n \leq \left(\left(\frac{p}{a}\right)^a e^{a-p}\right)^n.$$

Show that for $0 < a < p$, the same upper bounds hold for $\mathbf{P}\{B \leq an\}$ (Karp (1988), see also Hagerup and Rüb (1990)).

2.12. Let B be binomially distributed with parameters (n, p). Show that if $p \geq 1/2$,

$$\mathbf{P}\{B - np \geq n\varepsilon\} < e^{-\frac{n\varepsilon^2}{2p(1-p)}},$$

and if $p \leq 1/2$,

$$\mathbf{P}\{B - np \leq -n\varepsilon\} < e^{-\frac{n\varepsilon^2}{2p(1-p)}}$$

(Okamoto (1958)).

2.13. Let B be binomially distributed with parameters (n, p). Prove that

$$\mathbf{P}\left\{\sqrt{B} - \sqrt{np} \geq \varepsilon\sqrt{n}\right\} < e^{-2n\varepsilon^2},$$

and

$$\mathbf{P}\left\{\sqrt{B} - \sqrt{np} \leq -\varepsilon\sqrt{n}\right\} < e^{-n\varepsilon^2}$$

(Okamoto (1958)).

2.14. Let D and n be positive integers with $1 \leq D \leq n$. Show that

$$\sum_{j=0}^{D} \binom{n}{j} \leq \left(\frac{en}{D}\right)^D.$$

Hint: observe that the left-hand side is 2^n times a tail probability of a symmetric binomial random variable and use Chernoff's inequality.

2.15. (ALTERNATIVE PROOF OF BENNETT'S INEQUALITY) As in the proof of Lemma 2.2, show that if Y is a centered random variable with finite variance v such that $Y \leq 1$,

$$\psi_Y(\lambda) = \log \mathbf{E}e^{\lambda Y} \leq \log\left(1 + v\left(e^{\lambda} - \lambda - 1\right)\right)$$

for $\lambda \in \mathbb{R}$ by solving a differential inequality. *Hint:* let P denote the distribution of Y and let P_λ be the probability distribution with density $e^{-\psi_Y(\lambda)}e^{\lambda x}$ with respect to P. Let Z have distribution P_λ. Check first that $\text{Var}(Z) \leq ve^{\lambda}$.

2.16. Prove that if X is a sub-Gaussian random variable with variance factor v then $\text{Var}(X) \leq v$.

2.17. Let $G_1, \ldots, G_N$ be independent standard normal random variables. Then

$$\lim_{N\to\infty} \frac{\boldsymbol{E} \max_{i=1,\ldots,N} G_i}{\sqrt{2\log N}} = 1.$$

(See Galambos (1987).)

2.18. (MAXIMUM OF INDEPENDENT POISSON RANDOM VARIABLES) Let $X_1, \ldots, X_n$ be independent Poisson random variables with expectation 1. The Lambert W function is defined over $[-1/e, \infty)$ by the equation $W(x)e^{W(x)} = x$. Prove that

$$\boldsymbol{E} \max_{i=1,\ldots,n} X_i \leq \frac{\log(n/e)}{W(\log(n/e)/e)}$$

Prove that for $z \geq e$, $W(z) \geq \log(z) - \log\log(z)$ and that for $n \geq e^{1+e^2}$,

$$\boldsymbol{E} \max_{i=1,\ldots,n} X_i \leq \frac{\log(n/e)}{\log(\log(n/e)/e) - \log(\log(\log(n/e)/e))}.$$

The following upper bound may be more manageable:

$$\boldsymbol{E} \max_{i=1,\ldots,n} X_i \leq \frac{2\log n}{\log(\log(en))}.$$

Hint: use Theorem 2.4.

2.19. (MAXIMUM OF INDEPENDENT BINOMIAL RANDOM VARIABLES) Let $X_1, \ldots, X_n$ be independent Binomial (m, p) random variables. Prove that

$$\boldsymbol{E} \max_{i=1,\ldots,n} X_i \leq mpe^{1+W((\log(n)-mp)/(emp))},$$

where W is defined in Exercise 2.18.

2.20. (RANDOM ALLOCATIONS) Suppose we throw m balls into n bins uniformly independently at random. Let M be the maximum number of balls in any bin. Prove that

$$\boldsymbol{E}M \leq \frac{m}{n}e^{1+W((\log(n)-m/n)/(em/n))}.$$

Deduce from this that if $m = cn\log(n)$ for $c > 0$,

$$\boldsymbol{E}M \leq c\log(n)e^{1+W((1-c)/(ce))}.$$

Conclude that if $m = cn$ for $c > 0$, if $\log n > c$,

$$EM \leq \frac{\log(n) - c}{\log((\log(n) - c)/(ec)) - \log(\log((\log(n) - c)/(ec)))}.$$

Hint: use Theorem 2.4 and Exercises 2.18 and 2.19. See Raab and Steger (1998) for related results and asymptotics.

2.21. (SUB-GAMMA RANDOM VARIABLES: ONE-SIDED BOUNDS) Let X be a centered random variable (i.e. $EX = 0$) in $\Gamma_+(v, c)$. Show that $\text{Var}(X) \leq v$ and there exists a constant C such that for every integer $q \geq 2$, $\left(E\left[X_+^q\right]\right)^{1/q} \leq C\left(\sqrt{qv} + cq\right)$. Conversely, suppose that X is a centered random variable such that there exist constants A and B such that $\text{Var}(X) \leq A$ and $\left(E\left[X_+^q\right]\right)^{1/q} \leq \sqrt{qA} + Bq$ for all $q \geq 2$. Show that there exists a constant C such that X is in $\Gamma_+(v, c)$ with $v = C(A + B^2)$ and $c = CB$.

2.22. (SUB-EXPONENTIAL RANDOM VARIABLES) A nonnegative random variable X has *exponential distribution* with parameter $a > 0$ if X has a density ae^{-ax}, $x \geq 0$. The moment-generating function of X is then $Ee^{\lambda X} = 1/(1 - \lambda/a)$ for $\lambda \in (0, a)$. Show that if q is a positive integer, the q-th moment of X equals $E[X^q] = q!/a^q$. We say that a nonnegative random variable X has a *sub-exponential* distribution if there exists a constant $a > 0$ such that for all $0 < \lambda < a$, $Ee^{\lambda X} \leq 1/(1 - \lambda/a)$. Show that if X is sub-exponential, then for every positive integer q,

$$E[X^q] \leq 2^{q+1}\frac{q!}{a^q}.$$

2.23. (SUB-EXPONENTIAL DISTRIBUTION–CONTINUED) Let X be a random variable such that there exists a constant $a > 0$ in order that for every positive integer q,

$$E[X^q] \leq \frac{q!}{a^q}.$$

Show that X is sub-exponential. More precisely, show that for any $0 < \lambda < a$, $Ee^{\lambda X} \leq 1/(1 - \lambda/a)$.

2.24. (A TAIL-COMPARISON INEQUALITY) Let X and Y be two real-valued random variables such that for any real a,

$$E\left[(X - a)_+\right] \leq E\left[(Y - a)_+\right]$$

while for some $\kappa \geq 1$ and $b > 0$, for all $t \geq 0$,

$$P\{Y \geq t\} \leq \kappa e^{-bt}.$$

Prove that for all $t \geq 0$,

$$P\{X \geq t\} \leq \kappa\, e^{1-bt}$$

(see Panchenko (2003)).

2.25. Consider the Gaussian chaos of order two Z defined in Example 2.12. Show that there exist positive constants c and C such that for all $q \geq 2$ and $n \geq 1$,

$$c\left(\sqrt{q}\|A\|_{\mathrm{HS}} + q\|A\|\right) \leq \left(\mathbf{E}\left[|Z|^q\right]\right)^{1/q} \leq C\left(\sqrt{q}\|A\|_{\mathrm{HS}} + q\|A\|\right)$$

(see Latała (1999)).

2.26. Let f be a nonnegative nonincreasing and g a nondecreasing real-valued function. Let h be a nonnegative function with finite expectation, such that $\mathbf{E}[h(X)f(X)] \leq \mathbf{E}[h(X)]$. Then

$$\mathbf{E}\left[f(X)g(X)h(X)\right] \leq \mathbf{E}\left[h(X)g(X)\right].$$

2.27. (BETWEEN SUB-GAMMA AND GAUSSIAN) Let $X_1, \ldots, X_n$ be identically distributed independent random variables such that

$$\mathbf{P}\left\{|X_i| \geq u\right\} \leq e^{-u^p}$$

for some $p \geq 1$. Let $q = p/(p-1)$ be the conjugate of p. Let $s = (s_1 \ldots, s_n) \in \mathbb{R}^n$. Let $Z = \sum_{i=1}^n s_i X_i$. Prove that there exists a constant L (that depends on p but not on n) such that

$$\mathbf{P}\{Z \geq t\} \leq L \exp\left(-\frac{1}{L}\min\left(\frac{t^2}{\|s\|_2^2}, \frac{t^p}{\|s\|_q^p}\right)\right)$$

where $\|s\|_p^p = \sum_{i=1}^n |s_i|^p$ (if $p < \infty$) or $\max\left(|s_i|\right)$ (if $p = \infty$).

2.28. (MOMENTS OF THE GUMBEL DISTRIBUTION) Let X be distributed according to the Gumbel distribution: $\mathbf{P}\{X \leq t\} = \exp(-\exp(-t))$. Prove that $\mathbf{E}X$ equals the Euler–Mascheroni constant $(\lim_{n\to\infty} \sum_{i=1}^n 1/i - \log n)$, that $\mathrm{Var}(X) = \pi^2/6 = \lim_{n\to\infty} \sum_{i=1}^n 1/i^2$, and that for all $\lambda \geq 0$, $\log \mathbf{E} \exp(\lambda(X - \mathbf{E}X)) \leq \mathrm{Var}(X)\lambda^2/(2(1-\lambda))$. *Hint:* use Rényi's representation of order statistics of samples of the exponential distribution, described as follows: if $Y_1, \ldots, Y_n$ are independent exponentially distributed random varibles, then $\max(X_1, \ldots, X_n)$ is distributed as $\sum_{i=1}^n Y_i/i$, and $\max(Y_1, \ldots, Y_n) - \log n$ converges in distribution to the Gumbel distribution.

3

Bounding the Variance

The purpose of this chapter is to introduce a simple, yet powerful, inequality which offers a useful upper bound for the variance of a general function of several independent random variables.

Formally, let $f : \mathcal{X}^n \to \mathbb{R}$ be a real-valued function of n variables, where $\mathcal{X}$ is some measurable space. If $X_1, \ldots, X_n$ are independent random variables taking values in $\mathcal{X}$, then we may define the real-valued random variable

$$Z = f(X_1, \ldots, X_n).$$

We emphasize that the X_i may have different distributions, the only essential assumption is independence. Throughout this chapter we assume that Z has a finite variance, and our purpose is to find general upper bounds.

The basic result – the Efron–Stein inequality–, presented in Section 3.1, provides a bound in terms of "local" variations of the function f. This inequality is a prologue to the numerous results presented in this book in which concentration properties may be controlled by studying the local behavior of the function at hand.

A large part of this chapter (Sections 3.2–3.5) is devoted to applications of the Efron–Stein inequality for bounding the variance of complex functions of independent random variables. We hope that the elementary arguments will convince the reader of the versatility and power of this simple inequality.

Once the variance of Z is controlled, one may use Chebyshev's inequality to derive upper bounds for the tail probabilities $\boldsymbol{P}\{Z > \boldsymbol{E}Z + t\}$. However, interestingly, by a simple trick shown in Section 3.6, the Efron–Stein inequality may also be used to derive exponential bounds for the tail probability. These bounds are not optimal in the sense that they do not capture the right exponential rate of decrease of the tails. In subsequent chapters we show how these tail inequalities can be significantly sharpened to prove tail bounds which cannot be revealed by looking at the variance only. However, the techniques used in this chapter present many of the main ideas used later in a simple, digestible form.

In Section 3.7 we show how the Efron–Stein inequality implies the Gaussian Poincaré inequality, a classical result for smooth functions of independent normal random variables.

We close this chapter by providing an alternative proof of the Efron–Stein inequality, based on a duality argument, which opens the door to generalizations presented in subsequent chapters, notably in Section 4.9.

3.1 The Efron–Stein Inequality

One of the main messages of this book is that, in a certain sense, sums of independent random variables have an extremal place in the world of general functions of independent random variables. Before deriving a bound for the variance of a general function Z of independent random variables (this problem, of course, only makes sense when Z is square integrable), we can gain some insight by considering first the very special case when the variables $X_1, \ldots, X_n$ are real-valued and $Z = X_1 + \cdots + X_n$. In this case we can use the exact formula

$$\mathrm{Var}\,(Z) = \sum_{i=1}^{n} \mathrm{Var}\,(X_i).$$

Of course, the proof of this formula uses independence only through the pairwise orthogonality (in $\mathbb{L}_2$) of the variables $X_i - \mathbf{E}X_i$. Now it is a natural idea to bound the variance of a general function by expressing $Z - \mathbf{E}Z$ as a sum of martingale differences for the Doob filtration and use the orthogonality of these differences. More precisely, if we denote by $\mathbf{E}_i$ the conditional expectation operator, conditioned on $(X_1, \ldots, X_i)$, and use the convention $\mathbf{E}_0 = \mathbf{E}$, then we may define

$$\Delta_i = \mathbf{E}_i Z - \mathbf{E}_{i-1} Z$$

for every $i = 1, \ldots, n$. Starting from the decomposition

$$Z - \mathbf{E}Z = \sum_{i=1}^{n} \Delta_i$$

one has

$$\mathrm{Var}\,(Z) = \mathbf{E}\left[\left(\sum_{i=1}^{n} \Delta_i\right)^2\right] = \sum_{i=1}^{n} \mathbf{E}\left[\Delta_i^2\right] + 2\sum_{j>i} \mathbf{E}\left[\Delta_i \Delta_j\right].$$

Now, if $j > i$, $\mathbf{E}_i \Delta_j = 0$ implies that

$$\mathbf{E}_i\left[\Delta_j \Delta_i\right] = \Delta_i \mathbf{E}_i \Delta_j = 0,$$

and, a fortiori, $\mathbf{E}\left[\Delta_j \Delta_i\right] = 0$. Thus, we obtain the following analog of the additivity formula of the variance:

$$\text{Var}(Z) = \mathbf{E}\left[\left(\sum_{i=1}^{n} \Delta_i\right)^2\right] = \sum_{i=1}^{n} \mathbf{E}\left[\Delta_i^2\right].$$

Until now, we have made no use of the fact that Z is a function of independent variables $X_1, \ldots, X_n$. Indeed, the above formula holds for any martingale. Independence may be applied in the following argument: for any integrable function $Z = f(X_1, \ldots, X_n)$ one may write, by Fubini's theorem,

$$\mathbf{E}_i Z = \int_{\mathcal{X}^{n-i}} f(X_1, \ldots, X_i, x_{i+1}, \ldots, x_n)\, d\mu_{i+1}(x_{i+1}) \ldots d\mu_n(x_n),$$

where, for every $j = 1, \ldots, n$, μ_j denotes the probability distribution of X_j. Also, if we denote by $\mathbf{E}^{(i)}$ the conditional expectation operator conditioned on $X^{(i)} = (X_1, \ldots, X_{i-1}, X_{i+1}, \ldots, X_n)$, we have

$$\mathbf{E}^{(i)} Z = \int_{\mathcal{X}} f(X_1, \ldots, X_{i-1}, x_i, X_{i+1}, \ldots, X_n)\, d\mu_i(x_i).$$

Then, again by Fubini's theorem,

$$\mathbf{E}_i\left[\mathbf{E}^{(i)} Z\right] = \mathbf{E}_{i-1} Z. \tag{3.1}$$

This observation is key in the proof of the main result of this chapter which we state next.

Theorem 3.1 (EFRON–STEIN INEQUALITY) *Let $X_1, \ldots, X_n$ be independent random variables and let $Z = f(X)$ be a square-integrable function of $X = (X_1, \ldots, X_n)$. Then*

$$\text{Var}(Z) \le \sum_{i=1}^{n} \mathbf{E}\left[\left(Z - \mathbf{E}^{(i)} Z\right)^2\right] \stackrel{\text{def}}{=} v.$$

Moreover, if $X_1', \ldots, X_n'$ are independent copies of $X_1, \ldots, X_n$ and if we define, for every $i = 1, \ldots, n$,

$$Z_i' = f(X_1, \ldots, X_{i-1}, X_i', X_{i+1}, \ldots, X_n),$$

then

$$v = \frac{1}{2} \sum_{i=1}^{n} \mathbf{E}\left[(Z - Z_i')^2\right] = \sum_{i=1}^{n} \mathbf{E}\left[(Z - Z_i')_+^2\right] = \sum_{i=1}^{n} \mathbf{E}\left[(Z - Z_i')_-^2\right]$$

where $x_+ = \max(x, 0)$ and $x_- = \max(-x, 0)$ denote the positive and negative parts of a real number x. Also,

$$v = \sum_{i=1}^{n} \inf_{Z_i} \mathbf{E}\left[(Z - Z_i)^2\right],$$

where the infimum is taken over the class of all $X^{(i)}$-measurable and square-integrable variables Z_i, $i = 1, \ldots, n$.

Proof We begin with the proof of the first statement. Note that, using (3.1), we may write

$$\Delta_i = \mathbf{E}_i\left[Z - \mathbf{E}^{(i)}Z\right].$$

By Jensen's inequality, used conditionally,

$$\Delta_i^2 \le \mathbf{E}_i\left[\left(Z - \mathbf{E}^{(i)}Z\right)^2\right].$$

Using $\mathrm{Var}(Z) = \sum_{i=1}^n \mathbf{E}\left[\Delta_i^2\right]$, we obtain the desired inequality. To prove the identities for v, denote by $\mathrm{Var}^{(i)}$ the conditional variance operator conditioned on $X^{(i)}$. Then we may write v as

$$v = \sum_{i=1}^n \mathbf{E}\left[\mathrm{Var}^{(i)}(Z)\right].$$

Now note that one may simply use (conditionally) the elementary fact that if X and Y are independent and identically distributed real-valued random variables, then $\mathrm{Var}(X) = (1/2)\mathbf{E}[(X-Y)^2]$. Since conditionally on $X^{(i)}$, Z_i' is an independent copy of Z, we may write

$$\mathrm{Var}^{(i)}(Z) = \frac{1}{2}\mathbf{E}^{(i)}\left[(Z - Z_i')^2\right] = \mathbf{E}^{(i)}\left[(Z - Z_i')_+^2\right] = \mathbf{E}^{(i)}\left[(Z - Z_i')_-^2\right]$$

where we used the fact that the conditional distributions of Z and Z_i' are identical. The last identity is obtained by recalling that, for any real-valued random variable X, $\mathrm{Var}(X) = \inf_{a\in\mathbb{R}} \mathbf{E}[(X-a)^2]$. Using this fact conditionally, we have, for every $i = 1, \ldots, n$,

$$\mathrm{Var}^{(i)}(Z) = \inf_{Z_i} \mathbf{E}^{(i)}\left[(Z - Z_i)^2\right].$$

Note that this infimum is achieved whenever $Z_i = \mathbf{E}^{(i)}Z$. □

Observe that in the case when $Z = \sum_{i=1}^n X_i$ is a sum of independent random variables (with finite variance), then the Efron–Stein inequality becomes an equality. Thus, the bound in the Efron–Stein inequality is, in a sense, not improvable.

Remark 3.2 (THE JACKKNIFE ESTIMATE) We should note here that the Efron–Stein inequality was first motivated by the study of the so-called *jackknife estimate* of statistics. To describe this estimate, assume that $X_1, \ldots, X_n$ are i.i.d. random variables and one wishes to estimate a functional θ of the distribution of the X_i by a function $Z = f(X_1, \ldots, X_n)$ of the data. The quality of the estimate is often measured by its *bias* $\mathbf{E}Z - \theta$ and its variance $\mathrm{Var}(Z)$. Since the distribution of the X_i's is unknown, one needs to estimate the bias and variance from the same sample. The jackknife estimate of the bias is defined by

$$(n-1)\left(\frac{1}{n}\sum_{i=1}^{n} Z_i - Z\right)$$

where Z_i is an appropriately defined function of $X^{(i)} = (X_1, \ldots, X_{i-1}, X_{i+1}, \ldots, X_n)$ (see Exercise 3.4). $X^{(i)}$ is often called the i-th *jackknife sample* while Z_i is the so-called *jackknife replication* of Z. In an analogous way, the jackknife estimate of the variance is defined by

$$\sum_{i=1}^{n} (Z - Z_i)^2.$$

(Sometimes this sum is multiplied by $(n-1)/n$.) Using this language, the Efron–Stein inequality simply states that the jackknife estimate of the variance is always positively biased. In fact, this is how Efron and Stein originally formulated their inequality.

In the next sections we illustrate the use of the Efron–Stein inequality for various prototypical examples. For many of these examples we will be able to prove much stronger exponential tail estimates. However, the bases of the methodology are laid down here and the arguments presented in this chapter will be of great use in establishing sharper bounds. Also, useful bounds for the variance can often be derived under significantly more general conditions than sharper tail bounds.

3.2 Functions with Bounded Differences

We say that a function $f : \mathcal{X}^n \to \mathbb{R}$ has the *bounded differences property* if for some nonnegative constants $c_1, \ldots, c_n$,

$$\sup_{\substack{x_1,\ldots,x_n,\\ x_i' \in \mathcal{X}}} \left| f(x_1, \ldots, x_n) - f(x_1, \ldots, x_{i-1}, x_i', x_{i+1}, \ldots, x_n) \right| \le c_i, \ 1 \le i \le n.$$

In other words, if we change the i-th variable of f while keeping all the others fixed, the value of the function cannot change by more than c_i. Then the Efron–Stein inequality implies the following:

Corollary 3.2 *If f has the bounded differences property with constants $c_1, \ldots, c_n$, then*

$$\mathrm{Var}(Z) \le \frac{1}{4}\sum_{i=1}^{n} c_i^2.$$

Proof From the Efron–Stein inequality,

$$\mathrm{Var}(Z) \le \inf_{Z_i} \sum_{i=1}^{n} \boldsymbol{E}\left[(Z - Z_i)^2\right],$$

where the infimum is taken over the class of all $X^{(i)}$-measurable and square-integrable variables Z_i. Here we choose

$$Z_i = \frac{1}{2}\left(\sup_{x_i' \in \mathcal{X}} f(X_1, \ldots, X_{i-1}, x_i', X_{i+1}, \ldots, X_n) + \inf_{x_i' \in \mathcal{X}} f(X_1, \ldots, X_{i-1}, x_i', X_{i+1}, \ldots, X_n)\right).$$

Hence

$$(Z - Z_i)^2 \le \frac{c_i^2}{4},$$

and the proposition follows. □

Next we list some interesting applications of this corollary. In all cases the bound for the variance is obtained effortlessly, while a direct estimation of the variance may be quite involved.

Example 3.3 (BIN PACKING) This is one of the basic operations research problems. Given n numbers $x_1, \ldots, x_n \in [0, 1]$, the question is the following: what is the minimal number of "bins" into which these numbers can be packed such that the sum of the numbers in each bin does not exceed one. Let $f(x_1, \ldots, x_n)$ be this minimum number. Clearly, changing one of the x_i's, the value of $f(x_1, \ldots, x_n)$ cannot change by more than one, so whenever $X_1, \ldots, X_n$ are independent, $Z = f(X_1, \ldots, X_n)$ satisfies

$$\mathrm{Var}\,(Z) \le \frac{n}{4}.$$

This upper bound is not improvable because if the X_i are symmetric Bernoulli random variables (i.e. $\boldsymbol{P}\{X_i = 0\} = \boldsymbol{P}\{X_i = 1\} = 1/2$), then Z is binomially distributed with parameters n and $1/2$ and therefore $\mathrm{Var}\,(Z) = n/4$. On the other hand, sharper bounds, which depend on the distribution of the X_i, may be proved using Talagrand's convex distance inequality discussed in Chapter 7.

Example 3.4 (LONGEST COMMON SUB-SEQUENCE) The simplest version of the longest common sub-sequence problem is as follows: let $X_1, \ldots, X_n$ and $Y_1, \ldots, Y_n$ be two sequences of coin flips. Define Z as the length of the longest sub-sequence which appears in both sequences, that is,

$$Z = \max\{k : X_{i_1} = Y_{j_1}, \ldots, X_{i_k} = Y_{j_k}, \text{ where } 1 \le i_1 < \cdots < i_k \le n \text{ and } 1 \le j_1 < \cdots < j_k \le n\}.$$

The behavior of $\boldsymbol{E}Z$ has been investigated in many papers. It is known that $\boldsymbol{E}Z/n$ converges to some number γ, whose value is unknown. It is conjectured to be $2/(1 + \sqrt{2})$, and it is known to fall between 0.75796 and 0.83763. Here we are concerned with the concentration of Z. A moment of thought reveals that changing one bit cannot change the length of the longest common subsequence by more than one, so Z satisfies the bounded differences property with $c_i = 1$. Consequently,

$$\mathrm{Var}(Z) \leq \frac{n}{2}.$$

Thus, by Chebyshev's inequality, with large probability, Z is within a constant times $\sqrt{n}$ of its expected value. In other words, it is strongly concentrated around the mean, which means that results on $\mathbf{E}Z$ faithfully describe the behavior of the longest common subsequence of two random strings.

Example 3.5 (KERNEL DENSITY ESTIMATION) Let $X_1, \ldots, X_n$ be i.i.d. samples drawn according to some (unknown) density ϕ on the real line. The density is estimated by the kernel estimate

$$\phi_n(x) = \frac{1}{nh_n} \sum_{i=1}^{n} K\left(\frac{x - X_i}{h_n}\right),$$

where $h_n > 0$ is a smoothing parameter, and K is a nonnegative function with $\int K = 1$. The performance of the estimate is typically measured by the L_1 error

$$Z(n) = f(X_1, \ldots, X_n) = \int |\phi(x) - \phi_n(x)| dx.$$

It is easy to see that

$$\begin{aligned} |f(x_1, \ldots, x_n) - f(x_1, \ldots, x_i', \ldots, x_n)| &\leq \frac{1}{nh_n} \int \left| K\left(\frac{x - x_i}{h_n}\right) - K\left(\frac{x - x_i'}{h_n}\right) \right| dx \\ &\leq \frac{2}{n}, \end{aligned}$$

so without further work we obtain

$$\mathrm{Var}(Z(n)) \leq \frac{1}{n}.$$

It is known that for every ϕ, $\sqrt{n}\,\mathbf{E}Z(n) \to \infty$, which implies, by Chebyshev's inequality, that for every $\varepsilon > 0$

$$\mathbf{P}\left\{ \left| \frac{Z(n)}{\mathbf{E}Z(n)} - 1 \right| \geq \varepsilon \right\} = \mathbf{P}\left\{ |Z(n) - \mathbf{E}Z(n)| \geq \varepsilon \mathbf{E}Z(n) \right\} \leq \frac{\mathrm{Var}(Z(n))}{\varepsilon^2 (\mathbf{E}Z(n))^2} \to 0$$

as $n \to \infty$. That is, $Z(n)/\mathbf{E}Z(n) \to 1$ in probability, or in other words, $Z(n)$ is *relatively stable*. This means that the random L_1-error essentially behaves like its expected value.

Example 3.6 (RADEMACHER AVERAGES) Rademacher averages and processes have played an important role in a large variety of applications ranging from empirical process theory through geometry to statistical learning theory. Here we derive bounds for the variance of the supremum of a Rademacher process, using the Efron–Stein inequality.

To define Rademacher averages, let $(\alpha_{i,t})$ be a collection of real numbers indexed by $i = 1, \ldots, n$ and $t \in \mathcal{T}$ where $\mathcal{T}$ is some set. If $X_1, \ldots, X_n$ are independent symmetric random signs (i.e. with $\mathbf{P}\{X_i = -1\} = \mathbf{P}\{X_i = 1\} = 1/2$), then one may define $Z = \sup_{t\in\mathcal{T}} \sum_{i=1}^n X_i\alpha_{i,t}$. The X_i are often called *Rademacher variables* and Z is a *Rademacher average*. The size of the expected value of Z depends, in a delicate manner, on the $\alpha_{i,t}$. However, it is immediate to see that by changing one X_i, Z can change by at most $2\sup_{t\in\mathcal{T}} |\alpha_{i,t}|$, so regardless of the behavior of $\mathbf{E}Z$, by Corollary 3.2 we always have

$$\mathrm{Var}\,(Z) \le \sum_{i=1}^n \sup_{t\in\mathcal{T}} \alpha_{i,t}^2.$$

Next we show how a closer look at the Efron–Stein inequality implies a significantly better bound for the variance of Z. Let $X_1', \ldots, X_n'$ be independent copies of $X_1, \ldots, X_n$. Then

$$Z_i' = \sup_{t\in\mathcal{T}} \left[\left(\sum_{j:j\neq i}^n X_j\alpha_{j,t} \right) + X_i'\alpha_{i,t} \right].$$

Let t^* be a (random) index such that $\sup_{t\in\mathcal{T}} \sum_{j=1}^n X_j\alpha_{j,t} = \sum_{j=1}^n X_j\alpha_{j,t^*}$. Then, for every $i = 1, \ldots, n$,

$$Z - Z_i' \le (X_i - X_i')\,\alpha_{i,t^*}$$

which implies

$$(Z - Z_i')_+^2 \le (X_i - X_i')^2\,\alpha_{i,t^*}^2.$$

By independence of X_i' and $(X_1, \ldots, X_n)$,

$$\mathbf{E}\left[(Z - Z_i')_+^2\right] \le \mathbf{E}\left[\mathbf{E}\left[((X_i - X_i')^2)\,\alpha_{i,t^*}^2 | X_1, \ldots, X_n\right]\right] = 2\mathbf{E}\left[\alpha_{i,t^*}^2\right].$$

Hence, the Efron–Stein inequality implies

$$\mathrm{Var}\,(Z) \le 2\mathbf{E}\left[\sum_{i=1}^n \alpha_{i,t^*}^2\right] \le 2\sigma^2,$$

where $\sigma^2 = \sup_{t\in\mathcal{T}} \sum_{i=1}^n \alpha_{i,t}^2$. Note that, while we lost a factor of 2, the supremum is now outside of the sum and this bound may be a significant improvement on what we obtained as an immediate corollary of the bounded differences property.

3.3 Self-Bounding Functions

Another simple property which is satisfied for many important examples is the so-called *self-bounding* property. We say that a nonnegative function $f : \mathcal{X}^n \to [0, \infty)$ has the self-bounding property if there exist functions $f_i : \mathcal{X}^{n-1} \to \mathbb{R}$ such that for all $x_1, \ldots, x_n \in \mathcal{X}$ and all $i = 1, \ldots, n$,

$$0 \le f(x_1, \ldots, x_n) - f_i(x_1, \ldots, x_{i-1}, x_{i+1}, \ldots, x_n) \le 1$$

and also

$$\sum_{i=1}^{n} (f(x_1, \ldots, x_n) - f_i(x_1, \ldots, x_{i-1}, x_{i+1}, \ldots, x_n)) \le f(x_1, \ldots, x_n).$$

For self-bounding functions we clearly have

$$\sum_{i=1}^{n} (f(x_1, \ldots, x_n) - f_i(x_1, \ldots, x_{i-1}, x_{i+1}, \ldots, x_n))^2 \le f(x_1, \ldots, x_n)$$

and therefore the last expression of v in Theorem 3.1 implies the following:

Corollary 3.7 *If f has the self-bounding property, then*

$$\mathrm{Var}(Z) \le \mathbf{E}Z.$$

Next we mention some applications of this simple corollary. In many cases the obtained bound is a significant improvement over that which we would obtain using simply Corollary 3.2.

Remark 3.3 (RELATIVE STABILITY) A sequence of nonnegative random variables $(Z(n))_{n\in\mathbb{N}}$ is said to be relatively stable if $Z(n)/\mathbf{E}Z(n) \to 1$ in probability. This property guarantees that the random fluctuations of $Z(n)$ around its expectation are of negligible size when compared to the expectation, and therefore most information about the size of $Z(n)$ is given by $\mathbf{E}Z(n)$. Bounding the variance of $Z(n)$ by its expected value implies, in many cases, the relative stability of $(Z(n))_{n\in\mathbb{N}}$. If $Z(n)$ has the self-bounding property, then, by Chebyshev's inequality, for all $\varepsilon > 0$,

$$\mathbf{P}\left\{\left|\frac{Z(n)}{\mathbf{E}Z(n)} - 1\right| > \varepsilon\right\} \le \frac{\mathrm{Var}(Z(n))}{\varepsilon^2(\mathbf{E}Z(n))^2} \le \frac{1}{\varepsilon^2\mathbf{E}Z(n)}.$$

Thus, for relative stability, it suffices to have $\mathbf{E}Z(n) \to \infty$.

An important class of functions satisfying the self-bounding property consists of the so-called *configuration functions.*

Assume that we have a property Π defined over the union of finite products of a set $\mathcal{X}$, that is, a sequence of sets $\Pi_1 \subset \mathcal{X}, \Pi_2 \subset \mathcal{X} \times \mathcal{X}, \ldots, \Pi_n \subset \mathcal{X}^n$. We say that

$(x_1, \ldots x_m) \in \mathcal{X}^m$ satisfies the property Π if $(x_1, \ldots x_m) \in \Pi_m$. We assume that Π is *hereditary* in the sense that if $(x_1, \ldots x_m)$ satisfies Π then so does any sub-sequence $(x_{i_1}, \ldots x_{i_k})$ of $(x_1, \ldots x_m)$. The function f that maps any vector $x = (x_1, \ldots x_n)$ to the size of a largest sub-sequence satisfying Π is the *configuration function* associated with property Π.

Corollary 3.7 implies the following result:

Corollary 3.8 *Let f be a configuration function, and let $Z = f(X_1, \ldots, X_n)$, where $X_1, \ldots, X_n$ are independent random variables. Then*

$$\mathrm{Var}(Z) \leq \mathbf{E}Z.$$

Proof By Corollary 3.7 it suffices to show that any configuration function is self-bounding. Let $Z_i = f(X^{(i)}) = f(X_1, \ldots, X_{i-1}, X_{i+1}, \ldots, X_n)$. The condition $0 \leq Z - Z_i \leq 1$ is trivially satisfied. On the other hand, assume that $Z = k$ and let $\{X_{i_1}, \ldots, X_{i_k}\} \subset \{X_1, \ldots, X_n\}$ be a sub-sequence of cardinality k such that $f_k(X_{i_1}, \ldots, X_{i_k}) = k$. (Note that by the definition of a configuration function such a sub-sequence exists.) Clearly, if the index i is such that $i \notin \{i_1, \ldots, i_k\}$ then $Z = Z_i$, and therefore

$$\sum_{i=1}^{n} (Z - Z_i) \leq Z$$

is also satisfied, which concludes the proof. □

To illustrate the fact that configuration functions appear rather naturally in various applications, we describe some examples originating from different fields.

Example 3.9 (NUMBER OF DISTINCT VALUES IN A DISCRETE SAMPLE) Let $X_1, \ldots, X_n$ be independent, identically distributed random variables taking their values in the set of positive integers such that $\mathbf{P}\{X_1 = k\} = p_k$, and let $Z(n)$ denote the number of distinct values taken by these n random variables. Then we may write

$$Z(n) = \sum_{i=1}^{n} \mathbb{1}_{\{\{X_i \neq X_1, \ldots, X_i \neq X_{i-1}\}\}},$$

so the expected value of $Z(n)$ may be computed easily:

$$\mathbf{E}Z(n) = \sum_{i=1}^{n} \sum_{j=1}^{\infty} (1 - p_j)^{i-1} p_j.$$

It is easy to see that $\mathbf{E}[Z(n)]/n \to 0$ as $n \to \infty$ (see Exercise 3.8). But how concentrated is the distribution of $Z(n)$? Clearly, $Z(n)$ satisfies the bounded differences property with $c_i = 1$, so Corollary 3.2 implies $\mathrm{Var}(Z(n)) \leq n/4$ and therefore $Z(n)/n \to 0$ in probability by Chebyshev's inequality. On the other hand, it is obvious that $Z(n)$ is a configuration function associated with the property of "distinctness," and by Corollary 3.8 we have

$$\text{Var}\,(Z(n)) \leq \mathbf{E}Z(n)$$

which is a significant improvement since $\mathbf{E}Z(n) = o(n)$.

Example 3.10 (VC DIMENSION) One of the central quantities in statistical learning theory is the *Vapnik–Chervonenkis dimension*. Let $\mathcal{A}$ be an arbitrary collection of subsets of $\mathcal{X}$, and let $x = (x_1, \ldots, x_n)$ be a vector of n points of $\mathcal{X}$. Define the *trace* of $\mathcal{A}$ on x by

$$\text{tr}(x) = \{A \cap \{x_1, \ldots, x_n\} : A \in \mathcal{A}\}.$$

The *shatter coefficient*, (or *Vapnik–Chervonenkis growth function*) of $\mathcal{A}$ in x is $T(x) = |\text{tr}(x)|$, the size of the trace. $T(x)$ is the number of different subsets of the n-point set $\{x_1, \ldots, x_n\}$ generated by intersecting it with elements of $\mathcal{A}$. A subset $\{x_{i_1}, \ldots, x_{i_k}\}$ of $\{x_1, \ldots, x_n\}$ is said to be *shattered* if $2^k = T(x_{i_1}, \ldots, x_{i_k})$. The VC *dimension* $D(x)$ of $\mathcal{A}$ (with respect to x) is the cardinality k of the largest shattered subset of x. From the definition it is obvious that $f(x) = D(x)$ is a configuration function (associated with the property of "shatteredness") and therefore if $X_1, \ldots, X_n$ are independent random variables, then

$$\text{Var}\,(D(X)) \leq \mathbf{E}D(X).$$

Example 3.11 (INCREASING SUB-SEQUENCES) Consider a vector $x = (x_1, \ldots, x_n)$ of n distinct numbers in $[0, 1]$. The positive integers $i_1 < i_2 < \cdots < i_m$ form an *increasing sub-sequence* if $x_{i_1} < x_{i_2} < \cdots < x_{i_m}$ (where $i_1 \geq 1$ and $i_m \leq n$). Let $L(x)$ denote the length of a longest increasing sub-sequence. Clearly, $L(x)$ is a configuration function (associated with the "increasing sequence" property) and therefore, if $X_1, \ldots, X_n$ are independent random variables such that they are different with probability one (this is warranted if every X_i has an absolutely continuous distribution) then $\text{Var}\,(L(X)) \leq \mathbf{E}L(X)$. If the X_i's are uniformly distributed in $[0, 1]$ then it is known that $\mathbf{E}L(X) \sim 2\sqrt{n}$. The obtained bound for the variance appears to be quite loose and the right order is $\text{Var}\,(L(X)) = O(n^{1/3})$, an apparently difficult result.

In a variation of the problem, $X_1, \ldots, X_n$ take their values in a finite set $\{1, \ldots, m\}$. Here we define $L^{(m)}(X)$ to be the length of the longest increasing sub-sequence of $X = (X_1, \ldots, X_n)$, that is, the largest positive integer k for which there exist $1 \leq i_1 < \cdots < i_k \leq n$ such that $X_{i_1} \leq X_{i_2} \leq \ldots \leq X_{i_k}$. The analysis of the variance remains unchanged, and as above, we have $\text{Var}\,(L^{(m)}(X)) \leq \mathbf{E}L^{(m)}(X)$. This estimate has the right order of magnitude as it is known that if the X_i are uniformly distributed, $(L^{(m)}(X) - n/m)/\sqrt{2n/m}$ converges in distribution to a random variable whose distribution depends on m.

Example 3.12 (CONDITIONAL RADEMACHER AVERAGES) An example of a self-bounding function which is not a configuration function is that of Rademacher averages. Let $X_1, \ldots, X_n$ be independent random variables taking values in $[-1, 1]^d$ and denote the components of X_i by $X_{i,1} \ldots, X_{i,d}$, $i = 1, \ldots, n$. If $\varepsilon_1, \ldots, \varepsilon_n$ denote independent symmetric $\{-1, 1\}$-valued random variables, independent of the X_i's (the so-called Rademacher random variables), then we define the *conditional Rademacher average* as

$$Z = \mathbf{E}\left[\max_{j=1,\ldots,d}\sum_{i=1}^{n}\varepsilon_i X_{i,j}|X_1,\ldots,X_n\right].$$

(Thus, the expected value is taken with respect to the Rademacher variables and Z is a function of the X_i's.) Quantities like Z have been known to measure effectively the complexity of model classes in statistical learning theory. It is immediate that Z has the bounded differences property and Corollary 3.2 implies $\text{Var}(Z) \leq n/4$. However, this bound may be improved by observing that Z also has the self-bounding property, and therefore $\text{Var}(Z) \leq \mathbf{E}Z$. Indeed, defining

$$Z_i = \mathbf{E}\left[\max_{j=1,\ldots,d}\sum_{\substack{k=1\\k\neq i}}^{n}\varepsilon_i X_{k,j}|X^{(i)}\right]$$

it is easy to see that $0 \leq Z - Z_i \leq 1$ and $\sum_{i=1}^{n}(Z - Z_i) \leq Z$ (the details are left as an exercise). The improvement provided by Lemma 3.7 is essential since it is well known in empirical process theory and statistical learning theory that in many circumstances, $\mathbf{E}Z$ may be bounded by $Cn^{1/2}$ where the constant C that does not depend on n (see for example Section 13.3).

3.4 More Examples and Applications

Example 3.13 (FIRST PASSAGE PERCOLATION) Consider a graph such that a weight X_i is assigned to each edge e_i so that the X_i are nonnegative independent random variables with second moment $\mathbf{E}X_i^2 = \sigma^2$. Let v_1 and v_2 be fixed vertices of the graph. We are interested in the total weight of the path from v_1 to v_2 with minimum weight. (The weight of a path is defined as the sum of the weights of the edges on the path.) Thus,

$$Z = \min_P \sum_{e_i \in P} X_i$$

where the minimum is taken over all paths P from v_1 to v_2. Denote an arbitrary optimal path by P^*. By replacing X_i with X_i', the total minimum weight can only increase if the edge e_i is on P^*, and therefore

$$(Z - Z_i')_-^2 \leq (X_i' - X_i)_+^2 \mathbb{1}_{\{e_i \in P^*\}} \leq X_i'^2 \mathbb{1}_{\{e_i \in P^*\}}.$$

Thus,

$$\text{Var}(Z) \leq \mathbf{E}\sum_i X_i'^2 \mathbb{1}_{\{e_i \in P^*\}} = \sigma^2 \mathbf{E}\sum_i \mathbb{1}_{\{e_i \in P^*\}},$$

that is, the variance of Z is bounded by σ^2 times the expected number of edges in the minimum-weight path. Under general conditions, this is bounded by a constant times the graph distance between v_1 and v_2 (see the exercises). This linear bound, however, is known to be loose in some important special cases such as percolation on $\mathbb{Z}^d$. To prove bounds of the correct order for this special case remains to be a challenge.

Example 3.14 (THE LARGEST EIGENVALUE OF A RANDOM SYMMETRIC MATRIX) Let A be a symmetric real matrix whose entries $X_{i,j}$, $1 \le i \le j \le n$ are independent random variables with absolute value bounded by 1. Let $Z = \lambda_1$ denote the largest eigenvalue of A. The property of the largest eigenvalue we need in order to bound the variance of Z is that if $v = (v_1, \dots, v_n) \in \mathbb{R}^n$ is an eigenvector corresponding to the largest eigenvalue λ_1 with $\|v\| = 1$, then

$$\lambda_1 = v^T A v = \sup_{u:\|u\|=1} u^T A u.$$

Using Theorem 3.1, consider the symmetric matrix $A'_{i,j}$ obtained by replacing $X_{i,j}$ in A by the independent copy $X'_{i,j}$, while keeping all other variables fixed. Let $Z'_{i,j}$ denote the largest eigenvalue of the obtained matrix. Then by the above-mentioned property of the largest eigenvalue,

$$\begin{aligned}(Z - Z'_{i,j})_+ &\le \left(v^T A v - v^T A'_{i,j} v\right) \mathbb{1}_{\{Z > Z'_{i,j}\}} \\ &= \left(v^T (A - A'_{i,j}) v\right) \mathbb{1}_{\{Z > Z'_{i,j}\}} \le 2(v_i v_j (X_{i,j} - X'_{i,j}))_+ \\ &\le 4|v_i v_j|.\end{aligned}$$

Therefore,

$$\sum_{1 \le i \le j \le n} (Z - Z'_{i,j})_+^2 \le \sum_{1 \le i \le j \le n} 16|v_i v_j|^2 \le 16 \left(\sum_{i=1}^n v_i^2\right)^2 = 16.$$

Taking expectations of both sides and using the Efron–Stein inequality, we have $\mathrm{Var}(Z) \le 16$. Thus, the variance is bounded by a constant regardless of the size of the matrix and the distribution of the entries. The only condition we need is independence and boundedness of the entries; they don't even need to have the same distribution. Note that the same proof also works for the smallest eigenvalue.

Example 3.15 (MINIMUM WEIGHT SPANNING TREE) Consider the random variable T_m defined as the sum of weights on the minimum spanning tree of the complete graph K_m with independent uniformly distributed (on $[0, 1]$) weights $X_{i,j}$ $(1 \le i < j \le m)$ on the edges. (Thus, T_m is a function of $n = \binom{m}{2}$ independent random variables.) It is well known that the expected value of T_m converges to a constant $\zeta(3) = \sum_{i=1}^{\infty} i^{-3}$. Here we bound the variance of T_m. Using the Efron–Stein inequality directly gives a loose bound, but a simple trick will lead us close to the truth. The idea is that the largest weight of any edge in the minimum spanning tree is small, at most of the order $\log m/m$, with high probability. This observation allows us to replace T_m with the related random variable $\overline{T}_m$ obtained when the $X_{i,j}$ are replaced by $\min(X_{i,j}, \delta_m)$, where $\delta_m > 0$ is a small positive number. Note that if $\delta_m = c \log m/m$ for some

$c > 1$ then $T_m = \overline{T}_m$ with high probability. In order to appreciate this simply observe that $T_m \neq \overline{T}_m$ implies that the largest edge weight in the minimum spanning tree is greater than δ_m. However, this is just the probability that the Erdős–Rényi random graph $G(m, \delta_m)$ (i.e. a graph on m vertices in which each of the possible $\binom{m}{2}$ edges is present independently with probability δ_m) is not connected, which is at most $2\left(e^{m^{(1-c)/2}} - 1\right) + 2^{m+1}m^{-(c-1)m/4}$ (see Exercise 3.12) which is bounded by $4m^{-c/4}$, if $c \geq 2$. Since

$$\begin{aligned}
\operatorname{Var}(T_m) &= \boldsymbol{E}\left[T_m^2\right] - (\boldsymbol{E}T_m)^2 \\
&= \boldsymbol{E}\left[T_m^2 \mathbb{1}_{\{T_m = \overline{T}_m\}}\right] + \boldsymbol{E}\left[T_m^2 \mathbb{1}_{\{T_m \neq \overline{T}_m\}}\right] - (\boldsymbol{E}\overline{T}_m)^2 \\
&\leq \operatorname{Var}(\overline{T}_m) + m^2 \boldsymbol{P}\left\{T_m \neq \overline{T}_m\right\} \\
&\leq \operatorname{Var}(\overline{T}_m) + 4m^{2-c/4},
\end{aligned}$$

it suffices to bound the variance of $\overline{T}_m$. Here it is advantageous to use the variant of the Efron–Stein inequality which states

$$\operatorname{Var}(\overline{T}_m) \leq \boldsymbol{E} \sum_{\substack{i,j=1 \\ i \neq j}}^{n} \left(\overline{T}_m - \overline{T}'_{m,(i,j)}\right)_-^2$$

where $\overline{T}'_{m,(i,j)}$ is obtained by replacing $X_{i,j}$ by an independent copy. Clearly, if one replaces the weight $X_{i,j}$ then $\overline{T}_m$ can only decrease if the edge (i,j) is in the minimum weight spanning tree. Since there are $m - 1$ such edges and the change cannot be more than δ_m,

$$\sum_{\substack{i,j=1 \\ i \neq j}}^{n} \left(\overline{T}_m - \overline{T}'_{m,(i,j)}\right)_-^2 \leq m\delta_m^2.$$

In summary,

$$\operatorname{Var}(T_m) \leq m\delta_m^2 + 4m^{2-c/4} = \frac{144\log^2 m}{m} + \frac{4}{m}$$

where we choose $c = 12$. This bound is not quite of the correct order, since it is known that, asymptotically, $\operatorname{Var}(T_m) \sim (6\zeta(4) - 4\zeta(3))/m$. (In fact, $\sqrt{m}(T_m - \zeta(3))$ converges, in distribution, to a centered normal random variable with variance $6\zeta(4) - 4\zeta(3)$.) However, this argument illustrates how the Efron–Stein inequality can be used in a simple way to obtain powerful nonasymptotic inequalities.

Example 3.16 (PACKET ROUTING IN PARALLEL COMPUTATION) Here we describe a simple routing problem for massive parallel computation. Let N be an integer and suppose that 2^N processors are arranged in a binary hypercube. More precisely, consider the graph with vertex set $\{-1, 1\}^N$ in which two vertices are joined by an edge if and only if the corresponding binary N-vectors differ in exactly one bit. Each vertex represents a processor and processors with neighboring vertices are joined with a direct communication link. During the execution of a parallel computing task, processors need to communicate with each other. In the simple model considered here, at a certain point in time, every processor $u \in \{-1, 1\}^N$ needs to send a packet to another processor $v = \sigma(u)$ where σ is a permutation over $\{-1, 1\}^N$. Thus, for each vertex u, a path from u to $v = \sigma(u)$ has to be found on the graph, and the packet is sent from u to v along the chosen path. In total, 2^N paths are chosen (one for each vertex), some of which may intersect in certain edges of the graph. If some edge is contained in various paths, then congestion occurs and computation suffers a delay proportional to the number of paths going through the edge. Thus, the routing problem is to assign paths so that the maximum number of paths going through any single edge is as small as possible. Formally, a routing strategy is a mapping p from pairs of vertices to paths. The length of a path is the number of edges it goes through.

The simplest routing strategy one may think of is the following "shortest-path" strategy: for a given pair (u, v) of vertices with Hamming distance $\rho = d(u, v)$ (with $v = \sigma(u)$), choose the path $u_0 = u, u_1, \ldots, u_\rho = v$ defined recursively for $i = 1, \ldots, \rho$ such that u_{i+1} differs from u_i according to the ith position in which u and v differ. It is easy to see (Exercise 3.14) that this strategy may have a maximal congestion that is exponentially large in N. Now consider the following randomized solution. Given a routing task σ, every vertex u chooses an intermediate vertex $W(u)$ independently, uniformly at random, among all possible 2^N vertices. Then u is first routed to $W(u)$ and then $W(u)$ to $\sigma(u)$ using the shortest path strategy described above. One may show that the expected value of the maximal congestion of this scheme is $O(N)$ (see Exercise 3.15). On the other hand, the random variable $Z(\sigma)$ defined as the maximal congestion of any edge, may be considered as a function of the 2^N independent random variables $W(u)$. Now it is easy to see that $Z(\sigma)$ is a configuration function and therefore $\text{Var}(Z(\sigma)) \leq \mathbf{E}Z(\sigma) = O(N)$. Thus, with high probability, $Z(\sigma)$ remains bounded by a linear function of N, an exponential improvement compared to the worst-case performance of a deterministic routing strategy.

3.5 A Convex Poincaré Inequality

In Section 3.7 below we will use the Efron–Stein inequality to prove a classical statement that any Lipschitz function of a canonical Gaussian vector has a standard deviation bounded by the Lipschitz constant. Here we point out an analogous bound for functions of n independent random variables taking values in $[0, 1]^n$. The price we have to pay for this generality is an extra convexity condition on the function.

We assume that $f : [0, 1]^n \to \mathbb{R}$ is a *separately convex* function, that is, for any $i = 1, \ldots, n$ and fixed $x_1, \ldots, x_{i-1}, x_{i+1}, \ldots, x_n, f$ is a convex function of its i-th variable. We also assume that the partial derivatives of f exist, though this last condition may be removed by a routine approximation argument which we do not detail here.

Theorem 3.17 *Let $X_1, \dots, X_n$ be independent random variables taking values in the interval $[0, 1]$ and let $f : [0, 1]^n \to \mathbb{R}$ be a separately convex function whose partial derivatives exist. Then $f(X) = f(X_1, \dots, X_n)$ satisfies*

$$\mathrm{Var}\,(f(X)) \le \mathbf{E}\left[\|\nabla f(X)\|^2\right].$$

Proof The proof is an easy consequence of the Efron–Stein inequality, because by Theorem 3.1 it suffices to bound the random variable $\sum_{i=1}^n (Z - Z_i)^2$ where $Z_i = \inf_{x_i'} f(X_1, \dots, x_i', \dots, X_n)$. Denote by X_i' the value of x_i' for which the minimum is achieved. This is guaranteed by continuity and the compactness of the domain of f. Then, writing $\overline{X}^{(i)} = (X_1, \dots, X_{i-1}, X_i', X_{i+1}, \dots, X_n)$, we have

$$\begin{aligned}
\sum_{i=1}^n (Z - Z_i)^2 &= \sum_{i=1}^n \left(f(X) - f\left(\overline{X}^{(i)}\right)\right)^2 \\
&\le \sum_{i=1}^n \left(\frac{\partial f}{\partial x_i}(X)\right)^2 (X_i - X_i')^2 \\
&\qquad \text{(by separate convexity)} \\
&\le \sum_{i=1}^n \left(\frac{\partial f}{\partial x_i}(X)\right)^2 \\
&= \|\nabla f(X)\|^2.
\end{aligned}$$

□

Example 3.18 (THE LARGEST SINGULAR VALUE OF A RANDOM MATRIX) Let A be an $m \times n$ matrix with entries $X_{i,j}$ ($i = 1, \dots, m, j = 1, \dots, n$) of independent random variables taking values in $[0, 1]$. We are interested in concentration of the largest singular value Z of A, defined as the square root of the largest eigenvalue of the symmetric $n \times n$ matrix $A^T A$. Thus,

$$Z = \sqrt{\lambda_1(A^T A)} = \sqrt{\sup_{u \in \mathbb{R}^n : \|u\|=1} u^T A^T A u} = \sup_{u \in \mathbb{R}^n : \|u\|=1} \|Au\|.$$

For each fixed vector u, $\|Au\|$ is a convex function of the mn-dimensional vector formed by the $X_{i,j}$ and since the supremum of convex functions is convex, we see that Z is a convex function of the $X_{i,j}$. In order to apply Theorem 3.17, we may use Lidskii's inequality, a classical result of linear algebra, which states that if $A = (x_{i,j})_{m \times n}$ and $B = (y_{i,j})_{m \times n}$ are two matrices then, denoting by $s_1(M) \ge \dots \ge s_n(M)$ the singular values of an $m \times n$ matrix M,

$$\begin{aligned}
(s_1(A) - s_1(B))^2 &\le \sum_{i=1}^n (s_i(A) - s_i(B))^2 \le \sum_{i=1}^n s_i(A - B)^2 \\
&= \mathrm{tr}((A - B)^T (A - B)) = \sum_{i=1}^m \sum_{j=1}^n (x_{i,j} - y_{i,j})^2
\end{aligned}$$

(see Exercise 3.16). Therefore, the largest singular value is a Lipschitz function with Lipschitz constant $L = 1$ and by Theorem 3.17,

$$\mathrm{Var}(Z) \leq 1.$$

3.6 Exponential Tail Bounds via the Efron–Stein Inequality

The purpose of this section is to show two different ways by which the Efron–Stein inequality may be used in a simple and elementary way to prove exponential bounds for the tail probabilities of functions with bounded differences. These bounds are suboptimal but the main ideas will be used later to prove sharper bounds. Also, our intention is to provide further evidence of the surprising power of the Efron–Stein inequality.

In the arguments, in fact, we need less than bounded differences, just the property that a positive constant v exists such that

$$\sum_{i=1}^{n} (Z - Z_i')_+^2 \leq v \tag{3.4}$$

holds with probability one. Recall that, for example, the largest eigenvalue of a random symmetric matrix satisfies this condition with $v = 16$ (see Example 3.14). We establish exponential tail inequalities by deriving upper bounds for the distance between quantiles of Z. Define, for any $\alpha \in (0,1)$, the α-quantile of $Z = f(X) = f(X_1, \ldots, X_n)$ by

$$Q_\alpha = \inf\{z : \mathbf{P}\{Z \leq z\} \geq \alpha\}.$$

In particular, we denote the median of Z by $\mathbf{M}Z = Q_{1/2}$.

The trick of the first method is to use the Efron–Stein inequality for the random variable $g_{a,b}(X) = g_{a,b}(X_1, \ldots, X_n)$ where $b \geq a$ and the function $g_{a,b} : \mathcal{X}^n \to \mathbb{R}$ is defined as

$$g_{a,b}(x) = \begin{cases} b & \text{if } f(x) \geq b \\ f(x) & \text{if } a < f(x) < b. \\ a & \text{if } f(x) \leq a \end{cases}$$

First observe that if $a \geq \mathbf{M}Z$, then $\mathbf{E}g_{a,b}(X) \leq (a+b)/2$ and therefore

$$\mathrm{Var}(g_{a,b}(X)) \geq \frac{\mathbf{P}\{g_{a,b}(X) = b\}}{4}(b-a)^2 = \frac{\mathbf{P}\{Z \geq b\}}{4}(b-a)^2.$$

On the other hand, we may use the Efron–Stein inequality to obtain an upper bound for the variance of $g_{a,b}(X)$. To this end, observe that if $f(x) \leq a$ then $g_{a,b}(\tilde{x}^{(i)}) \geq g_{a,b}(x)$, for $\tilde{x}^{(i)} = (x_1, \ldots, x_{i-1}, x_i', x_{i+1}, \ldots, x_n)$ and so

$$\begin{aligned}
\sum_{i=1}^{n} \mathbf{E}\left(g_{a,b}(X) - g_{a,b}\left(\tilde{X}^{(i)}\right)\right)^2 &= 2\sum_{i=1}^{n} \mathbf{E}\left(g_{a,b}(X) - g_{a,b}\left(\tilde{X}^{(i)}\right)\right)_+^2 \\
&\leq 2\mathbf{E}\left[\mathbb{1}_{\{Z>a\}} \sum_{i=1}^{n} \left(g_{a,b}(X) - g_{a,b}\left(\tilde{X}^{(i)}\right)\right)_+^2\right] \\
&\leq 2v\mathbf{P}\{Z > a\}
\end{aligned}$$

where, in the last step, we used the fact that condition (3.4) implies that

$$\sum_{i=1}^{n} \left(g_{a,b}(X) - g_{a,b}\left(\tilde{X}^{(i)}\right)\right)_+^2 \leq \sum_{i=1}^{n} \left(f(X) - f\left(\tilde{X}^{(i)}\right)\right)_+^2 \leq v.$$

Comparing the obtained upper and lower bounds for Var $(g_{a,b}(X))$, we get

$$b - a \leq \sqrt{8v\frac{\mathbf{P}\{Z > a\}}{\mathbf{P}\{Z \geq b\}}}.$$

We may use this inequality to bound the distance between quantiles of Z. To this end, let $0 < \delta < \gamma \leq 1/2$ and choose $a = Q_{1-\gamma}$ and $b = Q_{1-\delta}$. Then $\mathbf{P}\{Z > a\} \leq \gamma$ and $\mathbf{P}\{Z \geq b\} \geq \delta$ and therefore the distance between any two quantiles of Z (to the right of the median) can be bounded as

$$Q_{1-\delta} - Q_{1-\gamma} \leq \sqrt{\frac{8v\gamma}{\delta}}.$$

It is instructive to choose $\gamma = 2^{-k}$ and $\delta = 2^{-(k+1)}$ for some integer $k \geq 1$. Then, denoting $a_k = Q_{1-2^{-k}}$, we get

$$a_{k+1} - a_k \leq 4\sqrt{v},$$

so the difference between consecutive quantiles corresponding to exponentially decreasing tail probabilities is bounded by a constant. In particular, by summing this inequality for $k = 1, \ldots, m$, we have $a_{m+1} \leq \mathbf{M}Z + 4m\sqrt{v}$ which implies that for all $t > 0$,

$$\mathbf{P}\{Z > \mathbf{M}Z + t\} \leq 2^{-t/(4\sqrt{v})}.$$

In Chapter 6 we will be able to improve this tail bound by showing that the exponent is, in fact, of the order of $-t^2/v$, that is, tail probabilities of functions satisfying condition (3.4) decrease in a sub-Gaussian manner. We emphasize that we have derived more than just bounds for tail probabilities as we have obtained explicit, nonasymptotic bounds for the distance between quantiles. We may call these "local" tail bounds. In many cases, these local bounds can also be sharpened to reveal the sub-Gaussian nature of the tails. This will be shown in Section 9.3 building on hypercontractivity arguments.

An alternative route to obtain exponential bounds is by applying the Efron–Stein inequality to $\exp(\lambda Z/2)$ with $\lambda > 0$. Then, by the mean-value theorem,

$$\mathbf{E}e^{\lambda Z} - \left(\mathbf{E}\left[e^{\lambda Z/2}\right]^2\right) \le \mathbf{E}\left[\sum_{i=1}^{n}\left(e^{\lambda Z/2} - e^{\lambda Z_i'/2}\right)_+^2\right]$$
$$\le \frac{\lambda^2}{4}\mathbf{E}\left[\sum_{i=1}^{n} e^{\lambda Z}\left(Z - Z_i'\right)_+^2\right].$$

Now we may use our condition (3.4) to derive

$$\mathbf{E}e^{\lambda Z} - \left(\mathbf{E}\left[e^{\lambda Z/2}\right]\right)^2 \le \frac{v\lambda^2}{4}\mathbf{E}e^{\lambda Z}$$

or equivalently

$$\left(1 - \frac{v\lambda^2}{4}\right)F(\lambda) \le \left(F(\lambda/2)\right)^2,$$

where $F(\lambda) = \mathbf{E}e^{\lambda(Z-\mathbf{E}Z)}$. We may now use the above functional inequality to control the moment generating function. The solution is based on elementary calculus and is summarized in the following lemma.

Lemma 3.19 *Let* $g : (0,1) \to (0,\infty)$ *be a function such that* $\lim_{x\to 0}\left(g(x) - 1\right)/x = 0$. *If for every* $x \in (0,1)$

$$\left(1 - x^2\right)g(x) \le g(x/2)^2,$$

then

$$g(x) \le \left(1 - x^2\right)^{-2}.$$

Proof We easily derive, by induction, that

$$g(x) \le \left(g\left(x2^{-k}\right)\right)^{2^k}\prod_{j=0}^{k}\left(1 - \left(x2^{-j}\right)^2\right)^{-2^j}.$$

The assumption on the behavior of the function g at 0 ensures that $\lim_{k\to\infty}\left(g\left(x2^{-k}\right)\right)^{2^k} = 1$. Hence, the previous inequality implies that

$$\log g(x) \le \sum_{j=0}^{\infty} 2^j\left[-\log\left(1 - \left(x2^{-j}\right)^2\right)\right]. \tag{3.5}$$

Now by concavity of the logarithm, $-u^{-1}\log(1-u)$ is a nondecreasing function of $u \in (0,1)$ and therefore, for every integer j,

$$-\log\left(1-\left(x2^{-j}\right)^2\right) \leq 2^{-2j}\left[-\log\left(1-x^2\right)\right].$$

Plugging this inequality in (3.5) leads to

$$\log g(x) \leq \left[-\log\left(1-x^2\right)\right]\sum_{j=0}^{\infty} 2^{-j}$$

and the result follows. □

Since $F(0)=1$ and $F'(0)=0$, we may apply Lemma 3.19 to the function $x \to F\left(2xv^{-1/2}\right)$ and get, for every $\lambda \in \left(0, 2v^{-1/2}\right)$,

$$F(\lambda) \leq \left(1-\frac{\lambda^2 v}{4}\right)^{-2}. \tag{3.6}$$

Thus, the Efron–Stein inequality may be used to prove exponential integrability of Z. Moreover, since by (3.6) $F\left(v^{-1/2}\right) \leq 2$, by Markov's inequality, for every $t > 0$,

$$\mathbf{P}\{Z-\mathbf{E}Z \geq t\} \leq 2e^{-t/\sqrt{v}}.$$

This inequality has the same form as the one derived using the first method of this section but now we bound deviations from the mean instead of the median and the constants are somewhat better.

In Chapter 6 we derive Gaussian instead of exponential-like tail bounds. Another way of exploiting (3.6) is to bound $-\log(1-u)$ by $u(1-u)^{-1}$ and conclude that for every $\lambda \in \left(0, 2v^{-1/2}\right)$

$$\log F(\lambda) \leq \frac{\lambda^2 v}{2\left(1-\left(\lambda^2 v/4\right)\right)} \leq \frac{\lambda^2 v}{2\left(1-\left(\lambda\sqrt{v}/2\right)\right)}.$$

This bound for the moment-generating function means that $Z-\mathbf{E}Z$ is a sub-gamma random variable with variance factor v and scale parameter $c=\sqrt{v}/2$, as introduced in Section 2.4. The calculations of that section show that for all $t > 0$,

$$\mathbf{P}\left\{Z-\mathbf{E}Z \geq \sqrt{2vt}+ct\right\} \leq e^{-t}. \tag{3.7}$$

Since $c=\sqrt{v}/2$, we see that as soon as t is not too small (say, $t \geq 1$), the linear term in the expression $\sqrt{2vt}+ct$ dominates the other one. This is the reason why one cannot interpret (3.7) as a sub-Gaussian inequality. In subsequent chapters we establish inequalities like

(3.7) with much more interesting values for c. The case $c = 0$ is of course the most interesting one but in some circumstances we will get moderate values for c, typically depending on a uniform bound on the increments $Z - Z_i'$.

3.7 The Gaussian Poincaré Inequality

The Efron–Stein inequality can be successfully applied to prove a sharp bound for the variance of a smooth function of a standard Gaussian random vector, known as the Gaussian Poincaré inequality. This result is a prelude to various related inequalities discussed in Chapter 5.

Theorem 3.20 (GAUSSIAN POINCARÉ INEQUALITY) *Let $X = (X_1, \ldots, X_n)$ be a vector of i.i.d. standard Gaussian random variables (i.e. X is a Gaussian vector with zero mean vector and identity covariance matrix). Let $f : \mathbb{R}^n \to \mathbb{R}$ be any continuously differentiable function. Then*

$$\mathrm{Var}\,(f(X)) \leq \mathbf{E}\left[\left\|\nabla f(X)\right\|^2\right].$$

Proof We may assume that $\mathbf{E}\left\|\nabla f(X)\right\|^2 < \infty$ since otherwise the inequality is trivial. The proof is based on a double use of the Efron–Stein inequality. A first straightforward use of it reveals that it suffices to prove the theorem when the dimension n equals 1. Thus, the problem reduces to show that

$$\mathrm{Var}\,(f(X)) \leq \mathbf{E}\left[f'(X)^2\right], \tag{3.8}$$

where $f : \mathbb{R} \to \mathbb{R}$ is any continuously differentiable function on the real line and X is a standard normal random variable. First, notice that it suffices to prove this inequality when f has a compact support and is twice continuously differentiable. Now let $\varepsilon_1, \ldots, \varepsilon_n$ be independent Rademacher random variables and introduce

$$S_n = n^{-1/2} \sum_{j=1}^{n} \varepsilon_j.$$

Since for every i

$$\mathrm{Var}^{(i)}\,(f\,(S_n)) = \frac{1}{4}\left(f\left(S_n + \frac{1-\varepsilon_i}{\sqrt{n}}\right) - f\left(S_n - \frac{1+\varepsilon_i}{\sqrt{n}}\right)\right)^2,$$

applying the Efron–Stein inequality again, we obtain

$$\mathrm{Var}\,(f\,(S_n)) \leq \frac{1}{4}\sum_{i=1}^{n}\mathbf{E}\left[\left(f\left(S_n + \frac{1-\varepsilon_i}{\sqrt{n}}\right) - f\left(S_n - \frac{1+\varepsilon_i}{\sqrt{n}}\right)\right)^2\right]. \tag{3.9}$$

The central limit theorem implies that S_n converges in distribution to X, where X has the standard normal law. Hence $\mathrm{Var}\,(f\,(S_n))$ converges to $\mathrm{Var}\,(f(X))$. Let K denote the supremum of the absolute value of the second derivative of f. Taylor's theorem implies that, for every i,

$$\left|f\left(S_n + \frac{1-\varepsilon_i}{\sqrt{n}}\right) - f\left(S_n - \frac{1+\varepsilon_i}{\sqrt{n}}\right)\right| \le \frac{2}{\sqrt{n}}|f'(S_n)| + \frac{2K}{n}$$

and therefore

$$\begin{aligned}&\frac{n}{4}\left(f\left(S_n + \frac{1-\varepsilon_i}{\sqrt{n}}\right) - f\left(S_n - \frac{1+\varepsilon_i}{\sqrt{n}}\right)\right)^2 \\ &\le f'(S_n)^2 + \frac{2K}{\sqrt{n}}|f'(S_n)| + \frac{K^2}{n}.\end{aligned}$$

This and the central limit theorem imply that

$$\limsup_{n\to\infty} \frac{1}{4}\sum_{i=1}^{n} \mathbf{E}\left[\left(f\left(S_n + \frac{1-\varepsilon_i}{\sqrt{n}}\right) - f\left(S_n - \frac{1+\varepsilon_i}{\sqrt{n}}\right)\right)^2\right] = \mathbf{E}\left[f'(X)^2\right],$$

which means that (3.9) leads to (3.8) by letting n go to infinity. □

A straightforward consequence of the Gaussian Poincaré inequality is that, whenever $f : \mathbb{R}^n \to \mathbb{R}$ is Lipschitz, that is, for all $x, y \in \mathbb{R}^n$

$$|f(x) - f(y)| \le \|x - y\|$$

and X is a standard Gaussian random vector, then

$$\operatorname{Var}(f(X)) \le 1.$$

Indeed, using an approximation argument (like convolution with a smooth kernel) one may always assume that f is differentiable and if this is the case then $\sup_x \|\nabla f(x)\| \le 1$ and the inequality easily follows from Theorem 3.20.

3.8 A Proof of the Efron–Stein Inequality Based on Duality

The Efron–Stein inequality is the first example of various closely related concentration inequalities. In order to better prepare similar results in a more general context, we provide an alternative proof based on a duality, rather than an orthogonality, argument.

Consider first the following elementary duality formula:

Proposition 3.21 *If Y is a real-valued square-integrable random variable ($Y \in \mathbb{L}_2$ in short), then*

$$\mathrm{Var}\,(Y) = \sup_{T \in \mathbb{L}_2} \left(2\,\mathrm{Cov}\,(Y,T) - \mathrm{Var}\,(T)\right).$$

Proof The proof is simple: since $\mathrm{Var}\,(Y - T) \geq 0$, and

$$\mathrm{Var}\,(Y - T) = \mathrm{Var}\,(Y) - 2\,\mathrm{Cov}(Y,T) + \mathrm{Var}\,(T),$$

we have

$$\mathrm{Var}\,(Y) \geq 2\,\mathrm{Cov}\,(Y,T) - \mathrm{Var}\,(T)$$

and since this inequality becomes an equality whenever $T = Y$, the duality formula follows. □

Now we may consider the telescoping sum

$$Z^2 - (\mathbf{E}Z)^2 = \sum_{i=1}^{n} \left((\mathbf{E}_i Z)^2 - (\mathbf{E}_{i-1} Z)^2\right),$$

which leads to

$$\mathrm{Var}\,(Z) = \sum_{i=1}^{n} \mathbf{E}\left[(\mathbf{E}_i Z)^2 - (\mathbf{E}_{i-1} Z)^2\right].$$

Note that on the one hand, this decomposition does not require any orthogonality argument. On the other hand, it is equivalent to the identity $\mathrm{Var}\,(Z) = \sum_{i=1}^{n} \mathbf{E}\left[\Delta_i^2\right]$ which served as our starting point in proving Theorem 3.1. Indeed, for every $i = 1, \ldots, n$, by orthogonality between $\mathbf{E}_{i-1}Z$ and Δ_i, the Pythagorean theorem implies that

$$\mathbf{E}\left[\Delta_i^2\right] = \mathbf{E}\left[(\mathbf{E}_i Z)^2 - (\mathbf{E}_{i-1} Z)^2\right].$$

Similarly to our first proof of the Efron–Stein inequality, the independence of the variables $X_1, \ldots, X_n$ is used by noting that

$$\mathbf{E}_{i-1} Z = \mathbf{E}^{(i)}\left[\mathbf{E}_i Z\right]$$

and therefore

$$\mathbf{E}\left[(\mathbf{E}_i Z)^2 - (\mathbf{E}_{i-1} Z)^2\right] = \mathbf{E}\left[\mathrm{Var}^{(i)}\left(\mathbf{E}_i Z\right)\right].$$

In other words, we have proven the following alternative formulation (using independence but without using the orthogonality structure of the martingale differences):

$$\text{Var}\,(Z) = \sum_{i=1}^{n} \mathbf{E}\left[\text{Var}^{(i)}\,(\mathbf{E}_i Z)\right]. \tag{3.10}$$

It remains to commute the $\text{Var}^{(i)}$ and $\mathbf{E}_i$ operators and this is precisely the step where we use a duality argument.

Lemma 3.22 *For every $i = 1, \ldots, n$,*

$$\mathbf{E}\left[\text{Var}^{(i)}\,(\mathbf{E}_i Z)\right] \le \mathbf{E}\left[\text{Var}^{(i)}\,(Z)\right].$$

Proof Applying the duality formula of Proposition 3.21 conditionally on $X^{(i)}$, we show that for any square-integrable variable T,

$$2Cov^{(i)}(Z,T) - \text{Var}^{(i)}(T) \le \text{Var}^{(i)}(Z). \tag{3.11}$$

But if we take T to be $(X_1, \ldots, X_i)$-measurable, then

$$\mathbf{E}\left[Cov^{(i)}\,(Z,T)\right] = \mathbf{E}\left[Z\left(T - \mathbf{E}^{(i)}T\right)\right] = \mathbf{E}\left[\mathbf{E}_i Z\left(T - \mathbf{E}^{(i)}T\right)\right]$$
$$= \mathbf{E}\left[Cov^{(i)}\,(\mathbf{E}_i Z, T)\right].$$

Hence, choosing $T = \mathbf{E}_i Z$ leads to

$$\mathbf{E}\left[Cov^{(i)}\,(Z, \mathbf{E}_i Z)\right] = \mathbf{E}\left[\text{Var}^{(i)}\,(\mathbf{E}_i Z)\right],$$

and therefore, by (3.11),

$$\mathbf{E}\left[\text{Var}^{(i)}\,(\mathbf{E}_i Z)\right] \le \mathbf{E}\left[\text{Var}^{(i)}\,(Z)\right].$$

□

Combining Lemma 3.22 with the decomposition (3.10) leads to

$$\text{Var}\,(Z) \le \sum_{i=1}^{n} \mathbf{E}\left[\text{Var}^{(i)}\,(Z)\right]$$

which is equivalent to the Efron–Stein inequality.

Note that there is no measure-theoretic trap here since by Fubini's theorem, the conditional expectations that we are dealing with can all be defined from regular versions of conditional probabilities. Hence it is perfectly legal to use the duality formula for the conditional variance as we did above.

3.9 Bibliographical Remarks

The Efron–Stein inequality got its name from Efron and Stein (1981). While the original result of Efron and Stein had some extra conditions and came with a sub-optimal constant, Steele (1986) and Rhee and Talagrand (1986) obtained improved versions and the form presented in Theorem 3.1. The proof shown in Section 3.1 appears in Rhee and Talagrand (1986).

In statistics, the jackknife estimate is attributed to Quenouille (1949) and Tukey (1958). For surveys and related methods we refer to Efron and Tibshirani (1994), and Politis, Romano, and Wolf (1999).

The behavior of $Z = f(X_1, \ldots, X_n)$ in the bin packing problem, when $X_1, \ldots, X_n$ are independent random variables, has been extensively studied (see, for example, Rhee and Talagrand (1987), Rhee (1993), and Talagrand (1995)).

The longest common sub-sequence problem has now been studied intensively for about 30 years (see Chvátal and Sankoff (1975), Deken (1979), Dančík and Paterson (1994), and Steele (1982, 1996)). This was one of the first applications of the Efron–Stein inequality, see Steele (1986), in which the power and simplicity of the inequality was clearly demonstrated. The Efron–Stein inequality may also be used in a more general setup when the two independent strings are made of independent symbols, but not necessarily with balanced Bernoulli distribution. Determining the correct order of magnitude of the variance is a challenging problem but recent progress shows that, in many cases, the variance grows linearly with n and therefore the Efron–Stein bound is of the correct order of magnitude (see Houdré, Lember, and Matzinger (2006), Lember and Matzinger (2009), and Amsalu, Houdré, and Matzinger (2012)).

Configuration functions were defined by Talagrand (1995, Section 7). Our definition, taken from Boucheron, Lugosi and Massart (2000), is a slight modification of Talagrand's.

The relative stability of the L_1 error of the kernel density estimate is due to Devroye (1988, 1991). For more on the behavior of the L_1 error of the kernel density estimate we refer to Devroye and Györfi (1985), and Devroye and Lugosi (2000).

Concentration properties of self-bounding functions have been studied by Boucheron, Lugosi, and Massart (2000, 2009), Rio (2001), Bousquet (2002*a*), Maurer (2006), and McDiarmid and Reed (2006).

The Vapnik–Chervonenkis dimension and growth function were introduced in the pioneering work of Vapnik and Chervonenkis (1971, 1974).

The fact that the longest increasing sub-sequence in a random permutation of n numbers satisfies $\mathbb{E}L(X) \sim 2\sqrt{n}$ is due to Logan and Shepp (1977) (see also Hammersley (1972), Aldous and Diaconis (1995), and Groeneboom (2002)). The celebrated paper of Baik, Deift, and Johansson (2000) establishes the limit distribution of $L(X)$. This result implies that $\mathrm{Var}(L(X)) = O(n^{1/3})$. Ledoux (2005) obtains nonasymptotic exponential tail inequalities which have the best possible orders of magnitude. For early work on the concentration on $L(X)$ we refer to Frieze (1991), Bollobás and Brightwell (1992), and Talagrand (1995).

The limit theorem for the longest increasing sub-sequence in a random string over a finite alphabet mentioned in Example 3.11 is due to Tracy and Widom (2001) and

Johansson (2001) (see also Its, Tracy, and Widom (2001), and Houdré and Litherland (2009)).

Ever since the pioneering paper of Vapnik and Chervonenkis (1971), Rademacher averages have played a central role in the theory of empirical processes and statistical learning theory. For more information on the behavior of Rademacher averages and on their role in learning theory see, for example, Giné and Zinn (1984), Devroye, Györfi, and Lugosi (1996), Vapnik (1998), van der Vaart and Wellner (1996), Dudley (1999), Bartlett and Mendelson (2002), Koltchinskii (2001, 2006), and Boucheron, Bousquet, and Lugosi (2005*a*). For a modern account of the behavior of the expectation of Rademacher averages (and more general empirical processes) we refer to Talagrand (2005).

For the role of conditional Rademacher averages in probability in Banach spaces, see among others, Ledoux and Talagrand (1991) and Talagrand (1995). For the role in statistical learning theory, see, among others, Koltchinskii (2001), Koltchinskii and Panchenko (2000), Bartlett, Boucheron, and Lugosi (2002*a*), Bartlett and Mendelson (2002), Bartlett, Bousquet, and Mendelson (2002*b*), Boucheron, Bousquet, and Lugosi (2005*a*), and Massart (2006).

The problem of first passage percolation was introduced by Hammersley and Welsh (1965). The fact that the variance of first passage percolation in $\mathbb{Z}^d$ between the origin and ne_1 (where e_1 is the first canonical basis vector in $\mathbb{Z}^d$) is bounded by a linear function of n was first shown by Kesten (1993). Benjamini, Kalai and Schramm (2003) proved an upper bound of order $n/\log n$ for a certain distribution of the edge weights (see also Benaïm and Rossignol (2006) for more general results). However, it is conjectured that the correct order for the variance is $O(n^{2/3})$ (see, for example Bramson and Durrett (1999)).

The argument for bounding the variance of the largest eigenvalue of a random symmetric matrix is based on Alon, Krivelevich, and Vu (2002) who prove an exponential tail bound which we reproduce later. The phenomenon that the variance is asymptotically bounded was already discovered by Füredi and Komlós (1981), who also prove a limit theorem for the largest eigenvalue when the distributions of the entries are identical and have positive expectation. The case where the entries are centered has been settled by Soshnikov (1999).

The asymptotic value $\lim_{m\to\infty} ET_m = \zeta(3)$ of the expected weight of the minimum spanning tree was determined by Frieze (1985). The limit theorem mentioned in Example 3.15 is due to Janson (1995) and Wästlund (2005).

The randomized solution for the routing problem described in Example 3.16 was proposed by Valiant and Brebner (1981) (see also Valiant (1982) for related results).

Theorem 3.17 was proved independently by Bobkov (1996) and Ledoux (1997). Lidskii's inequality, used in Example 3.18, appears in Lidskii (1950).

The first argument of Section 3.6 is based on an idea sketched by Benjamini, Kalai and Schramm (2003) and elaborated by Devroye and Lugosi (2008). The moment-generating function approach was apparently developed first by Aida and Stroock (1994). Our calculations follow those of Bobkov and Ledoux (1997).

The proof of the Gaussian Poincaré inequality presented here is borrowed from Ané *et al.* (2000).

3.10 EXERCISES

3.1. Let Z be a nonnegative random variable such that Z^2 has a chi-square distribution with D degrees of freedom. Prove that

$$\sqrt{D} - 1 \leq \mathbf{E}Z \leq \sqrt{D}.$$

3.2. Assume that the random variables $X_1, \ldots, X_n$ are independent and binary $\{-1, 1\}$-valued with $\mathbf{P}\{X_i = 1\} = p_i$ and that $f : \{-1, 1\}^n \to \mathbb{R}$ has the bounded differences property with constants $c_1, \ldots, c_n$. Show that if $Z = f(X_1, \ldots, X_n)$,

$$\mathrm{Var}\,(Z) \leq \sum_{i=1}^{n} c_i^2 p_i(1 - p_i).$$

3.3. (ORDER STATISTICS) Assume that the random variables $X_1, \ldots, X_n$ are independent. Let $X_{(1)} \leq X_{(2)} \leq \cdots \leq X_{(n)}$ denote a nondecreasing rearrangement of $X_1, \ldots, X_n$. Prove that, no matter what the distribution of the X_i's is, $\mathrm{Var}\,(X_{(n)}) \leq \mathbf{E}[(X_{(n)} - X_{(n-1)})^2]$. Compute the left-hand side and the right-hand side when the X_i's are exponentially distributed with parameter 1 or when the X_i's are uniformly distributed on $[0, 1]$. (Use the fact that if the X_i's are exponentially distributed with parameter 1, the coordinates of the random vector $(X_{(1)}, X_{(2)} - X_{(1)}, \ldots, X_{(n)} - X_{(n-1)})$ are independent and exponentially distributed with parameters $1/n, 1/(n-1), \ldots, 1$.)

3.4. (JACKKNIFE ESTIMATE OF THE BIAS) Consider a sequence of estimates $Z = f_n(X_1, \ldots, X_n)$ of a parameter θ and assume that its bias satisfies $\mathbf{E}Z - \theta = c/n + O(n^{-2})$ for some constant c. By using the jackknife estimate of the bias defined by

$$B = (n-1)\left(\frac{1}{n}\sum_{i=1}^{n} Z_i - Z\right)$$

where $Z_i = f_{n-1}(X_1, \ldots, X_{i-1}, X_{i+1}, \ldots, X_n)$, one may define the bias-corrected estimate $\tilde{Z} = Z - B$. Show that the bias of $\tilde{Z}$ satisfies $\mathbf{E}\tilde{Z} - \theta = O(n^{-2})$. (Quenouille, 1949.)

3.5. (AMONG LIPSCHITZ FUNCTIONS THE SUM HAS THE LARGEST VARIANCE) Consider the class $\mathcal{F}$ of functions $f : \mathbb{R}^n \to \mathbb{R}$ that are Lipschitz with respect to the ℓ^1 distance, that is, if $x = (x_1, \ldots, x_n) \in \mathbb{R}^n$ and $y = (y_1, \ldots, y_n) \in \mathbb{R}^n$, then $|f(x) - f(y)| \leq \sum_{i=1}^n |x_i - y_i|$. Let $X = (X_1, \ldots X_n)$ be a vector of independent random variables with finite variance. Use the Efron–Stein inequality to show that the maximal value of $\mathrm{Var}\,(f(X))$ over $f \in \mathcal{F}$ is attained by the function $f(x) = \sum_{i=1}^n x_i$. (Bobkov and Houdré, 1996).

3.6. (JACKKNIFE ESTIMATE OF THE VARIANCE OF THE MEDIAN) Assume that the random variables $X_1, \ldots, X_n$ are independent and uniformly distributed on $[0, 1]$.

Let $X_{(1)} \leq X_{(2)} \leq \cdots \leq X_{(n)}$ denote a nondecreasing rearrangement of $X_1, \ldots, X_n$. Assume n is even. Check that

$$\mathrm{Var}\left(X_{(n/2)}\right) \leq \frac{n}{2} \boldsymbol{E}\left[\left(X_{(n/2)} - X_{(n/2-1)}\right)^2\right].$$

Compute the right-hand side and the left-hand side, as well as their limiting value when $n \to \infty$. Is the jackknife estimate of the variance of the median consistent? What is its limiting distribution?

3.7. Complete the proof of the fact that the conditional Rademacher average has the self-bounding property.

3.8. Consider the example of the number of distinct values in a discrete sample described in the text. Show that $\boldsymbol{E}Z/n \to 0$ as $n \to \infty$. Calculate explicitely $\mathrm{Var}(Z)$ and compare it with the upper bound obtained by the Efron–Stein inequality.

3.9. Let Z be the number of triangles in a random graph $\mathcal{G}(n,p)$. Calculate the variance of Z and compare it with the result obtained using the Efron–Stein inequality. (In the $\mathcal{G}(n,p)$ model for random graphs, the random graph $G = (V, E)$ with vertex set V ($|V| = n$) and edge set E is generated by starting from the complete graph with n vertices and deleting each edge independently from the others with probability $1 - p$. A triangle is a complete three-vertex subgraph.)

3.10. Consider the problem of first passage percolation on the d-dimensional integer lattice $\mathbb{Z}^d$ between the origin and a vertex $v \in \mathbb{Z}^d$. Show that if the distribution of the weights of the edges is such that X_i takes its values in the interval $[a, b]$ for some $0 < a < b < \infty$ then the number of edges on the minimum weight path is bounded by $(b/a)\|v\|_1$.

3.11. Consider the adjacency matrix $A = (X_{i,j})_{n\times n}$ of a random graph $\mathcal{G}(n,p)$. (That is, $X_{i,j} = 1$ if vertex i is connected to vertex j and $X_{i,j} = 0$ otherwise.) Show that the expected value of the largest eigenvalue of A is at least $(n-1)p$. (This simple lower bound is apparently asymptotically correct, see Füredi and Komlós (1981).)

3.12. Consider a random graph $\mathcal{G}(n,p)$ with $p = c\log n/n$, where $c > 1$. Show how the probability that the random graph is not connected is at most $2\left(e^{m^{(1-c)/2}} - 1\right) + 2^{m+1}m^{-(c-1)m/4}$ (Erdős and Rényi, 1960), and (Palmer, 1985).

3.13. (THE ASSIGNMENT PROBLEM) In the assignment problem, given an $m \times m$ array $\{X_{i,j}\}_{m\times m}$ of independent random variables distributed uniformly on $[0,1]$, one considers the random quantity

$$Z_m = \min_{\pi} \sum_{i=1}^{m} X_{i,\pi(i)}$$

where the minimum is taken over all permutations π of $\{1, \ldots, m\}$. Mimic the argument given for the minimum weight spanning tree to show that $\mathrm{Var}(Z_m) = O(\log^2 m/m)$. (A few samples from the vast literature on the assignment problem include Aldous (2001), Linusson and Wästlund (2004), Nair, Prabhakar and Sharma (2005), and Talagrand (1995).)

3.14. Show that if one employs the shortest-path routing strategy described in Example 3.16 then there exists a permutation σ such that the maximal congestion over any edge is at least $2^{N/2}/N$. In fact, much more is true: Valiant (1982) showed that no oblivious deterministic routing algorithm can have maximal congestion less than $\Omega(2^{N/2}/N)$ where a deterministic routing algorithm is said to be oblivious if the path from u to $\sigma(u)$ only depends on the value of $\sigma(u)$ and not on any other aspect of the permutation σ. Even if a routing algorithm chooses a shortest path between u and $\sigma(u)$ at random, it is bound to suffer a maximal congestion of order $\Omega\left(2^{\alpha N}\right)$ for some $\alpha > 0$.

3.15. Prove that in the randomized routing scheme defined in Example 3.16 the expected value of the maximal congestion, over any edge, is $O(N)$ (Valiant and Brebner, 1981).

3.16. (LIDSKII'S INEQUALITY) Let $A = (a_{i,j})_{i,j=1}^n$ and $B = (b_{i,j})_{i,j=1}^n$ be two symmetric matrices. Let $(\lambda_i(A))_{i=1,\dots,n}$ and $(\lambda_i(B))_{i=1,\dots,n}$ denote the nonincreasing rearrangements of their eigenvalues. Recall that $\sqrt{\operatorname{tr}(AA^T)}$ is the Hilbert–Schmidt (or Frobenius) norm of A. Prove the following version of Lidskii's inequality:

$$\sum_{i=1}^{n} (\lambda_i(A) - \lambda_i(B))^2 \le \|A - B\|_{\mathrm{HS}}^2 = \sum_{i,j=1}^{n} (a_{i,j} - b_{i,j})^2 .$$

Hint: prove that there exists an orthogonal matrix $Q = (q_{i,j})_{i,j=1}^n$ such that

$$\sum_{i,j=1}^{n} (a_{i,j} - b_{i,j})^2 = \left\| \operatorname{diag}(\lambda_i(A))Q - Q\operatorname{diag}(\lambda_i(B)) \right\|_{\mathrm{HS}}^2$$

$$= \sum_{i,j=1}^{n} q_{i,j}^2 (\lambda_i(A) - \lambda_j(B))^2 .$$

where $\operatorname{diag}(\lambda_i(A))$ and $\operatorname{diag}(\lambda_i(B))$ are two diagonal matrices with diagonal entries matching the eigenvalues of A and B. Let $P = (p_{i,j})_{i,j=1}^n$ be a doubly stochastic matrix, and prove that $\sum_{i,j=1}^n p_{i,j}^2 (\lambda_i(A) - \lambda_j(B))^2$ is minimized if P is the identity matrix. You may proceed by repeated exchanges. Assume that for some $k < \ell$, $p_{k,\ell} \neq 0$, check that there exists another doubly stochastic matrix $P' = (p'_{i,j})_{i,j=1}^n$ with $\sum_{i\neq j} P'^2_{i,j} < \sum_{i\neq j} P^2_{i,j}$ and $\sum_{i,j=1}^n p'^2_{i,j}(\lambda_i(A) - \lambda_j(B))^2 < \sum_{i,j=1}^n p_{i,j}^2(\lambda_i(A) - \lambda_j(B))^2$. This elementary proof is due to Wilkinson (see Marshall and Olkin (1979), Horn and Johnson (1990), Bhatia (1997), and Garling (2007) for more and related inequalities). This inequality is sometimes referred to as the Hoffman-Wielandt inequality. See also Terence Tao's blog <http://terrytao.wordpress.com>, course 254a.

3.17. Modify the argument of Section 3.6 to show that if f is such that there exists a constant v such that $\sum_{i=1}^n (Z - Z_i')_-^2 \le v$ and $B = \sup_{x,x_i'} |f(x) - f(x_i')|$ then for all $0 < \delta < \gamma \le 1/2$ such that $Q_{1-\gamma} \ge MZ + B$,

$$Q_{1-\delta} - Q_{1-\gamma} \le B + \sqrt{\frac{8v\gamma}{\delta}} .$$

3.18. Mimic the argument of Section 3.6 to show that if f is a self-bounding function then $\sqrt{a_{k+1}} - \sqrt{a_k} \leq ck$ for a universal constant c which implies $\mathbf{P}\{Z > \mathbf{M}Z + t\} \leq Ce^{-\sqrt{t}/C}$ for another constant C. This tail bound will also be sharpened considerably in Chapter 6.

3.19. (VARIANCE OF THE SQUARE ROOT) Let X be a nonnegative random variable such that for some $a > 0$, $\mathrm{Var}\,(X) \leq a\mathbf{E}X$. Prove that

$$\mathrm{Var}\,\left(\sqrt{X}\right) \leq a.$$

Hint: the method of Exercise 5.8 may be useful.

3.20. Assume that f is a nonnegative valued function defined on $\mathcal{X}^n$. Let $X_1, \ldots, X_n$ be independent random variables taking values in $\mathcal{X}$ and let $Z = f(X_1, \ldots, X_n)$. Let $Z_i = \inf_{x_i \in \mathcal{X}} f(X_1, \ldots, X_{i-1}, x_i, X_{i+1}, \ldots, X_n)$ and let $V = \sum_{i=1}^n (Z - Z_i)^2$. Assume that there exists a random variable W such that

$$V \leq WZ.$$

Prove that

$$\mathrm{Var}\,\left(\sqrt{Z}\right) \leq \mathbf{E}W.$$

3.21. (A POISSON POINCARÉ INEQUALITY) Let f be a real-valued function defined on the set of nonnegative integers and denote its "discrete derivative" by $Df(x) = f(x+1) - f(x)$. Let X be a Poisson random variable with parameter $\mathbf{E}X = \mu$. Prove that

$$\mathrm{Var}\,(f(X)) \leq \mu\mathbf{E}\left[(Df(X))^2\right].$$

Hint: use the Efron–Stein inequality and the infinite divisibility of the Poisson distribution. (See Klaassen (1985) and Kontoyiannis, Harremoës, and Johnson (2005) for more on this topic.)

3.22. (A POINCARÉ INEQUALITY FOR THE EXPONENTIAL DISTRIBUTION) Let X be a real-valued random variable with symmetric exponential distribution, that is with density $(1/2)e^{-|x|}$ for $x \in \mathbb{R}$. Prove that for any differentiable function f for which $\mathrm{Var}\,(f(X)) < \infty$,

$$\mathrm{Var}\,(f(X)) \leq 4\mathbf{E}\left[(f'(X))^2\right].$$

Hint: use the fact that

$$\mathbf{E}[f(X)] = f(0) + \mathbf{E}\left[\mathrm{sgn}(X)f'(X)\right].$$

See Ledoux (1999).

3.23. (VARIANCE OF THE SQUARE-ROOT OF A POISSON RANDOM VARIABLE) Prove that if X is a Poisson random variable, then

$$\mathrm{Var}\left(\sqrt{X}\right) \leq (\boldsymbol{E}X)\boldsymbol{E}\left[\frac{1}{4X+1}\right].$$

Hint: use the Poisson Poincaré inequality of Exercise 3.21. See van der Vaart (1998) for statistical applications of this inequality in the so-called method of "variance stabilization." For bounds on $\boldsymbol{E}[1/(X+1)]$, see Arlot (2007).

3.24. (VARIANCE OF SUPREMA OF GAUSSIAN PROCESSES) Let $\mathcal{T}$ be a finite index set and let $(X_t)_{t\in\mathcal{T}}$ be a centered Gaussian vector. Let $Z = \max_{t\in\mathcal{T}} X_t$. Show that $\mathrm{Var}(Z) \leq \max_{t\in\mathcal{T}} \mathrm{Var}(X_t)$.

4

Basic Information Inequalities

This chapter introduces a series of inequalities which have their origin in different fields, such as geometry, combinatorics, and information theory. These elementary results, which, for historical reasons, we call information inequalities, will be the basis of exponential concentration inequalities for functions of various independent random variables.

In the first seven sections we concentrate on discrete random variables. This simplified setting allows us to present the main ideas in an elementary and transparent way. First we introduce the concepts of Shannon entropy and relative entropy. After summarizing their most basic properties, we prove a simple elementary entropy inequality, called Han's inequality, which has surprisingly far-reaching consequences. In Section 4.4 we show how some basic isoperimetric inequalities on the binary hypercube follow as simple applications of Han's inequality. In Section 4.5, as another combinatorial application of Han's inequality, we see that combinatorial entropies satisfy the self-bounding property, leading to interesting concentration properties of such functions.

For the purposes of this book, the perhaps most important application of Han's inequality is the sub-additivity of entropy proved in Section 4.7. This inequality is at the core of the so-called "entropy method" for proving concentration inequalities (see Chapters 5, 6 and 12).

In Section 4.8 we abandon the restricted world of discrete random variables and introduce the notion of relative entropy in a general, measure-theoretic framework. The key tool is a duality formula for entropy, shown in Section 4.9, which allows us to derive a simple "transportation cost" lemma (see Section 4.10). We also describe a fundamental result known as Pinsker's inequality which is at the basis of a successful method for proving concentration inequalities called the "transportation method" (see Chapter 8). The duality formula for relative entropy may also be used to establish a variety of properties of relative entropy (see the exercises) and to investigate the maximum error probability in multiple hypothesis testing (see Section 4.12). We also present a proof of the sub-additivity of entropy for general random variables.

The chapter is concluded by the Brunn–Minkowski inequality, a fundamental result that lies at the intersection of analysis, convex geometry, and information theory.

4.1 Shannon Entropy and Relative Entropy

Let X be a random variable taking values in the countable set $\mathcal{X}$ with distribution defined by

$$\mathbf{P}\{X = x\} = p(x) \quad \text{for all } x \in \mathcal{X}.$$

The *Shannon entropy* (or simply *entropy*) of X is defined by

$$H(X) = \mathbf{E}[-\log p(X)] = -\sum_{x \in \mathcal{X}} p(x) \log p(x)$$

(where log denotes natural logarithm and we agree on the convention $0 \log 0 = 0$).

Here we use the traditional notation $H(X)$ for the entropy of a random variable X. This notation may be somewhat misleading since $H(X)$ is not a function of the random variable X but rather a functional of the distribution of X.

The entropy is obviously nonnegative. A direct consequence of the fact that $x \mapsto -x \log x$ is a concave function on $[0, \infty)$ is that the entropy is a concave functional in the sense that if the distribution of X is a mixture of two probability distributions, then the entropy of X is at least as large as the corresponding convex combination of the entropies of the two distributions.

A closely related important concept is that of relative entropy. Let P and Q be two probability distributions over a countable set $\mathcal{X}$ with probability mass functions p and q. Then the *Kullback–Leibler divergence* or *relative entropy* of P and Q is

$$D(P\|Q) = \sum_{x \in \mathcal{X}} p(x) \log \frac{p(x)}{q(x)}$$

if P is absolutely continuous with respect to Q and infinite otherwise.

A basic property is that the relative entropy between P and Q is nonnegative, and equals zero if and only if $P = Q$. This follows simply by observing that if P is absolutely continuous with respect to Q, since $\log x \leq x - 1$ for all $x > 0$,

$$D(P\|Q) = -\sum_{x \in \mathcal{X}: p(x)>0} p(x) \log \frac{q(x)}{p(x)} \geq -\sum_{x \in \mathcal{X}: p(x)>0} p(x) \left(\frac{q(x)}{p(x)} - 1 \right) \geq 0.$$

This observation has some interesting consequences. The simplest of these follows by taking Q to be the uniform distribution over a finite set $\mathcal{X}$. If X is a random variable with distribution P, then

$$D(P\|Q) = \log |\mathcal{X}| - H(X).$$

The nonnegativity of the relative entropy implies that

$$H(X) \leq \log |\mathcal{X}|$$

and equality holds if and only if X is uniformly distributed over $\mathcal{X}$.

The entropy has a key role in information theory. Exercises 4.1 and 4.2 sketch some of the basic ideas.

4.2 Entropy on Product Spaces and the Chain Rule

As our primary interest is in functions of several independent random variables, we pay special attention to the Shannon entropy of distributions on product spaces. If (X, Y) is a pair of discrete random variables taking values in $\mathcal{X} \times \mathcal{Y}$ then the *joint entropy* $H(X, Y)$ of X and Y is defined as the entropy of the pair (X, Y).

Let the probability mass function of the joint distribution P of (X, Y) be defined by $(p(x, y))_{x,y \in \mathcal{X} \times \mathcal{Y}}$. The probability mass functions of the marginal distributions of X and Y are denoted by p_X and p_Y. Then

$$H(X) + H(Y) - H(X, Y) = \sum_{x,y} p(x, y) \log \frac{p(x, y)}{p_X(x) p_Y(y)}.$$

The latter expression is the relative entropy between the joint distribution P and the product of marginal distributions $P_X \otimes P_Y$ and therefore, it is nonnegative and equals zero if and only if X and Y are independent. This implies the sub-additivity of the Shannon entropy:

$$H(X, Y) \leq H(X) + H(Y)$$

and equality holds if and only if X and Y are independent.

Remark 4.1 The quantity $H(X) + H(Y) - H(X, Y)$ is usually called the *mutual information* between X and Y. The Shannon entropy of a random variable may be defined as the mutual information between a random variable and itself.

The *conditional entropy* $H(X|Y)$ is defined as

$$H(X|Y) = H(X, Y) - H(Y).$$

Observe that if we write the joint probability mass function $p(x, y) = \mathbf{P}\{X = x, Y = y\}$ and the conditional probability mass function $p(x|y) = \mathbf{P}\{X = x|Y = y\}$, then

$$\begin{aligned} H(X|Y) &= - \sum_{x \in \mathcal{X}, y \in \mathcal{Y}} p(x, y) \log p(x|y) \\ &= \sum_{y \in \mathcal{Y}} p_Y(y) \left(- \sum_{x \in \mathcal{X}} p(x|y) \log p(x|y) \right) \\ &= \mathbf{E}\left[- \log p(X|Y) \right]. \end{aligned}$$

As the conditional entropy is the expected value of the Shannon entropy of conditional distributions, we see that $H(X|Y) \geq 0$.

Consider a pair of random variables X, Y with joint distribution $P_{X,Y}$ and marginal distributions P_X and P_Y. Noting that

$$D(P_{X,Y} \| P_X \otimes P_Y) = H(X) - H(X|Y),$$

the nonnegativity of the relative entropy implies that $H(X) \geq H(X|Y)$, or in other words, conditioning decreases entropy.

It is similarly easy to see that this fact also remains true for conditional entropies, that is,

$$H(X|Y) \geq H(X|Y, Z).$$

It is easy to see that the defining identity of the conditional entropy remains true conditionally, that is, for any three (discrete) random variables X, Y, Z,

$$H(X, Y|Z) = H(Y|Z) + H(X|Y, Z).$$

Just add $H(Z)$ to both sides and use the definition of the conditional entropy. A repeated application of this yields the *chain rule for entropy*: for arbitrary discrete random variables $X_1, \ldots, X_n$,

$$\begin{aligned} H(X_1, \ldots, X_n) &= H(X_1) + H(X_2|X_1) + H(X_3|X_1, X_2) \\ &\quad + \cdots + H(X_n|X_1, \ldots, X_{n-1}). \end{aligned}$$

An analogous chain rule for relative entropies is given in Exercise 4.4.

4.3 Han's Inequality

Here we use the basic information-theoretic inequalities described in the previous sections to derive some simple and general inequalities for the joint entropy of several variables. Interestingly, these results have some immediate but nontrivial implications concerning the combinatorics of product spaces. We start with the simplest version.

Theorem 4.1 (HAN'S INEQUALITY) *Let $X_1, \ldots, X_n$ be discrete random variables. Then*

$$H(X_1, \ldots, X_n) \leq \frac{1}{n-1} \sum_{i=1}^{n} H(X_1, \ldots, X_{i-1}, X_{i+1}, \ldots, X_n).$$

Proof For any $i = 1, \ldots, n$, by the definition of the conditional entropy and the fact that conditioning reduces entropy,

$$\begin{aligned} &H(X_1, \ldots, X_n) \\ &= H(X_1, \ldots, X_{i-1}, X_{i+1}, \ldots, X_n) + H(X_i|X_1, \ldots, X_{i-1}, X_{i+1}, \ldots, X_n) \\ &\leq H(X_1, \ldots, X_{i-1}, X_{i+1}, \ldots, X_n) + H(X_i|X_1, \ldots, X_{i-1}). \end{aligned}$$

Summing these n inequalities and using the chain rule for entropy, we get

$$nH(X_1,\ldots,X_n) \leq \sum_{i=1}^{n} H(X_1,\ldots,X_{i-1},X_{i+1},\ldots,X_n) + H(X_1,\ldots,X_n)$$

which is what we wanted to prove. □

4.4 Edge Isoperimetric Inequality on the Binary Hypercube

In order to demonstrate the usefulness of Han's inequality, we show how it can be used to derive isoperimetric properties of the n-dimensional binary hypercube. It will be a recurring theme of this book that isoperimetric inequalities are intimately related to concentration of measure. This is the first and simplest illustration of the phenomenon.

Consider the binary hypercube $\{-1,1\}^n$ and for any $x, x' \in \{-1,1\}^n$, define the Hamming distance

$$d_H(x,x') = \sum_{i=1}^{n} \mathbb{1}_{\{x_i \neq x'_i\}}.$$

The elements x of the binary n-cube may be considered as vertices of a graph in which two elements x and x' of $\{-1,1\}^n$ are adjacent if and only if their Hamming distance is 1. The graph structure has $N = 2^n$ vertices and $n2^{n-1}$ undirected edges. Its density (the ratio between the number of edges and the number of vertices) is thus $n/2 = (\log_2 N)/2$.

A remarkable property of the binary n-cube is that for any subset $A \subseteq \{-1,1\}^n$, the density of the subgraph induced by A is at most $(\log_2 |A|)/2$. This is the message of the next statement which may be considered as an isoperimetric theorem for the binary hypercube. Note that equality is achieved if the graph induced by A is a lower-dimensional hypercube, since if A is a hypercube of dimension $d \leq n$, then the subgraph induced by A has 2^d vertices and $E(A) = d2^{d-1}$ edges.

Theorem 4.2 *Let A be a subset of $\{-1,1\}^n$. Let $E(A)$ denote the set of edges of the subgraph induced by A, that is, the collection of (unordered) pairs (x,x') with $x,x' \in A$ such that $d_H(x,x') = 1$. Then*

$$|E(A)| \leq \frac{|A|}{2} \times \log_2 |A|.$$

Proof Define the random vector $X = (X_1,\ldots,X_n)$ taking values in $\{-1,1\}^n$ such that X has the uniform distribution over A. Denote by p the probability mass function of X. The Shannon entropy of X is clearly $\log|A|$. Writing $X^{(i)} = (X_1,\ldots,X_{i-1},X_{i+1},\ldots,X_n)$, and using the definition of conditional entropy, we have

$$H(X) - H\left(X^{(i)}\right) = H\left(X_i|X^{(i)}\right) = -\sum_{x\in A} p(x)\log p\left(x_i|x^{(i)}\right).$$

By definition, $p(x) = 1/|A|$ for all $x \in A$. On the other hand, for $x \in A$,

$$p\left(x_i|x^{(i)}\right) = \begin{cases} 1/2 \text{ if } \overline{x}^{(i)} \in A \\ 1 \quad \text{otherwise} \end{cases}$$

where $\overline{x}^{(i)} = (x_1, \ldots, x_{i-1}, -x_i, x_{i+1}, \ldots, x_n)$ is obtained by flipping the i-th bit of x. Thus,

$$H(X) - H\left(X^{(i)}\right) = \frac{\log 2}{|A|} \sum_{x \in A} \mathbb{1}_{\{x, \overline{x}^{(i)} \in A\}}$$

and therefore

$$\sum_{i=1}^{n} \left(H(X) - H\left(X^{(i)}\right)\right) = \frac{\log 2}{|A|} \sum_{x \in A} \sum_{i=1}^{n} \mathbb{1}_{\{x, \overline{x}^{(i)} \in A\}} = \frac{|E(A)|}{|A|} 2 \log 2.$$

Thus, Han's inequality implies

$$\frac{|E(A)|}{|A|} 2 \log 2 = \sum_{i=1}^{n} \left(H(X) - H\left(X^{(i)}\right)\right) \leq H(X) = \log |A|.$$

This is precisely what we wanted to prove. □

Next we show how Theorem 4.2 can be turned into an inequality for the edge-perimeter of A, or equivalently for the total influence of the n variables. Let the binary random vector $X = (X_1, \ldots, X_n)$ be uniformly distributed over $\{-1, 1\}^n$ and denote by $\overline{X}^{(i)} = (X_1, \ldots, X_{i-1}, -X_i, X_{i+1}, \ldots, X_n)$ the vector obtained by flipping the i-th bit of X. For any $A \subset \{-1, 1\}^n$, the *influence* of the i-th variable is defined by

$$I_i(A) = \mathbf{P}\left\{\mathbb{1}_{\{X \in A\}} \neq \mathbb{1}_{\{\overline{X}^{(i)} \in A\}}\right\}.$$

If $\mathbb{1}_{\{X \in A\}} \neq \mathbb{1}_{\{\overline{X}^{(i)} \in A\}}$, then the i-th variable is said to be *pivotal* for A. Thus, the influence $I_i(A)$ is just the probability that the i-th variable is pivotal for A. The *total influence* is defined by the sum of individual influences

$$I(A) = \sum_{i=1}^{n} I_i(A).$$

Clearly, $I(A) = 2|\partial_E(A)|/2^n$ where $\partial_E(A)$ is the *edge boundary* of A defined by

$$\partial_E(A) = \left\{(x, x') : x \in A, x' \in A^c, d_H(x, x') = 1\right\}.$$

The following bound for the total influence is a simple corollary of Theorem 4.2.

Theorem 4.3 *For any $A \subset \{-1,1\}^n$, let $P(A)$ denote $\mathbf{P}\{X \in A\} = |A|/2^n$. Then*

$$I(A) \geq 2P(A)\log_2 \frac{1}{P(A)}.$$

Proof Since A is a subset of the n-cube, every point in A belongs to exactly n edges, so

$$n|A| = 2|E(A)| + |\partial_E(A)|$$

(since every edge with both endpoints in A is counted twice), and by Theorem 4.2,

$$|\partial_E(A)| \geq \left(n - \log_2 |A|\right) \times |A| = \log_2 \frac{2^n}{|A|} \times |A|$$

which is equivalent to the statement of the theorem. □

Remark 4.2 The random variable $Z = \mathbb{1}_{\{X\in A\}}$ may be considered as a function of the n independent random variables $X_1,\ldots,X_n$. Then $\mathrm{Var}(Z) = P(A)(1-P(A))$ and the Efron–Stein inequality immediately implies

$$P(A)(1-P(A)) \leq \frac{I(A)}{4}.$$

When $P(A)$ is small, Theorem 4.3 gives a much better bound.

Influences of subsets of the binary hypercube are basic in the study of threshold phenomena, percolation, game theory, complexity theory, and many other areas. In Chapters 9 and 10 we devote more effort to the understanding of this fundamental quantity.

4.5 Combinatorial Entropies

In Section 3.3 we considered functions satisfying a special property—the so-called self-bounding property—that have interesting concentration properties. In particular, Corollary 3.7 shows that if f is self-bounding and $X_1,\ldots,X_n$ are independent random variables, then $Z = f(X_1,\ldots,X_n)$ satisfies $\mathrm{Var}(Z) \leq \mathbf{E}Z$.

In Section 3.3 several examples of such functions are discussed. The purpose of this section is to show a whole new class of self-bounding functions that we call *combinatorial entropies*. The self-bounding property of these functions may be seen as an easy consequence of Han's inequality. The basic idea is quite similar to that of the proof of Theorem 4.2. We start by describing a simple example. The general case, shown below, mimics the same argument.

Example 4.4 (VC ENTROPY) In this first example we consider the so-called Vapnik–Chervonenkis (or VC) entropy, a quantity closely related to the VC dimension discussed in Section 3.3. Let $\mathcal{A}$ be an arbitrary collection of subsets of $\mathcal{X}$, and let

$x = (x_1, \ldots, x_n)$ be a vector of n points of $\mathcal{X}$. Recall that the *shatter coefficient* is defined as the size of the trace of $\mathcal{A}$ on x, that is,

$$T(x) = |\mathrm{tr}(x)| = \left|\{A \cap \{x_1, \ldots, x_n\} : A \in \mathcal{A}\}\right|.$$

The VC *entropy* is defined as the logarithm of the shatter coefficient, that is,

$$h(x) = \log_2 T(x).$$

Lemma 4.5 *The* VC *entropy has the self-bounding property.*

Proof We need to show that there exists a function h' of $n-1$ variables such that for all $i = 1, \ldots, n$, writing $x^{(i)} = (x_1, \ldots, x_{i-1}, x_{i+1}, \ldots, x_n)$, $0 \le h(x) - h'(x^{(i)}) \le 1$ and

$$\sum_{i=1}^{n} \left(h(x) - h'\left(x^{(i)}\right)\right) \le h(x).$$

We define h' in the natural way, that is, as the VC entropy based on the $n-1$ points in its arguments. Then, for any i, $h'(x^{(i)}) \le h(x)$, and the difference cannot be more than one. The nontrivial part of the proof is to show the second property. We do this using Han's inequality (Theorem 4.1).

Consider the uniform distribution over the set $\mathrm{tr}(x)$. This defines a random binary vector $Y = (Y_1, \ldots, Y_n) \in \{0,1\}^n$. Then

$$h(x) = \log_2 |\mathrm{tr}(x)| = \frac{1}{\log 2} H(Y_1, \ldots, Y_n),$$

where $H(Y_1, \ldots, Y_n)$ is the (joint) Shannon entropy of $Y_1, \ldots, Y_n$. Since the uniform distribution maximizes the Shannon entropy, we also find, for all $i \le n$, that

$$h'\left(x^{(i)}\right) \ge \frac{1}{\log 2} H(Y_1, \ldots, Y_{i-1}, Y_{i+1}, \ldots, Y_n).$$

Since by Han's inequality

$$H(Y_1, \ldots, Y_n) \le \frac{1}{n-1} \sum_{i=1}^{n} H(Y_1, \ldots, Y_{i-1}, Y_{i+1}, \ldots, Y_n),$$

we obtain

$$\sum_{i=1}^{n} \left(h(x) - h'\left(x^{(i)}\right)\right) \le h(x)$$

as desired. □

The above lemma, together with Corollary 3.7 immediately implies the following.

Corollary 4.6 *Let $X_1, \ldots, X_n$ be independent random variables taking their values in some set $\mathcal{X}$ and let $\mathcal{A}$ be an arbitrary collection of subsets of $\mathcal{X}$. If $Z = h(X)$ denotes the random* VC *entropy, then* $\text{Var}(Z) \leq \mathbf{E}Z$.

In Chapter 6 we extend this result to exponential inequalities.

The proof of concentration of the VC entropy may be generalized, in a straightforward way, to a class of functions we call *combinatorial entropies* defined as follows.

Let $x = (x_1, \ldots, x_n)$ be an n-vector of elements with $x_i \in \mathcal{X}_i$ to which we associate a set $\text{tr}(x) \subset \mathcal{Y}^n$ of n-vectors whose components are elements of a possibly different set $\mathcal{Y}$. We assume that for each $x \in \mathcal{X}^n$ and $i \leq n$, the set $\text{tr}(x^{(i)}) = \text{tr}(x_1, \ldots, x_{i-1}, x_{i+1}, \ldots, x_n)$ is the projection of $\text{tr}(x)$ along the ith coordinate, that is,

$$\text{tr}(x^{(i)}) = \Big\{ y^{(i)} = (y_1, \ldots, y_{i-1}, y_{i+1}, \ldots, y_n) \in \mathcal{Y}^{n-1} : \\ \exists y_i \in \mathcal{Y} \text{ such that } (y_1, \ldots, y_n) \in \text{tr}(x) \Big\}.$$

The associated combinatorial entropy is $h(x) = \log_b |\text{tr}(x)|$ where b is an arbitrary positive number.

As in the case of VC entropy, combinatorial entropies may be shown to have the self-bounding property. (The details are left as an exercise.) Then we immediately obtain the following generalization.

Theorem 4.7 *Assume that $h(x) = \log_b |tr(x)|$ is a combinatorial entropy such that for all $x \in \mathcal{X}^n$ and $i \leq n$,*

$$h(x) - h(x^{(i)}) \leq 1.$$

If $X = (X_1, \ldots, X_n)$ is a vector of n independent random variables taking values in $\mathcal{X}$, then the random combinatorial entropy $Z = h(X)$ satisfies $\text{Var}(Z) \leq \mathbf{E}Z$.

Example 4.8 (INCREASING SUB-SEQUENCES) Recall the setup of the example of increasing sub-sequences of Section 3.3, and let $N(x)$ denote the number of different increasing sub-sequences of x. Observe that $\log_2 N(x)$ is a combinatorial entropy. This is easy to see by considering $\mathcal{Y} = \{0, 1\}$ and by assigning, to each increasing sub-sequence $i_1 < i_2 < \cdots < i_m$ of x, a binary n-vector $y_1^n = (y_1, \ldots, y_n)$ such that $y_j = 1$ if and only if $j = i_k$ for some $k = 1, \ldots, m$ (i.e. the indices appearing in the increasing sequence are marked by 1). Now the conditions of Theorem 4.7 are obviously met and therefore $Z = \log_2 N(X)$ satisfies $\text{Var}(Z) \leq \mathbf{E}Z$.

4.6 Han's Inequality for Relative Entropies

In this section we derive an inequality which may be regarded as a version of Han's inequality for relative entropies. This inequality is fundamental in deriving a "sub-additivity" inequality (see Section 4.7) which, in turn, is at the basis of many exponential concentration inequalities.

Let $\mathcal{X}$ be a countable set, and let P and Q be probability distributions on $\mathcal{X}^n$ such that $P = P_1 \otimes \cdots \otimes P_n$ is a product measure. We denote the elements of $\mathcal{X}^n$ by $x = (x_1, \ldots, x_n)$ and write $x^{(i)} = (x_1, \ldots, x_{i-1}, x_{i+1}, \ldots, x_n)$ for the $(n-1)$-vector obtained by leaving out the i-th component of x. Denote by $Q^{(i)}$ and $P^{(i)}$ the marginal distributions of Q and P. Let $p^{(i)}$ and $q^{(i)}$ denote the corresponding probability mass function, that is,

$$q^{(i)}\left(x^{(i)}\right) = \sum_{y \in \mathcal{X}} q(x_1, \ldots, x_{i-1}, y, x_{i+1}, \ldots, x_n)$$

and

$$\begin{aligned} p^{(i)}\left(x^{(i)}\right) &= \sum_{y \in \mathcal{X}} p(x_1, \ldots, x_{i-1}, y, x_{i+1}, \ldots, x_n) \\ &= p_1(x_1) \cdots p_{i-1}(x_{i-1}) p_{i+1}(x_{i+1}) \cdots p_n(x_n). \end{aligned}$$

Then we have the following.

Theorem 4.9 (HAN'S INEQUALITY FOR RELATIVE ENTROPIES)

$$D(Q\|P) \geq \frac{1}{n-1} \sum_{i=1}^{n} D\left(Q^{(i)} \| P^{(i)}\right)$$

or equivalently,

$$D(Q\|P) \leq \sum_{i=1}^{n} \left(D(Q\|P) - D\left(Q^{(i)} \| P^{(i)}\right)\right).$$

Proof The statement is a straightforward consequence of Han's inequality. Indeed, Han's inequality states that

$$\sum_{x \in \mathcal{X}^n} q(x) \log q(x) \geq \frac{1}{n-1} \sum_{i=1}^{n} \sum_{x^{(i)} \in \mathcal{X}^{n-1}} q^{(i)}\left(x^{(i)}\right) \log q^{(i)}\left(x^{(i)}\right).$$

Since

$$D(Q\|P) = \sum_{x \in \mathcal{X}^n} q(x) \log q(x) - \sum_{x \in \mathcal{X}^n} q(x) \log p(x)$$

and

$$D\left(Q^{(i)} \| P^{(i)}\right) = \sum_{x^{(i)} \in \mathcal{X}^{n-1}} \left(q^{(i)}\left(x^{(i)}\right) \log q^{(i)}\left(x^{(i)}\right) - q^{(i)}\left(x^{(i)}\right) \log p^{(i)}\left(x^{(i)}\right)\right),$$

it suffices to show that

$$\sum_{x \in \mathcal{X}^n} q(x) \log p(x) = \frac{1}{n-1} \sum_{i=1}^{n} \sum_{x^{(i)} \in \mathcal{X}^{n-1}} q^{(i)}\left(x^{(i)}\right) \log p^{(i)}\left(x^{(i)}\right).$$

This may be seen easily by noting that by the product property of P, we have $p(x) = p^{(i)}(x^{(i)})p_i(x_i)$ for all i, and also $p(x) = \prod_{i=1}^{n} p_i(x_i)$, and therefore

$$\begin{aligned}
\sum_{x \in \mathcal{X}^n} q(x) \log p(x) &= \frac{1}{n} \sum_{i=1}^{n} \sum_{x \in \mathcal{X}^n} q(x) \left(\log p^{(i)}\left(x^{(i)}\right) + \log p_i(x_i) \right) \\
&= \frac{1}{n} \sum_{i=1}^{n} \sum_{x \in \mathcal{X}^n} q(x) \log p^{(i)}\left(x^{(i)}\right) + \frac{1}{n} \sum_{x \in \mathcal{X}^n} q(x) \log p(x).
\end{aligned}$$

Rearranging, we obtain

$$\begin{aligned}
\sum_{x \in \mathcal{X}^n} q(x) \log p(x) &= \frac{1}{n-1} \sum_{i=1}^{n} \sum_{x \in \mathcal{X}^n} q(x) \log p^{(i)}\left(x^{(i)}\right) \\
&= \frac{1}{n-1} \sum_{i=1}^{n} \sum_{x^{(i)} \in \mathcal{X}^{n-1}} q^{(i)}\left(x^{(i)}\right) \log p^{(i)}\left(x^{(i)}\right)
\end{aligned}$$

where we used the defining property of $q^{(i)}$. □

4.7 Sub-Additivity of the Entropy

We are now prepared to prove an inequality which will serve as the basis of the so-called "entropy method" for proving concentration inequalities. In Chapter 14 we give a much more general version with further important consequences. The reason we give this simple version here is that it is an easy corollary of Han's inequality for relative entropies, and it is sufficiently powerful to derive many interesting exponential concentration inequalities.

As in Section 3.1, we let $X_1, \ldots, X_n$ be independent random variables, and investigate concentration properties of $Z = f(X_1, \ldots, X_n)$. The basis of the entropy method is a powerful extension of the Efron–Stein inequality. Recall that the Efron–Stein inequality states that

$$\mathrm{Var}(Z) \leq \sum_{i=1}^{n} \boldsymbol{E}\left[\boldsymbol{E}^{(i)}[Z^2] - \left(\boldsymbol{E}^{(i)}Z\right)^2 \right],$$

where $\mathbf{E}^{(i)}$ denotes expectation with respect to the variable X_i only, that is, conditional expectation conditioned on $X^{(i)} = (X_1, \ldots, X_{i-1}, X_{i+1}, \ldots, X_n)$, or, putting $\Phi(x) = x^2$,

$$\mathbf{E}\Phi(Z) - \Phi(\mathbf{E}Z) \leq \sum_{i=1}^{n} \mathbf{E}\left[\mathbf{E}^{(i)}\Phi(Z) - \Phi\left(\mathbf{E}^{(i)}Z\right)\right].$$

In fact, this inequality remains true for a large class of convex functions Φ (see Chapter 14).

The case of interest in this section is when $\Phi(x) = x \log x$. For a nonnegative random variable Z, the quantity $\mathbf{E}\Phi(Z) - \Phi(\mathbf{E}Z)$ is often called the *entropy* of Z, denoted by $\text{Ent}(Z)$. This notion of entropy is not to be confused with the Shannon entropy introduced earlier in this chapter. Nevertheless, there is a close relationship between the two notions of entropy. As seen in the proof below, $\text{Ent}(Z)$ may be written as the relative entropy between the distribution induced by Z on $\mathcal{X}^n$ and the distribution of $X = (X_1, \ldots, X_n)$.

Theorem 4.10 (SUB-ADDITIVITY OF THE ENTROPY) *Let $\Phi(x) = x \log x$ for $x > 0$ and $\Phi(0) = 0$. Let $X_1 \ldots, X_n$ be independent random variables taking values in a countable set $\mathcal{X}$ and let $f : \mathcal{X}^n \to [0, \infty)$. Letting $Z = f(X_1, \ldots, X_n)$, we have*

$$\mathbf{E}\Phi(Z) - \Phi(\mathbf{E}Z) \leq \sum_{i=1}^{n} \mathbf{E}\left[\mathbf{E}^{(i)}\Phi(Z) - \Phi\left(\mathbf{E}^{(i)}Z\right)\right].$$

Introducing the notation $\text{Ent}^{(i)}(Z) = \mathbf{E}^{(i)}\Phi(Z) - \Phi(\mathbf{E}^{(i)}Z)$, this can be re-written as

$$\text{Ent}(Z) \leq \mathbf{E}\left[\sum_{i=1}^{n} \text{Ent}^{(i)}(Z)\right].$$

Here we only state the result for discrete random variables $X_1 \ldots, X_n$. However, the result may be extended to the general case as is shown below in Section 4.8 (see also the more general Theorem 14.1 in Chapter 14).

Proof The theorem is a direct consequence of Han's inequality for relative entropies. First note that if the inequality is true for a random variable Z then it is also true for cZ where c is a positive constant. Hence we may assume that $\mathbf{E}Z = 1$. Now define the probability measure Q on $\mathcal{X}^n$ by its probability mass function q given by

$$q(x) = f(x)p(x) \quad \text{for all } x \in \mathcal{X}^n$$

where p denotes the probability mass function of $X = (X_1, \ldots, X_n)$ and P the corresponding distribution. Then,

$$\mathbf{E}\Phi(Z) - \Phi(\mathbf{E}Z) = \mathbf{E}[Z \log Z] = D(Q\|P)$$

which, by Theorem 4.9, does not exceed $\sum_{i=1}^{n}\left(D(Q\|P) - D\left(Q^{(i)}\|P^{(i)}\right)\right)$. However, straightforward calculation shows that

$$\sum_{i=1}^{n}\left(D(Q\|P) - D\left(Q^{(i)}\|P^{(i)}\right)\right) = \sum_{i=1}^{n}\mathbf{E}\left[\mathbf{E}^{(i)}\Phi(Z) - \Phi\left(\mathbf{E}^{(i)}Z\right)\right]$$

and the statement follows. □

As a first application of the sub-additivity of the entropy, we derive a generalization of the edge isoperimetric inequality Theorem 4.3 when the distribution of the random vector $X = (X_1, \ldots, X_n)$ is such that $X_1, \ldots, X_n$ are independent binary random variables with $\mathbf{P}\{X_i = 1\} = 1 - \mathbf{P}\{X_i = -1\} = p$ where $p \in (0, 1)$.

For an index $i \leq n$, introduce the notation

$$X_i^+ = (X_1, \ldots, X_{i-1}, 1, X_{i+1}, \ldots, X_n) \quad \text{and} \quad X_i^- = (X_1, \ldots, X_{i-1}, -1, X_{i+1}, \ldots, X_n).$$

Let $A \subset \{-1, 1\}^n$ be an arbitrary set. The *positive* and *negative* influences of the i-th variable are defined as

$$I_i^+(A) = \mathbf{P}\left\{X_i^+ \in A \text{ and } X_i^- \notin A\right\}$$

and

$$I_i^-(A) = \mathbf{P}\left\{X_i^- \in A \text{ and } X_i^+ \notin A\right\}.$$

The influence $I_i(A) = \mathbf{P}\left\{\mathbb{1}_{\{X\in A\}} \neq \mathbb{1}_{\{\overline{X}^{(i)}\in A\}}\right\}$ is just the sum $I_i^+(A) + I_i^-(A)$ of positive and negative influences. The total positive and negative influences are defined as $I^+(A) = \sum_{i=1}^{n} I_i^+(A)$ and $I^-(A) = \sum_{i=1}^{n} I_i^-(A)$, respectively.

Theorem 4.11 *Let $A \in \{-1, 1\}^n$ be any set and let the random vector be distributed as described above. Then*

$$P(A)\log\frac{1}{P(A)} \leq I^+(A)p\log\frac{1}{p} + I^-(A)(1-p)\log\frac{1}{1-p}$$

where $P(A) = \mathbf{P}\{X \in A\}$.

Proof The proof is a simple application of Theorem 4.10 for the random variable $Z = \mathbb{1}_{\{X\in A\}}$. Since $\Phi(1) = \Phi(0) = 0$, we always have $\Phi(Z) = 0$, and the left-hand side of the sub-additivity inequality is simply $P(A)\log(1/P(A))$. On the other hand, for each i,

$$\Phi\left(\mathbf{E}^{(i)}Z\right) = \begin{cases} p\log p & \text{if } X_i^+ \in A \text{ and } X_i^- \notin A \\ (1-p)\log(1-p) & \text{if } X_i^- \in A \text{ and } X_i^+ \notin A \\ 0 & \text{if the } i\text{-th variable is not pivotal.} \end{cases}$$

Therefore, the right-hand side of the sub-additivity inequality becomes

$$I^+(A)p \log \frac{1}{p} + I^-(A)(1-p) \log \frac{1}{1-p},$$

proving the statement. □

Note that in the symmetric case, that is, when $p = 1/2$, the statement reduces to Theorem 4.3. Another important special case is when the set A is a monotone subset of $\{-1, 1, \}^n$. A set $A \subset \{-1, 1\}^n$ is said to be *monotone* if $\mathbb{1}_{\{x \in A\}} \geq \mathbb{1}_{\{y \in A\}}$ for all $x = (x_1, \ldots, x_n)$ and $y = (y_1, \ldots, y_n)$ in $\{-1, 1\}^n$ such that $x_i \geq y_i$ for all i. If A is monotone, the negative influence $I^-(A)$ equals zero, implying $I(A) = I^+(A)$, and we immediately obtain the following.

Corollary 4.12 *If A is a monotone subset of $\{-1, 1\}^n$, then the total influence is bounded as*

$$I(A) \geq \frac{P(A) \log \frac{1}{P(A)}}{p \log \frac{1}{p}}.$$

4.8 Entropy of General Random Variables

Up to this point we have only considered the entropy of discrete random variables. This is convenient as the main ideas can be explained in a more transparent way in this simple setting. However, in order to establish general concentration inequalities, we need to handle entropy of all kinds of random variables, not only those of a discrete distribution. In this section we introduce a general notion of entropy and the rest of the chapter is dedicated to describing some properties of this notion. In particular, in Section 4.9 we present a duality formula for entropy which allows us to derive a simple "transportation cost" lemma (see Section 4.10). In Chapter 8 we explore how this transportation lemma and its variants can be used to establish concentration inequalities. Finally, in Section 4.13 we prove, in its full generality, the sub-additivity of entropy that we already proved for discrete random variables in Theorem 4.10.

Luckily, the general framework does not require sophisticated measure theoretic arguments at all. We begin with a formal definition of relative entropy within a general framework and elementary properties of entropy. These properties are essentially the same as in the discrete case but their proofs are a little different.

As in Section 4.7, Φ denotes the function $\Phi(x) = x \log x$, defined on $[0, \infty)$ (where $0 \log 0$ is defined as 0). Let $(\Omega, \mathcal{A}, \boldsymbol{P})$ be a probability space and let Y be a nonnegative random variable defined on it such that Y is integrable, that is, $\boldsymbol{E}Y = \int_\Omega Y(\omega) d\boldsymbol{P}(\omega) < \infty$. As before, we define the *entropy* of Y by

$$\operatorname{Ent}(Y) = \boldsymbol{E}\Phi(Y) - \Phi(\boldsymbol{E}Y).$$

Note that since Φ is bounded from below by $-e^{-1}$, the expression $E\Phi(Y)$ is meaningful even if $\Phi(Y)$ is not integrable. Hence $\text{Ent}(Y)$ is well defined for all nonnegative random variables. Since Φ is a convex function, by Jensen's inequality, $\text{Ent}(Y)$ is a nonnegative (possibly infinite) quantity. Moreover $\text{Ent}(Y) < \infty$ if and only if $\Phi(Y)$ is integrable.

We may use this definition of entropy to introduce a general notion of the Kullback–Leibler divergence as follows. If Y is a nonnegative random variable with $EY = 1$, we may define another probability measure Q on $(\Omega, \mathcal{A})$ by $Q(A) = \int_A Y(\omega)dP(\omega) = E[Y\mathbb{1}_{\{A\}}]$ for all $A \in \mathcal{A}$. We write $Q = YP$ for such a probability measure. The *Kullback–Leibler divergence* (or *relative entropy*) of Q with respect to P, is defined by

$$D(Q\|P) = \text{Ent}(Y).$$

To see that this definition is a generalization of the one introduced in Section 4.1 for discrete probability distributions, observe that when Ω is at most countable and Q is absolutely continuous with respect to P, then we may write $Q = YP$ where the random variable Y is defined by

$$Y(\omega) = \begin{cases} q(\omega)/p(\omega) & \text{if } p(\omega) > 0 \\ 0 & \text{otherwise} \end{cases}$$

and therefore

$$D(Q\|P) = \sum_{\omega\in\Omega, p(\omega)>0} q(\omega) \log \frac{q(\omega)}{p(\omega)}.$$

More generally, if $Q \ll P$, that is, if Q is absolutely continuous with respect to P, one may always write $Q = YP$ with Y defined by the expression above where $p(x) = dP/d\lambda$ and $q(x) = dQ/d\lambda$ denote the densities of P and Q with respect to a common dominating measure λ.

4.9 Duality and Variational Formulas

The next result gives an alternative characterization of the relative entropy, close in spirit to the duality formula for the variance given in Proposition 3.21.

Theorem 4.13 (DUALITY FORMULA OF ENTROPY) *Let Y be a nonnegative random variable defined on a probability space $(\Omega, \mathcal{A}, P)$ such that $E\Phi(Y) < \infty$. Then we have the duality formula*

$$\text{Ent}(Y) = \sup_{U\in\mathcal{U}} E[UY]$$

where the supremum is taken over the set $\mathcal{U}$ of all random variables $U : \Omega \to \overline{\mathbb{R}}$ with $Ee^U = 1$.

Moreover, if U is such that $\mathbf{E}[UY] \leq \text{Ent}(Y)$ for all nonnegative random variables Y such that $\Phi(Y)$ is integrable and $\mathbf{E}Y = 1$, then $\mathbf{E}e^U \leq 1$.

Remark 4.3 By elementary calculations one sees that for all $u \in \mathbb{R}$,

$$\sup_{x>0} (xu - \Phi(x)) = e^{u-1},$$

so if $\Phi(Y)$ is integrable and $\mathbf{E}e^U = 1$, we have

$$UY \leq \Phi(Y) + \frac{1}{e}e^U.$$

Therefore U_+Y is integrable and one can always define $\mathbf{E}[UY]$ as $\mathbf{E}[U_+Y] - \mathbf{E}[U_-Y]$ (where U_+ and U_- denote the positive and negative parts of U). Thus, the right-hand side of the duality formula of Theorem 4.13 is always well defined.

Remark 4.4 (ALTERNATIVE FORMULATION OF THE DUALITY FORMULA) One may re-write the duality formula of Theorem 4.13 as

$$\text{Ent}(Y) = \sup_T \mathbf{E}\left[Y(\log T - \log(\mathbf{E}T))\right]$$

where the supremum is taken over all nonnegative and integrable random variables.

Proof To prove the duality formula simply observe that, for any random variable U with $\mathbf{E}e^U = 1$, we have

$$\text{Ent}(Y) - \mathbf{E}[UY] = \text{Ent}_{e^U\mathbf{P}}\left[Ye^{-U}\right]$$

where $\text{Ent}_{e^U\mathbf{P}}$ is defined as Ent with the only difference that expectations are taken with respect to the probability measure $e^U\mathbf{P}$ (instead of $\mathbf{P}$). This shows that $\text{Ent}(Y) - \mathbf{E}[UY] \geq 0$ with equality whenever $e^U = Y/\mathbf{E}Y$. This proves the duality formula.

Let U be such that $\mathbf{E}[UY] \leq \text{Ent}(Y)$ for all nonnegative random variables Y such that $\Phi(Y)$ is integrable. If $\mathbf{E}e^U = 0$, then there is nothing to prove. Otherwise, given a positive integer n large enough to ensure that $x_n = \mathbf{E}e^{\min(U,n)} > 0$, one may define $Y_n = e^{\min(U,n)}/x_n$, which leads to

$$\mathbf{E}[UY_n] \leq \text{Ent}(Y_n),$$

and therefore

$$\frac{1}{x_n}\mathbf{E}\left[Ue^{\min(U,n)}\right] \leq \frac{1}{x_n}\left[\mathbf{E}\left[(\min(U,n))\, e^{\min(U,n)}\right] - \log x_n\right].$$

Hence

$$\log x_n \leq 0$$

and taking the limit when $n \to \infty$, we show by monotone convergence that $\mathbf{E}e^U \leq 1$, which finishes the proof of the theorem. □

The previous theorem makes it possible to establish a duality between entropy and moment-generating functions.

Corollary 4.14 *Let Z be a real-valued integrable random variable. Then for every $\lambda \in \mathbb{R}$,*

$$\log \mathbf{E}e^{\lambda(Z-\mathbf{E}Z)} = \sup_{\mathbf{Q} \ll \mathbf{P}} \left[\lambda\left(\mathbf{E}_{\mathbf{Q}}Z - \mathbf{E}Z\right) - D(\mathbf{Q}\|\mathbf{P})\right]$$

where the supremum is taken over all probability measures $\mathbf{Q}$ absolutely continuous with respect to $\mathbf{P}$, and $\mathbf{E}_{\mathbf{Q}}$ denotes integration with respect to the measure $\mathbf{Q}$ (recall that $\mathbf{E}$ is integration with respect to $\mathbf{P}$).

As in Chapter 2, the logarithmic moment-generating function of a real-valued random variable Z is denoted by $\psi_Z(\lambda) = \log \mathbf{E}e^{\lambda Z}$ for $\lambda \in \mathbb{R}$.

Proof Let $\mathbf{Q}$ be a probability measure absolutely continuous with respect to $\mathbf{P}$. Taking $Y = d\mathbf{Q}/d\mathbf{P}$ and choosing $U = \lambda(Z - \mathbf{E}Z) - \psi_{Z-\mathbf{E}Z}(\lambda)$, it follows from the duality formula of Theorem 4.13 that

$$D(\mathbf{Q}\|\mathbf{P}) = \text{Ent}(Y) \geq \mathbf{E}[UY] = \lambda(\mathbf{E}_{\mathbf{Q}}Z - \mathbf{E}Z) - \psi_{Z-\mathbf{E}Z}(\lambda),$$

or equivalently that

$$\psi_{Z-\mathbf{E}Z}(\lambda) \geq \lambda\left(\mathbf{E}_{\mathbf{Q}}Z - \mathbf{E}Z\right) - D(\mathbf{Q}\|\mathbf{P}),$$

and therefore

$$\log \mathbf{E}e^{\lambda(Z-\mathbf{E}Z)} \geq \sup_{\mathbf{Q}' \ll \mathbf{P}} \left[\lambda\left(\mathbf{E}_{\mathbf{Q}'}Z - \mathbf{E}Z\right) - D(\mathbf{Q}'\|\mathbf{P})\right].$$

Conversely, setting

$$U = \lambda(Z - \mathbf{E}Z) - \sup_{\mathbf{Q}' \ll \mathbf{P}} \left[\lambda\left(\mathbf{E}_{\mathbf{Q}'}Z - \mathbf{E}Z\right) - D(\mathbf{Q}'\|\mathbf{P})\right]$$

for every nonnegative random variable Y such that $\mathbf{E}Y = 1$,

$$\mathbf{E}\left[UY\right] \leq \text{Ent}(Y).$$

Hence, by Theorem 4.13, $\mathbf{E}e^{U} \leq 1$ which means that

$$\log \mathbf{E}e^{\lambda(Z-\mathbf{E}Z)} \leq \sup_{\mathbf{Q}' \ll \mathbf{P}} \left[\lambda\left(\mathbf{E}_{\mathbf{Q}'}Z - \mathbf{E}Z\right) - D(\mathbf{Q}'\|\mathbf{P})\right]. \qquad \square$$

The duality formula implies the following property of the Kullback–Leibler divergence.

Corollary 4.15 *Let* $\mathbf{P}$ *and* $\mathbf{Q}$ *be two probability distributions on the same space. Then*

$$D(\mathbf{Q}\|\mathbf{P}) = \sup_Z \left[\mathbf{E}_{\mathbf{Q}} Z - \log \mathbf{E}e^Z\right]$$

where the supremum is taken over all random variables such that $\mathbf{E}e^Z < \infty$.

This corollary asserts that if $\mathbf{P}$ remains fixed, $D(\mathbf{Q}\|\mathbf{P})$ is the convex dual of the functional $Z \to \log \mathbf{E}e^Z$.

Proof If $\mathbf{Q} \ll \mathbf{P}$, $D(\mathbf{Q}\|\mathbf{P}) = \mathrm{Ent}\,(d\mathbf{Q}/d\mathbf{P})$ and the corollary follows from the alternative formulation of the duality formula. If $\mathbf{Q} \not\ll \mathbf{P}$, there exists an event A such that $\mathbf{Q}(A) > 0 = \mathbf{P}(A)$, $D(\mathbf{Q}\|\mathbf{P}) = \infty$, and choosing $Z_n = n\mathbb{1}_{\{A\}}$ and letting n tend to infinity, we observe that the supremum on the right-hand side is infinite. □

The duality formula for entropy and its corollaries have many useful consequences (see Exercises 4.10, 4.11, and 4.13).

The last results in this section will be useful when developing the entropy method in Chapters 5 and 6. It is well known that the expected value minimizes the average squared Euclidean distance to a random point. This is an instance of a more general statement.

Theorem 4.16 (THE EXPECTED VALUE MINIMIZES EXPECTED BREGMAN DIVERGENCE) *Let* $I \subseteq \mathbb{R}$ *be an open interval and let* $f : I \to \mathbb{R}$ *be convex and differentiable. For any* $x, y \in I$, *the Bregman divergence of* f *from* x *to* y *is* $f(y) - f(x) - f'(x)(y - x)$. *Let* X *be an* I*-valued random variable. Then*

$$\mathbf{E}\left[f(X) - f(\mathbf{E}X)\right] = \inf_{a\in I} \mathbf{E}\left[f(X) - f(a) - f'(a)(X - a)\right].$$

Taking $f(x) = x\log x$, we obtain the following variational formula for entropy.

Corollary 4.17 *Let* Y *be a nonnegative random variable such that* $\mathbf{E}\Phi(Y) < \infty$. *Then*

$$\mathrm{Ent}(Y) = \inf_{u>0} \mathbf{E}\left[Y(\log Y - \log u) - (Y - u)\right].$$

Proof Let $a \in I$. The difference between the expected Bregman divergence from a and the expected Bregman divergence from $\mathbf{E}X$

$$\mathbf{E}\left[f(X) - f(\mathbf{E}X) - f'(\mathbf{E}X)(X - \mathbf{E}X)\right] = \mathbf{E}\left[f(X) - f(\mathbf{E}X)\right]$$

satisfies

$$\begin{aligned}
&\mathbf{E}\left[f(X) - f(a) - f'(a)(X - a)\right] - \mathbf{E}\left[f(X) - f(\mathbf{E}X)\right] \\
&= \mathbf{E}\left[-f(a) - f'(a)(X - a) + f(\mathbf{E}X)\right] \\
&= f(\mathbf{E}X) - f(a) - f'(a)(\mathbf{E}X - a).
\end{aligned}$$

The last expression is the Bregman divergence of f from a to $\mathbf{E}X$. As f is convex, it is nonnegative. □

Theorem 4.13 and Corollary 4.17 relate to the convexity of two different functions: Theorem 4.13 is about the convexity of the entropy functional while Corollary 4.17 is about the convexity of $\Phi(x) = x \log x$.

4.10 A Transportation Lemma

The duality formula of Corollary 4.14 allows one to relate the concentration property of a random variable Z around its expectation to the so-called *transportation cost*, that is, the "price" one has to pay when one computes the expectation of Z under $\mathbf{Q}$ rather than under the original probability measure $\mathbf{P}$.

To render this simple but subtle connection more explicit, the following transportation lemma may be illuminating.

Lemma 4.18 *Let Z be a real-valued integrable random variable. Let ϕ be a convex and continuously differentiable function on a (possibly unbounded) interval $[0, b)$ and assume that $\phi(0) = \phi'(0) = 0$. Define, for every $x \geq 0$, $\phi^*(x) = \sup_{\lambda \in (0,b)} (\lambda x - \phi(\lambda))$, and let, for every $t \geq 0$, $\phi^{*-1}(t) = \inf\{x \geq 0 : \phi^*(x) > t\}$. Then the following two statements are equivalent:*
(i) for every $\lambda \in (0, b)$,

$$\log \mathbf{E}e^{\lambda(Z-\mathbf{E}Z)} \leq \phi(\lambda);$$

(ii) for any probability measure $\mathbf{Q}$ absolutely continuous with respect to $\mathbf{P}$ such that $D(\mathbf{Q}\|\mathbf{P}) < \infty$,

$$\mathbf{E}_{\mathbf{Q}}Z - \mathbf{E}Z \leq \phi^{*-1}\left[D(\mathbf{Q}\|\mathbf{P})\right].$$

In particular, given $v > 0$,

$$\log \mathbf{E}e^{\lambda(Z-\mathbf{E}Z)} \leq \frac{v\lambda^2}{2}$$

for every $\lambda > 0$ if and only if for any probability measure $\mathbf{Q}$ absolutely continuous with respect to $\mathbf{P}$ and such that $D(\mathbf{Q}\|\mathbf{P}) < \infty$,

$$\mathbf{E}_{\mathbf{Q}}Z - \mathbf{E}Z \leq \sqrt{2vD(\mathbf{Q}\|\mathbf{P})}.$$

Proof As a direct consequence of Corollary 4.14 we see that (i) holds if and only if for every distribution $\mathbf{Q}$ which is absolutely continuous with respect to $\mathbf{P}$,

$$\mathbf{E}_{\mathbf{Q}}Z - \mathbf{E}Z \leq \inf_{\lambda \in (0,b)} \left(\frac{\phi(\lambda) + D(\mathbf{Q}\|\mathbf{P})}{\lambda}\right).$$

However, it follows from Lemma 2.4 that

$$\phi^{*-1}(D(Q\|P)) = \inf_{\lambda\in(0,b)}\left(\frac{\phi(\lambda)+D(Q\|P)}{\lambda}\right),$$

which shows that (i) is equivalent to (ii). Applying the previous result with $\phi(\lambda) = \lambda^2 v/2$ for every $\lambda > 0$ leads to the stated special case of equivalence since then $\phi^{*-1}(t) = \sqrt{2vt}$. □

The last inequality of Lemma 4.18 is related to what is usually termed a *quadratic transportation cost inequality*. If Ω is a metric space, the probability measure P is said to satisfy a quadratic transportation cost inequality if the last inequality of Lemma 4.18 holds for every Z which is Lipschitz on Ω with Lipschitz norm at most 1. The link between quadratic transportation cost inequalities and sub-Gaussian concentration inequalities is studied in greater detail in Chapter 8, devoted to transportation inequalities.

4.11 Pinsker's Inequality

Pinsker's inequality relates the relative entropy of two probability distributions to their variational distance. Let P and Q be two probability measures on a measurable space $(\Omega, \mathcal{A})$. The *total variation* or *variational distance* between P and Q is defined by

$$V(P,Q) = \sup_{A\in\mathcal{A}} |P(A) - Q(A)|.$$

It is a well-known and simple fact that the total variation is half the L_1-distance, that is, if λ is a common dominating measure of P and Q and $p(x) = dP/d\lambda$ and $q(x) = dQ/d\lambda$ denote their respective densities, then

$$V(P,Q) = P(A^*) - Q(A^*) = \frac{1}{2}\int |p(x) - q(x)|d\lambda(x),$$

where $A^* = \{x : p(x) \geq q(x)\}$. We note that another important interpretation of the variational distance is related to the best coupling of the two measures

$$V(P,Q) = \min P\{X \neq Y\},$$

where the minimum is taken over all pairs of joint distributions for the random variables (X, Y) whose marginal distributions are $X \sim P$ and $Y \sim Q$. (The proof of these well-known facts is left as Exercise 4.5).

The importance of Pinsker's inequality in statistics stems from the fact that it provides a lower bound for the error of certain hypothesis testing problems. We use Pinsker's inequality for a completely different purpose, namely for establishing a transportation cost inequality that may be used to prove concentration inequalities. The proof of Pinsker's inequality derives easily from Hoeffding's inequality via the transportation cost bound of Lemma 4.18.

Theorem 4.19 (PINSKER'S INEQUALITY) *Let* $\mathbf{P}$ *and* $\mathbf{Q}$ *be probability distributions on* $(\Omega, \mathcal{A})$ *such that* $\mathbf{Q} \ll \mathbf{P}$. *Then*

$$V(\mathbf{P}, \mathbf{Q})^2 \leq \frac{1}{2} D(\mathbf{Q} \| \mathbf{P}).$$

Proof Define the random variable Y such that $\mathbf{Q} = Y\mathbf{P}$ and let $A^* = \{Y \geq 1\}$ be the set achieving the maximum in the definition of the total variation between $\mathbf{P}$ and $\mathbf{Q}$. Then, setting $Z = \mathbb{1}_{\{A^*\}}$,

$$V(\mathbf{P}, \mathbf{Q}) = \mathbf{Q}\{A^*\} - \mathbf{P}\{A^*\} = \mathbf{E}_{\mathbf{Q}} Z - \mathbf{E} Z.$$

It follows from Hoeffding's lemma (Lemma 2.2) that for any $\lambda > 0$,

$$\psi_{Z-\mathbf{E}Z}(\lambda) \leq \frac{\lambda^2}{8}$$

which, by Lemma 4.18, leads to

$$\mathbf{E}_{\mathbf{Q}} Z - \mathbf{E} Z \leq \sqrt{\frac{1}{2} D(\mathbf{Q} \| \mathbf{P})},$$

concluding the proof. □

4.12 Birgé's Inequality

Next we show how the ideas already used in the proof of Pinsker's inequality may be used to prove a sharper version. Then we use this inequality for deriving a lower bound for the probability of error in multiple testing problems. Let $h(q, p) = q \log(q/p) + (1 - q) \log((1 - q)/(1 - p))$ be the relative entropy between two Bernoulli distributions, with parameters q and p. Then we have the following strengthened version of Theorem 4.19.

Theorem 4.20 *Let* $\mathbf{P}$ *and* $\mathbf{Q}$ *be probability distributions on* $(\Omega, \mathcal{A})$ *such that* $\mathbf{Q} \ll \mathbf{P}$. *Then*

$$\sup_{A \in \mathcal{A}} h(\mathbf{Q}\{A\}, \mathbf{P}\{A\}) \leq D(\mathbf{Q} \| \mathbf{P}).$$

Proof For any $p \in [0, 1]$, let

$$\phi_p(\lambda) = \log\left(p\left(e^{\lambda} - 1\right) + 1\right)$$

denote the logarithm of the moment-generating function of the Bernoulli(p) distribution where $\lambda \in \mathbb{R}$. By Corollary 4.15, for any $A \in \mathcal{A}$, and $\lambda \geq 0$,

$$D(Q\|P) \geq \mathbf{E}_Q[\lambda \mathbb{1}_{\{A\}}] - \log \mathbf{E}_P e^{\lambda \mathbb{1}_{\{A\}}},$$

and therefore

$$D(Q\|P) \geq \sup_{\lambda \geq 0} (\lambda Q\{A\} - \phi_{P\{A\}}(\lambda)).$$

The theorem follows by noting that for any $a \in [0,1]$,

$$\sup_{\lambda > 0} \left(\lambda a - \phi_p(\lambda)\right) = h(a,p). \qquad \square$$

Since $h(Q\{A\}, P\{A\}) \geq 2(Q\{A\} - P\{A\})^2$, Theorem 4.20 implies Pinsker's inequality. Note also that Theorem 4.20 can be derived as a simple consequence of the so-called data processing lemma (see Exercise 4.10).

The variational representation of relative entropy (Corollary 4.15) may be used to establish lower bounds for the probability of error in multiple testing problems. The next result is a sharper version of Fano's inequality, a classical tool from information theory.

Theorem 4.21 (BIRGÉ'S INEQUALITY) *Let $P_0, P_1, \ldots, P_N$ be probability distributions over $(\Omega, \mathcal{A})$ and let $A_0, A_1, \ldots, A_N \in \mathcal{A}$ be pairwise disjoint events. If $a = \min_{i=0,\ldots,N} P_i(A_i) \geq 1/(N+1)$,*

$$h\left(a, \frac{1-a}{N}\right) \leq \frac{1}{N} \sum_{i=1}^{N} D(P_i\|P_0).$$

Proof By the variational representation of relative entropy (Corollary 4.15), for any $i = 1, \ldots, N$,

$$\sup_{\lambda > 0} \mathbf{E}_{P_i}\left[\lambda \mathbb{1}_{\{A_i\}}\right] - \log \mathbf{E}_{P_0} e^{\lambda \mathbb{1}_{\{A_i\}}} \leq D(P_i\|P_0).$$

Observe that $\sum_{i=1}^{N} P_0(A_i) \leq 1 - P_0(A_0) \leq 1 - a$. For any fixed $\lambda \geq 0$

$$\begin{aligned}
\frac{1}{N}\sum_{i=1}^{N} D(P_i\|P_0) &\geq \frac{1}{N}\sum_{i=1}^{N} \left(\lambda P_i(A_i) - \log\left[P_0(A_i)(e^\lambda - 1) + 1\right]\right) \\
&\geq \lambda a - \log\left(\frac{1 - P_0(A_0)}{N}(e^\lambda - 1) + 1\right) \\
&\geq \lambda a - \log\left(\frac{1-a}{N}(e^\lambda - 1) + 1\right),
\end{aligned}$$

where the second inequality follows from the concavity of the logarithm and Jensen's inequality. We may choose λ such that it satisfies $h(a,(1-a)/N) = \lambda a - \log\left(\frac{1-a}{N}(e^\lambda - 1) + 1\right)$. □

4.13 Sub-Additivity of Entropy: The General Case

We now turn to the sub-additivity of entropy in a general measure theoretic framework. We proved this inequality in Section 4.7 in the restricted setting of discrete random variables as an easy consequence of Han's inequality. In the general case our proof relies on the duality formula of Theorem 4.13.

In Chapter 14 we present an even more general version of the sub-additivity of entropy. As we will see there, it is deeply connected to the convexity of the entropy functional. In the proof below we start from a decomposition that we already used to prove the Efron–Stein inequality and then use the duality formula of Theorem 4.13.

Theorem 4.22 (SUB-ADDITIVITY OF ENTROPY) *Let $X_1,\ldots,X_n$ be independent random variables and let $Y = f(X_1,\ldots,X_n)$ be a nonnegative measurable function of these variables such that $\Phi(Y) = Y\log Y$ is integrable. For every $1 \le i \le n$, denote by $\mathbf{E}^{(i)}$ the expectation operator conditioned on $X^{(i)} = (X_1,\ldots,X_{i-1},X_{i+1},\ldots,X_n)$. Denote by $\mathrm{Ent}^{(i)}(Y)$ the conditional entropy of Y, given $X^{(i)}$, defined by*

$$\mathrm{Ent}^{(i)}(Y) = \mathbf{E}^{(i)}\Phi(Y) - \Phi\left(\mathbf{E}^{(i)}Y\right).$$

Then

$$\mathrm{Ent}(Y) \le \mathbf{E}\sum_{i=1}^{n}\mathrm{Ent}^{(i)}(Y).$$

Proof Introduce the conditional expectation operator $\mathbf{E}_i[\cdot] = \mathbf{E}\left[\cdot|X_1,\ldots,X_i\right]$ for $i = 1,\ldots,n$ and the convention $\mathbf{E}_0 = \mathbf{E}$. Noting that the operator $\mathbf{E}_n$ is just the identity when restricted to the set of $(X_1,\ldots,X_n)$-measurable and integrable random variables, we have the decomposition

$$Y(\log Y - \log(\mathbf{E}Y)) = \sum_{i=1}^{n} Y(\log(\mathbf{E}_iY) - \log(\mathbf{E}_{i-1}Y)).$$

Now the duality formula given in Remark 4.4 yields

$$\mathbf{E}^{(i)}\left[Y\left(\log(\mathbf{E}_iY) - \log\left(\mathbf{E}^{(i)}[\mathbf{E}_iY]\right)\right)\right] \le \mathrm{Ent}^{(i)}(Y).$$

Since $X_1,\ldots,X_n$ are independent, we have $\mathbf{E}^{(i)}[\mathbf{E}_iY] = \mathbf{E}_{i-1}Y$ and therefore taking expectations on both sides of the decomposition above yields

$$E\left[Y\left(\log Y-\log(EY)\right)\right] = \sum_{i=1}^{n} E\left[E^{(i)}\left[Y\left(\log(E_iY)-\log\left(E^{(i)}\left[E_iY\right]\right)\right)\right]\right]$$
$$\leq \sum_{i=1}^{n} E\left[\mathrm{Ent}^{(i)}(Y)\right]$$

and Theorem 4.22 follows. □

Recall that in Section 4.7 the sub-additivity of entropy for discrete probability distributions is derived from Han's inequality (Theorem 4.1). The alternative proof given here has the advantage that it works in a more general measure-theoretic framework. It is interesting to notice that Han's inequality itself can be derived from the sub-additivity of entropy. In other words, for discrete probability distributions, the sub-additivity of entropy and Han's inequality are equivalent. Indeed, let $\mathcal{X}$ be a finite set of cardinality k and consider a random variable X with values in $\mathcal{X}^n$. Setting $X^{(i)} = (X_1, \ldots, X_{i-1}, X_{i+1}, \ldots, X_n)$ for every $i = 1, \ldots, n$, recall that Han's inequality tells us that

$$H(X) \leq \frac{1}{n-1}\sum_{i=1}^{n} H\left(X^{(i)}\right).$$

Define Q as the distribution of X and let P be the uniform distribution on $\mathcal{X}^n$. Denote by q the probability mass function of Q, that is, for every $x \in \mathcal{X}^n$, $q(x) = P\{X = x\}$. Setting $Y = dQ/dP$, we have $Y(x) = q(x)k^n$ and

$$\mathrm{Ent}(Y) = D(Q\|P) = -H(X) + n\log k.$$

The inequality of Theorem 4.22 can be written in this case as

$$\mathrm{Ent}(Y) \leq E\sum_{i=1}^{n} \mathrm{Ent}^{(i)}(Y).$$

Now

$$E\left[\mathrm{Ent}^{(1)}(Y)\right] = \mathrm{Ent}(Y) - \sum_{x\in\mathcal{X}^{n-1}}\left(\sum_{t\in\mathcal{X}} q(t,x)\right)\log\left(k^{n-1}\sum_{t\in\mathcal{X}} q(t,x)\right)$$
$$= \mathrm{Ent}(Y) + H\left(X^{(1)}\right) - (n-1)\log k$$

and similarly, for all i,

$$E\left[\mathrm{Ent}^{(i)}(Y)\right] = \mathrm{Ent}(Y) - (n-1)\log k + H\left(X^{(i)}\right).$$

Putting the pieces together, Han's inequality follows.

4.14 The Brunn–Minkowski Inequality

Next we present a classical result of convex geometry that is of fundamental importance in a wide variety of areas, including analysis and information theory. We include it here because it provides a short proof of the classical isoperimetric inequality (see Chapter 7). To describe the basic inequality, consider sets $A, B \subset \mathbb{R}^n$ and define the *Minkowski sum* of A and B as the set of all vectors in $\mathbb{R}^n$ formed by sums of elements of A and B:

$$A + B = \{x + y : x \in A, y \in B\}.$$

Similarly, for $c \in \mathbb{R}$, let $c \cdot A = \{cx : x \in A\}$. Denote by $\mathrm{Vol}(A)$ the Lebesgue measure of a (measurable) set $A \subset \mathbb{R}^n$.

Theorem 4.23 (BRUNN–MINKOWSKI INEQUALITY) *Let $A, B \subset \mathbb{R}^n$ be non-empty compact sets. Then for all $\lambda \in [0, 1]$,*

$$\mathrm{Vol}((1-\lambda)A + \lambda B)^{1/n} \geq (1-\lambda)\mathrm{Vol}(A)^{1/n} + \lambda \mathrm{Vol}(B)^{1/n}.$$

Note that it is not necessary to assume compactness of A and B. We do it to avoid having to worry about measurability of the Minkowski sum set (see Exercise 4.9). Many different proofs of the Brunn–Minkowski inequality are known. Here we present possibly the simplest one, based on a powerful functional generalization known as the Prékopa–Leindler inequality. Before stating this, let us consider the special one-dimensional case of Theorem 4.23. To see why the theorem is true in this case, notice first that if $A \subset \mathbb{R}$ and $c \geq 0$, then $\mathrm{Vol}(cA) = c\mathrm{Vol}(A)$ and therefore it suffices to prove that for any compact sets $A, B \subset \mathbb{R}$,

$$\mathrm{Vol}(A + B) \geq \mathrm{Vol}(A) + \mathrm{Vol}(B).$$

To see this, observe that none of the three volumes involved changes if the sets A and B are translated arbitrarily. Now we may translate A to $A' = \{a\} + A$ and B to $B' = \{b\} + B$ such that $A' \subset (-\infty, 0]$, $B' \subset [0, \infty)$, and $A' \cap B' = \{0\}$ (simply pick $a = -\sup A$ and $b = -\inf B$). However, $A' \cup B' \subset A' + B'$ and therefore $\mathrm{Vol}(A' + B') \geq \mathrm{Vol}(A' \cup B') = \mathrm{Vol}(A') + \mathrm{Vol}(B')$, proving the one-dimensional Brunn–Minkowski inequality.

The next inequality may be regarded as a functional generalization of the Brunn–Minkowski inequality.

Theorem 4.24 (PRÉKOPA–LEINDLER INEQUALITY) *Let $\lambda \in (0, 1)$, and let $f, g, h : \mathbb{R}^n \to [0, \infty)$ be nonnegative measurable functions such that for all $x, y \in \mathbb{R}^n$,*

$$h((1-\lambda)x + \lambda y) \geq f(x)^{1-\lambda} g(y)^{\lambda}.$$

Then

$$\int_{\mathbb{R}^n} h(x)dx \geq \left(\int_{\mathbb{R}^n} f(x)dx\right)^{1-\lambda} \left(\int_{\mathbb{R}^n} g(x)dx\right)^{\lambda}.$$

Proof The proof goes by induction with respect to the dimension n. To prove the one-dimensional case, consider measurable nonnegative functions f, g, h satisfying the condition of the theorem. By the monotone convergence theorem, it suffices to prove the statement for *bounded* functions f and g. Now observe that we may assume, without loss of generality, that $\sup_{x\in\mathbb{R}^n} f(x) = \sup_{x\in\mathbb{R}^n} g(x) = 1$. Then

$$\int_{\mathbb{R}} f(x)dx = \int_0^1 \mathrm{Vol}(\{x : f(x) \geq t\})dt$$

and

$$\int_{\mathbb{R}} g(x)dx = \int_0^1 \mathrm{Vol}(\{x : g(x) \geq t\})dt.$$

For any fixed $t \in [0, 1]$, if $f(x) \geq t$ and $g(y) \geq t$, then by the hypothesis of the theorem, $h((1-\lambda)x + \lambda y) \geq t$. This implication may be re-written as

$$(1-\lambda)\{x : f(x) \geq t\} + \lambda\{x : g(x) \geq t\} \subseteq \{x : h(x) \geq t\}.$$

Thus,

$$\begin{aligned}
\int_{\mathbb{R}} h(x)dx &= \int_0^\infty \mathrm{Vol}(\{x : h(x) \geq t\})dt \\
&\geq \int_0^1 \mathrm{Vol}(\{x : h(x) \geq t\})dt \\
&\geq \int_0^1 \mathrm{Vol}\left((1-\lambda)\{x : f(x) \geq t\} + \lambda\{x : g(x) \geq t\}\right) dt \\
&\quad \text{(by the inclusion above)} \\
&\geq (1-\lambda)\int_0^1 \mathrm{Vol}(\{x : f(x) \geq t\})dt + \lambda \int_0^1 \mathrm{Vol}(\{x : g(x) \geq t\})dt \\
&\quad \text{(by the one-dimensional Brunn–Minkowski inequality)} \\
&= (1-\lambda)\int_{\mathbb{R}} f(x)dx + \lambda \int_{\mathbb{R}} g(x)dx \\
&\geq \left(\int_{\mathbb{R}} f(x)dx\right)^{1-\lambda} \left(\int_{\mathbb{R}} g(x)dx\right)^{\lambda} \\
&\quad \text{(by the arithmetic-geometric mean inequality)}
\end{aligned}$$

and this proves the one-dimensional case.

For the induction step, assume that the theorem holds for all dimensions $1, \ldots, n-1$ and let $f, g, h : \mathbb{R}^n \to [0, \infty)$, $\lambda \in (0, 1)$ be such that they satisfy the assumption of the theorem. Now let $x, y \in \mathbb{R}^{n-1}$ and $a, b \in \mathbb{R}$. Then

$$h(((1-\lambda)x + \lambda y, (1-\lambda)a + \lambda b)) = h((1-\lambda)(x, a) + \lambda(y, b)) \geq f(x, a)^{1-\lambda} g(y, b)^{\lambda}$$

so by the inductive hypothesis,

$$\int_{\mathbb{R}^{n-1}} h(x,(1-\lambda)a+\lambda b))dx \geq \left(\int_{\mathbb{R}^{n-1}} f(x,a)dx\right)^{1-\lambda}\left(\int_{\mathbb{R}^{n-1}} g(x,b)dx\right)^{\lambda}.$$

In other words, introducing

$$F(a) = \int_{\mathbb{R}^{n-1}} f(x,a)dx, \quad G(a) = \int_{\mathbb{R}^{n-1}} g(x,a)dx,$$
$$\text{and} \quad H(a) = \int_{\mathbb{R}^{n-1}} h(x,a)dx,$$

we have

$$H((1-\lambda)a+\lambda b) \geq F(a)^{1-\lambda}G(b)^{\lambda},$$

so by Fubini's theorem and the one-dimensional inequality, we have

$$\begin{aligned}\int_{\mathbb{R}^n} h(x)dx &= \int_{\mathbb{R}} H(a)da \\ &\geq \left(\int_{\mathbb{R}} F(a)da\right)^{1-\lambda}\left(\int_{\mathbb{R}} G(a)da\right)^{\lambda} \\ &= \left(\int_{\mathbb{R}^n} f(x)dx\right)^{1-\lambda}\left(\int_{\mathbb{R}^n} g(x)dx\right)^{\lambda}\end{aligned}$$

as desired. □

Corollary 4.25 (A WEAKER BRUNN–MINKOWSKI INEQUALITY) *Let $A, B \subset \mathbb{R}^n$ be compact sets. Then for all $\lambda \in [0,1]$,*

$$\text{Vol}((1-\lambda)A+\lambda B) \geq \text{Vol}(A)^{1-\lambda}\text{Vol}(B)^{\lambda}.$$

Proof We apply the Prékopa–Leindler inequality with $f(x) = \mathbb{1}_{\{x\in A\}}$, $g(x) = \mathbb{1}_{\{x\in B\}}$, and $h(x) = \mathbb{1}_{\{x\in(1-\lambda)A+\lambda B\}}$. To confirm that these functions satisfy the hypothesis of Theorem 4.24, observe that $f(x)^{1-\lambda}g(y)^{\lambda} = \mathbb{1}_{\{x\in A, y\in B\}} \leq h((1-\lambda)x+\lambda y)$. □

Observe that Corollary 4.25 is weaker than Theorem 4.23 because by the arithmetic-geometric mean inequality

$$(1-\lambda)\text{Vol}(A)^{1/n} + \lambda\text{Vol}(B)^{1/n} \geq \text{Vol}(A)^{(1-\lambda)/n}\text{Vol}(B)^{\lambda/n}.$$

Interestingly, however, one may deduce the Brunn–Minkowski inequality starting from this weaker form as follows.

Proof of Theorem 4.23. First observe that it suffices to prove that for all nonempty compact sets A and B,

$$\mathrm{Vol}(A+B)^{1/n} \geq \mathrm{Vol}(A)^{1/n} + \mathrm{Vol}(B)^{1/n}$$

because by replacing A by $(1-\lambda)A$ and B by λB we obtain the original statement. Also notice that we may assume that $\mathrm{Vol}(A), \mathrm{Vol}(B) > 0$ because otherwise the inequality holds trivially. Defining $A' = \mathrm{Vol}(A)^{-1/n}A$ and $B' = \mathrm{Vol}(B)^{-1/n}B$, we have $\mathrm{Vol}(A') = \mathrm{Vol}(B') = 1$. Therefore, by Corollary 4.25, for all $\lambda \in (0,1)$,

$$\mathrm{Vol}((1-\lambda)A' + \lambda B') \geq 1.$$

Finally, we apply this inequality with the choice

$$\lambda = \frac{\mathrm{Vol}(B)^{1/n}}{\mathrm{Vol}(A)^{1/n} + \mathrm{Vol}(B)^{1/n}},$$

obtaining

$$\begin{aligned} 1 &\leq \mathrm{Vol}\left(\frac{1}{\mathrm{Vol}(A)^{1/n} + \mathrm{Vol}(B)^{1/n}}A + \frac{1}{\mathrm{Vol}(A)^{1/n} + \mathrm{Vol}(B)^{1/n}}B\right) \\ &= \frac{\mathrm{Vol}(A+B)}{\left(\mathrm{Vol}(A)^{1/n} + \mathrm{Vol}(B)^{1/n}\right)^n}, \end{aligned}$$

proving the Brunn–Minkowski inequality. □

The Brunn–Minkowski inequality can be used to show that the uniform distribution over convex bodies exhibits the concentration of measure phenomenon (see Exercise 4.17).

4.15 Bibliographical Remarks

Information theory originated from Shannon's celebrated paper (Shannon, 1948) which introduced a general mathematical theory of communication. It was Shannon who defined the basic notions of entropy, relative entropy, and mutual information, and proved their significance in data compression and coding problems. However, very soon it became apparent that the significance of Shannon's techniques reached far beyond the engineering problems he had in mind, and today the toolbox of information theory is routinely used in a wide variety of mathematical problems. For some excellent textbooks on the topic we refer to Gallager (1968), Csiszár and Körner (1981), Cover and Thomas (1991), MacKay (2003), and Richardson and Urbanke (2008).

A geometric version of Han's inequality appears as early as in 1948 in a paper by Loomis and Whitney (1949). Han's inequality, as described in Theorems 4.1 and 4.9, was derived by Han (1978).

Different versions of the discrete isoperimetric inequalities of Theorems 4.2 and 4.3 go back to Harper (1966), Loomis and Whitney (1949), and Hart (1976). The subsets of the n-cube that maximize the edge-perimeter or the sum of average influences for a given cardinality and achieve equality in Theorem 4.2 have been described by Harper (see Bollobás (1986)). These combinatorial inequalities have also been derived without resorting to Han's inequality (see for example Bollobás (1986)).

The fact that combinatorial entropies satisfy the self-bounding property was shown by Boucheron, Lugosi, and Massart (2000).

The sub-additive property of entropy, often called *tensorization inequality* for entropy appears in Gross (1975) (and see also Ledoux (1997), Bobkov and Ledoux (1997)). Related inequalities may be found in Beckner (1989), Latała and Oleszkiewicz (2000), Chafaï (2002), and Boucheron *et al.* (2005*b*). The proof of the general result of Theorem 4.22 is borrowed from Ané *et al.* (2000).

The notion of relative entropy also plays an important role in the theory of large deviations which dates back to Cramér (1938) (see also Varadhan (1984), Deuschel and Stroock (1989), Dembo and Zeitouni (1998), and Dupuis and Ellis (1997)). The variational formulation of relative entropy (Theorem 4.13) is also frequently used in large deviations theory.

The link between quadratic transportation cost inequalities and Gaussian type concentration is well known (and see for example Marton (1996*a*), Dembo (1997), Bobkov and Götze (1999)). In particular, Lemma 4.18 is inspired by a related result on quadratic transportation cost inequalities in Bobkov and Götze (1999).

Theorem 4.19 was first proved by Pinsker (1964) with the worse constant 1, while Csiszár (1967) established it with the optimal constant 1/2. For some sharper versions we refer to Ordentlich and Weinberger (2005).

Theorem 4.21 is due to Birgé (2005). It improves on Fano's lemma (see, e.g., Cover and Thomas 1991) originally proved to estimate the probability of error in channel coding theory. Beginning with the work of Ibragimov and Khasminskii (1981), Fano's lemma has been proved to be a fundamental tool in deriving minimax lower bounds in statistics.

For the history of the Brunn–Minkowski inequality, its connection to numerous other inequalities, and various applications, we refer the reader to the comprehensive text of Schneider (1993) and to the survey by Gardner (2002). The Prékopa–Leindler inequality is established by Prékopa (1971, 1973) and Leindler (1972). The proof presented here is from Brascamp and Lieb (1976). For nice surveys of applications of the Brunn–Minkowski inequalities to concentration and convex geometry, see Ball (1997), Ledoux (2001), Schechtman (2003), and Barthe (2003). The connection between the Brunn–Minkowski inequality and concentration of measure was shown by Borell (1975). We recommend the survey of Giannopoulos and Milman (2001).

4.16 EXERCISES

4.1. (KRAFT–MCMILLAN INEQUALITY) Let $\mathcal{X}$ denote a countable set, P a probability distribution on $\mathcal{X}$. Let $\mathcal{Y}$ denote a finite set called the encoding alphabet. A uniquely decodable encoding of $\mathcal{X}$ using alphabet $\mathcal{Y}$, is a mapping ϕ from $\mathcal{X}$ to the set $\mathcal{Y}^*$ of sequences of finite length over the encoding alphabet, such that for any two sequences $x_1, \ldots, x_n$ and $x'_1, \ldots, x'_p$ of elements of $\mathcal{X}$, if the concatenations of $\phi(x_1), \ldots, \phi(x_n)$ and $\phi(x'_1), \ldots, \phi'(x'_n)$ are equal, then $n = p$, and $x_i = x'_i$ for $i \leq n$. If $x \in \mathcal{X}$, $\phi(x)$ is the codeword associated with x and $|\phi(x)|$ denotes the length of the codeword. The *Kraft–McMillan inequality* asserts that for any uniquely-decodable coding ϕ of $\mathcal{X}$ on alphabet $\mathcal{Y}$

$$\sum_{x \in \mathcal{X}} |\mathcal{Y}|^{-|\phi(x)|} \leq 1.$$

Prove the Kraft–McMillan inequality. Use the Kraft–MacMillan inequality to prove that the Shannon entropy with base $|\mathcal{Y}|$ is a lower bound on the average codeword length under P:

$$\frac{H(X)}{\log|\mathcal{Y}|} = \mathbf{E}\left[-\log_{|\mathcal{Y}|} p(X)\right] \leq \mathbf{E}\,|\phi(X)|\,.$$

4.2. (CONVERSE OF THE KRAFT–MCMILLAN INEQUALITY) Let $\ell : \mathcal{X} \to \{1, 2, \ldots\}$ be such that

$$\sum_{x \in \mathcal{X}} |\mathcal{Y}|^{-\ell(x)} \leq 1.$$

Prove the *converse of the Kraft–McMillan inequality*: there exists a uniquely decodable encoding ϕ such that for all $x \in \mathcal{X}$, $|\phi(x)| = \ell(x)$. Use the converse of the Kraft–McMillan inequality to prove that there exists a uniquely decodable encoding ϕ such that

$$\mathbf{E}\,|\phi(X)| \leq \mathbf{E}\left[-\log_{|\mathcal{Y}|} p(X)\right] + 1 = \frac{H(X)}{\log|\mathcal{Y}|} + 1.$$

4.3. (LOG–SUM INEQUALITY) Let $a_1, \ldots, a_n$ and $b_1, \ldots, b_n$ denote two sequences of positive integers. Prove that

$$\sum_i a_i \log \frac{a_i}{b_i} \geq \left(\sum_i a_i\right) \log \frac{\sum_i a_i}{\sum_i b_i}.$$

4.4. (CHAIN RULE FOR THE RELATIVE ENTROPY) Let P and Q denote two joint distributions for $X_1, \ldots, X_n$, let $P_{1:i}$ and $Q_{1:i}$ denote the marginal distributions of $X_1, \ldots, X_i$ under P and Q, respectively. Let $P_{X_i|1\ldots i-1}$ and $Q_{X_i|1\ldots i-1}$ denote the conditional distribution of X_i with respect to $X_1, \ldots, X_{i-1}$ under P and under Q. Show that

$$D(P\|Q) = \sum_{i=1}^{n} \mathbf{E}_{P_{1:i-1}}\left[D\left(P_{X_i|1\ldots i-1}\|Q_{X_i|1\ldots i-1}\right)\right].$$

4.5. (PROPERTIES OF THE VARIATIONAL DISTANCE) Let P and Q be two probability distributions on the same discrete set $\mathcal{X}$. Prove that the total variation distance $V(P,Q)$ satisfies

$$V(P,Q) = P(A^*) - Q(A^*) = \frac{1}{2}\sum_{x\in\mathcal{X}} |P(x) - Q(x)|,$$

where $A^* = \{x : P(x) \geq Q(x)\}$. (This identity is sometimes referred to as Scheffé's theorem (Scheffé, 1947).) Show that

$$V(P,Q) = \min \mathbf{P}\{X \neq Y\},$$

where the minimum is taken over all pairs of random variables (X,Y) whose marginal distributions are $X \sim P$ and $Y \sim Q$.

4.6. (DISCRETE LOOMIS–WHITNEY INEQUALITY) This exercise and the next illustrate the fact that Han's inequality has something simple to say about the combinatorics of product spaces. Let A denote a finite subset of $\mathbb{Z}^d$ and let A_i denote the projection of A along the i-th coordinate. Show that

$$|A|^{d-1} \leq \prod_{i=1}^{d} |A_i|$$

(see Loomis and Whitney (1949)).

4.7. (DISCRETE ISOPERIMETRIC INEQUALITY IN $\mathbb{Z}^d$) Let A denote a finite subset of $\mathbb{Z}^d$. Let B denote the canonical basis of $\mathbb{Z}^d$. Prove that the set ∂A defined by

$$\partial A = \left\{(x,y,s) : x \in A, y \in B, s \in \{-1,1\}, x + sy \notin A\right\}$$

has cardinality bounded as

$$|\partial A| \geq 2d|A|^{\frac{d-1}{d}}.$$

4.8. Assume that $h(x) = \log_b |\mathrm{tr}(x)|$ is a combinatorial entropy such that for all $x \in \mathcal{X}^n$ and $i \leq n$,

$$h(x) - h\left(x^{(i)}\right) \leq 1.$$

Show that h has the self-bounding property.

4.9. Prove that the Minkowski sum of two compact sets is compact.

4.10. (KULLBACK–LEIBLER DIVERGENCE AND SUB-σ-ALGEBRAS, DATA PROCESSING LEMMA) If $\mathcal{G}$ is a σ-algebra of subsets of $\mathcal{X}$, $A \in \mathcal{G}$ is said to be an atom in $\mathcal{G}$ if $B \subset A$ and $B \in \mathcal{G} \setminus \{\emptyset\}$ then $B = A$. Let atom($\mathcal{G}$) denote the set of atoms of $\mathcal{G}$. If $\mathbf{P}$ and $\mathbf{Q}$ are probability measures over $\mathcal{X}$, and $\mathcal{G}$ has countably many atoms, let

$$D(\mathbf{P}\|\mathbf{Q}|\mathcal{G}) = \sum_{A \in \text{atom}(\mathcal{G})} \mathbf{P}(A) \log \frac{\mathbf{P}(A)}{\mathbf{Q}(A)}.$$

Show that if $\mathcal{H} \subset \mathcal{G}$ where $\mathcal{H}$ is a σ-algebra, then $D(\mathbf{P}\|\mathbf{Q}|\mathcal{H}) \leq D(\mathbf{P}\|\mathbf{Q}|\mathcal{G})$. Show that

$$D(\mathbf{P}\|\mathbf{Q}) = \sup \left\{ D(\mathbf{P}\|\mathbf{Q}|\mathcal{G}) : \mathcal{G} \text{ has finitely many atoms.} \right\}.$$

Use this statement to prove Theorem 4.10 from Theorem 4.9 (that is, by checking that Theorem 4.9 still holds when $\mathcal{X}$ is not countable). *Hint:* the first part follows easily from the duality formula.

4.11. (CONVEXITY OF KULLBACK–LEIBLER DIVERGENCE) Prove that for any fixed probability measure $\mathbf{P}$ on $\mathcal{X}$, the function $\mathbf{Q} \to D(\mathbf{Q}\|\mathbf{P})$ is convex on the set of probability distributions over $\mathcal{X}$. *Hint:* use the duality representation.

4.12. (KULLBACK–LEIBLER DIVERGENCE WITH RESPECT TO A PRODUCT DISTRIBUTION) Let $\mathbf{P}$ denote a probability distribution over $\mathcal{X} \times Y$. Let $\mathbf{P}_X$ and $\mathbf{P}_Y$ denote its two marginal distributions and let $\mathbf{Q}_X$ and $\mathbf{Q}_Y$ denote two probability distributions over $\mathcal{X}$ and $\mathcal{Y}$. Prove that

$$D(\mathbf{P}\|\mathbf{Q}_X \otimes \mathbf{Q}_Y) = D(\mathbf{P}\|\mathbf{P}_X \otimes \mathbf{P}_Y) + D(\mathbf{P}_X\|\mathbf{Q}_X) + D(\mathbf{P}_Y\|\mathbf{Q}_Y).$$

4.13. (KULLBACK–LEIBLER DIVERGENCE AND LEGENDRE TRANSFORM OF LOGARITHMIC MOMENT-GENERATING FUNCTION) Let Z be a real-valued random variable. Recall that $\psi_Z(\lambda) = \log \mathbf{E}e^{\lambda Z}$ for $\lambda \in \mathbb{R}$. Let $\psi^*(t) = \sup_{\lambda \in \mathbb{R}}[\lambda t - \psi_{Z-\mathbf{E}Z}(\lambda)]$. Prove that for all $t > 0$,

$$\psi^*(t) = \inf \left\{ D(\mathbf{Q}\|\mathbf{P}) : \mathbf{E}_{\mathbf{Q}}Z - \mathbf{E}Z \geq t \right\}.$$

4.14. (LAW OF RARE EVENTS) Let $\mathbf{P}$ be the probability distribution of a sum of n independent Bernoulli random variables $X_1, \ldots, X_n$ with parameters $p_1, \ldots, p_n$. Let Po(μ) be the Poisson distribution with expectation $\mu = \sum_{i=1}^n p_i$. Prove that $V(\mathbf{P}, \text{Po}(\mu)) \leq \sum_{i=1}^n p_i^2$. Interpret this inequality by considering the Poisson approximation of the binomial distribution with parameters n and μ/n. *Hint:* use the infinite divisibility of the Poisson distribution and a coupling argument (see Exercise 4.5). (See Barbour, Holst and Janson (1992) for a thorough treatment of this topic.)

4.15. (LAW OF RARE EVENTS AND KULLBACK–LEIBLER DIVERGENCE) Let $X_1, \ldots, X_n$ be (not necessarily independent) Bernoulli random variables, with $\mathbf{E}X_i = p_i$ for $i \leq n$.

Let $S_n = \sum_{i=1}^n X_i$, and $\mu = \mathbf{E}S_n$. Let $\mathbf{P}$ denote the probability distribution of S_n and let $\text{Po}(\mu)$ be the Poisson distribution with expectation μ. Prove that

$$D(\mathbf{P}\|\text{Po}(\mu)) \le \sum_{i=1}^n p_i^2 + \sum_{i=1}^n H(X_i) - H(X_1, \ldots, X_n).$$

Hint: use the infinite divisibility of the Poisson distribution, the data processing lemma (Exercise 4.10), and the previous exercise. (Note that this result can be combined with Pinsker's inequality in order to derive a sub-optimal upper bound on the total variation distance between a binomial distribution and a Poisson distribution with the same expectation.)

4.16. (PRÉKOPA–LEINDLER INEQUALITY ON $\mathbb{R}$) Let $\lambda \in (0,1)$, and let $f, g, h : \mathbb{R} \to [0, \infty)$ be nonnegative measurable functions such that for all $x, y \in \mathbb{R}$,

$$h((1-\lambda)x + \lambda y) \ge f(x)^{1-\lambda} g(y)^{\lambda}.$$

Prove that

$$\int_{\mathbb{R}} h(x)dx \ge \left(\int_{\mathbb{R}} f(x)dx\right)^{1-\lambda} \left(\int_{\mathbb{R}} g(x)dx\right)^{\lambda}$$

without resorting to the Brunn–Minkowski inequality on $\mathbb{R}$. *Hint:* prove that it is possible to define two functions x and y by

$$\int_{-\infty}^{x(t)} f(u)du = t \int_{-\infty}^{\infty} f(u)du \quad \text{and} \quad \int_{-\infty}^{y(t)} f(u)du = t \int_{-\infty}^{\infty} g(u)du$$

and let $z(t) = (1-\lambda)x(t) + \lambda y(t)$. Verify that all three functions are differentiable and that $z'(t) \ge (x'(t))^{1-\lambda} (y'(t))^{\lambda}$. Use change of variables to finish the proof. See Barthe (2003).

4.17. (BORELL'S LEMMA) Let C be a convex body (a compact convex set with non-empty interior) in $\mathbb{R}^n$ and let $\mathbf{P}$ be the uniform probability distribution over C. Prove Borell's lemma that states the following: if A is a symmetric convex subset of C with $\mathbf{P}\{A\} > 1/2$, then for any $t > 1$,

$$\mathbf{P}\left\{(tA)^c\right\} \le \mathbf{P}\{A\} \left(\frac{1-\mathbf{P}\{A\}}{\mathbf{P}\{A\}}\right)^{(t+1)/2}.$$

Hint: prove first that for $t > 1$,

$$\frac{2}{t+1}(tA)^c + \frac{t-1}{t+1}A \subseteq A^c,$$

where A^c is the complement of A with respect to C. Then, use the Brunn–Minkowski inequality. Does the statement remain true if the convexity and symmetry assumptions on A are relaxed? Borell's lemma provides an example of the concentration of measure phenomenon. It asserts that, regardless of the dimension of the ambient space n and the convex body C, if A is a symmetric convex subset of C with $\boldsymbol{P}\{A\} > 1/2$, then $\boldsymbol{P}\{(tA)^c\}$ decreases exponentially fast as t increases. See Giannopoulos and Milman (2001).

4.18. (A CONSEQUENCE OF BORELL'S LEMMA) Let C be a convex body in $\mathbb{R}^n$ and let $\boldsymbol{P}$ be the uniform probability distribution over C. Assume $X = (X_1, \ldots, X_n)$ is distributed according to $\boldsymbol{P}$. Assume that $\boldsymbol{E}X = 0$. Prove that there exists a universal constant κ such that for $p \geq 2$ and for all $y \in \mathbb{R}^n$,

$$\boldsymbol{E}\left[\left|\sum_{i=1}^{n} y_i X_i\right|^p\right]^{1/p} < \kappa p \boldsymbol{E}\left[\left|\sum_{i=1}^{n} y_i X_i\right|^2\right]^{1/2}.$$

Is it possible to tighten this inequality for some special convex bodies? *Hint:* use Borell's lemma (see Exercise 4.17). It is enough to check that $\boldsymbol{P}\{\left|\sum_{i=1}^{n} y_i X_i\right| > t\}$ decays exponentially fast with t.

4.19. (AN ELEMENTARY VERSION OF THE BRUNN–MINKOWSKI INEQUALITY) Assume that A and B are axis-parallel hyper-rectangles in $\mathbb{R}^n$. Use the arithmetic-geometric mean inequality to verify that

$$\mathrm{Vol}(A+B)^{1/n} \geq \mathrm{Vol}(A)^{1/n} + \mathrm{Vol}(B)^{1/n}.$$

This statement is the first step of some proofs of the Brunn–Minkowski inequality (see Stein and Shakarchi 2005).

5

Logarithmic Sobolev Inequalities

In this chapter we prove a few inequalities known as *logarithmic Sobolev inequalities*. The simplest such result, stated and proved in Section 5.1 below, may be regarded as an extension of the edge isoperimetric inequality on the binary hypercube shown in the previous chapter (Theorem 4.2). This inequality is surprisingly powerful. The application most interesting to us in this chapter shows how this simple result can be used to prove a general exponential concentration inequality for functions defined on the binary hypercube. The passage between the logarithmic Sobolev inequality and the concentration bound is achieved by a clever trick, the so-called *Herbst argument* (see Section 5.2). This is the first instance of a general methodology that we explore in this book in detail. The proof technique, called the *entropy method,* is based on various modifications of the logarithmic Sobolev inequality and the Herbst argument. In Chapters 6 and 12 we explore this technique in detail, and derive concentration inequalities for general functions of independent random variables, not only those defined over the binary hypercube.

In Sections 5.3 and 5.4 we extend the arguments from Bernoulli to Gaussian random variables, obtaining a remarkably useful Gaussian concentration inequality whose use is illustrated in Section 5.5 in proving a concentration inequality for the supremum of a Gaussian process. We will return to Gaussian concentration in Chapter 10 where a sharp form is presented. The Gaussian logarithmic Sobolev inequality shown here has applications in a variety of areas of mathematics.

As an application of the Gaussian logarithmic Sobolev inequality, in Section 5.6 we derive a more general version of the Johnson–Lindenstrauss theorem of Section 2.9.

In Section 5.7, we describe some statistical applications: a bound for the performance of LASSO, and an ℓ_1-penalized least squares estimator. Gaussian concentration proves to be a convenient tool for such model selection problems in quite general Gaussian models.

In Sections 5.8 and 5.9, we establish a collection of closely related results, starting from the so-called Bonami–Beckner inequality. This so-called *hypercontractive* inequality has its origins in harmonic analysis and has countless applications in a variety of areas.

In Section 5.10, we close this chapter by invoking Gaussian hypercontractive inequalities to prove a challenging tail bound for the largest eigenvalue of random matrices from the Gaussian unitary ensemble.

5.1 Symmetric Bernoulli Distributions

The purpose of this section is to prove the simplest of a large family of inequalities, generally referred to as *logarithmic Sobolev inequalities*. For this simplest instance we consider real-valued functions defined on the binary hypercube $f : \{-1, 1\}^n \to \mathbb{R}$. Consider a uniformly distributed binary vector $X = (X_1, \ldots, X_n)$ on the hypercube $\{-1, 1\}^n$. In other words, the components of X are independent, identically distributed random sign (Rademacher) variables with $\mathbf{P}\{X_i = -1\} = \mathbf{P}\{X_i = 1\} = 1/2$. Consider the induced real-valued random variable $Z = f(X)$. The logarithmic Sobolev inequality presented here relates two functionals of f that have already appeared in Chapters 3 and 4. One of them is the *entropy*

$$\mathrm{Ent}(f) = \mathbf{E}\left[f(X)\log(f(X))\right] - \mathbf{E}f(X)\log \mathbf{E}f(X),$$

defined for nonnegative functions $f \geq 0$. We use either $\mathrm{Ent}(f)$ or $\mathrm{Ent}(Z)$ interchangeably to denote the entropy of $Z = f(X)$. The other functional is a quantity familiar from the Efron–Stein inequality,

$$\mathcal{E}(f) = \frac{1}{2}\mathbf{E}\left[\sum_{i=1}^{n}\left(f(X) - f\left(\tilde{X}^{(i)}\right)\right)^2\right]$$

where $\tilde{X}^{(i)} = (X_1, \ldots, X_{i-1}, X_i', X_{i+1}, \ldots, X_n)$ is obtained by replacing the i-th component of X by an independent copy X_i'. Recall that by the Efron–Stein inequality, $\mathrm{Var}\left(f(X)\right) \leq \mathcal{E}(f)$. Since X is uniformly distributed, $\mathcal{E}(f)$ may be written in a slightly more convenient form

$$\mathcal{E}(f) = \frac{1}{4}\mathbf{E}\left[\sum_{i=1}^{n}\left(f(X) - f\left(\overline{X}^{(i)}\right)\right)^2\right] = \frac{1}{2}\mathbf{E}\left[\sum_{i=1}^{n}\left(f(X) - f\left(\overline{X}^{(i)}\right)\right)_+^2\right]$$

where the random binary vector $\overline{X}^{(i)} = (X_1, \ldots, X_{i-1}, -X_i, X_{i+1}, \ldots, X_n)$ is obtained by flipping the i-th component of X while leaving the others intact. One may think about $\nabla_i f(x) = (f(x) - f(\overline{x}^{(i)}))/2$ as the i-th component of the *discrete gradient* vector $\nabla f(x) = (\nabla_1 f(x), \ldots, \nabla_n f(x))$. With this notation, the Efron–Stein estimate of the variance is just the expected squared norm of the discrete gradient: $\mathcal{E}(f) = \mathbf{E}\|\nabla f(X)\|^2$.

Theorem 5.1 (LOGARITHMIC SOBOLEV INEQUALITY FOR THE SYMMETRIC BERNOULLI DISTRIBUTION) *Let $f : \{-1, 1\}^n \to \mathbb{R}$ be an arbitrary real-valued function defined on the n-dimensional binary hypercube and assume that X is uniformly distributed over $\{-1, 1\}^n$. Then*

$$\mathrm{Ent}\left(f^2\right) \leq 2\mathcal{E}(f).$$

Before proving the inequality, we point out that Theorem 5.1 is a common generalization of the edge isoperimetric inequality of Theorem 4.2 and of the Efron–Stein inequality for the binary hypercube. Indeed, let $A \subset \{-1,1\}^n$ be any subset of the binary hypercube. Defining $f(x) = \mathbb{1}_{\{x \in A\}}$, we have, writing $P(A) = \mathbf{P}\{X \in A\}$, that $\mathrm{Ent}(f^2) = -P(A)\log P(A)$ and $4\mathcal{E}(f) = I(A)$ is just the total influence of A. In Chapter 9 we point out other deep connections between influences and logarithmic Sobolev inequalities.

On the other hand, note that if f is nonnegative, $\mathrm{Var}(f(X)) \le \mathrm{Ent}(f^2)$ (see Exercise 5.1). One can also show (see Exercise 5.2) that for any function $f : \{-1,1\}^n \to \mathbb{R}$ (not necessarily nonnegative), Theorem 5.1 implies $\mathrm{Var}(f(X)) \le \mathcal{E}(f)$ and therefore it is stronger than the Efron–Stein inequality (for the binary hypercube).

Proof The key to the proof is the sub-additivity property of entropy derived in Theorem 4.10. This property implies that, writing $Z = f(X)$,

$$\mathrm{Ent}(Z^2) \le \mathbf{E}\left[\sum_{i=1}^n \mathrm{Ent}^{(i)}(Z^2)\right]$$

where $\mathrm{Ent}^{(i)}(Z^2) = \mathbf{E}^{(i)}[Z^2\log(Z^2)] - \mathbf{E}^{(i)}[Z^2]\log(\mathbf{E}^{(i)}[Z^2])$. (Recall that $\mathbf{E}^{(i)}$ denotes conditional expectation conditioned on $X^{(i)} = (X_1,\ldots,X_{i-1},X_{i+1},\ldots,X_n)$.) Therefore, it suffices to show that for all $i = 1,\ldots,n$,

$$\mathrm{Ent}^{(i)}(Z^2) \le \frac{1}{2}\mathbf{E}^{(i)}\left[\left(f(X) - f\left(\overline{X}^{(i)}\right)\right)^2\right]. \tag{5.1}$$

Given any fixed realization of $X^{(i)}$, Z can take two different values with equal probability. Call these values a and b. Then the desired inequality (5.1) takes the form

$$\frac{a^2}{2}\log a^2 + \frac{b^2}{2}\log b^2 - \frac{a^2+b^2}{2}\log\frac{a^2+b^2}{2} \le \frac{1}{2}(a-b)^2.$$

Thus, it remains to prove that this elementary inequality holds for any $a, b \in \mathbb{R}$. As $(|a|-|b|)^2 \le (a-b)^2$, we may assume, without loss of generality, that both a and b are nonnegative. By symmetry, we may also assume that $a \ge b$. For any fixed value of $b \ge 0$, define the function

$$h(a) = \frac{a^2}{2}\log a^2 + \frac{b^2}{2}\log b^2 - \frac{a^2+b^2}{2}\log\frac{a^2+b^2}{2} - \frac{1}{2}(a-b)^2$$

for $a \in [b,\infty)$. Since $h(b) = 0$, it suffices to check that $h'(b) = 0$ and that h is concave on $[b,\infty)$. However, elementary calculus shows that

$$h'(a) = a \log \frac{2a^2}{a^2 + b^2} - (a - b)$$

from which $h'(b) = 0$ is clear, while using $\log x - x \leq -1$,

$$h''(a) = 1 + \log \frac{2a^2}{a^2 + b^2} - \frac{2a^2}{a^2 + b^2} \leq 0.$$

□

Theorem 5.1 is possibly the simplest in the family of logarithmic Sobolev inequalities. It is outside of the scope of this book to offer a general account of these inequalities and we even avoid the general definition of what a logarithmic Sobolev inequality is. We merely mention a few of them that are important for our purposes. The obvious next step after having Theorem 5.1 is to ask what happens if the distribution of X is not uniform but rather a product of i.i.d. Bernoulli distributions with parameter different from $1/2$. This is the setup we consider in the remaining part of this section. More precisely, we still consider functions $f : \{-1, 1\}^n \to \mathbb{R}$ defined on the binary hypercube, but we now assume that the components of the random vector $X = (X_1, \ldots, X_n) \in \{-1, 1\}^n$ are independent, identically distributed random bits with distribution $\mathbf{P}\{X_i = 1\} = 1 - \mathbf{P}\{X_i = -1\} = p$ where $p \in [0, 1]$. With the same notation as before, we now have

$$\begin{aligned} \mathcal{E}(f) &= \frac{1}{2} \mathbf{E}\left[\sum_{i=1}^{n} \left(f(X) - f\left(\tilde{X}^{(i)}\right)\right)^2\right] \\ &= p(1-p)\,\mathbf{E}\left[\sum_{i=1}^{n} \left(f(X) - f\left(\overline{X}^{(i)}\right)\right)^2\right]. \end{aligned}$$

Then Theorem 5.1 may be generalized as follows to include the case of asymmetric Bernoulli distributions.

Theorem 5.2 *For any function $f : \{-1, 1\}^n \to \mathbb{R}$,*

$$\operatorname{Ent}(f^2) \leq c(p)\mathcal{E}(f)$$

where

$$c(p) = \frac{1}{1 - 2p} \log \frac{1 - p}{p}.$$

It is easy to see that $\lim_{p \to 1/2} c(p) = 2$, thus recovering the case of the symmetric distribution. In Chapter 9 we will see several interesting applications of this inequality. The function $c(p)$ is plotted in Fig. 5.1.

Theorem 5.2 is a special case of a more general result that we prove in Section 14.3, but the reader may attempt a direct proof (see Exercise 5.4).

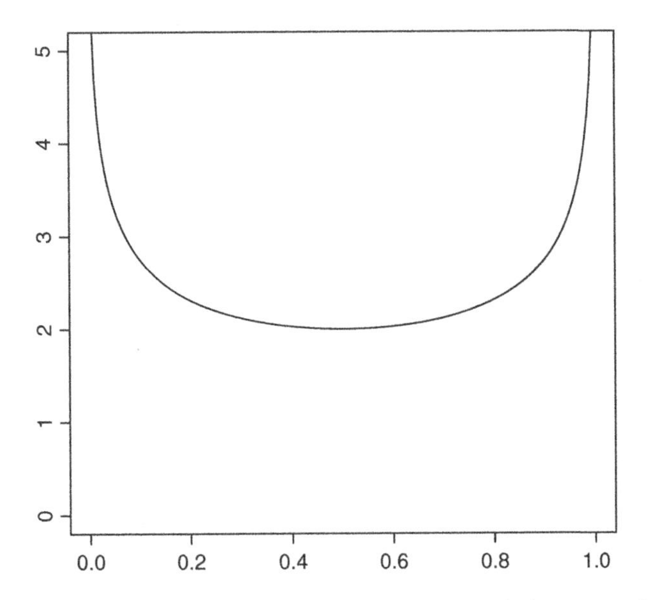

Figure 5.1 The constant $c(p)$ in the logarithmic Sobolev inequality of Theorem 5.2 for asymmetric Bernoulli distributions

5.2 Herbst's Argument: Concentration on the Hypercube

Simple as it is, the logarithmic Sobolev inequality of Theorem 5.1 has many interesting consequences. The most important from the point of view of this book is an exponential concentration inequality for functions defined on the binary hypercube. This is the first and simplest of a series of exponential inequalities which we expose. Many of them are based on generalizations and modifications of the argument presented here.

We consider an arbitrary function $f : \{-1, 1\}^n \to \mathbb{R}$ defined on the binary hypercube. Let $X = (X_1, \ldots, X_n)$ be a uniformly distributed random vector in $\{-1, 1\}^n$. We are interested in the concentration properties of the random variable $Z = f(X)$.

The following argument, attributed to Herbst, provides an exponential concentration inequality for Z. The main trick is to use the logarithmic Sobolev inequality for the nonnegative function $g(x) = e^{\lambda f(x)/2}$ where $\lambda \in \mathbb{R}$ is a parameter whose value we optimize later. Then the entropy of g^2 becomes

$$\mathrm{Ent}(g^2) = \mathrm{Ent}\left(e^{\lambda f}\right) = \lambda \boldsymbol{E}\left[Ze^{\lambda Z}\right] - \boldsymbol{E}e^{\lambda Z} \log \boldsymbol{E}e^{\lambda Z}.$$

The key observation is that if we introduce $F(\lambda) = \boldsymbol{E}e^{\lambda Z}$ for the moment generating function of Z, its derivative is $F'(\lambda) = \boldsymbol{E}\left[Ze^{\lambda Z}\right]$, and therefore the expression above may be written as

$$\mathrm{Ent}(g^2) = \lambda F'(\lambda) - F(\lambda) \log F(\lambda).$$

The idea of Herbst's argument is that by bounding $\text{Ent}(g^2)$ using the logarithmic Sobolev inequality of Theorem 5.1, a differential inequality for $F(\lambda)$ is achieved. By solving the differential inequality we obtain an upper bound for the moment-generating function which, in turn, may be converted into an exponential tail inequality by Cramér–Chernoff bounding (see Section 2.2).

By Theorem 5.1, we have

$$\begin{aligned}\text{Ent}(g^2) &\leq \frac{1}{2}\sum_{i=1}^{n} \mathbf{E}\left[\left(e^{\lambda f(X)/2} - e^{\lambda f\left(\overline{X}^{(i)}\right)/2}\right)^2\right] \\ &= \sum_{i=1}^{n} \mathbf{E}\left[\left(e^{\lambda f(X)/2} - e^{\lambda f\left(\overline{X}^{(i)}\right)/2}\right)_+^2\right]\end{aligned}$$

where we used the fact that X and $\overline{X}^{(i)}$ have the same distribution.

By convexity of the exponential function, for any real numbers $z > y$, $e^{z/2} - e^{y/2} \leq (z-y)e^{z/2}/2$, so we have

$$\begin{aligned}\text{Ent}(g^2) &\leq \frac{\lambda^2}{4}\sum_{i=1}^{n} \mathbf{E}\left[\left(f(X) - f\left(\overline{X}^{(i)}\right)\right)_+^2 e^{\lambda f(X)}\right] \\ &= \frac{\lambda^2}{4}\mathbf{E}\left[e^{\lambda f(X)} \sum_{i=1}^{n}\left(f(X) - f\left(\overline{X}^{(i)}\right)\right)_+^2\right].\end{aligned}$$

Recalling that for any $x = (x_1, \ldots, x_n) \in \{-1, 1\}^n$ we denote by $\overline{x}^{(i)}$ the vector $(x_1, \ldots, x_{i-1}, -x_i, x_{i+1}, \ldots, x_n)$, and introducing the quantity

$$v = \max_{x \in \{-1,1\}^n} \sum_{i=1}^{n} \left(f(x) - f\left(\overline{x}^{(i)}\right)\right)_+^2,$$

we obtain

$$\text{Ent}(e^{\lambda f}) \leq \frac{v\lambda^2}{4}\mathbf{E}e^{\lambda f(X)}.$$

Expressing the obtained inequality in terms of the moment-generating function F, we have

$$\lambda F'(\lambda) - F(\lambda)\log F(\lambda) \leq \frac{v\lambda^2}{4}F(\lambda).$$

This is the promised differential inequality. To solve it, divide both sides by the positive number $\lambda^2 F(\lambda)$. Defining $G(\lambda) = \log F(\lambda)$, we observe that the left-hand side is just the derivative of $G(\lambda)/\lambda$. Thus, we obtain the inequality

$$\left(\frac{G(\lambda)}{\lambda}\right)' \leq \frac{v}{4}.$$

By l'Hospital's rule we note that $\lim_{\lambda\to 0} G(\lambda)/\lambda = F'(0)/F(0) = \mathbf{E}Z$. If $\lambda > 0$, by integrating the inequality between 0 and λ, we get $G(\lambda)/\lambda \leq \mathbf{E}Z + \lambda v/4$, or in other words,

$$F(\lambda) \leq e^{\lambda \mathbf{E}Z + \lambda^2 v/4}.$$

Finally, by Markov's inequality,

$$\mathbf{P}\{Z > \mathbf{E}Z + t\} \leq \inf_{\lambda>0} F(\lambda) e^{-\lambda \mathbf{E}Z - \lambda t} \leq \inf_{\lambda>0} e^{\lambda^2 v/4 - \lambda t} = e^{-t^2/v}$$

where $\lambda = 2t/v$ minimizes the obtained upper bound. Similarly, if $\lambda < 0$, we may integrate the obtained upper bound for the derivative of $G(\lambda)/\lambda$ between $-\lambda$ and 0 to obtain

$$F(\lambda) \leq e^{\lambda \mathbf{E}Z + \lambda^2 v/4}$$

which implies the left-tail inequality

$$\mathbf{P}\{Z < \mathbf{E}Z - t\} \leq \inf_{\lambda<0} F(\lambda) e^{-\lambda \mathbf{E}Z + \lambda t} \leq \inf_{\lambda<0} e^{\lambda^2 v/4 + \lambda t} = e^{-t^2/v}.$$

The following theorem summarizes what we have just proved.

Theorem 5.3 *Let $f : \{-1,1\}^n \to \mathbb{R}$ and assume that X is uniformly distributed on $\{-1,1\}^n$. Let $v > 0$ be such that*

$$\sum_{i=1}^{n} \left(f(x) - f\left(\overline{x}^{(i)}\right)\right)_+^2 \leq v$$

for all $x \in \{-1,1\}^n$. Then the random variable $Z = f(X)$ satisfies, for all $t > 0$,

$$\mathbf{P}\{Z > \mathbf{E}Z + t\} \leq e^{-t^2/v} \quad \text{and} \quad \mathbf{P}\{Z < \mathbf{E}Z - t\} \leq e^{-t^2/v}.$$

Recall that by the Efron–Stein inequality, $\mathrm{Var}(Z) \leq v/2$. The theorem states much more: tail probabilities decrease similarly to the tail probabilities of a Gaussian random variable with variance $v/2$. The price we pay for such an improved inequality is that a pointwise control of $\sum_{i=1}^{n}(f(x) - f(\overline{x}^{(i)}))_+^2$ is required, while to bound the variance it suffices to keep its expected value under control. Recall that in Section 3.6, using the Efron–Stein estimate, we could derive the weaker exponential bound $\mathbf{P}\{Z > \mathbf{E}Z + t\} \leq 2e^{-t/\sqrt{v}}$. In Exercise 5.5 we describe a variant of Theorem 5.3 that allows the recovery of Hoeffding's inequality (with the right constant) in the special case of symmetric binomial distributions.

As we already mentioned, this is the first in a series of exponential inequalities that we prove in this book. It will be generalized and strengthened in several ways. For example, in Section 6 we show that this inequality holds for all functions of independent random variables, not only for those defined on the binary hypercube. However, the skeleton of several proofs to come is similar to the one of Theorem 5.3: a logarithmic Sobolev inequality (or one of its modifications) is used for the random variable $e^{\lambda Z}$ which leads to a differential inequality involving the moment generating function. Once the differential inequality is solved, the Cramér–Chernoff bound yields a concentration inequality. (See Exercise 5.6 for some simple extensions.)

5.3 A Gaussian Logarithmic Sobolev Inequality

In this section we use the logarithmic Sobolev inequality for the balanced Bernoulli distribution (Theorem 5.1) to derive an analog result under the canonical Gaussian distribution in $\mathbb{R}^n$. Even though the logarithmic Sobolev inequalities for the binary hypercube are interesting in their own right, at the inception of the theory, they were merely considered as intermediate results on the way to proving the Gaussian logarithmic Sobolev inequality and a series of related results.

Theorem 5.4 (GAUSSIAN LOGARITHMIC SOBOLEV INEQUALITY) *Let $X = (X_1, \ldots, X_n)$ be a vector of n independent standard normal random variables and let $f : R^n \to \mathbb{R}$ be a continuously differentiable function. Then*

$$\operatorname{Ent}\left(f^2\right) \leq 2\mathbf{E}\left[\left\|\nabla f(X)\right\|^2\right].$$

Note that the Gaussian logarithmic Sobolev inequality is an improvement on the Gaussian Poincaré inequality (see Exercise 5.2). The proof is based on Theorem 5.1 and follows the same pattern as the proof of the Gaussian Poincaré inequality in Section 3.7.

Proof We first prove the theorem for $n = 1$, that is, when $f : \mathbb{R} \to \mathbb{R}$ is a continuously differentiable function on the real line and X is a standard normal random variable. If $\mathbf{E}\left[f'(X)^2\right] = \infty$, there is nothing to prove, so assume $\mathbf{E}\left[f'(X)^2\right] < \infty$. By a standard density argument, it suffices to prove the theorem for twice differentiable functions with bounded support (Exercise 5.12).

Let $\varepsilon_1, \varepsilon_2, \ldots$ be independent Rademacher random variables. Recall from the proof of the Gaussian Poincaré inequality (Theorem 3.20) that

$$\lim_{n\to\infty} \mathbf{E}\left[\sum_{j=1}^{n}\left|f\left(\frac{1}{\sqrt{n}}\sum_{i=1}^{n}\varepsilon_i\right) - f\left(\frac{1}{\sqrt{n}}\sum_{i=1}^{n}\varepsilon_i - \frac{2\varepsilon_j}{\sqrt{n}}\right)\right|^2\right] = 4\mathbf{E}\left[f'(X)^2\right].$$

On the other hand, for any continuous uniformly bounded function f, by the central limit theorem, we have

$$\lim_{n\to\infty} \operatorname{Ent}\left[f^2\left(\frac{1}{\sqrt{n}}\sum_{i=1}^{n}\varepsilon_i\right)\right] = \operatorname{Ent}\left[f(X)^2\right].$$

The proof is then completed by invoking the logarithmic Sobolev inequality for balanced Bernoulli random variables (Theorem 5.1) which asserts that, for each n,

$$\mathrm{Ent}\left[f^2\left(\frac{1}{\sqrt{n}}\sum_{i=1}^{n}\varepsilon_i\right)\right]$$
$$\leq \frac{1}{2}\boldsymbol{E}\left[\sum_{j=1}^{n}\left|f\left(\frac{1}{\sqrt{n}}\sum_{i=1}^{n}\varepsilon_i\right)-f\left(\frac{1}{\sqrt{n}}\sum_{i=1}^{n}\varepsilon_i-\frac{2\varepsilon_j}{\sqrt{n}}\right)\right|^2\right].$$

The extension of the result to dimension $n \geq 1$ follows easily from the sub-additivity of entropy seen in Theorem 4.22 which states that

$$\mathrm{Ent}\left(f^2\right) \leq \sum_{i=1}^{n}\boldsymbol{E}\left[\boldsymbol{E}^{(i)}\left[f(X)^2\log f(X)^2\right]-\boldsymbol{E}^{(i)}\left[f(X)^2\right]\log\boldsymbol{E}^{(i)}\left[f(X)^2\right]\right]$$

where $\boldsymbol{E}^{(i)}$ denotes integration with respect to the i-th variable X_i only. The result for $n = 1$ proved above implies that

$$\boldsymbol{E}^{(i)}\left[f(X)^2\log f(X)^2\right]-\boldsymbol{E}^{(i)}\left[f(X)^2\right]\log\boldsymbol{E}^{(i)}\left[f(X)^2\right] \leq 2\boldsymbol{E}^{(i)}\left[(\partial_i f(X))^2\right].$$

Since $\|\nabla f(X)\|^2 = \sum_{i=1}^{n}(\partial_i f(X))^2$, the proof is complete. □

5.4 Gaussian Concentration: The Tsirelson–Ibragimov–Sudakov Inequality

In the same way that Theorem 5.1 led to exponential concentration inequalities for functions on the binary hypercube via Herbst's argument, if we start from the Gaussian logarithmic Sobolev inequality, the same proof leads to exponential tail inequalities for smooth functions of independent Gaussian random variables. The result is the following classical Gaussian concentration inequality.

Theorem 5.5 *Let $X = (X_1, \ldots, X_n)$ be a vector of n independent standard normal random variables. Let $f : \mathbb{R}^n \to \mathbb{R}$ denote an L-Lipschitz function, that is, there exists a constant $L > 0$ such that for all $x, y \in \mathbb{R}^n$,*

$$|f(x) - f(y)| \leq L\|x - y\|.$$

Then, for all $\lambda \in \mathbb{R}$,

$$\log \boldsymbol{E}e^{\lambda(f(X)-\boldsymbol{E}f(X))} \leq \frac{\lambda^2}{2}L^2.$$

Proof By a standard density argument we may assume that f is differentiable with gradient uniformly bounded by L. We may also assume, without loss of generality, that $\mathbf{E}f(X) = 0$. The argument is the same as that given in Section 5.2, except that Theorem 5.4 is used instead of Theorem 5.1. Using the Gaussian logarithmic Sobolev inequality for the function $e^{\lambda f/2}$, we obtain

$$\begin{aligned}\operatorname{Ent}\left(e^{\lambda f}\right) &\leq 2\mathbf{E}\left\|\nabla e^{\lambda f(X)/2}\right\|^2 \\ &= \frac{\lambda^2}{2}\mathbf{E}\left[e^{\lambda f(X)}\|\nabla f(X)\|^2\right] \\ &\leq \frac{\lambda^2 L^2}{2}\mathbf{E}e^{\lambda f(X)}.\end{aligned}$$

Writing $F(\lambda) = \mathbf{E}e^{\lambda f(X)}$, we obtain the differential inequality

$$\lambda F'(\lambda) - F(\lambda)\log F(\lambda) \leq \frac{\lambda^2 L^2}{2}F(\lambda)$$

which can be solved exactly as in Section 5.2 to obtain $\log F(\lambda) \leq \lambda^2 L^2/2$, as desired. □

The sub-Gaussian bound obtained for the moment-generating function of course implies an exponential tail inequality in the standard way by Markov's inequality. More precisely, we have derived the following.

Theorem 5.6 (GAUSSIAN CONCENTRATION INEQUALITY) *Let $X = (X_1, \ldots, X_n)$ be a vector of n independent standard normal random variables. Let $f : \mathbb{R}^n \to \mathbb{R}$ denote an L-Lipschitz function. Then, for all $t > 0$,*

$$\mathbf{P}\left\{f(X) - \mathbf{E}f(X) \geq t\right\} \leq e^{-t^2/(2L^2)}.$$

An important feature of the theorem is that the right-hand side does not depend on the dimension n. This inequality has served as a benchmark for the development of concentration inequalities during the last three decades. An important and prototypical application is described in the following example.

Example 5.7 (NORM OF A GAUSSIAN VECTOR) Let $X = (X_1, \ldots, X_n)$ be a jointly Gaussian vector with zero expectation and covariance matrix Γ. Let $p \geq 1$ and consider the real-valued random variable defined by the p-norm of X, that is,

$$Z = \|X\|_p = \left(\sum_{i=1}^{n}|X_i|^p\right)^{1/p}.$$

Since Γ is positive semidefinite, there exists an $n \times n$ matrix A satisfying $A^T A = \Gamma$. Then the Gaussian vector X is distributed as AY where $Y = (Y_1, \ldots, Y_n)$ is distributed according to the canonical Gaussian distribution, that is, the components of Y are independent standard normal random variables. Then $f(y) = \|Ay\|_p$ is a Lipschitz function from $\mathbb{R}^n$ to $\mathbb{R}$ with Lipschitz constant L equal to the operator norm of A mapping ℓ_2 to ℓ_p, that is,

$$L = \|A\|_{\ell_2 \to \ell_p} \stackrel{\text{def}}{=} \sup_{y \in \mathbb{R}^n : \|y\|_2 = 1} \|Ay\|_p .$$

Then by Theorems 3.20 and 5.6, $\mathrm{Var}(Z) \le L^2$ and for all $t > 0$, $\mathbf{P}\{|Z - \mathbf{E}Z| \ge t\} \le 2e^{-t^2/(2L^2)}$.

5.5 A Concentration Inequality for Suprema of Gaussian Processes

We illustrate the Gaussian concentration inequality of Theorem 5.6 by showing how it implies, in a simple way, a concentration inequality for the supremum of a Gaussian process. A key feature of the Gaussian concentration inequality is that the upper bound does not depend on the dimension n. This allows us to extend it easily to an infinite-dimensional setting which is described next.

Let $\mathcal{T}$ be a metric space and let $(X_t)_{t \in \mathcal{T}}$ be a Gaussian process indexed by $\mathcal{T}$. (This means simply that a random variable X_t is assigned to every $t \in \mathcal{T}$ and for any finite collection $\{t_1, \ldots, t_n\} \subset \mathcal{T}$, the vector $(X_{t_1} \ldots X_{t_n})$ has a jointly Gaussian distribution with mean zero.) In addition, we assume that $\mathcal{T}$ is totally bounded (i.e. for every $t > 0$ it can be covered by finitely many balls of radius t) and that the Gaussian process is almost surely continuous, that is, with probability 1, X_t is a continuous function of t.

Theorem 5.8 *Let $(X_t)_{t \in \mathcal{T}}$ be an almost surely continuous centered Gaussian process indexed by a totally bounded set $\mathcal{T}$. If*

$$\sigma^2 = \sup_{t \in \mathcal{T}} \mathbf{E}\left[X_t^2\right],$$

then $Z = \sup_{t \in \mathcal{T}} X_t$ satisfies $\mathrm{Var}(Z) \le \sigma^2$, and for all $u > 0$,

$$\mathbf{P}\{Z - \mathbf{E}Z \ge u\} \le e^{-u^2/(2\sigma^2)}$$

and

$$\mathbf{P}\{\mathbf{E}Z - Z \ge u\} \le e^{-u^2/(2\sigma^2)}.$$

Proof We assume that $\mathcal{T}$ is a finite set. The extension to arbitrary totally bounded $\mathcal{T}$ is based on a separability argument and monotone convergence, whose details are left to the reader (see Exercise 5.14). We may assume, for simplicity, that $\mathcal{T} = \{1, \ldots, n\}$. Let Γ be the covariance matrix of the centered Gaussian vector $X = (X_1, \ldots, X_n)$. Denote by A the square root of the positive semidefinite matrix Γ. If $Y = (Y_1, \ldots, Y_n)$ is a vector of independent standard normal random variables, then

$$f(Y) = \max_{i=1,\ldots,n} (AY)_i$$

has the same distribution as $\max_{i=1,\ldots,n} X_i$. Hence, we can apply the Gaussian concentration inequality by bounding the Lipschitz constant of f. By the Cauchy–Schwarz inequality, for all $u, v \in \mathbb{R}^n$ and $i = 1, \ldots, n$,

$$|(Au)_i - (Av)_i| = \left| \sum_j A_{i,j} (u_j - v_j) \right| \leq \left(\sum_j A_{i,j}^2 \right)^{1/2} \|u - v\| .$$

Since $\sum_j A_{i,j}^2 = \mathrm{Var}(X_i)$, we get

$$|f(u) - f(v)| \leq \max_{i=1,\ldots,n} |(Au)_i - (Av)_i| \leq \sigma \|u - v\| .$$

Therefore, f is Lipschitz with constant σ and the tail bounds follow from the Gaussian concentration inequality. The variance bound follows from the Gaussian Poincaré inequality. □

Exercise 5.37 describes an example where Theorem 5.8 is not tight.

5.6 Gaussian Random Projections

In this section we return to the Johnson–Lindenstrauss problem studied in Section 2.9. Recall that we showed that if A is a finite subset of $\mathbb{R}^D$ with cardinality n, and we defined the random projection $W : \mathbb{R}^D \to \mathbb{R}^d$ by assigning, to each $\alpha = (\alpha_1, \ldots, \alpha_D) \in \mathbb{R}^D$, the vector

$$W(\alpha) = \left(\frac{1}{\sqrt{d}} W_i(\alpha) \right)_{i=1}^{d}$$

with

$$W_i(\alpha) = \sum_{j=1}^{D} \alpha_j X_{i,j}$$

where the $X_{i,j}$ are sub-Gaussian random variables with zero mean and unit variance, then, with high probability, W is an ε-isometry on A provided that $d \geq \kappa \varepsilon^{-2} \log n$ for an absolute constant κ.

The purpose of the present section is to generalize this result to the case when A is not necessarily a finite set. We concentrate on the case where the $X_{i,j}$ are i.i.d. standard normal random variables. This allows us to use the Gaussian logarithmic Sobolev inequality which, together with a modification of Herbst's argument (see Section 5.2), serves as our main tool to improve on the crude bounds established in Section 2.9.

Our goal is to introduce a sharper measure for the complexity of the set A than its cardinality. This is interesting even if A is a finite set and allows us to extend the Johnson–Lindenstrauss lemma to possibly infinite sets A. As in Section 2.9, the results may be generalized in a straightforward way to the case when A is a subset of a general Hilbert space, but to avoid technicalities, we assume $A \subset \mathbb{R}^D$ for some finite D. As in Section 2.9, we set

$$T = \left\{ \frac{a - a'}{\|a - a'\|}, (a, a') \in A \times A \text{ with } a \neq a' \right\}.$$

Recall that since W is linear, for every $\alpha \in T$, $\mathbf{E}\|W(\alpha)\|^2 = 1$ and W is an ε-isometry on A if and only if

$$\sup_{\alpha \in T} \left| \|W(\alpha)\|^2 - 1 \right| \leq \varepsilon.$$

A way to guarantee that this happens with large probability is to show that the expected value of the random variable $\sup_{\alpha \in T} \left| \|W(\alpha)\|^2 - 1 \right|$ is significantly less than ε and that this random variable is highly concentrated around its mean. In this section we address the issue of concentration. In Section 13.6 we return to the problem and show techniques to bound the expected value of the supremum above in terms of the "size" of the set T, measured by the so-called metric entropy.

When one considers a supremum of possibly uncountably many random variables, some care should be taken to ensure that the supremum is measurable. Luckily, measurability is guaranteed here since W is continuous on the totally bounded set T. Hence, there exists an at most countable subset $T' \subset T$ such that $\sup_{\alpha \in T} \left| \|W(\alpha)\|^2 - 1 \right| = \sup_{\alpha \in T'} \left| \|W(\alpha)\|^2 - 1 \right|$ which is measurable. The following concentration inequality is the key to the main result of this section.

Theorem 5.9 *Define Z by either*

$$Z = d \sup_{\alpha \in T} \|W(\alpha)\|^2 \quad \text{or} \quad Z = d \inf_{\alpha \in T} \|W(\alpha)\|^2.$$

Then, for all $t > 0$,

$$\mathbf{P}\left\{ Z - \mathbf{E}Z \geq 2\sqrt{2t\mathbf{E}Z} + 2t \right\} \leq e^{-t}$$

and for all $t > 1/2$,

$$\mathbf{P}\left\{ Z - \mathbf{E}Z \leq -2\sqrt{2t\mathbf{E}Z} \right\} \leq e^{-t}.$$

Proof First observe that it suffices to prove the statement when T is a finite set. This is because, as observed above, without loss of generality, we may assume that T is a countable set. However, the supremum may be written as the limit of a sequence of suprema taken over finite subsets. Once the inequalities stated in the theorem are proved for finite sets T, the monotone convergence theorem implies that they also hold for countable sets T. So assume that T is a finite set.

The proof is based on the Gaussian concentration inequality (Theorem 5.6). In order to apply it, we write Z as a function of the vector of $d \times D$ independent standard normal random variables $X = (X_{i,j})_{i=1,\dots,d,j=1,\dots,D}$. To this end, write $x = (x_{i,j})_{i=1,\dots,d,j=1,\dots,D} \in \mathbb{R}^{dD}$ and define the function $f : \mathbb{R}^{dD} \to \mathbb{R}$ either by

$$f(x) = \sup_{\alpha \in T} \sum_{i=1}^{d} \left(\sum_{j=1}^{D} \alpha_j x_{i,j} \right)^2$$

or by

$$f(x) = \inf_{\alpha \in T} \sum_{i=1}^{d} \left(\sum_{j=1}^{D} \alpha_j x_{i,j} \right)^2 .$$

Thus, $Z = f(X)$. The crucial property is that $\sqrt{f}$ is 1-Lipschitz. Hence, by the Gaussian concentration inequality,

$$\begin{aligned} \mathbf{P}\left\{Z \geq \mathbf{E}Z + 2\sqrt{2t\mathbf{E}Z} + 2t\right\} &\leq \mathbf{P}\left\{Z \geq \left(\mathbf{E}\sqrt{Z}\right)^2 + 2\sqrt{2t}\mathbf{E}\sqrt{Z} + 2t\right\} \\ &\leq \mathbf{P}\left\{\sqrt{Z} \geq \mathbf{E}\sqrt{Z} + \sqrt{2t}\right\} \\ &\leq e^{-t}. \end{aligned}$$

Meanwhile, by the Gaussian Poincaré inequality, we have

$$\mathrm{Var}\left(\sqrt{Z}\right) = \mathbf{E}Z - \left(\mathbf{E}\sqrt{Z}\right)^2 \leq 1,$$

and thus, invoking again the Gaussian concentration inequality, for $t > 1/2$,

$$\begin{aligned} \mathbf{P}\left\{Z - \mathbf{E}Z \leq -2\sqrt{2t\mathbf{E}Z}\right\} &\leq \mathbf{P}\left\{Z \leq \mathbf{E}Z - 1 - 2\sqrt{2t\mathbf{E}Z} + 2t\right\} \\ &\leq \mathbf{P}\left\{Z \leq \left(\mathbf{E}\sqrt{Z}\right)^2 - 2\sqrt{2t}\mathbf{E}\sqrt{Z} + 2t\right\} \\ &\leq \mathbf{P}\left\{\sqrt{Z} \leq \mathbf{E}\sqrt{Z} - \sqrt{2t}\right\} \\ &\leq e^{-t}. \end{aligned}$$

□

Now we are ready to use Theorem 5.9 to derive a generalized version of the Johnson–Lindenstrauss lemma. Introducing the random variables

$$Z = d \sup_{\alpha \in T} \|W(\alpha)\|^2 \quad \text{and} \quad Z' = d \inf_{\alpha \in T} \|W(\alpha)\|^2,$$

we have

$$V \stackrel{\text{def}}{=} \sup_{\alpha \in T} \left(\|W(\alpha)\|^2 - 1 \right) = \frac{Z}{d} - 1$$

and

$$V' \stackrel{\text{def}}{=} \sup_{\alpha \in T} \left(- \|W(\alpha)\|^2 + 1 \right) = -\frac{Z'}{d} + 1.$$

Now, for any $t > 1/2$, with a double application of Theorem 5.9, we obtain, with probability at least $1 - 2e^{-t}$,

$$\sup_{\alpha \in T} \left| \|W(\alpha)\|^2 - 1 \right| = \max(V, V') \leq \max(\boldsymbol{E}V, \boldsymbol{E}V') + 2\sqrt{\frac{2(1 + \boldsymbol{E}V)t}{d}} + \frac{2t}{d}.$$

The quantity $\Delta = d \max\left(\boldsymbol{E}V, \boldsymbol{E}V'\right)^2$ may be regarded as a measure of "complexity" of the set T (or of the set A). Using this notation, the previous inequality implies

$$\sup_{\alpha \in T} \left| \|W(\alpha)\|^2 - 1 \right| \leq 2\sqrt{\frac{\Delta}{d}} + 2\sqrt{\frac{2t}{d}} + \frac{4t}{d},$$

which holds with probability at least $1 - 2e^{-t}$. As a consequence of this and some straightforward computation we derive the following structural result which provides a fairly general answer to the Johnson–Lindenstrauss problem in the Gaussian case.

Theorem 5.10 *Consider the random projection $W : \mathbb{R}^D \to \mathbb{R}^d$ based on i.i.d. standard normal variables $X_{i,j}$, $i = 1, \ldots, d$, $j = 1, \ldots, D$ and let $A \subset \mathbb{R}^d$. If*

$$T = \left\{ \frac{a - a'}{\|a - a'\|}, (a, a') \in A \times A \text{ with } a \neq a' \right\}$$

and

$$\Delta = d \max\left(\boldsymbol{E} \sup_{\alpha \in T} \left(\|W(\alpha)\|^2 - 1 \right), \boldsymbol{E} \sup_{\alpha \in T} \left(- \|W(\alpha)\|^2 + 1 \right) \right)^2,$$

then there exists an absolute constant κ ($\kappa = 20$ works) such that, for every $\varepsilon, \delta \in (0, 1)$, if $d \geq \kappa \left(\Delta + \log(2/\delta)\right) \varepsilon^{-2}$, W is an ε-isometry on A, with probability larger than $1 - \delta$.

The main message of the theorem is that as long as d is larger than $20(\Delta + 1)/\varepsilon^2$, with positive probability, W is an ε-isometry on A and therefore there exists a linear embedding of A in $\mathbb{R}^d$ that is an ε-isometry. The key quantity here is Δ which, in a sense, measures the richness of the set A. One may bound Δ in terms of metric entropies of the set T. We return to this problem in Section 13.6. Here we merely point out that if A is a finite set, one may recover Theorem 2.13. Indeed, since each variable $d \sup_{\alpha \in T} \|W(\alpha)\|^2$ follows a chi-square distribution with d degrees of freedom, the inequality obtained in Example 2.7 in Chapter 2 implies that

$$\Delta \leq 4 \log N \left(1 + \sqrt{\frac{\log N}{d}}\right)^2,$$

where $N \leq \binom{n}{2}$ is the cardinality of the set T. We may assume that $\kappa \geq 20$ (otherwise we change κ to $\max(\kappa, 20)$ and Theorem 5.10 still holds). Assuming that $d \geq 10\kappa\varepsilon^{-2} \log(n/\sqrt{\delta}) \geq 100 \log N$, we derive that $\Delta \leq 4(1.1)^2 \log N \leq 10 \log n - 4 \log 2$ and therefore the condition $d \geq \kappa(\Delta + \log(2/\delta))\varepsilon^{-2}$ is satisfied whenever $d \geq 10\kappa\varepsilon^{-2} \log\left(n/\sqrt{\delta}\right)$. This means that the conclusion of Theorem 5.10 holds provided that $d \geq 10\kappa\varepsilon^{-2} \log\left(n/\sqrt{\delta}\right)$. In other words, we recover Theorem 2.13 up to the absolute constant involved in the constraint on the dimension d.

5.7 A Performance Bound for the Lasso

Numerous applications for concentration inequalities have been found in mathematical statistics and statistical learning theory. In this section we describe an application of the Gaussian concentration inequality to a general model selection problem and show how it can be used in the analysis of one of the popular methods of regression function estimation, the so-called LASSO.

First, we describe the generalized linear Gaussian model we work with. To this end, we need the notion of an *isonormal* Gaussian process. Let $\mathbb{H}$ be a separable Hilbert space and let $(W(t))_{t \in \mathbb{H}}$ be a centered Gaussian process on $\mathbb{H}$. The process is called isonormal if its covariance is given by $\boldsymbol{E}\left[W(t)W(u)\right] = \langle t, u \rangle$ for all $t, u \in \mathbb{H}$ where $\langle t, u \rangle$ denotes the inner product of t and u. In our generalized linear Gaussian model, one observes, for all $t \in \mathbb{H}$,

$$Y(t) = \langle s, t \rangle + \varepsilon W(t), \tag{5.2}$$

where $\varepsilon > 0$ is a fixed parameter and W is an isonormal process.

The statistical problem we consider in this section is as follows: upon observing the process $Y(t), t \in \mathbb{H}$, find, or at least approximate, the element $s \in \mathbb{H}$ generating $Y(t)$ according to (5.2).

This framework is convenient to cover both finite-dimensional linear models and the infinite-dimensional white noise model as described in the following examples.

Example 5.11 (CLASSICAL LINEAR GAUSSIAN REGRESSION MODEL) In the classical Gaussian linear regression model, a random vector $Y = (Y_1, \ldots, Y_n)$ is observed, given by

$$Y_j = s_j + \sigma X_j.$$

$X = (X, \ldots, X_n)$ is a vector of independent standard normal random variables, $\sigma > 0$, and $s = (s_1, \ldots, s_n) \in \mathbb{R}^d$ is a fixed unknown vector. Setting

$$W(t) = \sqrt{n} \langle X, t \rangle$$

with the scalar product $\langle u, v \rangle = (1/n) \sum_{j=1}^{n} u_j v_j$, we see that W is an isonormal process on $\mathbb{R}^n$ and that $Y(t) = \langle Y, t \rangle$ satisfies (5.2) with $\varepsilon = \sigma / \sqrt{n}$.

Example 5.12 (WHITE NOISE MODEL) In this case a realization of the stochastic process $\zeta(x)$ for $x \in [0, 1]$ is observed, given by the stochastic differential equation

$$d\zeta(x) = s(x)dx + \varepsilon dB(x) \text{ with } \zeta(0) = 0$$

where B is a standard Brownian motion, s is a square-integrable function, and $\varepsilon > 0$. If we define $W(t) = \int_0^1 t(x)dB(x)$ for every square-integrable function $t \in \mathbb{L}_2([0, 1])$, then W is indeed an isonormal process on $\mathbb{H} = \mathbb{L}_2([0, 1])$ and $Y(t) = \int_0^1 t(x)d\zeta(x)$ satisfies the definition (5.2), provided that $\mathbb{H}$ is equipped with its usual scalar product $\langle s, t \rangle = \int_0^1 s(x)t(x)dx$. Typically, s is a signal and $d\zeta(x)$ represents the noisy signal received at time x. This framework easily extends to a d-dimensional setting if one considers a multivariate Brownian sheet B on $[0, 1]^d$ and takes $\mathbb{H} = \mathbb{L}_2([0, 1]^d)$.

Example 5.13 (FIXED DESIGN GAUSSIAN REGRESSION) The model of fixed-design Gaussian regression is a special case of the classical Gaussian linear model for which $s_j = s(j/n)$, $j = 1, \ldots, n$, where $s : [0, 1] \to \mathbb{R}$ is a fixed unknown function. The observed values Y_j represent the "noisy" version of the "signal" s observed at "time" j/n. It may be considered as a discretized version of the white noise model. Indeed, if one observes $\zeta(x)$, $x \in [0, 1]$ such that

$$d\zeta(x) = s(x)dx + \varepsilon dB(x)$$

only at the points j/n for $j = 1 \ldots, n$, then with $\sigma = \varepsilon\sqrt{n}$ and

$$X_j = \sqrt{n}\left(B(j/n) - B((j-1)/n)\right) \text{ for all } j \in [1, n],$$

the noisy signal at time j/n is

$$Y_j = n\left(\zeta(j/n) - \zeta((j-1)/n)\right) = n \int_{(j-1)/n}^{j/n} s(x)dx + \sigma X_j.$$

Since $X_1, \ldots, X_n$ are independent standard normal, we indeed obtain the fixed design Gaussian regression model with $s_j = s^{(n)}(j/n)$ where $s^{(n)}(x) = n \int_{(j-1)/n}^{j/n} s(y)\, dy$ whenever $x \in [(j-1)/n, j/n)$. $s^{(n)}$ is a piecewise constant approximation of s.

Next we describe a general way of addressing the statistical problem.

A *model* is a closed and convex set $S \subset \mathbb{H}$. If one wants to approximate $s \in \mathbb{H}$ by an element of the model S, it makes sense to choose the best approximating point of s in S by minimizing $\|t - s\|^2$ or, equivalently, $-2\langle s, t\rangle + \|t\|^2$ over $t \in S$. However, s is unknown, so it may be necessary to choose its "noisy" analog, the *least squares estimator* defined as a minimizer of *the least squares criterion* $\gamma(t) = -2Y(t) + \|t\|^2$ with respect to $t \in S$.

Such a minimizer may not exist so it may be necessary to resort to approximate minimization (as in Theorem 5.14 below). For now assume for simplicity that a least squares estimator exists and denote it by $\hat{s}$. The quality of the estimate $\hat{s}$ (and the model S) is measured by the *quadratic risk* $\mathbf{E}\left[\|\hat{s} - s\|^2\right]$.

The problem of *model selection* is to select a model from a collection such that the least squares estimator has a quadratic risk as small as possible. To describe the problem in mathematical terms, consider a finite or countable family of models $\{S_m : m \in \mathcal{M}\}$ where each S_m is a closed and convex subset of $\mathbb{H}$. For each $m \in \mathcal{M}$, we denote by $\hat{s}_m \in S_m$ the least squares estimator corresponding to model S_m. A model selection procedure uses the data to select a value $\hat{m} \in \mathcal{M}$ and chooses $\hat{s}_{\hat{m}}$ as the final estimator. Ideally, the risk of the resulting estimator $\mathbf{E}\left[\|\hat{s}_{\hat{m}} - s\|^2\right]$ is as close as possible to the minimal risk $\inf_{m\in\mathcal{M}} \mathbf{E}\left[\|\hat{s}_m - s\|^2\right]$.

A widely used principle for model selection is *penalized risk minimization*. In the context of this section, it may be defined as follows. Supposing that a nonnegative number $\text{pen}(m)$ is assigned to each model $m \in \mathcal{M}$, these are the so-called *penalties*. Then one selects $\hat{m} \in \mathcal{M}$ minimizing

$$\gamma(\hat{s}_m) + \text{pen}(m)$$

over $m \in \mathcal{M}$.

It is outside of the scope of this book to discuss how such a penalty function should be chosen. We merely present the following general bound which suggests some guidelines.

Theorem 5.14 *Let $\{S_m\}_{m\in\mathcal{M}}$ be a countable collection of convex and compact subsets of $\mathbb{H}$. Assume the existence of an almost surely continuous version of W on each set S_m. Define, for any $m \in \mathcal{M}$,*

$$\Delta_m = \mathbf{E} \sup_{t\in S_m} W(t)$$

and consider weights $x_m > 0$, $m \in \mathcal{M}$ such that

$$\sum_{m\in\mathcal{M}} e^{-x_m} \stackrel{\text{def}}{=} \Sigma < \infty.$$

Let $K > 1$ and assume that for any $m \in \mathcal{M}$,

$$\text{pen}(m) \geq 2K\varepsilon\left(\Delta_m + \varepsilon x_m + \sqrt{\Delta_m \varepsilon x_m}\right).$$

Given nonnegative numbers ρ_m, $m \in \mathcal{M}$, define a penalized approximate least squares estimator as any $\tilde{s} \in \cup_{m\in\mathcal{M}} S_m$ such that

$$\gamma(\tilde{s}) + \text{pen}(\hat{m}) \leq \inf_{m \in \mathcal{M}} \left(\inf_{t \in S_m} \gamma(t) + \text{pen}(m) + \rho_m \right)$$

where $\hat{m} = \arg\min_{m \in \mathcal{M}: \tilde{s} \in S_m} \text{pen}(m)$. *Then there is a constant* $C = C(K)$ *such that for all* $s \in \mathbb{H}$,

$$\boldsymbol{E}\left[\|\tilde{s} - s\|^2\right] \leq C\left[\inf_{m \in \mathcal{M}} \left(\inf_{t \in S_m} \|s - t\|^2 + \text{pen}(m) + \rho_m \right) + \varepsilon^2 (\Sigma + 1)\right].$$

Proof For each $m \in \mathcal{M}$, let s_m be the projection of s onto S_m, that is, the unique element of S_m such that $\|s - s_m\| = \inf_{t \in S_m} \|s - t\|$. Then, by the definition of $\tilde{s}$, for all $m \in \mathcal{M}$,

$$\gamma(\tilde{s}) + \text{pen}(\hat{m}) \leq \gamma(s_m) + \text{pen}(m) + \rho_m.$$

Since $\|s\|^2 + \gamma(t) = \|t - s\|^2 - 2\varepsilon W(t)$, this implies that

$$\|\tilde{s} - s\|^2 \leq \|s - s_m\|^2 + 2\varepsilon\left[W(\tilde{s}) - W(s_m)\right] - \text{pen}(\hat{m}) + \text{pen}(m) + \rho_m.$$

For all $m' \in \mathcal{M}$, let $y_{m'}$ be a positive number whose value will be specified below and define, for every $t \in S_{m'}$,

$$2w_{m'}(t) = \left[\|s - s_m\| + \|s - t\|\right]^2 + y_{m'}^2.$$

Finally, define the supremum of the weighted empirical process

$$V_{m'} = \sup_{t \in S_{m'}} \left[\frac{W(t) - W(s_m)}{w_{m'}(t)}\right].$$

Taking these definitions into account, the previous inequality implies

$$\|\tilde{s} - s\|^2 \leq \|s - s_m\|^2 + 2\varepsilon w_{\hat{m}}(\tilde{s}) V_{\hat{m}} - \text{pen}(\hat{m}) + \text{pen}(m) + \rho_m. \tag{5.3}$$

The proof mostly consists of controlling the fluctuations of the random variables $V_{m'}$. To this end, we may use the concentration inequality for suprema of Gaussian processes (Theorem 5.8) which ensures that, given $z > 0$, for all $m' \in \mathcal{M}$,

$$\boldsymbol{P}\left\{V_{m'} \geq \boldsymbol{E}V_{m'} + \sqrt{2v_{m'}(x_{m'} + z)}\right\} \leq e^{-x_{m'}} e^{-z} \tag{5.4}$$

where

$$v_{m'} = \sup_{t \in S_{m'}} \text{Var}\left(\frac{W(t) - W(s_m)}{w_{m'}(t)}\right) = \sup_{t \in S_{m'}} \frac{\|t - s_m\|^2}{w_{m'}^2(t)}.$$

Since $w_{m'}(t) \geq (\|s - s_m\| + \|s - t\|)\, y_{m'} \geq \|t - s_m\|\, y_{m'}$, we have $v_{m'} \leq y_{m'}^{-2}$. Therefore, summing the inequalities (5.4) over $m' \in \mathcal{M}$, we obtain for every $z > 0$, an event Ω_z with $\boldsymbol{P}\{\Omega_z\} > 1 - \Sigma e^{-z}$, such that on Ω_z, for all $m' \in \mathcal{M}$,

$$V_{m'} \leq \boldsymbol{E}V_{m'} + y_{m'}^{-1}\sqrt{2\left(x_{m'} + z\right)}. \tag{5.5}$$

Next we bound $\boldsymbol{E}V_{m'}$. We may write

$$\boldsymbol{E}V_{m'} \leq \boldsymbol{E}\left[\frac{\sup_{t\in S_{m'}}\left(W(t) - W\left(s_{m'}\right)\right)}{\inf_{t\in S_{m'}} w_{m'}(t)}\right] + \boldsymbol{E}\left[\frac{\left(W\left(s_{m'}\right) - W\left(s_m\right)\right)_+}{\inf_{t\in S_{m'}} w_{m'}(t)}\right]. \tag{5.6}$$

Since $2w_{m'}(t) \geq \left(\|s_{m'} - s\| + \|s_m - s\|\right)^2 + y_{m'}^2 \geq \|s_{m'} - s_m\|^2 + y_{m'}^2$ for all $t \in S_{m'}$, we have $2\inf_{t\in S_{m'}}\left[w_{m'}(t)\right] \geq \left(y_{m'}^2 \vee 2y_{m'}\|s_{m'} - s_m\|\right)$. Hence, on the one hand, by the definition of Δ_m,

$$\boldsymbol{E}\left[\frac{\sup_{t\in S_{m'}}\left(W(t) - W\left(s_{m'}\right)\right)}{\inf_{t\in S_{m'}} w_{m'}(t)}\right] \leq 2y_{m'}^{-2}\boldsymbol{E}\left[\sup_{t\in S_{m'}}\left(W(t) - W\left(s_{m'}\right)\right)\right] = 2\Delta_{m'}y_{m'}^{-2},$$

and on the other hand,

$$\boldsymbol{E}\left[\frac{\left(W\left(s_{m'}\right) - W\left(s_m\right)\right)_+}{\inf_{t\in S_{m'}} w_{m'}(t)}\right] \leq y_{m'}^{-1}\boldsymbol{E}\left[\frac{\left(W\left(s_{m'}\right) - W\left(s_m\right)\right)_+}{\|s_m - s_{m'}\|}\right].$$

Now, since $\left[W\left(s_{m'}\right) - W\left(s_m\right)\right] / \|s_m - s_{m'}\|$ is a standard Gaussian random variable,

$$\boldsymbol{E}\left[\frac{\left(W\left(s_{m'}\right) - W\left(s_m\right)\right)_+}{\inf_{t\in S_{m'}} w_{m'}(t)}\right] \leq y_{m'}^{-1}\left(2\pi\right)^{-1/2}$$

and collecting these inequalities, we obtain from (5.6), for all $m' \in \mathcal{M}$,

$$\boldsymbol{E}V_{m'} \leq 2\Delta_{m'}y_{m'}^{-2} + \left(2\pi\right)^{-1/2}y_{m'}^{-1}.$$

Hence, setting $\delta = \left(\left(4\pi\right)^{-1/2} + \sqrt{z}\right)^2$, (5.5) implies that on the event Ω_z, for all $m' \in \mathcal{M}$,

$$V_{m'} \leq y_{m'}^{-1}\left[2\Delta_{m'}y_{m'}^{-1} + \sqrt{2x_{m'}} + \left(2\pi\right)^{-1/2} + \sqrt{2z}\right]$$

or equivalently, for all $m' \in \mathcal{M}$,

$$V_{m'} \leq y_{m'}^{-1}\left[2\Delta_{m'}y_{m'}^{-1} + \sqrt{2x_{m'}} + \sqrt{2\delta}\right].$$

Defining

$$y_{m'}^2 = 2K\varepsilon^2\left[\left(\sqrt{x_{m'}} + \sqrt{\delta}\right)^2 + \varepsilon^{-1}K^{-1/2}\Delta_{m'} + \sqrt{\delta\varepsilon^{-1}K^{-1/2}\Delta_{m'}}\right],$$

the previous bound implies that on the event Ω_z, $\varepsilon V_{m'} \leq K^{-1/2}$ for all $m' \in \mathcal{M}$, which, in particular, implies that $\varepsilon V_{\hat{m}} \leq K^{-1/2}$ and therefore, by (5.3),

$$\|\tilde{s} - s\|^2 \leq \|s - s_m\|^2 + 2K^{-1/2} w_{\hat{m}}(\tilde{s}) - \text{pen}(\hat{m}) + \text{pen}(m) + \rho_m,$$

or equivalently,

$$\|\tilde{s} - s\|^2 \leq \|s - s_m\|^2 + K^{-1/2} \left[[\|s - s_m\| + \|s - \tilde{s}\|]^2 + y_{\hat{m}}^2 \right] - \text{pen}(\hat{m}) + \text{pen}(m) + \rho_m.$$

Using repeatedly the elementary inequality

$$2ab \leq \theta a^2 + \theta^{-1} b^2$$

for various values of $\theta > 0$, we derive that on the one hand,

$$K^{-1/2} y_{\hat{m}}^2 \leq 2K\varepsilon^2 \left[\varepsilon^{-1} \Delta_{\hat{m}} + x_{\hat{m}} + \sqrt{\varepsilon^{-1} \Delta_{\hat{m}} x_{\hat{m}}} + \frac{2}{\sqrt{K} - 1} \left(\frac{1}{2\pi} + 2z \right) \right],$$

and on the other hand,

$$[\|s - s_m\| + \|s - \tilde{s}\|]^2 \leq K^{1/4} \left(\|s - \tilde{s}\|^2 + \frac{\|s - s_m\|^2}{K^{1/4} - 1} \right).$$

Hence, setting $A' = \left(1 + K^{-1/4} \left(K^{1/4} - 1 \right)^{-1} \right)$, on the event Ω_z,

$$\begin{aligned} \|\tilde{s} - s\|^2 &\leq A' \|s - s_m\|^2 + K^{-1/4} \|s - \tilde{s}\|^2 \\ &+ 2K\varepsilon \left[\Delta_{\hat{m}} + \varepsilon x_{\hat{m}} + \sqrt{\varepsilon \Delta_{\hat{m}} x_{\hat{m}}} \right] - \text{pen}(\hat{m}) \\ &+ \text{pen}(m) + \rho_m + \frac{4K\varepsilon^2}{\sqrt{K} - 1} \left(\frac{1}{2\pi} + 2z \right). \end{aligned}$$

This, by condition on the penalty function, implies

$$\begin{aligned} \left(\frac{K^{1/4} - 1}{K^{1/4}} \right) \|\tilde{s} - s\|^2 &\leq A' \|s - s_m\|^2 + \text{pen}(m) + \rho_m \\ &+ \frac{2K\varepsilon^2}{\sqrt{K} - 1} \left(\frac{1}{2\pi} + 2z \right). \end{aligned}$$

Integrating this inequality with respect to z leads to the announced risk bound. □

In the rest of the section we apply the model selection theorem to the analysis of LASSO, a popular algorithm for regression function estimation.

Let $\Lambda = \{\varphi_1, \ldots, \varphi_N\} \subset \mathbb{H}$ be a finite set of (not necessarily linearly independent) vectors in $\mathbb{H}$. We seek estimates of s in the form of a linear combination of the vectors in Λ, often called the *dictionary*.

We may assume, without loss of generality, that $\|\varphi_i\| = 1$, for every $i = 1, \ldots, N$ (otherwise one may simply replace φ_i by $\varphi_i / \|\varphi_i\|$). Denote by $\mathcal{L}_1(\Lambda)$ the linear span of Λ equipped with the ℓ_1 norm

$$\|t\|_1 = \inf\left\{\sum_{i=1}^{N} |\beta_i| : \beta \in \mathbb{R}^N \text{ such that } \sum_{i=1}^{N} \beta_i \varphi_i = t\right\}.$$

Given a parameter $r > 0$ (called the regularization parameter), the LASSO estimator $\tilde{s}$ of s is defined as a minimizer of

$$\gamma(t) + r\|t\|_1$$

over all $t \in \mathcal{L}_1(\Lambda)$. Thus, the LASSO estimator is an ℓ_1-penalized least squares estimator. Here, we prove the following performance bound.

Theorem 5.15 *Consider the isonormal model introduced in (5.2). Let $\tilde{s}$ be a minimizer of $\gamma(t) + r\|t\|_1$ over $t \in \mathcal{L}_1(\Lambda)$. Assume that $r \geq 4\varepsilon\left(1 + \sqrt{\log N}\right)$. Then there exists an absolute constant $C \geq 1$ such that*

$$\mathbf{E}\left[\|\tilde{s} - s\|^2\right] \leq C\left[\inf_{t \in \mathcal{L}_1(\Lambda)} \left(\|s - t\|^2 + r\|t\|_1\right) + r\varepsilon\right].$$

The theorem states that, up to a constant factor, the "noisy" LASSO behaves as well as the deterministic LASSO. The discussion of the approximation-theoretic implications of this result goes beyond the scope of this book. The interested reader may find pointers in the section on bibliographical remarks below.

The proof is based on an application of Theorem 5.14. The basic idea is that LASSO can be considered as a penalized approximate least squares estimator over a properly defined sequence of models. The key observation that allows one to make this connection is the simple fact that the LASSO estimator $\tilde{s}$ satisfies

$$\gamma(\tilde{s}) + r\|\tilde{s}\|_1 = \inf_{R \geq 0} \inf_{\|t\|_1 \leq R} (\gamma(t) + rR).$$

To obtain a countable collection of models, we "discretize" the family of ℓ_1 balls by defining, for all $m = 1, 2, \ldots$, $S_m = \left\{t \in \mathcal{L}_1(\Lambda), \|t\|_1 \leq m\varepsilon\right\}$. We may define $\hat{m}$ as the smallest integer such that $\tilde{s} \in S_{\hat{m}}$ and notice that

$$\gamma(\tilde{s}) + r\hat{m}\varepsilon \leq \inf_{m \geq 1} \inf_{t \in S_m} (\gamma(t) + rm\varepsilon) + r\varepsilon.$$

This means that $\tilde{s}$ is equivalent to an approximate penalized least squares estimator over the sequence of models given by the collection of ℓ_1 balls $\{S_m, m \geq 1\}$.

Deriving Theorem 5.15 from Theorem 5.14 is now an exercise.

Proof of Theorem 5.15. Consider the ℓ_1 balls $S_m = \{t \in \mathcal{L}_1(\Lambda) : \|t\|_1 \leq m\varepsilon\}$ for $m = 1, 2, \ldots$, and choose the weights of the form $x_m = \theta m$, where $\theta > 0$ is a numerical constant specified later. Then $\sum_{m \geq 1} e^{-x_m} = \Sigma_\theta = e^\theta / \left(e^\theta - 1\right)$ and

$$\sup_{t \in S_m} W(t) \leq m\varepsilon \max_{i=1,\ldots,N} \left| W(\varphi_i) \right|.$$

Since the variables $W(\varphi_i)$, $i = 1, \ldots, N$ are standard normal, $\mathbf{E} \sup_{i=1,\ldots,N} \left| W(\varphi_i) \right| \leq \sqrt{2 \log(2N)}$ and therefore

$$\Delta_m = \mathbf{E} \sup_{t \in S_m} W(t) \leq m\varepsilon \sqrt{2 \log(2N)} \leq m\varepsilon \left(\sqrt{2 \log N} + \sqrt{2 \log 2} \right).$$

We may apply now Theorem 5.14 with $K = 4\sqrt{2}/5 > 1$, $\rho_m = r\varepsilon$, and $\text{pen}(m) = rm\varepsilon$. Defining $\theta = \left(1 - \sqrt{\log 2}\right)/K$, since

$$\begin{aligned} 2K\varepsilon \left(\Delta_m + \varepsilon x_m + \sqrt{\Delta_m \varepsilon x_m} \right) &\leq K\varepsilon \left(\frac{5}{2} \Delta_m + 4x_m \varepsilon \right) \\ &\leq m\varepsilon^2 \left(4\sqrt{\log N} + 4\sqrt{\log 2} + 4K\theta \right) \\ &= 4m\varepsilon^2 \left(\sqrt{\log N} + 1 \right), \end{aligned}$$

the constraint $r \geq 4\varepsilon \left(1 + \sqrt{\log N}\right)$ implies that the condition of Theorem 5.14 on the penalty function is satisfied. The risk bound of Theorem 5.14 becomes

$$\begin{aligned} \mathbf{E}\left[\|\tilde{s} - s\|^2 \right] &\leq C(K) \left[\inf_{m \geq 1} \left(\inf_{\|t\|_1 \leq m\varepsilon} \|s - t\|^2 + rm\varepsilon + r\varepsilon \right) + (1 + \Sigma_\theta) \varepsilon^2 \right] \\ &\leq C(K) \left[\inf_{t \in \mathcal{L}_1(\Lambda)} \left(\|s - t\|^2 + r\|t\|_1 \right) + 2r\varepsilon + (1 + \Sigma_\theta) \varepsilon^2 \right], \end{aligned}$$

hence the result. □

5.8 Hypercontractivity: The Bonami–Beckner Inequality

In this section we present a powerful concentration inequality for functions defined on the binary hypercube. This so-called *hypercontractive* inequality bounds higher-order moments of Boolean polynomials in terms of lower-order moments. The result, also known as the *Bonami–Beckner* inequality, has its origins in harmonic analysis and has countless applications and generalizations. The Bonami–Beckner inequality is closely related to the logarithmic Sobolev inequality presented in Section 5.1. In fact, the proof presented here is based on Theorem 5.1.

In this section we consider real-valued functions $f : \{-1, 1\}^n \to \mathbb{R}$. Every such function can be expressed in a unique way as

$$f(x) = \sum_{S \subset \{1,\ldots,n\}} \alpha_S u_S(x)$$

where the sum is over all 2^n subsets $S \subset \{1, \ldots, n\}$, and to each set S we assign the function

$$u_S(x) = \prod_{i \in S} x_i.$$

(If $S = \emptyset$, we define $u_S \equiv 1$.) The α_S are real-valued coefficients. To demonstrate why such a representation is unique, observe that if we define, for real-valued functions $f, g : \{-1, 1\}^n \to \mathbb{R}$, the inner product

$$\langle f, g \rangle = 2^{-n} \sum_{x \in \{-1,1\}^n} f(x) g(x),$$

then it can be seen immediately that for any $S, S' \subset \{1, \ldots, n\}$,

$$\langle u_S, u_{S'} \rangle = \begin{cases} 0 \text{ if } S \neq S' \\ 1 \text{ if } S = S' \end{cases}$$

and therefore the u_S form an orthonormal basis of the vector space of all functions $f : \{-1, 1\}^n$. This means that for all $S \subset \{1, \ldots, n\}$, $\alpha_S = \langle f, u_S \rangle$. The formula $f = \sum_S \alpha_S u_S$ is often called the *Fourier–Walsh expansion* of f and the α_S are the *Fourier coefficients* of f.

For any $q \geq 1$, we define the *norm* $\|f\|_q = \left(2^{-n} \sum_{x \in \{-1,1\}^n} |f(x)|^q\right)^{1/q}$.

The main result of this section, the Bonami–Beckner inequality, can be stated in various forms. Before stating the theorem in its full generality, we describe two of its corollaries as these are relatively simple to formulate and are the versions that we use in this book.

Corollary 5.16 *Let k be a positive integer and assume that $f : \{-1, 1\}^n \to \mathbb{R}$ has the form $f = \sum_{S:|S|=k} \alpha_S u_S$. Then for all $1 < p < q < \infty$,*

$$\|f\|_q \leq \left(\frac{q-1}{p-1}\right)^{k/2} \|f\|_p.$$

A function of the form $f = \sum_{S:|S|=k} \alpha_S u_S$ is sometimes referred to as a *homogeneous Rademacher chaos of order k*. If $X = (X_1, \ldots, X_n)$ is a vector of i.i.d. Rademacher random variables and we define the random variable

$$Z = f(X) = \sum_{S:|S|=k} \alpha_S u_S(X),$$

then Corollary 5.16 states that

$$\left(\mathbf{E}|Z|^q\right)^{1/q} \leq \left(\frac{q-1}{p-1}\right)^{k/2} \left(\mathbf{E}|Z|^p\right)^{1/p}.$$

Thus, higher-order moments of Z can be bounded by a constant multiple of lower-order moments. This is an important generalization of the so-called Kahane–Khinchine inequalities that deal with the special case when $k = 1$.

To state another useful formulation of the Bonami–Beckner inequality, we introduce, for any positive number γ, an operator T_γ that maps an arbitrary function $f = \sum_{S\subset\{1,\dots,n\}} \alpha_S u_S$ to another function

$$T_\gamma f = \sum_{S\subset\{1,\dots,n\}} \gamma^{|S|}\alpha_S u_S.$$

For $\gamma = 1$ this is just the identity operator. For $\gamma < 1$, the Fourier coefficients corresponding to a set S are shrunk by a factor that is exponential in the size of the set. For $\gamma > 1$, the Fourier coefficients are blown up similarly.

Corollary 5.17 *For any $f : \{-1,1\}^n \to \mathbb{R}$ and $\gamma \le 1$,*

$$\|T_\gamma f\|_2 \le \|f\|_{1+\gamma^2}.$$

The corollary above asserts that, considered as an operator from $L_{1+\gamma^2}$ to L_2, T_γ has an operator norm $\|T_\gamma\|_{op} \stackrel{\text{def}}{=} \sup_{f:\{-1,1\}^n\to\mathbb{R}} \|T_\gamma f\|_2/\|f\|_{1+\gamma^2}$ bounded by 1. (In fact, it equals 1; just consider the function $f \equiv 1$.) As the inequality involves different norms, the property is often called *hypercontractivity*. Next we formulate the general statement.

Theorem 5.18 (BONAMI–BECKNER INEQUALITY) *Let $1 < p < q < \infty$ and let $\beta > 0$. Define $\gamma = \sqrt{\beta/(q-1)}$ and $\delta = \sqrt{\beta/(p-1)}$. Then, for any function $f : \{-1,1\}^n \to \mathbb{R}$,*

$$\|T_\gamma f\|_q \le \|T_\delta f\|_p.$$

Observe that Corollary 5.16 follows simply by taking $\beta = 1$ while Corollary 5.17 is recovered by setting $q = 2$ and $\beta = p - 1$.

Proof The key idea of the proof is to define the function $q(t) = \beta e^{2t} + 1$ for $t \ge 0$. Then the statement of the theorem becomes

$$\|T_{e^{-t}} f\|_{q(t)} \le \|T_{e^{-s}} f\|_{q(s)}$$

where $t = \log\sqrt{(q-1)/\beta}$ and $s = \log\sqrt{(p-1)/\beta}$ (i.e., $s < t$). If $X = (X_1,\dots,X_n)$ is a uniformly distributed random vector on $\{-1,1\}^n$, then defining the random variable $Z_t = \sum_{S\subset\{1,\dots,n\}} e^{-t|S|}\alpha_S u_S(X)$, we may write

$$\|T_{e^{-t}} f\|_{q(t)} = \left(\mathbf{E}\left[|Z_t|^{q(t)}\right]\right)^{1/q(t)}$$

and therefore we need to prove that for all $0 \le s < t$,

$$\frac{1}{q(t)}\log\mathbf{E}\left[|Z_t|^{q(t)}\right] \le \frac{1}{q(s)}\log\mathbf{E}\left[|Z_s|^{q(s)}\right],$$

that is, that $(1/q(t))\log\mathbf{E}\left[|Z_t|^{q(t)}\right]$ is a nonincreasing function of $t \ge 0$. We do this by induction on n. The following lemma establishes the result for the case of $n = 1$ variable.

Lemma 5.19 *Let X be a Rademacher random variable, let $\alpha_0, \alpha_1 \in \mathbb{R}$ be real coefficients and let*

$$Z_t = \alpha_0 + e^{-t}\alpha_1 X.$$

Then $(1/q(t)) \log \mathbf{E}\left[|Z_t|^{q(t)}\right]$ is a nonincreasing function of $t \geq 0$ where $q(t) = \beta e^{2t} + 1$ with $\beta > 0$.

Proof Note first that for $s, t \geq 0$, $Z_t = e^{-(t-s)}Z_s + (1 - e^{-(t-s)})\mathbf{E}Z_s$. Now define, for all $t \geq 0$, $Y_t = e^{-(t-s)}|Z_s| + (1 - e^{-(t-s)})\mathbf{E}|Z_s|$. Then, for all $t \geq s$, $Y_s = |Z_s|$ and $|Z_t| \leq Y_t$ and thus,

$$\begin{aligned}
&\frac{1}{q(t)} \log \mathbf{E}\left[Y_t^{q(t)}\right] \leq \frac{1}{q(s)} \log \mathbf{E}\left[Y_s^{q(s)}\right] \\
\Longrightarrow \quad &\frac{1}{q(t)} \log \mathbf{E}\left[|Z_t|^{q(t)}\right] \leq \frac{1}{q(s)} \log \mathbf{E}\left[|Z_s|^{q(s)}\right].
\end{aligned}$$

In order to prove the lemma, it suffices to establish that $(1/q(t)) \log \mathbf{E}[Y_t^{q(t)}]$ is a nonincreasing function of $t \geq s$. However,

$$Y_t = e^{-(t-s)}|\alpha_0 + e^{-s}\alpha_1 X| + (1 - e^{-(t-s)})\frac{|\alpha_0 + e^{-s}\alpha_1| + |\alpha_0 - e^{-s}\alpha_1|}{2}.$$

If we exchange the roles of α_0 and $e^{-s}\alpha_1$, the distribution of Y_t does not change. Thus, without loss of generality, we may assume that $\alpha_0 \geq e^{-s}|\alpha_1|$. This means that $Z_t \geq 0$ for $t \geq s$. Summarizing, we have shown that in order to prove the lemma, it suffices to show that for $t \geq s$, if $Z_t \geq 0$ then

$$\frac{d}{dt}\left(\frac{1}{q(t)} \log \mathbf{E}\left[Z_t^{q(t)}\right]\right) \leq 0.$$

The derivative, by straightforward differentiation, may be written as

$$\begin{aligned}
&\frac{d}{dt}\left(\frac{1}{q(t)} \log \mathbf{E}\left[Z_t^{q(t)}\right]\right) \\
&= \frac{q'(t)}{q^2(t)} \frac{1}{\mathbf{E}\left[Z_t^{q(t)}\right]} \left(-\mathbf{E}\left[Z_t^{q(t)}\right] \log \mathbf{E}\left[Z_t^{q(t)}\right] + \mathbf{E}\left[Z_t^{q(t)} \log Z_t^{q(t)}\right]\right. \\
&\qquad \left. + \frac{q^2(t)}{q'(t)} \mathbf{E}\left[Z_t^{q(t)-1} \frac{dZ_t}{dt}\right]\right).
\end{aligned}$$

On the right-hand side we recognize the entropy $\text{Ent}(Z_t^{q(t)})$. Also, using the simple fact that $dZ_t/dt = \mathbf{E}Z_t - Z_t$, we have

$$
\begin{aligned}
&\frac{d}{dt}\left(\frac{1}{q(t)}\log \mathbf{E}\left[Z_t^{q(t)}\right]\right)\\
&= \frac{q'(t)}{q^2(t)}\frac{1}{\mathbf{E}\left[Z_t^{q(t)}\right]}\left(\mathrm{Ent}\left(Z_t^{q(t)}\right) + \frac{q^2(t)}{q'(t)}\mathbf{E}\left[Z_t^{q(t)-1}(\mathbf{E}Z_t - Z_t)\right]\right).
\end{aligned}
$$

Since $q'(t) > 0$, it suffices to show that the expression in parentheses is non-positive. To this end, we invoke the logarithmic Sobolev inequality of Theorem 5.1. We get

$$
\mathrm{Ent}\left(Z_t^{q(t)}\right) \leq \mathbf{E}\left[\left(Z_t^{q(t)/2} - Z_t'^{q(t)/2}\right)_+^2\right]
$$

with $Z_t' = \alpha_0 + e^{-t}\alpha_1 X'$ where X' is a Rademacher random variable, independent of X. In order to further bound the right-hand side, we observe that for $0 \leq a < b$,

$$
\begin{aligned}
\left(\frac{b^{q/2} - a^{q/2}}{b-a}\right)^2 &= \left(\frac{q}{2(b-a)}\int_a^b u^{\frac{q}{2}-1}du\right)^2\\
&\leq \frac{q^2}{4(b-a)}\int_a^b u^{q-2}du \quad \text{(by Cauchy–Schwarz)}\\
&= \frac{q^2}{4(q-1)}\frac{b^{q-1} - a^{q-1}}{b-a}.
\end{aligned}
$$

Using this inequality and the identical distribution of Z_t and Z_t', we obtain

$$
\begin{aligned}
\mathrm{Ent}\left(Z_t^{q(t)}\right) &\leq \mathbf{E}\left[\left(Z_t^{q(t)/2} - Z_t'^{q(t)/2}\right)_+^2\right]\\
&\leq \frac{q^2(t)}{4(q(t)-1)}\mathbf{E}\left[\left(Z_t^{q(t)-1} - Z_t'^{q(t)-1}\right)(Z_t - Z_t')\right]\\
&= \frac{q^2(t)}{2(q(t)-1)}\mathbf{E}\left[Z_t^{q(t)-1}(Z_t - Z_t')\right]\\
&= \frac{q^2(t)}{2(q(t)-1)}\mathbf{E}\left[Z_t^{q(t)-1}(Z_t - \mathbf{E}Z_t)\right].
\end{aligned}
$$

Using this bound, we finally have

$$
\begin{aligned}
&\mathrm{Ent}\left(Z_t^{q(t)}\right) + \frac{q^2(t)}{q'(t)}\mathbf{E}\left[Z_t^{q(t)-1}(\mathbf{E}Z_t - Z_t)\right]\\
&\leq \left(\frac{-q^2(t)}{2(q(t)-1)} + \frac{q^2(t)}{q'(t)}\right)\mathbf{E}\left[Z_t^{q(t)-1}(\mathbf{E}Z_t - Z_t)\right]\\
&= 0
\end{aligned}
$$

because the expression in parentheses involving $q(t)$ equals zero. □

With the proof of the case $n = 1$ completed, we proceed with the induction step to finish the proof of Theorem 5.18. Assume that the statement of the theorem holds for $n-1$ variables. The argument is based on the general form of Minkowski's inequality (Theorem 2.16). Recall that $X = (X_1, \ldots, X_n)$ is a uniformly distributed vector on $\{-1,1\}^n$ and $Z_t = \sum_{S\subset\{1,\ldots,n\}} e^{-t|S|}\alpha_S u_S(X)$. Introduce the random variables

$$V_t = \sum_{S\subset\{1,\ldots,n\}, n\notin S} e^{-t|S|}\alpha_S u_S(X)$$

and

$$W_t = \sum_{S\subset\{1,\ldots,n\}, n\in S} e^{-t(|S|-1)}\alpha_S u_{S\setminus\{n\}}(X)$$

so that $Z_t = V_t + e^{-t}X_n W_t$. Write $\overline{\mathbf{E}}_{n-1}$ for the conditional expectation operator conditioned on X_n (i.e. integration with respect to $X_1, \ldots, X_{n-1}$) and $\mathbf{E}^{(n)}$ for expectation taken with respect to X_n only (i.e. conditional on $X_1, \ldots, X_{n-1}$). Then

$$\begin{aligned}
\left(\mathbf{E}\|Z_t\|^{q(t)}\right)^{1/q(t)} &= \left(\overline{\mathbf{E}}_{n-1}\left[\mathbf{E}^{(n)}\left[\|V_t + e^{-t}X_n W_t\|^{q(t)}\right]\right]\right)^{1/q(t)} \\
&\le \left(\overline{\mathbf{E}}_{n-1}\left[\left(\mathbf{E}^{(n)}\left[\|V_t + e^{-s}X_n W_t\|^{q(s)}\right]\right)^{q(t)/q(s)}\right]\right)^{1/q(t)} \\
&\quad \text{(by Lemma 5.19)} \\
&\le \left(\mathbf{E}^{(n)}\left[\left(\overline{\mathbf{E}}_{n-1}\left[\|V_t + X_n W_t\|^{q(t)}\right]\right)^{q(s)/q(t)}\right]\right)^{1/q(s)} \\
&\quad \text{(by Minkowski's inequality; Theorem 2.16)} \\
&\le \left(\mathbf{E}^{(n)}\left[\overline{\mathbf{E}}_{n-1}\left[\|V_s + X_n W_s\|^{q(s)}\right]\right]\right)^{1/q(s)} \\
&\quad \text{(by the induction hypothesis)} \\
&= \left(\mathbf{E}\left[\|Z_s\|^{q(s)}\right]\right)^{1/q(s)},
\end{aligned}$$

where the last inequality is a consequence of the induction hypothesis. This completes the proof of Theorem 5.18. □

The Bonami–Beckner inequality may be extended to the case of vector-valued functions. For example, an extended version of Corollary 5.16 states that if $X = (X_1, \ldots, X_n)$ is a vector of i.i.d. Rademacher random variables and for each $S \subset \{1, \ldots, n\}$, α_S is an element of a normed vector space, then the random vector defined by

$$Z = \sum_{S:|S|=k} \alpha_S u_S(X)$$

satisfies

$$(\mathbf{E}\|Z\|^q)^{1/q} \le \left(\frac{q-1}{p-1}\right)^{k/2} (\mathbf{E}\|Z\|^p)^{1/p}$$

where $1 < p < q$ and $k \le n$ is a positive integer. The proof goes similarly to the case of real-valued coefficients, only the proof of Lemma 5.19 needs to be adjusted. We leave the details as an exercise (see Exercise 5.7).

The special case when $k = 1$ is a classical and thoroughly studied problem. In this case $f(x) = \sum_{i=1}^n b_i x_i$ and the inequality above is a version of the classical Kahane–Khinchine inequality. An especially interesting and important case is when $q = 2$ and $p = 1$. Unfortunately, in this case the Bonami–Beckner inequality is vacuous. However, the Bonami–Beckner inequality may be used to control the L_2-norm of Z by a constant multiple of the L_1 norm. To this end, just observe that by the Cauchy–Schwarz inequality,

$$\mathbf{E}\|Z\|^{3/2} \le \sqrt{\mathbf{E}\|Z\|}\sqrt{\mathbf{E}\|Z\|^2}$$

and therefore

$$\frac{(\mathbf{E}\|Z\|^2)^{1/2}}{\mathbf{E}\|Z\|} \le \left(\frac{(\mathbf{E}\|Z\|^2)^{1/2}}{(\mathbf{E}\|Z\|^{3/2})^{2/3}}\right)^3.$$

Thus, the Bonami–Beckner inequality implies

$$\left(\mathbf{E}\|Z\|^2\right)^{1/2} \le 2^{3k/2}\mathbf{E}\|Z\|.$$

However, the constant $2^{3k/2}$ is not optimal. For $k = 1$, an ancient and elementary argument shows that $2^{3/2}$ may be replaced by $3^{1/2}$ (see Exercise 5.8). Moreover the optimal constant is not difficult to determine. We close this section by a short and elegant proof of the Kahane–Khinchine inequality for $q = 2$ and $p = 1$ with the best possible constant. We prove the result for general, vector-valued coefficients as it does not require any additional effort.

Theorem 5.20 (SZAREK'S INEQUALITY) *Let $b_1, \ldots, b_n$ be elements of a normed vector space and let $X_1, \ldots, X_n$ be independent Rademacher random variables. If $Z = \| \sum_{i=1}^n b_i X_i \|$, then*

$$\sqrt{\mathbf{E}\,[Z^2]} \le \sqrt{2}\,\mathbf{E}Z.$$

Proof Let $f(x) = \| \sum_{i=1}^n b_i X_i \|$ for $x = (x_1, \ldots, x_n) \in \{-1, 1\}^n$ and denote its Fourier coefficients by $\alpha_S = \langle f, u_S \rangle$, $S \subset \{1, \ldots, n\}$. Recalling $\overline{x}^{(i)} = (x_1, \ldots, x_{i-1}, -x_i, x_{i+1}, \ldots, x_n)$, define $\overline{f}(x) = \sum_{i=1}^n f(\overline{x}^{(i)})$. Since

$$u_S(\overline{x}^{(i)}) = \begin{cases} u_S(x) & \text{if } i \notin S \\ -u_S(x) & \text{if } i \in S, \end{cases}$$

the Fourier coefficient of $\bar{f}$ corresponding to $S \subset \{1,\ldots,n\}$ equals $\alpha_S(n-2|S|)$. This means that

$$\begin{aligned}\langle f,\bar{f}\rangle &= \left\langle \sum_{S\subset\{1,\ldots,n\}} \alpha_S u_S, \sum_{S\subset\{1,\ldots,n\}} \alpha_S(n-2|S|)u_S \right\rangle \\ &= \sum_{S\subset\{1,\ldots,n\}} \alpha_S^2(n-2|S|).\end{aligned}$$

A key property of f is that if $|S|$ is odd then $\alpha_S = 0$. This simply follows because if $|S|$ is odd, $u_S(-x) = -u_S(x)$ and $f(-x) = f(x)$, so $\alpha_S = \sum_{x\in\{-1,1\}^n} f(x)u_S(x) = 0$. Using this fact implies

$$\begin{aligned}\langle f,\bar{f}\rangle &= \sum_{S\subset\{1,\ldots,n\}} \alpha_S^2(n-2|S|) \\ &\le n\alpha_\emptyset^2 + (n-4)\sum_{S\neq\emptyset}\alpha_S^2 \\ &= 4\alpha_\emptyset^2 + (n-4)\sum_{S\subset\{1,\ldots,n\}}\alpha_S^2 \\ &= 4\|f\|_1^2 + (n-4)\|f\|_2^2,\end{aligned}$$

where we used the simple facts that $\alpha_\emptyset = \|f\|_1$ and that $\sum_{S\subset\{1,\ldots,n\}}\alpha_S^2 = \|f\|_2^2$, known as Parseval's identity. We compare the upper bound obtained for $\langle f,\bar{f}\rangle$ by a simple lower bound derived as follows. Note that, for every $x \in \{-1,1\}^n$,

$$\bar{f}(x) = \sum_{i=1}^n \left\| \sum_{j=1}^n b_j \bar{x}_j^{(i)} \right\| \ge \left\| \sum_{i=1}^n \sum_{j=1}^n b_j \bar{x}_j^{(i)} \right\| = (n-2)f(x).$$

Thus, since f is nonnegative, $\langle f,\bar{f}\rangle \ge (n-2)\|f\|_2^2$. Comparing the upper and lower bounds obtained for $\langle f,\bar{f}\rangle$, we get $\|f\|_2^2 \le 2\|f\|_1^2$ which is precisely what we wanted to show. □

To confirm that the constant $\sqrt{2}$ is the best possible, just consider the case $Z = X_1 + X_2$.

5.9 Gaussian Hypercontractivity

The hypercontractivity property of the symmetric Bernoulli distribution given by the Bonami–Beckner inequality also has its Gaussian analog, called Nelson's theorem, which we do not detail here (see, however, Exercises 5.18, 5.19, and 5.20). However, we point out a simple consequence of the Bonami–Beckner inequality for moments of polynomials of a Gaussian variable.

Corollary 5.21 *Let $f(x) = \sum_{i=0}^{k} a_i x^i$ be a polynomial of degree k of a real variable and let X be a standard normal random variable. Then for any $q > 2$,*

$$\left(\mathbf{E}\left[|f(X)|^q\right]\right)^{1/q} \leq (q-1)^{k/2} \left(\mathbf{E}\left[|f(X)|^2\right]\right)^{1/2}.$$

Proof Let $\varepsilon = (\varepsilon_1, \ldots, \varepsilon_n)$ be a vector of n i.i.d. Rademacher random variables. By the central limit theorem, it suffices to prove that for all n,

$$\left(\mathbf{E}\left[\left|f\left(\frac{1}{\sqrt{n}}\sum_{i=1}^{n}\varepsilon_i\right)\right|^q\right]\right)^{1/q} \leq (q-1)^{k/2}\left(\mathbf{E}\left[\left|f\left(\frac{1}{\sqrt{n}}\sum_{i=1}^{n}\varepsilon_i\right)\right|^2\right]\right)^{1/2}.$$

Introducing $g(\varepsilon_1, \ldots, \varepsilon_n) = f\left(\frac{1}{\sqrt{n}}\sum_{i=1}^{n}\varepsilon_i\right)$, we observe that g is a (nonhomogeneous) Rademacher chaos of order d, that is, $g : \{-1,1\}^n \to \mathbb{R}$ may be expressed as

$$g(\varepsilon) = \sum_{S \subset \{1,\ldots,n\}: |S| \leq k} \alpha_S u_S(\varepsilon).$$

We may then apply the Bonami–Beckner inequality (Theorem 5.18) with $\beta = q - 1$ and $p = 2$ to get, with $\delta = \sqrt{q-1} > 1$,

$$\|g\|_q^2 \leq \|T_\delta g\|_2^2 = \sum_{S \subset \{1,\ldots,n\}: |S| \leq k} \alpha_S^2 \delta^{2|S|} \leq \delta^{2k} \sum_{S \subset \{1,\ldots,n\}: |S| \leq k} \alpha_S^2 = \delta^{2k} \|g\|_2^2$$

which is exactly what we wanted to prove. □

5.10 The Largest Eigenvalue of Random Matrices

In this section we investigate concentration properties of the largest eigenvalue of random Hermitian matrices with Gaussian entries. This is just one example from the vast literature on tail bounds for eigenvalues of random matrices. We present it to illustrate how the Gaussian hypercontractive inequality (Corollary 5.21) may be used to obtain powerful results in a nontrivial example.

Recall that in Example 3.14 (see also Example 6.8) we studied the random fluctuations of the largest eigenvalue of random symmetric matrices with independent bounded entries. However, the Gaussian assumption used here allows us to obtain significantly sharper results.

A complex $n \times n$ matrix H is called *Hermitian* if $H = H^*$ where H^* is the transposed conjugate of H (i.e. $H^*_{i,j} = \overline{H}_{j,i}$ for all $1 \leq i,j \leq n$). The set of $n \times n$ Hermitian matrices is denoted by $\mathcal{H}_n$ while the set of $n \times n$ unitary matrices is denoted by $\mathcal{U}_n$. The spectral decomposition theorem for Hermitian matrices asserts that any Hermitian matrix H can be written as $H = UDU^*$ where $U \in \mathcal{U}_n$ and D is a diagonal matrix with real entries. The entries of D are the eigenvalues of H, denoted by $\lambda_1 \geq \lambda_2 \geq \ldots \geq \lambda_n$.

In this section we consider a special random matrix model, called the *Gaussian unitary ensemble* (GUE). A random matrix H is said to belong to the GUE if H is a Hermitian matrix whose diagonal entries $(H_{i,i})_{i \leq n}$ are independent real Gaussian variables with variance $\sigma^2 = 1/(4n)$ and whose off-diagonal entries $(H_{i,j})_{1 \leq i < j \leq n}$ are independent complex Gaussian random variables with independent real and imaginary parts, both with variances $\sigma^2/2 = 1/(8n)$.

The distribution of a random $n \times n$ matrix from the GUE may be described by its density with respect to the Lebesgue measure over $\mathbb{R}^{n^2}$ (using the straightforward one-to-one mapping between the set of $n \times n$ Hermitian matrices and $\mathbb{R}^{n^2}$). This density is proportional to $\exp(-\|H\|_{\mathrm{HS}}^2/(2\sigma^2))$. Recall that the *Hilbert–Schmidt norm* of a complex $n \times n$ matrix A is defined by $\|A\|_{\mathrm{HS}}^2 = \sum_{1 \leq i,j \leq n} |A_{i,j}|^2$.

Here we study the largest eigenvalue $\lambda_1(H)$ of a random matrix H from the GUE. One may use Lidskii's inequality (see Exercise 3.16) to show that $\lambda_1(H)$ is a Lipschitz function of n^2 independent standard Gaussian random variables with Lipschitz constant $n^{-1/2}$. The Gaussian concentration inequality (Theorem 5.6) implies that the fluctuations of $\lambda_1(H)$ around its expectation are of an order of at most $n^{-1/2}$, with high probabiliy. The main focus of this section is the following theorem. It shows that the upper tail is significantly lighter than implied by the Gaussian concentration inequality, as typical deviations are of the order of $n^{-2/3}$.

Theorem 5.22 *Let $Z = \lambda_1(H)$ be the largest eigenvalue of a random $n \times n$ matrix H, distributed according to the* GUE. *Then for all $0 \leq t \leq 1$,*

$$\boldsymbol{P}\{Z \geq 1 + t\} \leq \frac{1}{2t^{1/2}} e^{-nt^{3/2}}.$$

By a more involved analysis, one may show that the factor $1/(2t^{1/2})$ can be replaced by a universal constant.

Note that $\boldsymbol{E}Z \neq 1$, so Theorem 5.22 is not about deviations from the mean. Nevertheless, it follows from Wigner's theorem (Theorem 5.23 below) that $\liminf_{n\to\infty} \boldsymbol{E}Z \geq 1$. This, combined with Theorem 5.22, implies that, in fact, $\lim_{n\to\infty} \boldsymbol{E}Z = 1$. One may also show that there exists a universal constant $\kappa > 0$ such that for all n, $|\boldsymbol{E}Z - 1| \leq \kappa n^{-2/3}$.

Wigner's celebrated *semi-circular law* determines the asymptotic distribution of the eigenvalues of random matrices from the GUE. In order to state Wigner's theorem, we define the *spectral measure* L_n of an $n \times n$ Hermitian matrix H as the discrete probability measure on the real line that assigns weight $1/n$ to each eigenvalue of H. In other words, for any function f defined over $\mathbb{R}$, we let

$$L_n f = \frac{1}{n} \sum_{i=1}^{n} f(\lambda_i).$$

The *semi-circular density* is defined, for $x \in \mathbb{R}$, by $\phi(x) = (2/\pi)\sqrt{1 - x^2}\,\mathbb{1}_{\{x \in [-1,1]\}}$.

Theorem 5.23 (WIGNER'S THEOREM) *Let L_n denote the spectral measure of a random $n \times n$ matrix from the GUE. Then the sequence L_n converges weakly in probability, to the semi-circular distribution. This means that for all $\varepsilon > 0$,*

$$\lim_{n\to\infty} P\left\{\sup_{f\in\mathcal{B}} \left|L_n f - \int f(x)\phi(x)dx\right| > \varepsilon\right\} = 0,$$

where $\mathcal{B}$ denotes the set of 1-Lipschitz functions $f : \mathbb{R} \to [-1, 1]$.

Wigner's theorem may be proved by solving Exercises 5.32–5.35.

In preparation for the proof of Theorem 5.22, we need to introduce the so-called *Hermite polynomials*. For every $k = 1, 2, \ldots$, the normalized Hermite polynomial of degree k is defined by

$$h_k(x) = \frac{1}{\sqrt{k!}} \frac{d^k e^{\lambda x - \lambda^2/2}}{d\lambda^k}\Bigg|_{\lambda=0}.$$

An important property of Hermite polynomials is that they form an orthonormal family in the space of square-integrable functions under the standard Gaussian distribution. That is, if X is a standard Gaussian random variable, then $\boldsymbol{E}h_i(X) = 0$ for all $i > 1$, and

$$\boldsymbol{E}\left[h_i(X)h_j(X)\right] = \begin{cases} 1 & \text{if } i = j \\ 0 & \text{otherwise.} \end{cases}$$

A proof of this well-known fact and some other useful properties of Hermite polynomials are suggested in Exercise 5.22.

The proof of Theorem 5.22 starts from the determinantal description of the joint distribution of the eigenvalues as shown in the next lemma. Deriving this lemma requires a substantial amount of work (see Exercises 5.24–5.30). Observe that with probability 1, a random matrix from the GUE has pairwise distinct eigenvalues.

Henceforth, let $\Delta(x_1, \ldots, x_n)$ be the *Vandermonde determinant* defined by $x_1, \ldots, x_n$:

$$\Delta(x_1, \ldots, x_n) = \prod_{1\le i<j\le n} (x_i - x_j) = \det \begin{pmatrix} 1 & x_1 & x_1^2 & \cdots & x_1^{n-1} \\ 1 & x_2 & x_2^2 & \cdots & x_2^{n-1} \\ \vdots & \vdots & \vdots & & \vdots \\ 1 & x_n & x_n^2 & \cdots & x_n^{n-1} \end{pmatrix}.$$

Lemma 5.24 *The joint density of the eigenvalues of an $n \times n$ random matrix from the GUE at $\lambda_1 > \lambda_2 > \cdots > \lambda_n$ equals*

$$\frac{\prod_{j=0}^{n-1} 1/j!}{(2\pi)^{n/2}\sigma^n} \Delta\left(\frac{\lambda_1}{\sigma}, \ldots, \frac{\lambda_n}{\sigma}\right)^2 \exp\left(-\frac{\sum_{i=1}^n \lambda_i^2}{2\sigma^2}\right),$$

where $\sigma = 1/\sqrt{4n}$.

The density of the unordered sequence $(\lambda_1, \ldots, \lambda_n) \in \mathbb{R}^n$ of eigenvalues of a random $n \times n$ matrix from the GUE is obtained by dividing the formula above by $n!$.

Starting from Lemma 5.24, the proof of Theorem 5.22 has two main steps. The first relates expectations under the mean spectral measure with sums of Gaussian integrals. This lemma is also used to establish Theorem 5.23.

Lemma 5.25 *Let H be an $n \times n$ random matrix from the GUE and let L_n denote the (random) spectral measure of H. For any $f : \mathbb{R} \to \mathbb{R}$,*

$$\mathbf{E}L_n f = \frac{1}{n} \sum_{i=0}^{n-1} \mathbf{E}\left[f(\sigma X) h_i(X)^2\right],$$

where X is a standard Gaussian random variable and $\sigma = 1/\sqrt{4n}$.

Proof As h_k is of degree k and as the leading coefficient of $\sqrt{k!}h_k$ is 1, the Vandermonde determinant may be rewritten in terms of the Hermite polynomials as

$$\Delta(\lambda_1, \ldots, \lambda_n) = \left(\prod_{j=0}^{n-1} \sqrt{j!}\right) \det\Big(h_j(\lambda_i)\Big)_{\substack{1 \le i \le n \\ 0 \le j < n}}.$$

The determinant may also be written by summing, over the set S_n of all permutations of $\{1, \ldots, n\}$, the signed product of diagonal elements:

$$\Delta(\lambda_1, \ldots, \lambda_n) = \prod_{j=0}^{n-1} \sqrt{j!} \sum_{\tau \in S_n} \operatorname{sgn}(\tau) \prod_{i=1}^{n} h_{\tau(i)-1}(\lambda_i)$$

where $\operatorname{sgn}(\tau) = 1$ (resp. -1) if τ is the product of an even (resp. odd) number of transpositions. In the sequel, $\tau \circ \tau'$ is the composition of permutations τ and τ', that is, $\tau \circ \tau'(x) = \tau(\tau'(x))$ and $\operatorname{sgn}(\tau \circ \tau') = \operatorname{sgn}(\tau)\operatorname{sgn}(\tau')$.

For each $i \in \{1, \ldots, n\}$ and for any measurable function f,

$$\begin{aligned}
&\mathbf{E}f(\lambda_i(H)) \\
&= \frac{1}{n!} \sum_{\tau, \tau' \in S_n} \operatorname{sgn}(\tau \circ \tau') \\
&\qquad \int_{\mathbb{R}^n} f(\lambda_i) \prod_{j=1}^{n} h_{\tau(j)-1}\left(\frac{\lambda_j}{\sigma}\right) h_{\tau'(j)-1}\left(\frac{\lambda_j}{\sigma}\right) \frac{e^{-\sum_{k=1}^{n} \lambda_k^2/(2\sigma^2)}}{(2\pi\sigma^2)^{n/2}} d\lambda_1 \cdots d\lambda_n \\
&= \frac{1}{n!} \sum_{\tau, \tau' \in S_n} \operatorname{sgn}(\tau \circ \tau') \\
&\qquad \int_{\mathbb{R}^n} f(\sigma\lambda_i) \prod_{j=1}^{n} h_{\tau(j)-1}(\lambda_j)\, h_{\tau'(j)-1}(\lambda_j) \frac{e^{-\sum_{k=1}^{n} \lambda_k^2/2}}{(2\pi)^{n/2}} d\lambda_1 \cdots d\lambda_n.
\end{aligned}$$

For all $\tau, \tau' \in S_n$, by Fubini's theorem,

$$\int_{\mathbb{R}^n} f(\sigma\lambda_i) \prod_{j=1}^{n} h_{\tau(j)-1}(\lambda_j) h_{\tau'(j)-1}(\lambda_j) e^{-\sum_{k=1}^{n} \lambda_k^2/2} d\lambda_1 \cdots d\lambda_n$$
$$= \left(\int_{\mathbb{R}} f(\sigma\lambda_i) h_{\tau(i)-1}(\lambda_i) h_{\tau'(i)-1}(\lambda_i) e^{-\lambda_i^2/2} d\lambda_i \right)$$
$$\times \prod_{j \neq i} \left(\int_{\mathbb{R}} h_{\tau(j)-1}(\lambda_j) h_{\tau'(j)-1}(\lambda_j) e^{-\lambda_j^2/2} d\lambda_j \right).$$

By the orthogonality property of the Hermite polynomials, the last factor on the right-hand side vanishes unless $\tau(j) = \tau'(j)$ for all $j \neq i$, that is unless $\tau = \tau'$. Hence,

$$\boldsymbol{E} f(\lambda_i(H)) = \frac{1}{n!} \sum_{\tau \in S_n} \int_{\mathbb{R}} f(\sigma\lambda_i) h_{\tau(i)-1}(\lambda_i)^2 \frac{e^{-\lambda_i^2/2}}{\sqrt{2\pi}} d\lambda_i$$
$$= \frac{1}{n} \int_{\mathbb{R}} f(\sigma\lambda_i) \left(\sum_{k=0}^{n-1} h_k(\lambda_i)^2 \right) \frac{e^{-\lambda_i^2/2}}{\sqrt{2\pi}} d\lambda_i.$$

The lemma follows by simplifying the expansion of $\boldsymbol{E} L_n$. □

Proof of Theorem 5.22. We may combine Lemma 5.25 and the simple bound

$$\mathbb{1}_{\{\max_{i=1,\ldots,n} \lambda_i \geq 1+t\}} \leq \sum_{i=1}^{n} \mathbb{1}_{\{\lambda_i \geq 1+t\}},$$

that is, we choose $f(\lambda) = \mathbb{1}_{\{\lambda \geq 1+t\}}$ to obtain

$$\boldsymbol{P}\{Z \geq 1+t\} \leq n\boldsymbol{E}\left[\frac{1}{n} \sum_{i=1}^{n} \mathbb{1}_{\{\lambda_i(H) \geq 1+t\}} \right]$$
$$\leq \sum_{i=0}^{n-1} \boldsymbol{E}\left[\mathbb{1}_{\{X \geq 2\sqrt{n}(1+t)\}} h_i(X)^2 \right].$$

By Hölder's inequality, for any $r > 1$, letting $r^* = r/(r-1)$,

$$\boldsymbol{E}\left[\mathbb{1}_{\{X \geq 2\sqrt{n}(1+t)\}} h_i(X)^2 \right] \leq \left(\boldsymbol{P}\{X \geq 2\sqrt{n}(1+t)\} \right)^{1/r^*} \left(\boldsymbol{E}\left[h_i(X)^{2r} \right] \right)^{1/r}$$
$$\leq e^{-2n(1+t)^2/r^*} \|h_i\|_{2r}^2$$
$$\leq e^{-2n(1+t)^2/r^*} (2r-1)^i \|h_i\|_2^2$$
$$= e^{-2n(1+t)^2/r^*} (2r-1)^i,$$

where the last inequality follows from Corollary 5.21. Summing the n upper bounds,

$$P\{Z \geq 1 + t\} \leq e^{-2n(1+t)^2/r^*} \sum_{i=0}^{n-1} (2r - 1)^i$$
$$= e^{-2n(1+t)^2/r^*} \frac{(2r - 1)^n}{2r - 2}.$$

The theorem now follows by choosing $r = 1 + \sqrt{t}$. □

5.11 Bibliographical Remarks

It is outside the scope of this book to offer an exhaustive account of logarithmic Sobolev inequalities. Instead, we refer the interested reader to the excellent book of Ané *et al.* (2000) for an extensive survey, with connections to other functional inequalities, Markov chains, information theory, etc. The investigation of logarithmic Sobolev inequalities, Poincaré inequalities and hypercontractivity was initially motivated by an analysis of the mixing properties of Markov processes and Markov chains. We refer the reader to the survey of Diaconis and Saloff-Coste (1998), the lecture notes by Saloff-Coste (1997), and Martinelli (1997) for a presentation of the role of functional inequalities in that field.

The logarithmic Sobolev inequalities for the balanced Bernoulli and Gaussian distributions were first derived by Gross (1975). It was Gross who determined the optimal constant in the logarithmic Sobolev inequality for the balanced Bernoulli distribution. The case of general Bernoulli distributions (Theorem 5.2) was clarified 20 years later by Higuchi and Yoshida (1995) and independently by Diaconis and Saloff-Coste (1996). The proof of Theorem 5.2 suggested in Exercise 5.4 is attributed to Bobkov, as it is presented in the lecture notes by Saloff-Coste (1997) and in Ané *et al.* (2000, Chapter 1). The logarithmic Sobolev constants for Bernoulli distributions can also be recovered from the more general result of Latała and Oleszkiewicz (2000).

The argument, attributed to Herbst, to derive concentration inequalities based on logarithmic Sobolev inequalities appears first in Davies and Simon (1984) (see also Aida, Masuda, and Shigekawa (1994)). The method was greatly generalized and popularized by Ledoux (1997, 1996, 1999, 2001) (and see Chapters 6 and 12).

The story of the Kahane–Khinchine inequalities date back to Khinchine (1923), Littlewood (1930), and Paley and Zygmund (1930), who proved it in the case of one-dimensional coefficients with different constants. It was extended to vector-valued coefficients by Kahane (1964). (For Littlewood's argument see Exercise 5.8.) The optimal constant $\sqrt{2}$ for real Rademacher sums in Theorem 5.20 was established by Szarek (1976). It was further generalized by Haagerup (1981) for comparing any q-th moment of a real Rademacher sum to the second moment. The optimal comparison between the first and second moments for norms of vector valued Rademacher sums is due to Latała and Oleszkiewicz (1994). The proof of Theorem 5.20 given here was inspired by the proof given by de la Penã and Giné (1999) who attribute it to Kwapień, Latała, and Oleszkiewicz (1996). We refer to de la Penã and Giné (1999) for many related results.

Theorem 5.6 was originally proved by Tsirelson, Ibragimov, and Sudakov (1976) using arguments different to the ones given here, based on stochastic calculus. A sharper form of this inequality is given in Section 10.4. For a thorough account of Gaussian concentration inequalities see Ledoux (1996).

The generalized Johnson–Lindenstrauss problem in Section 5.6 was investigated by Klartag and Mendelson (2005) whose results essentially contain Theorem 5.10 and also the bounds on Δ derived in Section 13.6.

The generalized linear Gaussian model discussed in Section 5.7 was introduced in Birgé and Massart (2001). For a detailed account of Gaussian model selection and related problems we refer the reader to Massart (2006). The LASSO estimator was introduced by Tibshirani (1996) and has become an important tool for high-dimensional regression problems. We refer the interested reader to Barron *et al.* (2008), Bickel, Ritov, and Tsybakov (2009), Bunea, Tsybakov, and Wegkamp (2007), Candès and Tao (2005, 2007), Donoho (2006*b*, 2006*c*), Huang, Cheang and Barron (2010), Koltchinskii (2009*a*, 2009*b*), and van de Geer (2008) for a variety of theoretical results. Theorem 5.15 and the argument presented here are borrowed from Massart and Meynet (2010). Related results were obtained by Bartlett, Mendelson and Neeman (2012).

The Bonami–Beckner inequality (Theorem 5.18) is due to Bonami (1970) and Beckner (1975). The Gaussian analog of Theorem 5.16 described in Exercises 5.18 and 5.19 is from Nelson (1973). Gross (1975) established the equivalence between hypercontractivity and logarithmic Sobolev inequalities in a general framework that we do not discuss here. Our proof of the Bonami–Beckner inequality is based on some of these ideas (see again Ané *et al.* (2000)) and see also Exercise 5.18 for another aspect in this connection. Starting with an important paper by Kahn, Kalai, and Linial (1988), the Bonami–Beckner inequality has found many interesting applications in the geometry of the binary hypercube and in the study of threshold phenomena. Several of these applications are described in Chapter 9 (though we prove most of these results using logarithmic Sobolev inequalities).

Wishart (1928) initiated the analysis of random matrices, namely the analysis of empirical covariance matrices of multivariate Gaussian samples. A survey of recent developments in the non-asymptotic analysis of random covariance matrices can be found in Rudelson and Vershynin (2010), and see also Section 13.4.

Nowadays, eigenvalues and singular values of random matrices are a major topic of study in mathematical physics, multivariate statistics, combinatorics, and information theory, to name but a few. The interested reader is referred to Mehta (2004) or Anderson, Guionnet, and Zeitouni (2010) for a thorough presentation (see also Tao (2012)). Theorem 5.23 was proved by Wigner (1955) who actually proved the weak convergence of the mean spectral measure to the semi-circular distribution. The convergence of the empirical spectral measure to the semi-circular distribution has been established for many other matrix ensembles using a variety of proof techniques (see Anderson, Guionnet, and Zeitouni (2010)). It holds for random real symmetric Hermitian matrices with independent entries under some mild tail assumptions on the distribution of the entries. Götze and Tikhomirov (2003, 2005) provide upper bounds on the rate of convergence of the spectral measure to the semi-circular distribution. Refer also to Meckes and Meckes (2012) and references therein for recent progress on the rate of convergence for spectral measures of a variety of matrix ensembles.

The asymptotic distribution of the largest eigenvalue of random matrices from the Gaussian unitary ensemble was characterized by Tracy and Widom (1994). The Tracy–Widom asymptotics for the largest eigenvalue has been extended to other ensembles of random matrices, including some non-Gaussian ensembles (see Soshnikov (1999)). Erdős and Yau (2012) survey universality issues raised by spectra of random matrices. In particular, the Tracy–Widom asymptotics holds for ensembles of random symmetric matrices with Rademacher entries, suggesting that there is room for improvement in the variance bound described in Example 3.14.

The largest eigenvalue of a random matrix from the GUE has interesting connections outside the random matrix theory. For example, for large n, once properly centered and standardized, the length of the longest increasing sequence in a random permutation over $\{1, \ldots, n\}$ behaves like the largest eigenvalue of a random matrix distributed as the GUE (Baik, Deift and Johansson (1999)).

The proof of Theorem 5.22 is taken from Ledoux (2003) but see also Aubrun (2005) for an alternative approach. Using more involved arguments, Ledoux (2003) proves that the polynomial factor in the tail bound is not necessary. The survey by Ledoux (2007) provides an accessible account of a wide range of non-asymptotic as well as asymptotic results on eigenvalues of random matrices.

5.12 EXERCISES

5.1. Show that for any nonnegative random variable Z, $\operatorname{Var}(Z) \leq \operatorname{Ent}(Z^2)$ (Latała and Oleszkiewicz (2000).) Show by example that the inequality is not necessarily true if Z is not required to be nonnegative. *Hint:* introduce, for $p \in [1, 2)$, the functional $\Psi_p(Z) = \mathbf{E}[Z^2] - (\mathbf{E}[Z^p])^{2/p}$. Show that $\lim_{p\uparrow 2} \Psi(Z) = \operatorname{Ent}(Z^2)/2$. Moreover, show that $\Psi_p(Z)/((1/p) - (1/2))$ is nondecreasing in p.

5.2. Show that Theorem 5.1 implies that for any function $f : \{-1, 1\}^n \to \mathbb{R}$, $\operatorname{Var}(f(X)) \leq \mathcal{E}(f)$. Prove also similarly that the Gaussian logarithmic Sobolev inequality (Theorem 5.4) implies the Gaussian Poincaré inequality (Theorem 3.20). *Hint:* let $\varepsilon > 0$ be small and use the logarithmic Sobolev inequality for $1 + \varepsilon f$. Show that $\operatorname{Ent}((1 + \varepsilon f)^2) = 2\varepsilon^2 \operatorname{Var}(f(X)) + O(\varepsilon^3)$.

5.3. (OPTIMALITY OF THE CONSTANT IN THE LOGARITHMIC SOBOLEV INEQUALITY) Prove that Theorem 5.1 does not hold if the constant 2 is replaced by any smaller constant.

5.4. Prove Theorem 5.2. Prove also that $c(p)$ is the best possible constant. *Hint:* by subadditivity of the entropy it suffices to prove the theorem for $n = 1$. Start with the duality formula of the entropy (Theorem 4.13). Show first that it suffices to prove the statement for strictly positive functions f.

5.5. Prove the following variant of Theorem 5.3. Let $f : \{-1,1\}^n \to \mathbb{R}$ and let X be uniformly distributed on $\{-1,1\}^n$. Let $v > 0$ be such that

$$\sum_{i=1}^{n} \left(f(x) - f\left(\overline{x}^{(i)}\right)\right)^2 \leq v$$

for all $x \in \{-1,1\}^n$. (Note that, as opposed to the statement of Theorem 5.3, the positive part is omitted in the definition of v.) Prove that, for all $t > 0$, $Z = f(X)$ satisfies

$$\mathbf{P}\{Z > \mathbf{E}Z + t\} \leq e^{-2t^2/v}.$$

Hint: proceed as in the proof of the theorem, but instead of using the simple convexity argument, establish first that for real numbers $z \geq y$,

$$\left(e^{z/2} - e^{y/2}\right)^2 \leq \frac{(z-y)^2}{8}\left(e^z + e^y\right).$$

Use this to show that

$$\operatorname{Ent}\left(e^{\lambda f(X)}\right) \leq \frac{1}{2}\sum_{i=1}^{n} \mathbf{E}\left[\left(e^{\lambda f(X)/2} - e^{\lambda f\left(\overline{X}^{(i)}\right)/2}\right)^2\right] \leq \mathbf{E}\left[\frac{\lambda^2 v}{8} e^{\lambda f(X)}\right].$$

5.6. Prove the following version of Theorem 5.3 for asymmetric Bernoulli distributions. Let $f : \{-1,1\}^n \to \mathbb{R}$ and assume that $X = (X_1, \ldots, X_n)$ has i.i.d. components with distribution $\mathbf{P}\{X_i = 1\} = 1 - \mathbf{P}\{X_i = -1\} = p$. Let $v > 0$ be such that

$$\sum_{i=1}^{n} \left(f(x) - f\left(\overline{x}^{(i)}\right)\right)_+^2 \leq v$$

for all $x \in \{-1,1\}^n$. Show that if f is nondecreasing in all of its components then for all $t > 0$,

$$\mathbf{P}\left\{f(X) > \mathbf{E}f(X) + t\right\} \leq \exp\left(\frac{-t^2}{(1-p)c(p)v}\right).$$

If f is nonincreasing then

$$\mathbf{P}\left\{f(X) > \mathbf{E}f(X) + t\right\} \leq \exp\left(\frac{-t^2}{pc(p)v}\right).$$

Hint: use Theorem 5.2 together with Herbst's argument.

5.7. (EXTENSION OF BONAMI–BECKNER TO VECTOR-VALUED FUNCTIONS) This exercise extends Lemma 5.19 to vector-valued functions. Let X be a Rademacher random variable and let $Z = \alpha_0 + \alpha_1 X$ where α_0, α_1 belong to a normed vector space. For $t \geq 0$, let $q(t) = \beta e^{2t} + 1$ and define $Z_t = \alpha_0 + e^{-t}\alpha_1 X$. Show that for all $0 \leq s < t$,

$$\left(\mathbf{E}\left[\|Z_t\|^{q(t)}\right]\right)^{1/q(t)} \leq \left(\mathbf{E}\left[\|Z_s\|^{q(s)}\right]\right)^{1/q(s)}.$$

Hint: if $v = \alpha_0 + \alpha_1$ and $w = \alpha_0 - \alpha_1$, notice that $\alpha_0 + e^{-t}\alpha_1 = v(1+e^t)/2 + w(1-e^{-t})/2$ and $\alpha_0 - e^{-t}\alpha_1 = v(1-e^t)/2 + w(1+e^{-t})/2$. By the convexity of the norm,

$$\left(\mathbf{E}\left[|Z_t|^{q(t)}\right]\right)^{1/q(t)}$$

$$\leq \left(\frac{\left(\frac{1+e^t}{2}|v| + \frac{1-e^{-t}}{2}|w|\right)^{q(t)} + \left(\frac{1-e^t}{2}|v| + \frac{1+e^{-t}}{2}|w|\right)^{q(t)}}{2}\right)^{1/q(t)}.$$

Write $\beta_0 = (|v| + |w|)/2$, $\beta_1 = (|v| - |w|)$, and use Lemma 5.19.

5.8. (LITTLEWOOD'S INEQUALITY FOR REAL RADEMACHER SUMS) Let $Z = \left|\sum_{i=1}^n b_i X_i\right|$ be a real-valued Rademacher sum where $b_1, \ldots, b_n \in \mathbb{R}$ are fixed coefficients and $X_1, \ldots, X_n$ are i.i.d. Rademacher random variables. Show first by elementary arguments that $\mathbf{E}[Z^4] \leq 3(\mathbf{E}[Z^2])^2$. Next use Hölder's inequality to derive $\mathbf{E}[Z^2] \leq (\mathbf{E}Z)^{2/3}(\mathbf{E}[Z^4])^{1/3}$. Conclude that $\mathbf{E}[Z^2] \leq 3(\mathbf{E}Z)^2$. This is a slightly weaker version of Theorem 5.20. Is the comparison between the fourth and the second moments improvable?

5.9. (MARCINKIEWICZ'S INEQUALITIES) Let $Y_1, \ldots, Y_n$ be independent random variables with finite variance and let $X_1, \ldots, X_n$ be independent Rademacher variables. Prove that

$$\mathbf{E}\left[\left(\sum_{i=1}^n Y_i^2\right)^{1/2}\right] \leq \sqrt{2}\mathbf{E}\left[\left|\sum_{i=1}^n X_i Y_i\right|\right] \leq 2\sqrt{2}\mathbf{E}\left[\left|\sum_{i=1}^n Y_i\right|\right].$$

Hint: use Theorem 5.20 and symmetrization.

5.10. (KHINCHINE'S INEQUALITY) Let $\varepsilon_1, \ldots, \varepsilon_n$, be a sequence of independent Rademacher random variables. Let $\alpha_1, \ldots, \alpha_n$ be n fixed real numbers. Prove that for $p = 1, 2, \ldots$,

$$\mathbf{E}\left[\left|\sum_{i=1}^n \varepsilon_i \alpha_i\right|^{2p}\right] \leq \frac{(2p)!}{2^p p!}\left(\sum_{i=1}^n \alpha_i^2\right)^p.$$

Using the central limit theorem and the known values for the moments of the standard Gaussian distribution, check that the dimension-free coefficients $\frac{(2p)!}{2^p p!}$ cannot be improved. *Hint:* if we are ready to replace the constants $(2p)!/(2^p p!)$ by $(2p-1)^p$, the above inequalities follow from the Bonami–Beckner inequalities. Another version can be derived from Hoeffding's inequality.

5.11. Show that the constant 2 on the right-hand side of the Gaussian logarithmic Sobolev inequality (Theorem 5.4) is the best possible. *Hint:* the bound for the moment generating function in the Gaussian concentration inequality is an equality if f is linear.

5.12. (THEOREM 5.4) Work out the details of the density argument used in the proof of Theorem 5.4.

5.13. (POINCARÉ AND LOGARITHMIC SOBOLEV INEQUALITIES FOR GENERAL GAUSSIAN DISTRIBUTIONS) Assume that the random vector $X \in \mathbb{R}^n$ has centered Gaussian distribution with covariance matrix Γ. Show that for any continuously differentiable function $f : \mathbb{R}^n \to \mathbb{R}$,

$$\mathrm{Var}\,(f(X)) \le \boldsymbol{E}\left[\langle \Gamma \nabla f(X), \nabla f(X)\rangle\right]$$

and

$$\mathrm{Ent}\,(f^2) \le 2\boldsymbol{E}\left[\langle \Gamma \nabla f(X), \nabla f(X)\rangle\right].$$

5.14. Detail the first step of the proof of Theorem 5.8. *Hint:* by total boundedness and sample path continuity, $Z = \sup_{t\in\mathcal{D}} X_t$ where $\mathcal{D}$ is a dense countable subset of $\mathcal{T}$. Use the Gaussian Poincaré inequality for finite subsets and monotone convergence to show that Z has an expected value (by relating it to the median of Z). Then again, use monotone convergence and the theorem for finite sets to conclude.

5.15. (NON-CENTERED CHI-SQUARED RANDOM VARIABLES) If $X_1, \ldots, X_D$ are independent standard normal random variables, then $Z^2 = (X_1 + \delta)^2 + \sum_{i=2}^{D} X_i^2$ has chi-square distribution with D degrees of freedom and non-centrality parameter δ^2. Compute the expected value and the variance of Z^2. Show that Z^2 is sub-gamma with variance factor $v = 2\boldsymbol{E}Z^2 + 2\delta^2$ and scale factor 2. Use the Gaussian Poincaré inequality and the Gaussian concentration inequality to show that the variance of Z is less than 1, and that Z is sub-Gaussian with variance factor 1. Show how this implies that Z^2 is sub-gamma with variance factor $4\boldsymbol{E}Z^2$ and scale factor 2.

5.16. (ADAPTING HERBST'S ARGUMENT) Let $X_1, \ldots, X_n$ be independent standard Gaussian random variables. Let f denote a differentiable function on $\mathbb{R}^n$ such that $\boldsymbol{E}\left[\exp(\lambda \|\nabla f(X_1, \ldots, X_n)\|^2)\right] < \infty$ for $\lambda < \lambda_0$ where λ_0 may be ∞. Let $Z = f(X_1, \ldots, X_n)$. Prove that for λ, θ satisfying $\lambda/\theta < \lambda_0$ and $\lambda\theta < 2$,

$$\log \boldsymbol{E}\left[\exp\left(\lambda(Z - \boldsymbol{E}Z)\right)\right] \le \frac{\lambda\theta}{2(1 - \lambda\theta/2)} \log \boldsymbol{E}\left[\exp\left(\lambda \|\nabla F\|^2/\theta\right)\right].$$

Hint: starting from Gaussian logarithmic Sobolev inequality, use Corollary 4.15 to upper bound $\boldsymbol{E}\left[\|\nabla f\|^2 \exp(\lambda Z)\right]$. Apply this result when f is the squared norm of the orthogonal projection of X on some linear subspace of $\mathbb{R}^n$.

5.17. (SZAREK'S INEQUALITY FOR GAUSSIAN SUMS) Let $b_1, \ldots, b_n$ be elements of a normed vector space and let $X_1, \ldots, X_n$ be independent standard Gaussian random variables. Let $Z = \|\sum_{i=1}^{n} b_i X_i\|$. Prove that

$$\sqrt{\boldsymbol{E}\,[Z^2]} \le \sqrt{2}\,\boldsymbol{E}Z.$$

Hint: start from Theorem 5.20 and use the central limit theorem as in the proof of the Gaussian Poincaré inequality or as in the proof of the Gaussian logarithmic Sobolev inequality. The factor $\sqrt{2}$ is not optimal and can be improved to $\sqrt{\pi/2}$. The best constants in comparison of moments of Gaussian vectors can be found in Latała and Oleszkiewicz (1999).

5.18. (NELSON'S THEOREM) Let X be a standard Gaussian random variable. For any $0 < \gamma \leq 1$, let the operator T_γ map any function f with $E[f(X)^2] < \infty$ to another function $T_\gamma f$ defined by

$$T_\gamma f(y) = E\left[f\left(\gamma y + \sqrt{1-\gamma^2}X\right)\right].$$

Check first that for all $\gamma \leq 1$, T_γ is a contraction, that is, $E\left[(T_\gamma f(X))^2\right] \leq E[f(X)^2]$. Let $t \geq 0$, $1 < p < \infty$, $q(t) = 1 + e^{2t}(p-1)$, and let the function f be such that $E[|f(X)|^p] < \infty$. Prove that

$$\left(E\left[|T_{e^{-t}}f(X)|^{q(t)}\right]\right)^{1/q(t)} \leq \left(E\left[|f(X)|^p\right]\right)^{1/p}.$$

Check that this is enough to establish the property for nonnegative twice differentiable functions. Define the differential operator L by $Lf(x) = f''(x) - xf'(x)$. Check first that for any nonnegative twice differentiable function g, $dT_{e^{-t}}g/dt = LT_{e^{-t}}g$ and that, for any $r > 1$,

$$\text{Ent}\left(g(X)^r\right) \leq -\frac{r^2}{2(r-1)}E\left[g(X)^{r-1}Lg(X)\right].$$

Hint: this follows from the Gaussian logarithmic Sobolev inequality by rewriting $E\left[h(X)^2\right]$ using integration by parts, where $h(x) = \partial_x g^{r/2}(x)$. Prove that $(1/q(t)) \log E\left[|T_{e^{-t}}f(X)|^{q(t)}\right]$ is a nonincreasing function of t. The argument parallels the proof of Lemma 5.19. The collection of operators $(T_{e^{-t}})_{t\geq 0}$ is known as the Ornstein–Uhlenbeck semigroup. The hypercontractivity of the Ornstein–Uhlenbeck semigroup was first proved by Nelson (1973).

5.19. (GAUSSIAN HYPERCONTRACTIVITY IN SEVERAL DIMENSIONS) Suppose $X_1, \ldots, X_n$ are independent standard Gaussian random variables. For any $0 < \gamma \leq 1$, let the operator T_γ map any function $f: \mathbb{R}^n \to \mathbb{R}$ such that $E[f(X_1, \ldots, X_n)^2] < \infty$ to another function

$$T_\gamma f(y_1, \ldots, y_n) = E\left[f\left(\gamma y_1 + \sqrt{1-\gamma^2}X_1, \ldots, \gamma y_n + \sqrt{1-\gamma^2}X_n\right)\right].$$

Let $t \geq 0$, $1 < p < \infty$, $q(t) = 1 + e^{2t}(p-1)$, and let the function f be such that $E[|f(X_1, \ldots, X_n)|^p] < \infty$. Prove that

$$\left(E\left[|T_{e^{-t}}f(X_1, \ldots, X_n)|^{q(t)}\right]\right)^{1/q(t)} \leq \left(E\left[|f(X_1, \ldots, X_n)|^p\right]\right)^{1/p}.$$

Hint: use the results of Exercise 5.18 and imitate the last part of the proof of Theorem 5.16.

5.20. (GAUSSIAN HYPERCONTRACTIVITY AND HERMITE POLYNOMIALS) Recall the definition of Hermite polynomials h_n from Section 5.9. Let the operator T_γ with $0 < \gamma \leq 1$ be defined as in Exercise 5.18. Prove that the Hermite polynomials are eigenfunctions of T_γ, for all $n = 1, 2, \ldots$, that is,

$$T_\gamma h_n = \gamma^n h_n.$$

Hint: recall the definition of the differential operator $Lf(x) = f''(x) - xf'(x)$ from Exercise 5.18. Use the relation $dT_{e^{-t}}g/dt = LT_{e^{-t}}g$ established in Exercise 5.18, the fact that $L \circ T_{e^{-t}} = T_{e^{-t}} \circ L$, and the fact that Hermite polynomials satisfy $nh_n = -Lh_n$. For a vector $\overline{k} = (k_1, \ldots, k_n)$ of nonnegative integers, let $|\overline{k}| = \sum_{i=1}^n k_i$. Define $f(x_1, \ldots, x_n) = \sum_{\overline{k} \in \mathbb{N}^n} \alpha_{\overline{k}} \prod_{i=1}^n h_{k_i}(x_i)$, where $\sum_{\overline{k} \in \mathbb{N}^n} \alpha_{\overline{k}}^2 < \infty$. Show that

$$T_\gamma f = \sum_{\overline{k}} \gamma^{\overline{k}} \alpha_{\overline{k}} \prod_{i=1}^n h_{k_i}.$$

This is the exact Gaussian analog of Theorem 5.16. The Hermite polynomials form an orthonormal basis of the Hilbert space of square integrable functions of a vector of independent standard Gaussian random variables, and they are the eigenfunctions of the hypercontractive operator T_γ.

5.21. (TIGHTNESS OF HYPERCONTRACTIVE BOUNDS) For $\lambda \geq 0$, define $f_\lambda(x) = \exp(\lambda x - \lambda^2/2)$. Let the operator T_γ (for $\gamma \in [0, 1)$ be defined as in Exercise 5.18. Compute $T_\gamma f_\lambda$ and $\mathbf{E}[|f_\lambda(X)|^p]$, where X is a standard Gaussian random variable and $p > 1$. Check that if $q > 1 + e^{2t}$,

$$\sup_{f:\mathbf{E}[|f(X)|^2]<\infty} \frac{\mathbf{E}[|T_{e^{-t}}f(X)|^q]^{1/q}}{\mathbf{E}[|f(X)|^2]^{1/2}} = \infty.$$

This proves that the hypercontractive bounds of Exercises 5.18 and 5.19 are tight.

5.22. (HERMITE POLYNOMIALS) Recall the definition of the Hermite polynomials from Section 5.9. Prove that for $\lambda, x \in \mathbb{R}$,

$$e^{\lambda x - \lambda^2/2} = \sum_{k=0}^{\infty} \frac{\lambda^k}{\sqrt{k!}} h_k(x).$$

Prove that if (X_1, X_2) is a Gaussian vector where X_1 and X_2 are standard Gaussian random variables, then

$$\mathbf{E}\left[\exp\left(\lambda X_1 - \frac{\lambda^2}{2}\right)\exp\left(\mu X_2 - \frac{\mu^2}{2}\right)\right] = \exp\left(\lambda\mu\mathbf{E}[X_1 X_2]\right).$$

Combine the two statements in order to establish that the Hermite polynomials form an orthonormal family, that is,

$$\boldsymbol{E}\left[h_i(X)h_j(X)\right] = \begin{cases} 1 & \text{if } i = j \\ 0 & \text{otherwise,} \end{cases}$$

where X is standard Gaussian. Prove the following three-term recurrences for normalized Hermite polynomials:

$$xh_n(x) = \sqrt{n+1}h_{n+1}(x) + h_n'(x)$$
$$xh_n(x) = \sqrt{n+1}h_{n+1}(x) + \sqrt{n}h_{n-1}(x)$$

for all $n = 0, 1, 2, \ldots$ and $x \in \mathbb{R}$. Note that the three-term recurrences entail $h_n'(x) = \sqrt{n}h_{n-1}(x)$. From the recurrences, deduce the Christoffel–Darboux formula: for $x \neq y$, for $n = 1, 2, \ldots,$

$$\sum_{i=0}^{n-1} h_i(x)h_i(y) = \sqrt{n}\frac{h_n(x)h_{n-1}(y) - h_{n-1}(x)h_n(y)}{(x-y)}.$$

The Hermite polynomials form an orthonormal basis of the space $L_2(\gamma)$ of square-integrable functions under the standard Gaussian distribution γ. This can be checked by invoking the density of bounded continuous functions in $L_2(\gamma)$ and the density of polynomials in the set of continuous functions with respect to the supremum norm over compact sets.

5.23. (INVARIANCE OF GUE) Prove that the GUE is invariant under unitary transformations: if $W \in \mathcal{U}_n$, and the random matrix H is distributed according to the GUE, then so is WH.

5.24. (ZEROS OF MULTIVARIATE POLYNOMIAL) Prove that if p is a nonzero n-variate polynomial, then $\{x \in \mathbb{R}^n : p(x) = 1\}$ has Lebesgue measure 0 over $\mathbb{R}^n$.
Hint: use induction over n and the Tonelli–Fubini theorem.

5.25. (MULTIPLE ROOTS AND DISCRIMINANT) Let $P(x) = \sum_{i=0}^m a_i x^i$ and $Q(x) = \sum_{j=0}^n b_j x^j$. The *Sylvester matrix* $S_{P,Q}$ is the $(n+m) \times (n+m)$ matrix defined by stacking $n-1$ circular shifts of

$$(a_m, a_{m-1}, \ldots, a_0, \underbrace{0, \ldots, 0}_{n-1 \text{ times}})$$

and $m-1$ circular shifts of

$$(b_n, b_{n-1}, \ldots, b_0, \underbrace{0, \ldots, 0}_{m-1 \text{ times}}):$$

$$S_{P,Q} = \begin{pmatrix} a_m & a_{m-1} & \dots & & a_0 & 0 & \dots & \dots & 0 \\ 0 & a_m & a_{m-1} & \dots & \dots & a_0 & 0 & \dots & 0 \\ 0 & 0 & a_m & a_{m-1} & \dots & \dots & a_0 & 0 \dots & 0 \\ 0 & 0 & \dots & \dots & \dots & \dots & \dots & \dots & 0 \\ 0 & 0 & 0 & \dots & \dots & a_m & a_{m-1} & \dots & a_0 \\ b_n & b_{n-1} & \dots & \dots & \dots & b_0 & 0 & \dots & 0 \\ 0 & b_n & b_{n-1} & \dots & \dots & \dots & b_0 & 0 \dots & 0 \\ \dots & 0 & b_n & b_{n-1} & \dots & \dots & \dots & b_0 \dots & 0 \\ 0 & \dots & \dots & \dots & \dots & \ddots & \dots & b_0 & 0 \\ 0 & 0 & \dots & 0 & b_n & b_{n-1} & \dots & \dots & b_0 \end{pmatrix}.$$

The determinant of $S_{P,Q}$ is called the *discriminant* of P and Q and it is denoted by $D(P, Q)$. Prove that if P and Q have a common root, then $D(P, Q) = 0$. Prove that if P has multiple roots, then $D(P, P') = 0$. Prove that there exists an n^2-variate polynomial P such that if an $n \times n$ matrix A has eigenvalues with multiplicity larger than 1, then P vanishes on the vector defined by the coefficients of A. See Lang (1965) for details about discriminants.

In Exercises 5.26–5.35, we denote by $\mathcal{H}_n^d$ the subset of $n \times n$ Hermitian matrices with pairwise distinct eigenvalues. Let $\mathcal{D}_n^d$ denote the subset of $n \times n$ real diagonal matrices with decreasing diagonal entries. Let $\mathcal{U}_n^g$ denote the subset of $n \times n$ unitary matrices with real positive diagonal entries. If A is an $n \times n$ matrix, then the i, j minor of A, $A^{(i,j)}$ is obtained by deleting the ith row and the jth column of A. We denote $A^{(k)} = A^{(k,k)}$. A matrix from $\mathcal{U}_n^g$ belongs to $\mathcal{U}_n^{vg}$ if all its minors are invertible. Let the set of "good" Hermitian matrices $\mathcal{H}_n^{d,g}$ be the subset of $n \times n$ Hermitian matrices that admit a decomposition UDU^* where $U \in \mathcal{U}_n^{vg}$ and $D \in \mathcal{D}_n^d$.

5.26. Prove that, almost surely, a random matrix from the GUE has pairwise distinct eigenvalues, that is, $\mathcal{H}_n \setminus \mathcal{H}_n^d$ has Lebesgue measure 0. *Hint:* the coefficients of the characteristic polynomial of a matrix are polynomials of the entries of the matrix. Use Exercises 5.24 and 5.25.

5.27. Prove that if $H \in \mathcal{H}_n^d$ and for all $1 \le k \le n$, H and $H^{(k)}$ do not have common eigenvalues then if $H = UDU^*$ with $U \in \mathcal{U}_n$ and $D \in \mathcal{D}_n$, U has nonzero entries. Prove that $\mathcal{H}_n \setminus \mathcal{H}_n^{d,g}$ has Lebesgue measure 0. *Hint:* the *adjugate* $\text{Adj}(H)$ of H is defined by $\text{Adj}(H)_{i,j} = (-1)^{i+j}\det(H^{(j,i)})$. Recall that $H\text{Adj}(H) = \text{Adj}(H)H = \det(H)I_n$ (see, e.g. Apostol (1969, Theorem 3.12)). Let λ be an eigenvalue of $H \in \mathcal{H}_n^d$. Let $A = H - \lambda\text{Id}_n$. Use the assumption $H \in \mathcal{H}_n^d$ to check that the columns of $\text{Adj}(A)$ are scalar multiples of a column of U. Finally, use the assumption that H and $H^{(k)}$ do not have common eigenvalues to verify that $\text{Adj}(A)$ has nonzero entries. To prove the last statement, use results from Exercises 5.24 and 5.25. (See the proof of Lemma 2.5.5 in Anderson, Guionnet and Zeitouni (2010).)

5.28. (DENSITY OF EIGENVALUES I) Prove the existence of a diffeomorphism (i.e. a bijective differentiable map whose inverse is differentiable) between $\mathcal{H}_n^{d,g}$ and $\mathcal{D}_n^d \times \mathbb{R}^{n(n-1)}$ where $\mathcal{D}_n^d$ is the set of $n \times n$ real diagonal matrices with decreasing diagonal coefficients. *Hint:* let T be the operator that maps $U \in \mathcal{U}_n^{vg}$ to the vector

$$T(U) = \left(\frac{U_{1,2}}{U_{1,1}}, \frac{U_{1,3}}{U_{1,1}}, \ldots, \frac{U_{1,n}}{U_{1,1}}, \frac{U_{2,3}}{U_{2,2}}, \ldots, \frac{U_{2,n}}{U_{2,2}}, \ldots, \frac{U_{n-1,n}}{U_{n-1,n-1}} \right).$$

Each $U_{i,j}/U_{i,i}$ $(1 \leq i < j \leq n)$ should be considered as a pair of real numbers corresponding to the real and imaginary part. Check that T is one-to-one on $\mathcal{U}_n^{vg}$ and that $T(\mathcal{U}_n^{vg})$ is open in $\mathbb{R}^{n(n-1)}$. (See Anderson, Guionnet, and Zeitouni (2010, Lemma 2.5.5).)

5.29. (DENSITY OF EIGENVALUES II) Let T be defined as in Exercise 5.28. Let $J : \mathcal{D}_n^d \times T(\mathcal{U}_n^g) \to \mathcal{H}_n^{d,g}$ be the inverse of the mapping defined by

$$\mathcal{H}_n^{d,g} \to \mathcal{D}_n^d \times T(\mathcal{U}_n^g)$$
$$H = U\text{diag}(\lambda_1, \ldots, \lambda_n)U^* \mapsto (\text{diag}(\lambda_1, \ldots, \lambda_n), T(U)).$$

Let $p = (p_1, \ldots, p_{n(n-1)}) \in T(\mathcal{U}_n^g)$. Define a one-to-one mapping $r : \{(i,j) : 1 \leq i < j \leq n\} \to \{1, n(n-1)/2\}$ by $r(i,j) = \sum_{k=1}^{i-1}(n-k) + j - i + 1$ (this is the rank of (i,j) when traversing the upper-triangle in a row-wise fashion). The purpose of this exercise is to outline a collection of equations satisfied by the $n^2 \times n^2$ matrix of partial derivatives of J (the Jacobian matrix of J). Exercise 5.30 takes advantage of these equations to establish the fact that the determinant of the Jacobian matrix (the Jacobian determinant) is the product of the square of the Vandermonde determinant defined by $(\lambda_1, \ldots, \lambda_n)$ (that is $\prod_{1 \leq i<j \leq n}(\lambda_i - \lambda_j)$) and of a function of $(p_1, \ldots, p_{n(n-1)}) \in T(\mathcal{U}_n^g)$. This observation is an essential part of the proof of Lemma 5.24. For each $1 \leq \ell \leq n(n-1)$, let $\partial U/\partial p_\ell$ be the $n \times n$ complex matrix of partial derivatives of U with respect to p_ℓ. Verify that the complex matrix $S_\ell = U^* \frac{\partial U}{\partial p_\ell}$ is skew Hermitian, that is,

$$S_\ell^* = \frac{\partial U^*}{\partial p_\ell} U = -S_\ell.$$

Now letting $B_\ell = U^* \partial H/\partial p_\ell U$ for each $1 \leq \ell \leq n(n-1)$, verify that

$$U^* \frac{\partial H}{\partial p_\ell} U = S_\ell \times \text{diag}(\lambda_1, \ldots, \lambda_n) - \text{diag}(\lambda_1, \ldots, \lambda_n) \times S_\ell,$$

or equivalently that

$$B_\ell[i,j] = S_\ell[i,j](\lambda_j - \lambda_i), \tag{5.7}$$

for $i, j \leq n$. Verify that for each $1 \leq i \leq n$, $\partial H/\partial \lambda_i$ is the matrix of the orthogonal projection on the line generated by the ith column of U. Verify that for each $1 \leq i \leq n$,

$$U^* \frac{\partial H}{\partial \lambda_i} U = \operatorname{diag}\Big(\underbrace{0, \ldots 0}_{i-1 \text{ times}}, 1, \underbrace{0, \ldots, 0}_{n-i \text{ times}} \Big),$$

which implies

$$\left(U^* \frac{\partial H}{\partial \lambda_i} U \right) [j,k] = \sum_{j'=1}^{n} \sum_{k'=1}^{n} \frac{\partial H}{\partial \lambda_i} [j', k'] \overline{U}[j', j] U[k', k] = \mathbb{1}_{\{i=j=k\}}, \tag{5.8}$$

for $1 \leq j, k \leq n$.

5.30. (DENSITY OF EIGENVALUES III) The Jacobian matrix of the mapping J defined in Exercise 5.29 is an $n^2 \times n^2$ real matrix $\operatorname{Jac}(J)$ which may be described in partitioned form by

$$\operatorname{Jac}(J) = \begin{pmatrix} \overbrace{\dfrac{\partial H_{j,j}}{\partial \lambda_i}}^{1 \leq j \leq n} & \overbrace{\dfrac{\operatorname{Re} \partial H_{j,k}}{\partial \lambda_i}}^{1 \leq j < k \leq n} & \overbrace{\dfrac{\operatorname{Im} \partial H_{j,k}}{\partial \lambda_i}}^{1 \leq j < k \leq n} \\ \hdashline \dfrac{\partial H_{j,j}}{\partial p_\ell} & \dfrac{\operatorname{Re} \partial H_{j,k}}{\partial p_\ell} & \dfrac{\operatorname{Im} \partial H_{j,k}}{\partial p_\ell} \end{pmatrix} \begin{matrix} 1 \leq i \leq n \\ \\ 1 \leq \ell \leq n(n-1). \end{matrix}$$

The key step in the proof of Lemma 5.24 consists of showing that $\det(\operatorname{Jac}(J))$ is the product of the square of a Vandermonde determinant and of an expression that only depends on U. Define the matrix M in partitioned form as

$$M = \begin{pmatrix} \overbrace{\overline{U}[i,j]U[i,j]}^{1 \leq j \leq n} & \overbrace{\operatorname{Re} \overline{U}[i,j]U[i,k]}^{1 \leq j < k \leq n} & \overbrace{\operatorname{Im} \overline{U}[i,j]U[i,k]}^{1 \leq j < k \leq n} \\ \hdashline \operatorname{Re} \overline{U}[j',j]U[k',j] & 2\operatorname{Re} \overline{U}[j',j]U[k',k] & 2\operatorname{Im} \overline{U}[j',j]U[k',k] \\ \hdashline -\operatorname{Im} \overline{U}[j',j]U[k',j] & -2\operatorname{Im} \overline{U}[j',j]U[k',k] & 2\operatorname{Re} \overline{U}[j',j]U[k',k] \end{pmatrix} \begin{matrix} 1 \leq i \leq n \\ \\ 1 \leq j' < k' \leq n \\ \\ 1 \leq j' < k' \leq n. \end{matrix}$$

Write $C = \mathrm{Jac}(J) \times M$ in partitioned form as

$$C = \begin{pmatrix} C^{1,1} & C^{1,2} & C^{1,3} \\ C^{2,1} & C^{2,2} & C^{2,3} \end{pmatrix}$$

The exercise mostly consists of checking that

$$C = \left(\begin{array}{ccc} \overbrace{}^{1\le j\le n} & \overbrace{}^{1\le j<k\le n} & \overbrace{}^{1\le j<k\le n} \\ \mathrm{Id}_n & 0 & 0 \\ \hdashline ? & \mathrm{Re}B_\ell[j,k] & \mathrm{Im}B_\ell[j,k] \end{array} \right) \begin{array}{l} \\ 1 \le i \le n \\ 1 \le \ell \le n(n-1) \end{array}$$

where B_ℓ is defined as in Exercise 5.29.

1. Check that $C^{1,1} = \mathrm{Id}_n$ while $C^{1,2} = C^{1,3} = 0$. *Hint:* verify that $C^{1,1}[i,j] = U^* \frac{\partial H}{\partial \lambda_i} U[j,j]$, while for $1 \le j < k \le n$, $m = r(j,k)$, $C^{1,2}[i,m]$ (resp. $C^{1,3}[i,m]$) is the real (resp. imaginary) part of the $U^* \frac{\partial H}{\partial \lambda_i} U[j,k]$. Use (5.8) from Exercise 5.29.
2. Check that for $\ell \in \{1,\ldots,n(n-1)\}$ and $m \in \{1,\ldots,n(n-1)/2\}$, $C^{2,2}[\ell,m] = \mathrm{Re}B_\ell[i,j] = \mathrm{Re}S_\ell[i,j](\lambda_i - \lambda_j)$ and $C^{2,3}[\ell,m] = \mathrm{Im}B_\ell[i,j] = \mathrm{Im}S_\ell[i,j](\lambda_i - \lambda_j)$ where $1 \le i < j \le n$, $m = r(i,j)$. *Hint:* use (5.7) from Exercise 5.29.
3. Check that the determinant of the $n(n-1)$ by $n(n-1)$ real matrix $\left(C^{2,2}\ C^{2,3}\right)$ is the product of $\Delta(\lambda_1,\ldots,\lambda_n)^2$ and of a quantity that only depends on U, where $\Delta(\lambda_1,\ldots,\lambda_n)$ is the Vandermonde determinant $\prod_{1\le i<j\le n}(\lambda_i - \lambda_j)$. Deduce from this that the Jacobian determinant $\det(\mathrm{Jac}(J))$ can be written as the product of $\Delta(\lambda_1,\ldots,\lambda_n)^2$ and of a quantity that only depends on the coefficients of U.
4. Conclude the proof of Lemma 5.24 by combining the results of Exercises 5.24–5.29 and the change-of-variables formula in multiple integrals.

(This argument is from Mehta (2004, Chapter 3) and Anderson, Guionnet, and Zeitouni (2010, Chapter 2); see also Tao (2012). It can be tailored to other ensembles of Gaussian random matrices.)

5.31. Using the notation of Theorem 5.23, prove that

$$\int_{\mathbb{R}^n} \Delta(x_1,\ldots,x_n)^2 \frac{e^{-\sum_{i=1}^n \frac{x_i^2}{2}}}{\sqrt{2\pi}^n} = \prod_{j=0}^{n-1} j!\,.$$

Hint: use the pattern of proof of Lemma 5.24.

5.32. (MOMENTS OF THE SEMI-CIRCULAR DISTRIBUTION) The semi-circular distribution has density $2/\pi\sqrt{1-x^2}\mathbb{1}_{\{|x|\le 1\}}$. Let m_{2k} denote its $2k^{\text{th}}$ moment for $k = 1, 2, \ldots$. Prove that

$$m_{2k} = \frac{C_k}{2^{2k}}$$

where $C_k = \binom{2k}{k}/(k+1)$ is the kth Catalan number. *Hint:* prove that $m_{2k} = 2/(\pi(2k+1))\int_{-\pi/2}^{\pi/2}\sin(\theta)^{2k+2}d\theta$ and also that $m_{2k} = 2/(\pi(2k+2))\int_{-\pi/2}^{\pi/2}\sin(\theta)^{2k}d\theta$.

5.33. (CONCENTRATION OF THE SPECTRAL MEASURE) Let H be a random $n \times n$ matrix from the GUE with eigenvalues $\lambda_1 \geq \cdots \geq \lambda_n$ and spectral measure L_n. Let $\mathcal{B}$ denote the set of functions f on $\mathbb{R}$ with $\sup_{x\in\mathcal{R}} |f(x)| \leq 1$ and Lipschitz constant not larger than 1. Prove that for $t \geq 0$

$$\sup_{f\in\mathcal{B}} \boldsymbol{P}\left\{\left|L_n(f) - \boldsymbol{E}[L_n(f)]\right| \geq t\right\} \leq 2e^{-\frac{n^2t^2}{2}}.$$

Hint: use Lidskii's inequality (see Exercise 3.16) and Theorem 5.6. (See Anderson, Guionnet, and Zeitouni (2010, Theorem 2.3.5).)

5.34. (CONCENTRATION OF THE SPECTRAL MEASURE OF MATRICES FROM THE GUE, CONTINUED) Let H be a random $n \times n$ random matrix from the GUE with eigenvalues $\lambda_1 \geq \cdots \geq \lambda_n$ and spectral measure L_n. Let $\mathcal{B}$ denote the set of functions f on $\mathbb{R}$ with $\sup_{x\in\mathcal{R}} |f(x)| \leq 1$ and Lipschitz constant not larger than 1. Let $Z = \sup_{f\in\mathcal{B}} \left|L_n(f) - \boldsymbol{E}[L_n(f)]\right|$ be the bounded Lipschitz distance between the empirical spectral measure and the average spectral measure. Prove that for $t \geq 0$,

$$\boldsymbol{P}\{Z \geq \boldsymbol{E}Z + t\} \leq 2e^{-\frac{n^2t^2}{2}}.$$

Prove that there exists a universal constant κ such that

$$\boldsymbol{E}Z \leq \frac{\kappa}{\sqrt{n}}.$$

Hint: use again Theorem 5.6 as in Exercise 5.33 to establish the tail bound. (See Anderson, Guionnet, and Zeitouni (2010, Theorem 2.3.5).) A proof of the upper bound for $\boldsymbol{E}Z$ can be derived from Götze and Tikhomirov (2003, 2005) who state similar bounds for the uniform distance between the empirical spectral distribution function and the semi-circular distribution function and the uniform distance between the average empirical spectral distribution function and the semi-circular distribution function. By standard results (see Dudley, 2002), the bounded-Lipschitz distance to the semi-circular distribution is within a constant factor of the uniform distance to the semi-circular distribution. Note that this upper bound holds under rather general moment conditions on the entries of the random Hermitian matrices.

5.35. (MOMENT-GENERATING FUNCTION OF THE SPECTRAL MEASURE OF RANDOM MATRICES FROM THE GUE) Let H be an $n \times n$ random matrix from the GUE. The aim of this exercise is to compute the expected moment-generating function of the spectral measure L_n of H, that is,

$$F_n(s) = \mathbf{E}\left[\sum_{i=1}^{n} \frac{1}{n} e^{s\lambda_i}\right] \quad \text{for } s \in \mathbb{R},$$

and then to check the pointwise convergence of F_n to the moment-generating function of the semi-circular distribution, that is,

$$\lim_{n\to\infty} F_n(s) = \sum_{k\in\mathbb{N}} \frac{m_{2k}s^{2k}}{(2k)!} \quad \text{for all } s \in \mathbb{R}.$$

The even moments $(m_{2k})_{k\in\mathbb{N}}$ of the semi-circular distribution are determined in Exercise 5.32. As in Exercises (5.20–5.22), h_i denotes the ith normalized Hermite polynomial. Use the Christoffel–Darboux's identity (see Exercise 5.22) to establish that for all $x \in \mathbb{R}$ and $n = 1, 2, \ldots$, $\frac{1}{n}\sum_{i=0}^{n-1} h_i(x)^2 = \frac{1}{\sqrt{n}}\left(h_n'(x)h_{n-1}(x) - h_n(x)h_{n-1}'(x)\right)$. To lighten notation, denote $K_n(x,x) = \frac{e^{-x^2/2}}{\sqrt{2\pi n}}\left(h_n'(x)h_{n-1}(x) - h_n(x)h_{n-1}'(x)\right)$. Use Lemma 5.25 to prove that for any bounded continuous function f,

$$\mathbf{E}L_n f = \int_{\mathbb{R}} f(x/\sqrt{4n})K_n(x,x)dx.$$

Hint: prove and use the fact that

$$\frac{d}{dx}K_n(x,x) = -e^{-x^2/2}h_n(x)h_{n-1}(x) = -e^{-x^2/2}\frac{h_n(x)h_n'(x)}{\sqrt{n}}.$$

(See Anderson, Guionnet, and Zeitouni (2010, page 102).)

5.36. (GAUSSIAN ORTHOGONAL ENSEMBLE) A random real symmetric $n \times n$ matrix A belongs to the *Gaussian Orthogonal Ensemble* (GOE) if the entries $(A_{i,j})_{1\le i\le j\le n}$ are independent centered Gaussian random variables with variance $1/n$. Following the approach described in Exercises 5.24–5.30, prove the determinantal formula for the GOE: the joint density of the eigenvalues of a $n \times n$ random matrix from the GOE at $\lambda_1 > \lambda_2 > \cdots > \lambda_n$ is

$$\frac{D_n}{\sigma^{2n}}\Delta(\lambda_1, \ldots, \lambda_n)\exp\left(-\frac{\sum_{i=1}^{n}\lambda_i^2}{2\sigma^2}\right)$$

where D_n is a normalizing constant and $\sigma = 1/\sqrt{4n}$.

5.37. (TAIL INEQUALITY FOR MAXIMA OF GAUSSIAN RANDOM VECTORS) Let Z be the maximum of the absolute value of n independent standard Gaussian random variables, and for $t \geq 1$, let $U(t) = \inf\{x : \Phi(x) \geq 1 - 1/t\}$. Prove that for $t > 0$,

$$P\{Z - EZ \geq t + \delta_n\} \leq \exp\left(-\frac{t^2 U(2n)^2}{2(2 + tU(2n)/3)}\right),$$

where $\delta_n > 0$ and $\lim_n (2\log(2n))^{3/2}\delta_n = \pi^2/12$. *Hint:* represent Z as $U(2\exp(Y))$ where Y is the maximum of n independent exponential random variables with expected value 1 and use the fact that $U(e^x)$ is concave in x. The second part of the statement may be checked using standard results from extreme value theory (see de Haan and Ferreira (2006)).

6

The Entropy Method

In Chapter 3 we saw that the Efron–Stein inequality served as a powerful tool for bounding the variance of functions of several independent random variables. In many cases, however, it is reasonable to expect that, as in the case of sums of bounded random variables, the tail probabilities decrease at an exponential speed, a phenomenon the Efron–Stein inequality fails to capture. In Chapter 5 we have seen that logarithmic Sobolev inequalities, together with Herbst's argument, may be used to derive exponential concentration inequalities. However, the logarithmic Sobolev inequalities presented there are only valid for functions of either Bernoulli or Gaussian random variables and therefore the scope of the concentration inequalities obtained is significantly more limited than that of the Efron–Stein inequality.

The purpose of this chapter is to attempt to generalize the methodology based on logarithmic Sobolev inequalities that allows one to prove exponential concentration bounds that hold for functions of *arbitrary* independent random variables. A way to achieve this is by trying to mimic the procedure that worked for functions of Bernoulli and Gaussian random variables, that is, to start with a logarithmic Sobolev inequality and then, according to Herbst's trick, apply it to exponential functions of the random variable of interest. Since exact analogs of the Bernoulli and Gaussian logarithmic Sobolev inequalities do not always exist, we need to resort to appropriate modifications. Luckily, the sub-additivity of entropy (see Theorems 4.10 and 4.22) holds in a great generality and indeed, this inequality serves as our starting point. Then, by bounding the right-hand side of the inequality of Theorem 4.10, we obtain an appropriate *modified logarithmic Sobolev inequality* which, in turn, can be used via Herbst's argument to derive exponential concentration inequalities.

We term the proof method described above the *entropy method*, and the purpose of this chapter is to define its basis and to show some of the simplest powerful concentration bounds one can achieve using this method. In Chapters 11, 12, 14, and 15 we elaborate the entropy method and show various extensions.

As in Chapter 3, we investigate the concentration behavior of a real-valued random variable $Z = f(X_1, \ldots, X_n)$ where $X_1, \ldots, X_n$ are independent random variables taking values in a measurable space $\mathcal{X}$ and $f : \mathcal{X}^n \to \mathbb{R}$ is a function.

The main purpose of the entropy method for proving concentration inequalities is to apply the sub-additivity of entropy (Theorems 4.10 and 4.22) for the positive random variable $Y = e^{\lambda Z}$ where λ is a real number. Recall that by the sub-additivity of entropy,

$$\text{Ent}(Y) \le \mathbf{E} \sum_{i=1}^{n} \text{Ent}^{(i)}(Y)$$

or, equivalently,

$$\begin{aligned} &\mathbf{E}[Y \log Y] - (\mathbf{E}Y) \log(\mathbf{E}Y) \\ &\le \sum_{i=1}^{n} \mathbf{E}\left[\mathbf{E}^{(i)}[Y \log Y] - \left(\mathbf{E}^{(i)}Y\right) \log\left(\mathbf{E}^{(i)}Y\right)\right] \end{aligned} \tag{6.1}$$

where $\mathbf{E}^{(i)}$ denotes integration with respect to the distribution of X_i only. Then, normalizing by $\mathbf{E}e^{\lambda Z}$ and denoting the logarithmic moment-generating function of $Z - \mathbf{E}Z$ by $\psi(\lambda) = \log \mathbf{E}e^{\lambda(Z-\mathbf{E}Z)}$, the left-hand side of this inequality becomes

$$\frac{\text{Ent}\left(e^{\lambda Z}\right)}{\mathbf{E}e^{\lambda Z}} = \lambda\psi'(\lambda) - \psi(\lambda). \tag{6.2}$$

Our strategy is based on using (6.2) the sub-additivity of entropy and then univariate calculus to derive upper bounds for the derivative of $\psi(\lambda)$. By solving the obtained differential inequality, we obtain tail bounds via Chernoff's bounding.

To achieve this in a convenient way, we need some further bounds for the right-hand side of the inequality above. This is the purpose of Section 6.3 in which, relying on the sub-additivity of entropy, we prove some basic results which will serve as our starting point. These results are reminiscent of the classical logarithmic Sobolev inequalities discussed in Chapter 5, where it is shown that concentration inequalities follow from logarithmic Sobolev inequalities by *Herbst's argument.* Here we formalize this argument.

Proposition 6.1 (HERBST'S ARGUMENT) *Let Z be an integrable random variable such that for some $v > 0$, we have, for every $\lambda > 0$,*

$$\frac{\text{Ent}\left(e^{\lambda Z}\right)}{\mathbf{E}e^{\lambda Z}} \le \frac{\lambda^2 v}{2}.$$

Then, for every $\lambda > 0$,

$$\log \mathbf{E}e^{\lambda(Z-\mathbf{E}Z)} \le \frac{\lambda^2 v}{2}.$$

Proof The condition of the proposition means, via (6.2), that

$$\lambda\psi'(\lambda) - \psi(\lambda) \le \frac{\lambda^2 v}{2},$$

or equivalently,

$$\frac{1}{\lambda}\psi'(\lambda) - \frac{1}{\lambda^2}\psi(\lambda) \le \frac{v}{2}.$$

Setting $G(\lambda) = \lambda^{-1}\psi(\lambda)$, we see that the differential inequality becomes $G'(\lambda) \le v/2$. Since $G(\lambda) \to 0$ as $\lambda \to 0$, this implies $G(\lambda) \le \lambda v/2$, and the result follows. □

First, we present in Section 6.1 two simple direct methods to bound the right-hand side of the inequality of the sub-additivity of entropy and use Herbst's argument to conclude. This permits us to derive the celebrated *bounded differences inequality*, a simple prototypical exponential concentration inequality for functions of bounded differences that has found countless applications. We also present a sharper version in which the bounded differences assumption is significantly relaxed.

In Section 6.4 we present the first and simplest application of these modified logarithmic Sobolev inequalities. This first example is surprisingly powerful as it may be used to prove exponential concentration in many interesting cases. We describe some applications. The obtained inequalities reach further than the bounded differences inequality as they are able to handle much more general functions than just those having the bounded-differences property. A simple but useful application for convex Lipschitz functions of independent random variables is presented in Section 6.6.

In Section 6.7 we return to the class of self-bounding functions introduced in Section 3.3 and prove an exponential concentration inequality, thus providing a significant sharpening of Corollary 3.7. The notion of self-bounding function is generalized and further investigated in Section 6.11.

In Sections 6.8, 6.9, and 6.13 we use the entropy method to prove inequalities that may be considered as exponential versions of the Efron–Stein inequality. Various concentration results are shown here under different conditions with the purpose of demonstrating the flexibility of the entropy method.

We close the chapter by proving Janson's celebrated inequality for the lower tail probabilities of random Boolean polynomials. Even though Janson's inequality is not based on the entropy method, its proof shows some similarities with the techniques we use throughout the chapter.

6.1 The Bounded Differences Inequality

As a first illustration of the entropy method, we derive an exponential concentration inequality for functions of bounded differences. Unlike the Bernoulli and Gaussian concentration inequalities of Chapter 5, this inequality is distribution free: apart from independence, nothing else is required from the random variables $X_1, \ldots, X_n$.

Recall that a function $f : \mathcal{X}^n \to \mathbb{R}$ has the *bounded differences property* if for some nonnegative constants $c_1, \ldots, c_n$,

$$\sup_{\substack{x_1,\ldots,x_n,\\ x_i' \in \mathcal{X}}} |f(x_1,\ldots,x_n) - f(x_1,\ldots,x_{i-1},x_i',x_{i+1},\ldots,x_n)| \leq c_i\,,\ 1 \leq i \leq n.$$

In Chapter 3, as a corollary of the Efron–Stein inequality, we saw that if f has the bounded differences property, then $Z = f(X_1,\ldots,X_n)$ satisfies $\mathrm{Var}(Z) \leq (1/4)\sum_{i=1}^n c_i^2$ (see Corollary 3.2). The *bounded differences inequality* shows that such functions satisfy a sub-Gaussian tail inequality in which the role of the variance factor is played by the Efron–Stein upper bound of the variance $v = (1/4)\sum_{i=1}^n c_i^2$.

Theorem 6.2 (BOUNDED DIFFERENCES INEQUALITY) *Assume that the function f satisfies the bounded differences assumption with constants $c_1,\ldots,c_n$ and denote*

$$v = \frac{1}{4}\sum_{i=1}^n c_i^2 .$$

Let $Z = f(X_1,\ldots,X_n)$ where the X_i are independent. Then

$$\mathbf{P}\{Z - \mathbf{E}Z > t\} \leq e^{-t^2/(2v)}.$$

Note that since the bounded differences assumption is symmetric, Z also satisfies the lower-tail inequality

$$\mathbf{P}\{Z - \mathbf{E}Z < -t\} \leq e^{-t^2/(2v)}.$$

The proof combines sub-additivity of entropy, Hoeffding's lemma (Lemma 2.2) and Herbst's argument. The following way of looking at Hoeffding's lemma may illuminate the use of the sub-additivity of entropy: if Y is a random variable taking its values in $[a,b]$, then we know from Lemma 2.2 that $\psi''(\lambda) \leq (b-a)^2/4$ for every $\lambda \in \mathbb{R}$, where $\psi(\lambda) = \log \mathbf{E}e^{\lambda(Y-\mathbf{E}Y)}$. Hence,

$$\lambda\psi'(\lambda) - \psi(\lambda) = \int_0^\lambda \theta\psi''(\theta)d\theta \leq \frac{(b-a)^2\lambda^2}{8},$$

which means that

$$\frac{\mathrm{Ent}(e^{\lambda Y})}{\mathbf{E}e^{\lambda Y}} \leq \frac{(b-a)^2\lambda^2}{8}. \tag{6.3}$$

By Proposition 6.1, this inequality implies Hoeffding's inequality, that is, $\psi(\lambda) \leq (b-a)^2\lambda^2/8$ for all λ. Thus, (6.3) is a way of rephrasing Hoeffding's inequality, which is stronger than the usual one.

Proof Recall that by the sub-additivity of entropy (6.1),

$$\mathrm{Ent}(e^{\lambda Z}) \leq \mathbf{E}\sum_{i=1}^n \mathrm{Ent}^{(i)}\left(e^{\lambda Z}\right)$$

where $\mathrm{Ent}^{(i)}$ denotes conditional entropy, given $X^{(i)} = (X_1, \ldots, X_{i-1}, X_{i+1}, \ldots, X_n)$. By the bounded differences assumption, given $X^{(i)}$, Z is a random variable whose range is in an interval of length at most c_i, so by (6.3),

$$\frac{\mathrm{Ent}^{(i)}\left(e^{\lambda Z}\right)}{\mathbf{E}^{(i)} e^{\lambda Z}} \leq \frac{c_i^2 \lambda^2}{8}.$$

Hence, by the sub-additivity of entropy,

$$\mathrm{Ent}(e^{\lambda Z}) \leq \mathbf{E}\left[\sum_{i=1}^{n} \left(\frac{c_i^2 \lambda^2}{8}\right) \mathbf{E}^{(i)} e^{\lambda Z}\right] = \sum_{i=1}^{n} \frac{c_i^2 \lambda^2}{8} \mathbf{E} e^{\lambda Z},$$

or equivalently,

$$\frac{\mathrm{Ent}\left(e^{\lambda Z}\right)}{\mathbf{E} e^{\lambda Z}} \leq \frac{\lambda^2 v}{2}.$$

Proposition 6.1 allows us to conclude that

$$\psi(\lambda) = \log \mathbf{E} e^{\lambda(Z - \mathbf{E}Z)} \leq \frac{\lambda^2 v}{2}.$$

Finally, by Markov's inequality,

$$\mathbf{P}\{Z > \mathbf{E}Z + t\} \leq e^{\psi(\lambda) - \lambda t} \leq e^{\lambda^2 v/2 - \lambda t}.$$

Choosing $\lambda = t/v$, the upper bound becomes $e^{-t^2/(2v)}$. □

This extends Corollary 3.2 to an exponential concentration inequality. Thus, the applications of Corollary 3.2 in all examples of functions with bounded differences shown in Section 3.2 (such as bin packing, the length of the longest common subsequence, the L_1 error of the kernel density estimate, etc.) are improved in an essential way without further work.

Next we describe another application which is the simplest example of a concentration inequality for sums of independent vector-valued random variables.

Example 6.3 (A HOEFFDING-TYPE INEQUALITY IN HILBERT SPACE) As an illustration of the power of the bounded differences inequality, we derive a Hoeffding-type inequality for sums of random variables taking values in a Hilbert space. In particular, let $X_1, \ldots, X_n$ be independent zero-mean random variables taking values in a separable Hilbert space such that $\|X_i\| \leq c_i/2$ with probability one and denote $v = (1/4) \sum_{i=1}^{n} c_i^2$. Then, for all $t \geq \sqrt{v}$,

$$\mathbf{P}\left\{\left\|\sum_{i=1}^{n} X_i\right\| > t\right\} \leq e^{-(t-\sqrt{v})^2/(2v)}.$$

This follows simply by observing that, by the triangle inequality, $Z = \left\| \sum_{i=1}^n X_i \right\|$ satisfies the bounded differences property with constants c_i, and therefore

$$P\left\{\left\|\sum_{i=1}^n X_i\right\| > t\right\} = P\left\{\left\|\sum_{i=1}^n X_i\right\| - E\left\|\sum_{i=1}^n X_i\right\| > t - E\left\|\sum_{i=1}^n X_i\right\|\right\}$$
$$\leq \exp\left(-\frac{\left(t - E\left\|\sum_{i=1}^n X_i\right\|\right)^2}{2v}\right).$$

The proof is completed by observing that, by independence,

$$E\left\|\sum_{i=1}^n X_i\right\| \leq \sqrt{E\left\|\sum_{i=1}^n X_i\right\|^2} = \sqrt{\sum_{i=1}^n E\|X_i\|^2} \leq \sqrt{v}.$$

The next example illustrates a surprising application in which the bounded differences inequality is applied in a quite unexpected context.

Example 6.4 (SPECTRAL MEASURE OF RANDOM HERMITIAN MATRICES) Let $H = (H_{i,j})$ be an $n \times n$ random Hermitian matrix such that the vectors $(H_i)_{1 \leq i \leq n}$ are independent, where $H_i = (H_{i,j})_{1 \leq j \leq i}$. Let L_H denote the empirical spectral measure of H (i.e. the probability measure that gives mass r/n to an eigenvalue of H with multiplicity r). Given a bounded function $g : \mathbb{R} \to \mathbb{R}$ that has total variation $\|g\|_{TV} \leq 1$, we are interested in the concentration of the random variable $Z = \int g dL_H$. Recall that the total variation of a function $g : \mathbb{R} \to \mathbb{R}$ is defined by

$$\|g\|_{TV} = \sup_{n=1,2,\ldots} \sup_{x_1 < \cdots < x_n} \sum_{i=1}^{n-1} |f(x_{i+1}) - f(x_i)|.$$

Remarkably, much can be said about Z without imposing any moment assumption on the entries of the matrix. The argument is surprisingly simple. Indeed, for every $x = (x_1, \ldots, x_n)$ such that $x_i \in \mathbb{C}^{i-1} \times \mathbb{R}$ for all i, denote by $H(x)$ the Hermitian matrix given by $(H(x))_{i,j} = x_{i,j}$ for $1 \leq j \leq i \leq n$ and define the function f by

$$f(x) = \int g dL_{H(x)}.$$

The random variable of interest Z is just $f(H_1, \ldots, H_n)$ and it remains to establish the bounded differences property for f to get a concentration inequality of Z around its mean. To this end, we apply the following deterministic rank inequality for spectral measures (which relies on the Cauchy interlacing theorem, see Exercises 6.2 and 6.3 below). Let A and B denote Hermitian matrices. If one denotes by F_A and F_B the distribution functions related to the spectral measures L_A and L_B, then

$$\|F_A - F_B\|_\infty \leq \frac{\operatorname{rank}(A - B)}{n}.$$

Integrating by parts (noting that $F_A - F_B$ tends to 0 at $-\infty$ and $+\infty$), one has

$$\left|\int g dL_A - \int g dL_B\right| = \left|\int (F_A - F_B)\, dg\right| \leq \|F_A - F_B\|_\infty ,$$

where the last inequality comes from the fact that the absolute total mass of the Stieljes measure dg equals $\|g\|_{TV} \leq 1$. Combining the two inequalities above, we find that for every x and x',

$$|f(x) - f(x')| \leq \frac{\text{rank}(H(x) - H(x'))}{n}.$$

Now if x' differs from x only in the i-th coordinate, the matrix $H(x) - H(x')$ has all zero entries, except maybe for one row and one column which proves that $\text{rank}\,(H(x) - H(x')) \leq 2$. This shows that f satisfies the bounded differences condition with $c_i = 2/n$ for all i and, therefore, the bounded differences inequality tells us that Z is a sub-Gaussian random variable with variance factor $1/n$. Consequently $\mathbf{P}\left\{|Z - \mathbf{E}Z| \geq t\right\} \leq 2e^{-nt^2/2}$ for all $t > 0$.

6.2 More on Bounded Differences

Next we show a more flexible variant of the bounded differences inequality of Theorem 6.2. It relaxes the bounded differences condition in that differences need not be bounded by "hard" constants c_i but rather by quantities that are allowed to depend on x, as long as the sum of their squares are bounded. More precisely, we say that a function $f : \mathcal{X}^n \to \mathbb{R}$ has the *x-dependent bounded differences property* if there exists a constant $v > 0$ such that for all $x = (x_1, \ldots, x_n) \in \mathcal{X}^n$ there exist n functions of $n - 1$ variables $c_1, \ldots, c_n : \mathcal{X}^{n-1} \to [0, \infty)$, such that for $1 \leq i \leq n$,

$$\sup_{\substack{x_i' \in \mathcal{X} \\ x_i'' \in \mathcal{X}}} |f(x_1, \ldots, x_{i-1}, x_i'', x_{i+1}, \ldots, x_n) - f(x_1, \ldots, x_{i-1}, x_i', x_{i+1}, \ldots, x_n)|$$

$$\leq c_i\left(x^{(i)}\right),$$

and $(1/4)\sum_{i=1}^n c_i^2(x^{(i)}) \leq v$ for all $x \in \mathcal{X}^n$. Here $x^{(i)} = (x_1, \ldots, x_{i-1}, x_{i+1}, \ldots, x_n)$ stands for the $(n-1)$-vector obtained by dropping the i-th component of x.

Clearly, the Efron–Stein inequality still implies that if f has the x-dependent bounded differences property, then $Z = f(X_1, \ldots, X_n)$ satisfies $\text{Var}\,(Z) \leq v$. The next sub-Gaussian tail inequality extends Theorem 6.2 to such functions.

Theorem 6.5 *Assume that the function f satisfies the x-dependent bounded differences property with constant v. Let $Z = f(X_1, \ldots, X_n)$ where the X_i are independent. Then for all $t > 0$,*

$$\mathbf{P}\{Z - \mathbf{E}Z \geq t\} \leq e^{-t^2/(2v)}.$$

Proof Since the proof is a simple extension of that for bounded differences inequality, we will only sketch it. By the x-dependent bounded differences assumption, for fixed $X^{(i)}$, conditionally, Z is a random variable whose range is in an interval of length at most $c_i\left(X^{(i)}\right)$ so by (6.3),

$$\frac{\mathrm{Ent}^{(i)}\left(e^{\lambda Z}\right)}{\boldsymbol{E}^{(i)}e^{\lambda Z}} \leq \frac{c_i^2\left(X^{(i)}\right)\lambda^2}{8}$$

and by (6.1),

$$\mathrm{Ent}(e^{\lambda Z}) \leq \sum_{i=1}^{n} \boldsymbol{E}\left[\left(\frac{c_i^2\left(X^{(i)}\right)\lambda^2}{8}\right)\boldsymbol{E}^{(i)}e^{\lambda Z}\right] = \sum_{i=1}^{n} \boldsymbol{E}\left[\left(\frac{c_i^2\left(X^{(i)}\right)\lambda^2}{8}\right)e^{\lambda Z}\right].$$

Since $(1/4)\sum_{i=1}^{n} c_i^2(x^{(i)}) \leq v$, this inequality implies that

$$\frac{\mathrm{Ent}\left(e^{\lambda Z}\right)}{\boldsymbol{E}e^{\lambda Z}} \leq \frac{\lambda^2 v}{2}$$

and the announced inequality follows by using Herbst's argument as we did at the end of the proof of Theorem 6.2. □

6.3 Modified Logarithmic Sobolev Inequalities

In this section we present a simple inequality with the purpose of bringing sub-additivity of entropy into a more manageable form, providing a versatile tool for deriving exponential concentration inequalities. This tool will help us prove inequalities under much more flexible conditions than bounded differences. This is achieved by further developing the right-hand side of Eq. (6.1). The obtained inequalities are closely related to the *logarithmic Sobolev inequalities* that we met in Chapter 5, but there we were restricted to functions of Bernoulli or Gaussian random variables.

Our first modified logarithmic Sobolev inequality follows from the sub-additivity and the variational formulation of entropy. Throughout the entire chapter, we consider independent random variables $X_1, \ldots, X_n$ taking values in some space $\mathcal{X}$, a real-valued function $f : \mathcal{X}^n \to \mathbb{R}$, and the random variable $Z = f(X_1, \ldots, X_n)$. As in Section 3.1, we denote $Z_i = f_i(X^{(i)}) = f_i(X_1, \ldots, X_{i-1}, X_{i+1}, \ldots, X_n)$ where $f_i : \mathcal{X}^{n-1} \to \mathbb{R}$ is an arbitrary function.

Theorem 6.6 (A MODIFIED LOGARITHMIC SOBOLEV INEQUALITY) *Let* $\phi(x) = e^x - x - 1$. *Then for all* $\lambda \in \mathbb{R}$,

$$\lambda\boldsymbol{E}\left[Ze^{\lambda Z}\right] - \boldsymbol{E}\left[e^{\lambda Z}\right]\log\boldsymbol{E}\left[e^{\lambda Z}\right] \leq \sum_{i=1}^{n} \boldsymbol{E}\left[e^{\lambda Z}\phi\left(-\lambda(Z - Z_i)\right)\right].$$

Proof We bound each term on the right-hand side of the sub-additivity of entropy (6.1). To do this, recall that by the variational formula of entropy given in Corollary 4.17, for any nonnegative random variable Y and for any $u > 0$,

$$\mathbf{E}[Y \log Y] - (\mathbf{E}Y)\log(\mathbf{E}Y) \leq \mathbf{E}[Y\log Y - Y\log u - (Y - u)].$$

We use this bound conditionally. It implies that if Y_i is a positive function of the random variables $X_1, \ldots, X_{i-1}, X_{i+1}, \ldots, X_n$, then

$$\mathbf{E}^{(i)}[Y \log Y] - \left(\mathbf{E}^{(i)}Y\right)\log\left(\mathbf{E}^{(i)}Y\right) \leq \mathbf{E}^{(i)}\left[Y(\log Y - \log Y_i) - (Y - Y_i)\right].$$

Applying the above inequality to the variables $Y = e^{\lambda Z}$ and $Y_i = e^{\lambda Z_i}$, one obtains

$$\mathbf{E}^{(i)}[Y \log Y] - \left(\mathbf{E}^{(i)}Y\right)\log\left(\mathbf{E}^{(i)}Y\right) \leq \mathbf{E}^{(i)}\left[e^{\lambda Z}\phi(-\lambda(Z - Z_i))\right]$$

and the proof is completed by (6.1). □

6.4 Beyond Bounded Differences

Simplicity and generality make the bounded differences inequality attractive and it has become a universal tool as witnessed by its countless applications. However, it is possible to improve this simple inequality in various ways, and the entropy method provides a versatile tool. In this section we first give a simple example that is quite easy to obtain from the modified logarithmic Sobolev inequalities of the previous section yet has numerous interesting applications. Its proof is essentially identical to that of Theorem 5.3 but thanks to the generality of Theorem 6.6, we do not need to restrict ourselves to functions of Bernoulli random variables.

Here we consider a general real-valued function of n independent random variables $Z = f(X_1, \ldots, X_n)$ and Z_i denotes an $X^{(i)}$-measurable random variable defined by $Z_i = \inf_{x_i'} f(X_1, \ldots, x_i', \ldots, X_n)$.

Theorem 6.7 *Assume that Z is such that there exists a constant $v > 0$ such that, almost surely,*

$$\sum_{i=1}^{n}(Z - Z_i)^2 \leq v.$$

Then for all $t > 0$,

$$\mathbf{P}\{Z - \mathbf{E}Z > t\} \leq e^{-t^2/(2v)}.$$

Proof The result follows easily from the modified logarithmic Sobolev inequality proved in the previous section. Observe that for $x > 0$, $\phi(-x) \leq x^2/2$, and therefore, for all $\lambda > 0$, Theorem 6.6 implies

$$\lambda \mathbf{E}\left[Ze^{\lambda Z}\right] - \mathbf{E}\left[e^{\lambda Z}\right] \log \mathbf{E}\left[e^{\lambda Z}\right] \leq \mathbf{E}\left[e^{\lambda Z} \sum_{i=1}^{n} \frac{\lambda^2}{2}(Z - Z_i)^2\right]$$
$$\leq \frac{\lambda^2 v}{2} \mathbf{E} e^{\lambda Z},$$

where we used the assumption of the theorem. The obtained inequality has the same form as the one we already faced in the proof of Theorem 6.2 and the proof may be finished in an identical way. □

By replacing f by $-f$ in the theorem above, we see that if Z is such that

$$\sum_{i=1}^{n} (Z - Z_i)^2 \leq v$$

with $Z_i = \sup_{x_i'} f(X_1, \ldots, X_{i-1}, x_i', X_{i+1}, \ldots, X_n)$, then one obtains an analogous bound for the lower tail

$$\mathbf{P}\{Z < \mathbf{E}Z - t\} \leq e^{-t^2/(2v)}.$$

As a consequence, if the condition

$$\sum_{i=1}^{n} (Z - Z_i)^2 \leq v$$

is satisfied both for $Z_i = \inf_{x_i'} f(X_1, \ldots, X_{i-1}, x_i', X_{i+1}, \ldots, X_n)$ and for $Z_i = \sup_{x_i'} f(X_1, \ldots, X_{i-1}, x_i', X_{i+1}, \ldots, X_n)$, one has the two-sided inequality

$$\mathbf{P}\left\{|Z - \mathbf{E}Z| > t\right\} \leq 2e^{-t^2/(2v)}.$$

To understand why this inequality is a significant step forward in comparison with Theorem 6.2, simply observe that the conditions of Theorem 6.7 do not require that f should have bounded differences. All they require is that

$$\sup_{\substack{x_1,\ldots,x_n, \\ x_1',\ldots,x_n' \in \mathcal{X}}} \sum_{i=1}^{n} \left(f(x_1, \ldots, x_n) - f(x_1, \ldots, x_{i-1}, x_i', x_{i+1}, \ldots, x_n)\right)^2 \leq v.$$

The quantity v may be interpreted as an upper bound for the Efron–Stein estimate of the variance Var (Z). Many of the inequalities proved by the entropy method in this chapter have a similar flavor: a sub-Gaussian (or sometimes sub-gamma) tail bound where the role of the variance factor is played by a suitable upper bound based on the Efron–Stein inequality.

Note, however, that if f satisfies the bounded differences assumption (or the x-dependent bounded differences assumption), then Theorems 6.2 and 6.5 provide better constants in the exponent. To illustrate why Theorem 6.7 is an essential improvement, recall the example of the largest eigenvalue of a random symmetric matrix, as described in Example 3.14. For this example Theorem 6.5 fails to provide a meaningful inequality.

Example 6.8 (THE LARGEST EIGENVALUE OF A RANDOM SYMMETRIC MATRIX) As in Example 3.14, we consider a random symmetric real matrix A with entries $X_{i,j}$, $1 \leq i \leq j \leq n$ where the $X_{i,j}$ are independent random variables with absolute value bounded by 1. Let $Z = \lambda_1$ denote the largest eigenvalue of A. In Section 3.14, we have already seen that, almost surely,

$$\sum_{1 \leq i \leq j \leq n} (Z - Z_{i,j})^2 \leq 16.$$

We used this estimate and the Efron–Stein inequality to conclude that $\mathrm{Var}(Z) \leq 16$. Using Theorem 6.7, we get, without further work, the sub-Gaussian tail estimate

$$\boldsymbol{P}\{Z > \boldsymbol{E}Z + t\} \leq e^{-t^2/32}.$$

Clearly, the bounded differences inequality is useless here as it is impossible to handle the individual differences $Z - Z'_{i,j}$ in a meaningful way, while the sum of their squares is bounded by 16. In Section 8.2 we return to this example, re-prove the exponential tail inequality with a different method and derive a corresponding lower-tail inequality.

6.5 Inequalities for the Lower Tail

In the previous section we showed that the condition

$$\sum_{i=1}^{n} \left(f(X_1, \ldots, X_n) - \inf_{x'_i} f(X_1, \ldots, X_{i-1}, x'_i, X_{i+1}, \ldots, X_n) \right)^2 \leq v$$

guarantees a sub-Gaussian behavior for the upper tail probabilities $\boldsymbol{P}\{Z > \boldsymbol{E}Z + t\}$. To obtain an analogous bound for the lower tail probabilities $\boldsymbol{P}\{Z < \boldsymbol{E}Z - t\}$, however, one needs a condition of the form

$$\sum_{i=1}^{n} \left(f(X_1, \ldots, X_n) - \sup_{x'_i} f(X_1, \ldots, X_{i-1}, x'_i, X_{i+1}, \ldots, X_n) \right)^2 \leq v.$$

In many interesting cases, only one of the two quantities can be controlled easily, although one would like to handle both upper and lower tails. This is possible under an additional condition of bounded differences. Here we show a simple version of such a result. Note that it is not quite a sub-Gaussian but rather a sub-Poisson bound. As we point out in subsequent sections, there are some important applications in which sub-Gaussian lower tail bounds hold. In particular, in Section 6.11 below, we show a general sub-Gaussian lower tail inequality under some additional conditions (see Corollary 6.24). For more discussion and related results, we refer to Chapters 7, 9, and 15.

Theorem 6.9 *Assume that $X_1, \ldots, X_n$ are independent and $Z = f(X_1, \ldots, X_n)$ is such that there exists a constant $v > 0$ such that, almost surely,*

$$\sum_{i=1}^{n} (Z_i - Z)^2 \leq v$$

where $Z_i = \sup_{x'_i} f(X_1, \ldots, X_{i-1}, x'_i, X_{i+1}, \ldots, X_n)$. Assume also that $Z_i - Z \leq 1$ almost surely for all $i = 1, \ldots, n$. Then for all $t > 0$,

$$\mathbf{P}\{Z - \mathbf{E}Z > t\} \leq e^{-vh(t/v)} \leq e^{-t^2/(2(v+t/3))}$$

where $h(x) = (1 + x)\log(1 + x) - x$ for $x > -1$.

Proof Our starting point is, once again, the modified logarithmic Sobolev inequality of Theorem 6.6. In order to bound the right-hand side of that inequality, we need to bound $\mathbf{E}\left[e^{\lambda Z}\phi\left(-\lambda(Z - Z_i)\right)\right]$ with Z_i defined above. The key observation is that $\phi(x)/x^2 = (e^x - x - 1)/x^2$ is an increasing function of x and therefore, for any $\lambda > 0$,

$$\frac{\phi\left(-\lambda(Z - Z_i)\right)}{\lambda^2(Z - Z_i)^2} \leq \frac{\phi(\lambda)}{\lambda^2}$$

where we used the fact that $Z_i - Z \leq 1$. Thus, by Theorem 6.6, for $\lambda > 0$, we have

$$\begin{aligned}
\frac{d}{d\lambda}\left(\frac{1}{\lambda}\log \mathbf{E}e^{\lambda Z}\right) &\leq \frac{1}{\lambda^2 \mathbf{E}e^{\lambda Z}} \sum_{i=1}^{n} \mathbf{E}\left[e^{\lambda Z}\phi\left(-\lambda(Z - Z_i)\right)\right] \\
&\leq \frac{\phi(\lambda)}{\lambda^2 \mathbf{E}e^{\lambda Z}} \mathbf{E}\left[e^{\lambda Z}\sum_{i=1}^{n}(Z - Z_i)^2\right] \\
&\leq \frac{v\phi(\lambda)}{\lambda^2}
\end{aligned}$$

where we used the hypothesis of the theorem. The proof can now be finished as in Theorem 6.7, by integrating the bound above. We thus obtain

$$\mathbf{E}e^{\lambda(Z - \mathbf{E}Z)} \leq e^{\phi(\lambda)v}.$$

The upper bound is just the moment-generating function of a centered Poisson(v) random variable and the tail bounds follow from the calculations shown in Sections 2.2 and 2.7. □

Of course, by replacing f by $-f$, we get the analog result that if

$$\sum_{i=1}^{n}\left(f(X_1,\ldots,X_n)-\inf_{x_i'}f(X_1,\ldots,X_{i-1},x_i',X_{i+1},\ldots,X_n)\right)^2\leq v$$

(i.e. under the same condition as in Theorem 6.7) and also

$$f(X_1,\ldots,X_n)-\inf_{x_i'}f(X_1,\ldots,X_{i-1},x_i',X_{i+1},\ldots,X_n)\leq 1,$$

then for all $0 < t$,

$$\mathbf{P}\{Z < \mathbf{E}Z - t\} \leq e^{-t^2/(2(v+t/3))}.$$

This bound explains the title of the section.

6.6 Concentration of Convex Lipschitz Functions

In Section 5.4 we proved the fundamental result that any Lipschitz function of a canonical Gaussian vector has sub-Gaussian tails. The entropy method presented in the previous sections allows us to extend this to much more general product distributions, though we need an extra convexity condition on the Lipschitz function. This is analogous to the relationship of the "convex" Poincaré inequality of Section 3.5 to the Gaussian Poincaré inequality presented in Section 3.7. We state the result for functions of n independent random variables taking values in $[0,1]^n$. However, the same proof extends easily to functions of n independent vector-valued random variables under appropriate Lipschitz and convexity assumptions (see Exercise 6.5).

Recall that $f : [0,1]^n \to \mathbb{R}$ is said to be *separately convex* if, for every $i = 1,\ldots,n$, it is a convex function of i-th variable if the rest of the variables are fixed.

Theorem 6.10 *Let $X_1,\ldots,X_n$ be independent random variables taking values in the interval $[0,1]$ and let $f : [0,1]^n \to \mathbb{R}$ be a separately convex function such that $|f(x)-f(y)| \leq \|x-y\|$ for all $x,y \in [0,1]^n$. Then $Z = f(X_1,\ldots,X_n)$ satisfies, for all $t > 0$,*

$$\mathbf{P}\{Z > \mathbf{E}Z + t\} \leq e^{-t^2/2}.$$

Proof We may assume without loss of generality that the partial derivatives of f exist. (Otherwise one may approximate f by a smooth function via a standard argument.) Theorem 6.7 suffices to bound the random variable $\sum_{i=1}^n (Z-Z_i)^2$ where $Z_i = \inf_{x_i'} f(X_1,\ldots,x_i',\ldots,X_n)$. However, we have already shown in the proof of Theorem 3.17 that

$$\sum_{i=1}^{n}(Z-Z_i)^2 \leq \|\nabla(f(X))\|^2 \leq 1$$

where at the last step we used the Lipschitz property of f. Therefore, Theorem 6.7 is applicable with $v = 1$. □

Note that a naive bound using the Lipschitz condition would only give the bound $\sum_{i=1}^{n}\left(f(X)-f(\overline{X}^{(i)})\right)^2 \le 4n$. The convexity assumption provides an immense improvement over this simple bound.

Example 6.11 (THE LARGEST SINGULAR VALUE OF A RANDOM MATRIX) Consider again Example 3.18, that is, let Z be the largest singular value of an $m \times n$ matrix with independent entries $X_{i,j}$ $(i = 1,\ldots,m, j = 1,\ldots,n)$ taking values in $[0, 1]$. As we pointed out, Z is a convex function of the $X_{i,j}$, which is also Lipschitz, so Theorem 6.10 implies

$$\mathbf{P}\{Z > \mathbf{E}Z + t\} \le e^{-t^2/2}.$$

Here, we assumed that all entries of the matrix A are independent. This assumption may be weakened at the price of obtaining a weaker sub-Gaussian bound. The same argument may be used to establish concentration properties of the largest singular value of a matrix whose columns are independent vectors, but the components of these vectors are not necessarily independent (see Exercise 6.6).

6.7 Exponential Inequalities for Self-Bounding Functions

In this section we revisit self-bounding functions introduced in Section 3.3. Recall that a function $f : \mathcal{X}^n \to \mathbb{R}$ is said to have the self-bounding property if, for some functions $f_i : \mathcal{X}^{n-1} \to \mathbb{R}$, for all $x = (x_1,\ldots,x_n) \in \mathcal{X}^n$, and for all $i = 1,\ldots,n$,

$$0 \le f(x) - f_i\left(x^{(i)}\right) \le 1$$

and

$$\sum_{i=1}^{n}\left(f(x) - f_i\left(x^{(i)}\right)\right) \le f(x),$$

where, as usual, $x^{(i)} = (x_1,\ldots,x_{i-1},x_{i+1},\ldots,x_n)$. If $X_1,\ldots,X_n$ are independent random variables taking values in $\mathcal{X}$ and $Z = f(X_1,\ldots,X_n)$ for a self-bounding function f, then the Efron–Stein inequality implies $\mathrm{Var}(Z) \le \mathbf{E}Z$. We have seen several interesting examples of self-bounding functions, including various configuration functions, Rademacher averages (Section 3.3), and the combinatorial entropies introduced in Section 4.5. Here, building on the modified logarithmic Sobolev inequality of Theorem 6.6, we obtain exponential concentration bounds for self-bounding functions.

To state the main result of this section, recall the definition of the following two functions that we have already seen in Bennett's inequality and in the modified logarithmic Sobolev inequalities above:

$$h(u) = (1+u)\log(1+u) - u, \quad u \ge -1$$

and

$$\phi(v) = \sup_{u \geq -1} (uv - h(u)) = e^v - v - 1.$$

Theorem 6.12 *Assume that Z satisfies the self-bounding property. Then for every* $\lambda \in \mathbb{R}$,

$$\log \mathbf{E}e^{\lambda(Z-\mathbf{E}Z)} \leq \phi(\lambda)\mathbf{E}Z.$$

Moreover, for every $t > 0$,

$$\mathbf{P}\{Z \geq \mathbf{E}Z + t\} \leq \exp\left(-h\left(\frac{t}{\mathbf{E}Z}\right)\mathbf{E}Z\right)$$

and for every $0 < t \leq \mathbf{E}Z$,

$$\mathbf{P}\{Z \leq \mathbf{E}Z - t\} \leq \exp\left(-h\left(-\frac{t}{\mathbf{E}Z}\right)\mathbf{E}Z\right).$$

By recalling that $h(u) \geq u^2/(2 + 2u/3)$ for $u \geq 0$ (we have already used this in the proof of Bernstein's inequality; see Exercise 2.8) and observing that $h(u) \geq u^2/2$ for $u \leq 0$, we obtain the following immediate, perhaps more transparent, corollaries: for every $t > 0$,

$$\mathbf{P}\{Z \geq \mathbf{E}Z + t\} \leq \exp\left(-\frac{t^2}{2\mathbf{E}Z + 2t/3}\right)$$

and for every $0 < t \leq \mathbf{E}Z$,

$$\mathbf{P}\{Z \leq \mathbf{E}Z - t\} \leq \exp\left(-\frac{t^2}{2\mathbf{E}Z}\right).$$

In these sub-gamma tail bounds the variance factor $\mathbf{E}Z$ is the Efron–Stein upper bound of the variance $\mathrm{Var}(Z)$.

Proof We first invoke the modified logarithmic Sobolev inequality (Theorem 6.6). Since the function ϕ is convex with $\phi(0) = 0$, for any λ and any $u \in [0, 1]$, $\phi(-\lambda u) \leq u\phi(-\lambda)$. Thus, since $Z - Z_i \in [0, 1]$, we have, for every λ, $\phi(-\lambda(Z - Z_i)) \leq (Z - Z_i)\phi(-\lambda)$ and therefore, Theorem 6.6 and the condition $\sum_{i=1}^n (Z - Z_i) \leq Z$ imply that

$$\begin{aligned}\lambda\mathbf{E}\left[Ze^{\lambda Z}\right] - \mathbf{E}\left[e^{\lambda Z}\right]\log\mathbf{E}\left[e^{\lambda Z}\right] &\leq \mathbf{E}\left[\phi(-\lambda)e^{\lambda Z}\sum_{i=1}^n (Z - Z_i)\right] \\ &\leq \phi(-\lambda)\mathbf{E}\left[Ze^{\lambda Z}\right].\end{aligned}$$

Define, for $\lambda \in \mathbb{R}$, $F(\lambda) = \mathbf{E}e^{\lambda(Z-\mathbf{E}Z)}$. Then the inequality above becomes

$$\left[\lambda - \phi(-\lambda)\right] \frac{F'(\lambda)}{F(\lambda)} - \log F(\lambda) \leq \phi(-\lambda)\mathbf{E}Z,$$

which, writing $G(\lambda) = \log F(\lambda)$, implies

$$\left(1 - e^{-\lambda}\right) G'(\lambda) - G(\lambda) \leq \phi(-\lambda)\mathbf{E}Z.$$

For $\lambda \geq 0$ this inequality is equivalent to

$$\left(\frac{G(\lambda)}{e^{\lambda} - 1}\right)' \leq \mathbf{E}Z \cdot \left(\frac{-\lambda}{e^{\lambda} - 1}\right)'.$$

The last differential inequality is straightforward to solve and we obtain, for $\lambda > \lambda_0 > 0$,

$$G(\lambda) \leq \left(e^{\lambda} - 1\right)\left(\frac{G(\lambda_0)}{e^{\lambda_0} - 1} + \mathbf{E}Z\left(\frac{\lambda_0}{e^{\lambda_0} - 1} - \frac{\lambda}{e^{\lambda} - 1}\right)\right).$$

Letting λ_0 tend to 0 and observing that $\lim_{\lambda_0 \to 0} \lambda_0/(e^{\lambda_0} - 1) = 1$ and that, by l'Hospital's rule, $\lim_{\lambda_0 \to 0} G(\lambda_0)/(e^{\lambda_0} - 1) = \mathbf{E}[Z - \mathbf{E}Z] = 0$, for $\lambda \geq 0$, we get

$$G(\lambda) \leq \phi(\lambda)\mathbf{E}Z.$$

Proceeding in a similar way for $\lambda \leq 0$, we obtain the first inequality of the theorem.

On the right-hand side we recognize the moment-generating function of a centered Poisson random variable with parameter $\mathbf{E}Z$. The probability bounds are the corresponding Poisson tail inequalities and are obtained by Chernoff's bounding, as calculated in Section 2.2. □

Theorem 6.12 provides concentration inequalities for any function satisfying the self-bounding property. In Sections 3.3 and 4.5 several examples of such functions are discussed. Here we mention one more example.

Example 6.13 (MAXIMAL DEGREE IN A RANDOM GRAPH) Consider the Erdős–Rényi $G(n, p)$ model of a random graph. In this model a graph of n vertices is obtained if each one of the $m = \binom{n}{2}$ possible edges is selected, independently, with probability p. The *degree* of a vertex is the number of edges adjacent to that vertex. Note that the degree of any vertex is a binomial $(n - 1, p)$ random variable. Let D denote the maximal degree of any vertex in the graph. Clearly, D is a configuration function, so Theorem 6.12 applies. See Exercise 6.14 for properties of D.

Next we write out explicitly what the theorem implies for combinatorial entropies, defined in Section 4.5.

Theorem 6.14 *Assume that $h(x) = \log_b |tr(x)|$ is a combinatorial entropy such that for all $x \in \mathcal{X}^n$ and $i \leq n$,*

$$h(x) - h\left(x^{(i)}\right) \leq 1.$$

If $X = (X_1, \ldots, X_n)$ is a vector of n independent random variables taking values in $\mathcal{X}$, then the random combinatorial entropy $Z = h(X)$ satisfies

$$\mathbf{P}\{Z \geq \mathbf{E}Z + t\} \leq \exp\left(-\frac{t^2}{2\mathbf{E}Z + 2t/3}\right),$$

and

$$\mathbf{P}\{Z \leq \mathbf{E}Z - t\} \leq \exp\left(-\frac{t^2}{2\mathbf{E}Z}\right).$$

Moreover,

$$\mathbf{E}\log_b |tr(X)| \leq \log_b \mathbf{E}|tr(X)| \leq \frac{b-1}{\log b}\mathbf{E}\log_b |tr(X)|.$$

Note that the left-hand side of the last statement follows from Jensen's inequality, while the right-hand side follows by taking $\lambda = \log b$ in the first inequality of Theorem 6.12. One of the examples of combinatorial entropies, defined in Section 4.5, is VC entropy. For the random VC entropy $T(X)$, we obtain

$$\mathbf{E}\log_2 T(X) \leq \log_2 \mathbf{E}T(X) \leq (\log_2 e)\mathbf{E}\log_2 T(X).$$

This last statement shows that the expected VC entropy $\mathbf{E}\log_2 T(X)$ and the *annealed* VC entropy $\log_2 \mathbf{E}T(X)$ are tightly connected, regardless of the class of sets $\mathcal{A}$ and the distribution of the X_i's.

The same inequality holds for the logarithm of the number of increasing subsequences of a random permutation (see Section 4.5 for the definitions).

6.8 Symmetrized Modified Logarithmic Sobolev Inequalities

One of the most useful forms of the Efron–Stein inequality establishes an upper bound for the variance of $Z = f(X_1, \ldots, X_n)$ in terms of the behavior of the random variables $Z - Z_i'$ where $Z_i' = f(X_1, \ldots, X_i', \ldots, X_n)$ is obtained by replacing the variable X_i by an independent copy X_i' (see Theorem 3.1). The purpose of the next few sections is the search for exponential concentration inequalities involving the differences $Z - Z_i'$. The following symmetrized modified logarithmic Sobolev inequality is at the basis of such exponential tail inequalities.

Theorem 6.15 (SYMMETRIZED MODIFIED LOGARITHMIC SOBOLEV INEQUALITIES) *For all $\lambda \in \mathbb{R}$,*

$$\lambda \mathbf{E}\left[Ze^{\lambda Z}\right] - \mathbf{E}\left[e^{\lambda Z}\right] \log \mathbf{E}\left[e^{\lambda Z}\right] \leq \sum_{i=1}^{n} \mathbf{E}\left[e^{\lambda Z} \phi\left(-\lambda(Z - Z_i')\right)\right]$$

where $\phi(x) = e^x - x - 1$. Moreover, denoting $\tau(x) = x(e^x - 1)$, for all $\lambda \in \mathbb{R}$,

$$\lambda \mathbf{E}\left[Ze^{\lambda Z}\right] - \mathbf{E}\left[e^{\lambda Z}\right] \log \mathbf{E}\left[e^{\lambda Z}\right] \leq \sum_{i=1}^{n} \mathbf{E}\left[e^{\lambda Z} \tau(-\lambda(Z - Z_i')_+)\right],$$

$$\lambda \mathbf{E}\left[Ze^{\lambda Z}\right] - \mathbf{E}\left[e^{\lambda Z}\right] \log \mathbf{E}\left[e^{\lambda Z}\right] \leq \sum_{i=1}^{n} \mathbf{E}\left[e^{\lambda Z} \tau(\lambda(Z_i' - Z)_+)\right].$$

Proof The first inequality is proved exactly as for Theorem 6.6, simply by noting that, like Z_i, Z_i' is also independent of X_i. To prove the second and third inequalities, write

$$e^{\lambda Z} \phi\left(-\lambda(Z - Z_i')\right) = e^{\lambda Z} \phi\left(-\lambda(Z - Z_i')_+\right) + e^{\lambda Z} \phi\left(\lambda(Z_i' - Z)_+\right).$$

By symmetry, the conditional expectation of the second term, conditioned on $X_1, \ldots, X_{i-1}, X_{i+1}, \ldots, X_n$, may be written as

$$\begin{aligned} \mathbf{E}^{(i)}\left[e^{\lambda Z} \phi\left(\lambda(Z_i' - Z)_+\right)\right] &= \mathbf{E}^{(i)}\left[e^{\lambda Z_i'} \phi\left(\lambda(Z - Z_i')_+\right)\right] \\ &= \mathbf{E}^{(i)}\left[e^{\lambda Z} e^{-\lambda(Z - Z_i')} \phi\left(\lambda(Z - Z_i')_+\right)\right]. \end{aligned}$$

Summarizing, we have

$$\begin{aligned} &\mathbf{E}^{(i)}\left[e^{\lambda Z} \phi\left(-\lambda(Z - Z_i')\right)\right] \\ &= \mathbf{E}^{(i)}\left[\left(\phi\left(-\lambda(Z - Z_i')_+\right) + e^{-\lambda(Z - Z_i')} \phi\left(\lambda(Z - Z_i')_+\right)\right) e^{\lambda Z}\right]. \end{aligned}$$

The second inequality of the theorem follows simply by noting that $\phi(x) + e^x \phi(-x) = x(e^x - 1) = \tau(x)$. The last inequality follows similarly. □

6.9 Exponential Efron–Stein Inequalities

Recall that by the Efron–Stein inequality, if $X = (X_1, \ldots, X_n)$ is a vector of independent random variables, then the variance of $Z = f(X)$ is bounded as

$$\text{Var}(Z) \leq \frac{1}{2} \sum_{i=1}^{n} \mathbf{E}\left[(Z - Z_i')^2\right].$$

If we denote by $E'[\cdot] = E[\cdot|X]$ expectation with respect to the variables $X'_1, \ldots, X'_n$ only, then by introducing the random variables

$$V^+ = \sum_{i=1}^{n} E'\left[(Z - Z'_i)_+^2\right]$$

and

$$V^- = \sum_{i=1}^{n} E'\left[(Z - Z'_i)_-^2\right],$$

the Efron–Stein inequality can be written in either one of the equivalent forms

$$\mathrm{Var}\,(Z) \le EV^+ \quad \text{and} \quad \mathrm{Var}\,(Z) \le EV^-.$$

The message of the next theorem is that upper bounds for the moment-generating function of the random variables V^+ and V^- may be translated into exponential concentration inequalities for Z. In a sense, these may be understood as exponential versions of the Efron–Stein inequality.

Theorem 6.16 *Let $Z = f(X_1, \ldots, X_n)$ be a real-valued function of n independent random variables. Let $\theta, \lambda > 0$ be such that $\theta\lambda < 1$ and $Ee^{\lambda V^+/\theta} < \infty$. Then*

$$\log Ee^{\lambda(Z-EZ)} \le \frac{\lambda\theta}{1-\lambda\theta} \log Ee^{\lambda V^+/\theta}.$$

Next assume that Z is such that $Z'_i - Z \le 1$ for every $1 \le i \le n$. Then for all $\lambda \in (0, 1/2)$,

$$\log Ee^{\lambda(Z-EZ)} \le \frac{2\lambda}{1-2\lambda} \log Ee^{\lambda V^-}.$$

Proof The proof of the first statement is based on the second inequality of Theorem 6.15. To apply this inequality, we need to establish appropriate upper bounds for the quantity $\sum_{i=1}^{n} E\left[e^{\lambda Z}\tau(-\lambda(Z - Z'_i)_+)\right]$ appearing on the right-hand side. By noting that $\tau(-x) \le x^2$ for all $x \ge 0$, we see that it suffices to bound

$$\sum_{i=1}^{n} E\left[e^{\lambda Z}\lambda^2(Z - Z'_i)_+^2\right] = \lambda^2 E\left[V^+e^{\lambda Z}\right].$$

In previous applications of the entropy method, our strategy was to relate $E\left[V^+e^{\lambda Z}\right]$ to quantities expressed as a functional of the random variable Z. Here our approach is different: we bound the right-hand side by something that involves the moment-generating function of Z and a functional of V^+. In order to do this, we "decouple" the random variables $e^{\lambda Z}$ and V^+.

The duality formula of the entropy given in Theorem 4.13 serves as an ideal tool for this purpose. Recall that the duality formula implies that for any random variable W such that $\boldsymbol{E}e^W < \infty$,

$$\boldsymbol{E}\left[\left(W - \log \boldsymbol{E}e^W\right) e^{\lambda Z}\right] \leq \operatorname{Ent}(e^{\lambda Z}),$$

or equivalently,

$$\boldsymbol{E}\left[We^{\lambda Z}\right] \leq \boldsymbol{E}\left[e^{\lambda Z}\right] \log \boldsymbol{E}\left[e^{W}\right] + \operatorname{Ent}(e^{\lambda Z}).$$

A natural choice for W is λV^+ but it is advantageous to introduce a free parameter $\theta > 0$ and apply the "decoupling" inequality above with $W = \lambda V^+/\theta$. Now the symmetrized modified logarithmic Sobolev inequality becomes

$$\operatorname{Ent}(e^{\lambda Z}) \leq \lambda\theta \left(\boldsymbol{E}\left[e^{\lambda Z}\right] \log \boldsymbol{E}\left[e^{\lambda V^+/\theta}\right] + \operatorname{Ent}(e^{\lambda Z})\right).$$

Rearranging, and writing $\rho(\lambda) = \log \boldsymbol{E}e^{\lambda V^+}$ for the logarithmic moment generating function of V^+, we have

$$(1 - \lambda\theta) \operatorname{Ent}(e^{\lambda Z}) \leq \lambda\theta\rho(\lambda/\theta)\boldsymbol{E}e^{\lambda Z}$$

which, of course, is only meaningful if $\lambda\theta < 1$. If, as before, we let $G(\lambda) = \log \boldsymbol{E}e^{\lambda(Z-\boldsymbol{E}Z)}$, then the previous inequality becomes

$$\lambda G'(\lambda) - G(\lambda) \leq \frac{\lambda\theta}{1 - \lambda\theta}\rho(\lambda/\theta).$$

This differential inequality is of the form that we have already encountered and indeed, by Lemma 6.25,

$$G(\lambda) \leq \lambda\theta \int_0^\lambda \frac{\rho(u/\theta)}{u(1 - u\theta)} du.$$

Since $\rho(0) = 0$, the convexity of ρ implies that $\rho(u/\theta)/(u(1 - u\theta))$ is a non-decreasing function and therefore

$$G(\lambda) \leq \frac{\theta\lambda\rho(\lambda/\theta)}{1 - \lambda\theta},$$

and the first inequality of the theorem follows.

To prove the second statement of the theorem, we start with the last inequality of Theorem 6.15 which may be written as

$$\operatorname{Ent}\left(e^{\lambda Z}\right) \leq \sum_{i=1}^{n} \boldsymbol{E}\left[e^{\lambda Z}\lambda^2(Z_i' - Z)_+^2 \frac{e^{\lambda(Z_i'-Z)_+} - 1}{\lambda(Z_i' - Z)_+}\right].$$

Since $(e^x - 1)/x$ is an increasing function, the conditions $Z_i' - Z \leq 1$ and $\lambda < 1/2$ imply that

$$\mathrm{Ent}\left(e^{\lambda Z}\right) \leq \lambda^2 \sum_{i=1}^{n} \boldsymbol{E}\left[e^{\lambda Z}(Z_i' - Z)_+^2 2\left(e^{1/2} - 1\right)\right] \leq 2\lambda^2 \boldsymbol{E}\left[e^{\lambda Z} V^-\right].$$

The rest of the proof is the same as for the first inequality of the theorem. □

6.10 A Modified Logarithmic Sobolev Inequality for the Poisson Distribution

In the previous sections we derived modifications of the Gaussian logarithmic Sobolev inequality that allowed us to prove concentration inequalities for functions of independent random variables of arbitrary distribution. For certain specific distributions, apart from the normal distribution, sharper inequalities are available. Here we show such a "modified logarithmic Sobolev inequality" for Poisson random variables. Recall that X has a Poisson distribution with parameter $\mu > 0$ if X takes nonnegative integer values and for every $k = 0, 1, \ldots, \boldsymbol{P}\{X = k\} = \mu^k e^{-\mu}/k!$.

If f is a real-valued function defined on the set of nonnegative integers $\mathbb{N}$, then define the *discrete derivative* of f at $x \in \mathbb{N}$ by $Df(x) = f(x+1) - f(x)$. If one wanted to establish a "discrete" analog of the Gaussian logarithmic Sobolev inequality, one would hope to prove that all functions $f : \mathbb{N} \to \mathbb{R}$, $\mathrm{Ent}(f^2(X)) \leq \kappa \boldsymbol{E}[|Df(X)|^2]$ for some constant κ. Unfortunately, such a result is not true if X is Poisson because the supremum of $\mathrm{Ent}((f(X))^2)/\boldsymbol{E}[(Df(X))^2]$ is infinite.

However, Theorem 6.15 may be used to prove the following modified logarithmic Sobolev inequalities for Poisson distributions, which is a refinement of the Poisson Poincaré inequality of Exercise 3.21.

Theorem 6.17 (POISSON LOGARITHMIC SOBOLEV INEQUALITY) *Let X be a Poisson random variable and let $f : \mathbb{N} \to (0, \infty)$. Then*

$$\mathrm{Ent}(f(X)) \leq (\boldsymbol{E}X)\boldsymbol{E}\left[Df(X) D \log f(X)\right],$$

and

$$\mathrm{Ent}[f(X)] \leq (\boldsymbol{E}X)\boldsymbol{E}\left[\frac{|Df(X)|^2}{f(X)}\right].$$

The theorem may be proved in a way similar to that with which we proved the Gaussian logarithmic Sobolev inequality: first we establish an inequality for the Bernoulli distribution (see the lemma below) and then use the convergence of the binomial distribution to Poisson. We leave the details of the proof to the reader.

Lemma 6.18 (MODIFIED LOGARITHMIC SOBOLEV INEQUALITIES FOR BERNOULLI DISTRIBUTIONS) *For any function $f : \{0, 1\} \to (0, \infty)$, let $\nabla f(x) = f(1 - x) - f(x)$.*

Let $p \in (0,1)$, and let X be a Bernoulli random variable with parameter p (i.e., $\mathbf{P}\{X = 1\} = 1 - \mathbf{P}\{X = 0\} = p$). Then

$$\mathrm{Ent}(f(X)) \leq p(1-p)\mathbf{E}\left[\nabla f(X) \nabla \log f(X)\right]$$

and

$$\mathrm{Ent}(f(X)) \leq p(1-p)\mathbf{E}\left[\frac{|\nabla f(X)|^2}{f(X)}\right].$$

Proof We only prove the first inequality. The proof of the second is left as an exercise. Let X' be an independent copy of X. Let $q = 1 - p$. By the first inequality of Theorem 6.15, taking $\lambda = 1$ and $Z = \log f(X)$,

$$\begin{aligned}
\mathrm{Ent}(f(X)) &\leq \mathbf{E}\left[f(X)\phi(\log(f(X')/f(X)))\right] \\
&= \mathbf{E}\left[f(X') - f(X) - f(X)(\log(f(X')) - \log(f(X)))\right] \\
&= pq\left[-f(1)(\log(f(0) - \log f(1)))\right] + pq\left[-f(0)(\log(f(1) - \log f(0)))\right] \\
&= pq\mathbf{E}\left[\nabla f(X) \nabla \log f(X)\right].
\end{aligned}$$

□

It is easy to deduce from Theorem 6.17 that the square root of a Poisson random variable X satisfies

$$\log \mathbf{E} e^{\lambda(\sqrt{X} - \mathbf{E}\sqrt{X})} \leq v(e^{\lambda} - 1)$$

where $v = (\mathbf{E}X)\mathbf{E}[1/(4X+1)]$. This represents an improvement over what can be obtained from Theorem 6.29 below (see Exercise 6.12).

6.11 Weakly Self-Bounding Functions

Self-bounding functions, discussed in Section 6.7, appear naturally in numerous applications including configuration functions and combinatorial entropies. Theorem 6.12 is quite satisfactory as it cannot be improved in this generality and its proof is rather simple. However, one often faces functions that only satisfy slightly weaker conditions. A prime example, presented in Chapter 7, is the squared "convex distance." In order to handle this example, as well as various other naturally emerging cases, we generalize the definition of self-bounding functions in two different ways. This section is dedicated to inequalities for such generalized self-bounding functions. The proofs are variants of the entropy method, all based on the modified logarithmic Sobolev inequality of Theorem 6.6. However, the resulting differential inequality for the moment-generating function is not always as easy to solve as in Theorems 6.7 and 6.12, and most of our effort is devoted to the solution of these differential inequalities.

We distinguish two notions of generalized self-bounding functions. In both of the following definitions, a and b are nonnegative constants.

A nonnegative function $f : \mathcal{X}^n \to [0, \infty)$ is called *weakly* (a, b)*-self-bounding* if there exist functions $f_i : \mathcal{X}^{n-1} \to [0, \infty)$ such that for all $x \in \mathcal{X}^n$,

$$\sum_{i=1}^{n} \left(f(x) - f_i\left(x^{(i)}\right)\right)^2 \leq af(x) + b.$$

On the other hand, we say that a function $f : \mathcal{X}^n \to [0, \infty)$ is *strongly* (a, b)*-self-bounding* if there exist functions $f_i : \mathcal{X}^{n-1} \to [0, \infty)$ such that for all $i = 1, \ldots, n$ and all $x \in \mathcal{X}^n$,

$$0 \leq f(x) - f_i\left(x^{(i)}\right) \leq 1,$$

and

$$\sum_{i=1}^{n} \left(f(x) - f_i\left(x^{(i)}\right)\right) \leq af(x) + b.$$

Clearly, a self-bounding function is strongly $(1, 0)$-self-bounding and every strongly (a, b)-self-bounding function is weakly (a, b)-self-bounding. In both cases, the Efron–Stein inequality implies $\text{Var}(Z) \leq a\mathbf{E}Z + b$. Indeed, this quantity appears as a variance factor in the exponential bounds established below.

We present three inequalities. The simplest is an inequality for the upper tails of weakly (a, b)-self-bounding functions.

Theorem 6.19 *Let $X = (X_1, \ldots, X_n)$ be a vector of independent random variables, each taking values in a measurable set $\mathcal{X}$, let $a, b \geq 0$ and let $f : \mathcal{X}^n \to [0, \infty)$ be a weakly (a, b)-self-bounding function. Let $Z = f(X)$. If, in addition, $f_i(x^{(i)}) \leq f(x)$ for all $i \leq n$ and $x \in \mathcal{X}^n$, then for all $0 \leq \lambda \leq 2/a$,*

$$\log \mathbf{E}e^{\lambda(Z-\mathbf{E}Z)} \leq \frac{(a\mathbf{E}Z + b)\lambda^2}{2(1 - a\lambda/2)}$$

and for all $t > 0$,

$$\mathbf{P}\{Z \geq \mathbf{E}Z + t\} \leq \exp\left(-\frac{t^2}{2\,(a\mathbf{E}Z + b + at/2)}\right).$$

Proof Once again, our starting point is the modified logarithmic Sobolev inequality. Write $Z_i = f_i(X^{(i)})$. The main observation is that for $x \geq 0$, $\phi(-x) \leq x^2/2$. Since $Z - Z_i \geq 0$, for $\lambda > 0$, by further bounding the right-hand side of the inequality of Theorem 6.6, we obtain

$$\lambda \mathbf{E}\left[Ze^{\lambda Z}\right] - \mathbf{E}\left[e^{\lambda Z}\right] \log \mathbf{E}\left[e^{\lambda Z}\right] \leq \frac{\lambda^2}{2}\mathbf{E}\left[e^{\lambda Z}\sum_{i=1}^{n}(Z - Z_i)^2\right]$$
$$\leq \frac{\lambda^2}{2}\mathbf{E}\left[(aZ + b)e^{\lambda Z}\right]$$

where we use the assumption that f is weakly (a, b)-self-bounding. Introducing $G(\lambda) = \log \mathbf{E}e^{\lambda(Z-\mathbf{E}Z)}$, the inequality obtained above may be re-arranged to read

$$\left(\frac{1}{\lambda} - \frac{a}{2}\right) G'(\lambda) - \frac{G(\lambda)}{\lambda^2} \leq \frac{v}{2}$$

where we write $v = a\mathbf{E}Z + b$.

To finish the proof, simply observe that the left-hand side is just the derivative of the function $(1/\lambda - a/2)\, G(\lambda)$. Using the fact that $G(0) = G'(0) = 0$, and that $G'(\lambda) \geq 0$ for $\lambda > 0$, integrating this differential inequality leads to

$$G(\lambda) \leq \frac{v\lambda^2}{2(1 - a\lambda/2)} \quad \text{for all} \quad \lambda \in [0, 2/a).$$

This shows that $Z - \mathbf{E}Z$ is a sub-gamma random variable with variance factor $v = a\mathbf{E}Z + b$ and scale parameter $a/2$. The tail bound follows from the calculations shown is Section 2.4. □

The next theorem provides lower tail inequalities for weakly (a, b)-self-bounding functions. This will become essential for proving the convex distance inequality in Section 7.4.

Theorem 6.20 *Let $X = (X_1, \ldots, X_n)$ be a vector of independent random variables, each taking values in a measurable set $\mathcal{X}$, let $a, b \geq 0$ and let $f : \mathcal{X}^n \to [0, \infty)$ be a weakly (a, b)-self-bounding function. Let $Z = f(X)$ and define $c = (3a - 1)/6$. If, in addition, $0 \leq f(x) - f_i(x^{(i)}) \leq 1$ for each $i \leq n$ and $x \in \mathcal{X}^n$, then for $0 < t \leq \mathbf{E}Z$,*

$$\mathbf{P}\{Z \leq \mathbf{E}Z - t\} \leq \exp\left(-\frac{t^2}{2\left(a\mathbf{E}Z + b + c\_t\right)}\right).$$

Note that if $a \geq 1/3$, then the left tail is sub-Gaussian with variance proxy $a\mathbf{E}Z + b$, while for $a < 1/3$ we will only obtain a sub-gamma tail bound.

The proof of this theorem is shown below, together with the proof of the following upper tail inequality for strongly (a, b)-self-bounding functions.

Theorem 6.21 *Let $X = (X_1, \ldots, X_n)$ be a vector of independent random variables, each taking values in a measurable set $\mathcal{X}$, let $a, b \geq 0$ and let $f : \mathcal{X}^n \to [0, \infty)$ be a strongly (a, b)-self-bounding function. Let $Z = f(X)$ and define $c = (3a - 1)/6$. Then for all $\lambda \geq 0$,*

$$\log \mathbf{E}e^{\lambda(Z-\mathbf{E}Z)} \leq \frac{(a\mathbf{E}Z + b)\lambda^2}{2(1 - c_+\lambda)}$$

and for all $t > 0$,

$$\mathbf{P}\{Z \geq \mathbf{E}Z + t\} \leq \exp\left(-\frac{t^2}{2\,(a\mathbf{E}Z + b + c_+t)}\right).$$

In this upper tail bound we observe a similar phenomenon as in Theorem 6.20 but with a different sign. If $a \leq 1/3$, then the upper tail of a strongly (a, b)-self-bounding function is purely sub-Gaussian.

Our starting point is once again the modified logarithmic Sobolev inequality of Theorem 6.6.

If $\lambda \geq 0$ and f is strongly (a, b)-self-bounding, then, using $Z - Z_i \leq 1$ and the fact that for all $x \in [0, 1]$, $\phi(-\lambda x) \leq x\phi(-\lambda)$,

$$\begin{aligned}\lambda\mathbf{E}\left[Ze^{\lambda Z}\right] - \mathbf{E}\left[e^{\lambda Z}\right]\log\mathbf{E}\left[e^{\lambda Z}\right] &\leq \phi(-\lambda)\mathbf{E}\left[e^{\lambda Z}\sum_{i=1}^n (Z - Z_i)\right]\\ &\leq \phi(-\lambda)\mathbf{E}\left[(aZ + b)\,e^{\lambda Z}\right].\end{aligned}$$

For any $\lambda \in \mathbb{R}$, define $G(\lambda) = \log\mathbf{E}e^{(\lambda Z - \mathbf{E}Z)}$. Then the previous inequality may be written as the differential inequality

$$\left[\lambda - a\phi(-\lambda)\right] G'(\lambda) - G(\lambda) \leq v\phi(-\lambda), \tag{6.4}$$

where $v = a\mathbf{E}Z + b$.

On the other hand, if $\lambda \leq 0$ and f is weakly (a, b)-self-bounding, then since $\phi(x)/x^2$ is nondecreasing over $\mathbb{R}^+$, $\phi(-\lambda(Z - Z_i)) \leq \phi(-\lambda)(Z - Z_i)^2$ so

$$\begin{aligned}\lambda\mathbf{E}\left[Ze^{\lambda Z}\right] - \mathbf{E}\left[e^{\lambda Z}\right]\log\mathbf{E}\left[e^{\lambda Z}\right] &\leq \phi(-\lambda)\mathbf{E}\left[e^{\lambda Z}\sum_{i=1}^n (Z - Z_i)^2\right]\\ &\leq \phi(-\lambda)\mathbf{E}\left[(aZ + b)e^{\lambda Z}\right].\end{aligned}$$

This again leads to the differential inequality (6.4) but this time for $\lambda \leq 0$.

When $a = 1$, this differential inequality can be solved exactly as we saw it in the proof of Theorem 6.12, and one obtains the sub-Poissonian inequality

$$G(\lambda) \leq v\phi(\lambda).$$

However, when $a \neq 1$, it is not obvious what kind of bounds for G should be expected. If $a > 1$, then $\lambda - a\phi(-\lambda)$ becomes negative when λ is large enough. Since both $G'(\lambda)$ and $G(\lambda)$ are nonnegative when λ is nonnegative, (6.4) becomes trivial for large values of λ. Hence, at least when $a > 1$, there is no hope to derive Poissonian bounds from (6.4) for positive values of λ (i.e. for the upper tail).

The following lemma, proved in Section 6.12 below, is the key to the proof of both Theorems 6.20 and 6.21. It shows that if f satisfies a self-bounding property, then on the relevant interval, the logarithmic moment-generating function of $Z - \mathbf{E}Z$ is upper bounded by v times a function G_γ defined by

$$G_\gamma(\lambda) = \frac{\lambda^2}{2(1-\gamma\lambda)} \qquad \text{for every } \lambda \text{ such that } \gamma\lambda < 1$$

where $\gamma \in \mathbb{R}$ is a real-valued parameter. In the lemma below we mean $c_+^{-1} = \infty$ (resp. $c_-^{-1} = \infty$) when $c_+ = 0$ (resp. $c_- = 0$).

Lemma 6.22 *Let $a, v > 0$ and let G be a solution of the differential inequality*

$$[\lambda - a\phi(-\lambda)]\, H'(\lambda) - H(\lambda) \leq v\phi(-\lambda).$$

Define $c = (a - 1/3)/2$. Then, for every $\lambda \in (0, c_+^{-1})$

$$G(\lambda) \leq vG_{c_+}(\lambda)$$

and for every $\lambda \in (-\theta, 0)$

$$G(\lambda) \leq vG_{-c_-}(\lambda)$$

where $\theta = c_-^{-1}\left(1 - \sqrt{1 - 6c_-}\right)$ if $c_- > 0$ and $\theta = a^{-1}$ whenever $c_- = 0$.

The proof is given in the next section. Equipped with this lemma, it is now easy to obtain Theorems 6,20 and 6.21.

Proof of Theorem 6.20. We have to check that the condition $\lambda > -\theta$ is harmless. Since $\theta < c_-^{-1}$, by continuity, for every $t > 0$,

$$\sup_{u\in(0,\theta)}\left(tu - \frac{u^2 v}{2(1-c_- u)}\right) = \sup_{u\in(0,\theta]}\left(tu - \frac{u^2 v}{2(1-c_- u)}\right).$$

Note that we are only interested in values of t that are smaller than $\mathbf{E}Z \leq v/a$. Now the supremum of

$$tu - \frac{u^2 v}{2(1-c_- u)}$$

as a function of $u \in (0, c_-^{-1})$ is achieved either at $u_t = t/v$ (if $c_- = 0$) or at $u_t = c_-^{-1}\left(1 - (1 + (2tc_-/v))^{-1/2}\right)$ (if $c_- > 0$).

It is time to take into account the restriction $t \le v/a$. In the first case, when $u_t = t/v$, it implies that $u_t \le a^{-1} = \theta$, while in the second case, since $a = (1 - 6c_-)/3$ it implies that $1 + (2tc_-/v) \le (1 - 6c_-)^{-1}$ and therefore $u_t \le c_-^{-1}\left(1 - \sqrt{1 - 6c_-}\right) = \theta$. In both cases $u_t \le \theta$ which means that for every $t \le v/a$

$$\sup_{u \in (0,\theta]} \left(tu - \frac{u^2 v}{2(1 - c_- u)} \right) = \sup_{u \in (0, c_-^{-1})} \left(tu - \frac{u^2 v}{2(1 - c_- u)} \right)$$

and the result follows. □

Proof of Theorem 6.21. The upper-tail inequality for strongly (a, b)-self-bounding functions follows from Lemma 6.22 and Markov's inequality by routine calculations, exactly as in the proof of Bernstein's inequality when $c_+ > 0$, and it is straightforward when $c_+ = 0$. □

Example 6.23 (THE SQUARE OF A REGULAR FUNCTION) To illustrate the use of the results of this section, consider a function $g : \mathcal{X}^n \to \mathbb{R}$ and assume that there exists a constant $v > 0$ and that there are measurable functions $g_i : \mathcal{X}^{n-1} \to \mathbb{R}$ such that for all $x \in \mathcal{X}^n$, $g(x) \ge g_i(x^{(i)})$,

$$\sum_{i=1}^{n} \left(g(x) - g\left(x^{(i)}\right) \right)^2 \le v.$$

We term such a function *v-regular*. If $X = (X_1, \ldots, X_n) \in \mathcal{X}^n$ is a vector of independent $\mathcal{X}$-valued random variables, then by Theorem 6.7, for all $t > 0$,

$$\mathbf{P}\left\{ g(X) \ge \mathbf{E}g(X) + t \right\} \le e^{-t^2/(2v)}.$$

Even though Theorem 6.7 provides an exponential inequality for the lower tail, it fails to give an analogous sub-Gaussian bound for $\mathbf{P}\left\{ g(X) \le \mathbf{E}g(X) - t \right\}$. Here we show how Theorem 6.20 may be used to derive lower-tail bounds under an additional bounded-differences condition for the *square* of g.

Corollary 6.24 *Let $g : \mathcal{X}^n \to \mathbb{R}$ be a v-regular function such that for all $x \in \mathcal{X}^n$ and $i = 1, \ldots, n$, $g(x)^2 - g_i(x^{(i)})^2 \le 1$. Then for all $t \ge 0$,*

$$\mathbf{P}\left\{ g(X)^2 \le \mathbf{E}\left[g(X)^2\right] - t \right\} \le \exp\left(\frac{-t^2}{8v\mathbf{E}\left[g(X)^2\right] + t(4v - 1/3)_-} \right).$$

In particular, if g is nonnegative and $v \ge 1/12$, then for all $0 \le t \le \mathbf{E}g(X)$,

$$\mathbf{P}\left\{ g(X) \le \mathbf{E}g(X) - t \right\} \le e^{-t^2/(8v)}.$$

Proof Introduce $f(x) = g(x)^2$ and $f_i(x^{(i)}) = g_i(x^{(i)})^2$. Then

$$0 \le f(x) - f_i(x^{(i)}) \le 1.$$

Moreover,

$$\begin{aligned}\sum_{i=1}^{n}\left(f(x)-f_i\left(x^{(i)}\right)\right)^2 &= \sum_{i=1}^{n}\left(g(x)-g_i\left(x^{(i)}\right)\right)^2\left(g(x)+g_i\left(x^{(i)}\right)\right)^2\\ &= 4g(x)^2\sum_{i=1}^{n}\left(g(x)-g_i\left(x^{(i)}\right)\right)^2\\ &\leq 4vf(x)\end{aligned}$$

and therefore f is weakly $(4v,0)$-self-bounding. This means that Theorem 6.20 is applicable and this is how the first inequality is obtained.

The second inequality follows from the first by noting that

$$\begin{aligned}\mathbf{P}\left\{g(X)\leq \mathbf{E}g(X)-t\right\} &\leq \mathbf{P}\left\{g(X)\sqrt{\mathbf{E}\left[g(X)^2\right]}\leq \mathbf{E}\left[g(X)^2\right]-t\sqrt{\mathbf{E}\left[g(X)^2\right]}\right\}\\ &\leq \mathbf{P}\left\{g(X)^2\leq \mathbf{E}\left[g(X)^2\right]-t\sqrt{\mathbf{E}\left[g(X)^2\right]}\right\},\end{aligned}$$

and now the first inequality may be applied. □

For a more concrete class of applications, consider a nonnegative separately convex Lipschitz function g defined on $[0,1]^n$. If $X=(X_1,\ldots,X_n)$ are independent random variables taking values in $[0,1]$, then by Theorem 6.10,

$$\mathbf{P}\{g(X)-\mathbf{E}g(X)>t\}\leq e^{-t^2/2}.$$

Now we may derive a lower-tail inequality for g, under the additional assumption that g^2 takes its values in an interval of length 1. Indeed, without loss of generality we may assume that g is differentiable on $[0,1]^n$ because otherwise one may approximate g by a smooth function in a standard way. Then, denoting

$$g_i\left(x^{(i)}\right)=\inf_{x_i'\in\mathcal{X}} g(x_1,\ldots,x_{i-1},x_i',x_{i+1},\ldots,x_n),$$

by separate convexity,

$$g(x)-g_i\left(x^{(i)}\right)\leq\left|\frac{\partial g}{\partial x_i}(x)\right|.$$

Thus, for every $x\in[0,1]^n$,

$$\sum_{i=1}^{n}\left(g(x)-g_i\left(x^{(i)}\right)\right)^2\leq 1.$$

We return to the this problem in Section 7.5 where we will be able to drop the extra assumptions on the range of g^2.

For a concrete example, consider the ℓ_p norm $\|x\|_p$ for some $p \geq 2$. Then $g(x) = \|x\|_p$ is convex and Lipschitz, so we obtain that if $X = (X_1, \ldots, X_n)$ is a vector of independent random variables taking values in an interval of length 1, then for all $t > 0$,

$$\mathbf{P}\left\{\|X\|_p^2 \leq \mathbf{E}\|X\|_p^2 - t\right\} \leq e^{-t^2/(8\mathbf{E}\|X\|_p^2)}$$

and

$$\mathbf{P}\left\{\|X\|_p \leq \mathbf{E}\|X\|_p - t\right\} \leq e^{-t^2/8}.$$

6.12 Proof of Lemma 6.22

The key to the success of the entropy method is that the differential inequalities for the logarithmic moment-generating function of Z can be solved in many interesting cases. The cases considered so far were all easily solvable by lucky coincidences. Here we try to extract the essence of these circumstances and generalize them so that a large family of solvable differential inequalities can be dealt with. The next lemma establishes some simple sufficient conditions. Then Lemma 6.26 will allow us to use Lemma 6.25 to cope with more difficult cases, and this will lead to the proof of Lemma 6.22.

Lemma 6.25 *Let f be a nondecreasing continuously differentiable function on some interval I containing 0 such that $f(0) = 0$, $f'(0) > 0$ and $f(x) \neq 0$ for every $x \neq 0$. Let g be a continuous function on I and consider an infinitely many times differentiable function G on I such that $G(0) = G'(0) = 0$ and for every $\lambda \in I$,*

$$f(\lambda)G'(\lambda) - f'(\lambda)G(\lambda) \leq f^2(\lambda)g(\lambda).$$

Then, for every $\lambda \in I$, $G(\lambda) \leq f(\lambda)\int_0^\lambda g(x)dx$.

Note the special case when $f(\lambda) = \lambda$, and $g(\lambda) = L^2/2$ is the differential inequality obtained, for example, in Theorems 5.3 and 6.7 and is used to obtain sub-Gaussian concentration inequalities. If we choose $f(\lambda) = e^\lambda - 1$ and $g(\lambda) = -d(\lambda/e^\lambda - 1)/d\lambda$, we recover the differential inequality seen in the proof of Theorem 6.12.

Proof Define $\rho(\lambda) = G(\lambda)/f(\lambda)$ for every $\lambda \neq 0$ and $\rho(0) = 0$. Using the assumptions on G and f, we see that ρ is continuously differentiable on I with

$$\rho'(\lambda) = \frac{f(\lambda)G'(\lambda) - f'(\lambda)G(\lambda)}{f^2(\lambda)} \quad \text{for} \quad \lambda \neq 0 \quad \text{and} \quad \rho'(0) = \frac{G''(0)}{2f'(0)}.$$

Hence $f(\lambda)G'(\lambda) - f'(\lambda)G(\lambda) \leq f^2(\lambda)g(\lambda)$ implies that

$$\rho'(\lambda) \leq g(\lambda)$$

and therefore that the function $\Delta(\lambda) = \int_0^\lambda g(x)dx - \rho(\lambda)$ is nondecreasing on I. Since $\Delta(0) = 0$, Δ and f have the same sign on I, which means that $\Delta(\lambda)f(\lambda) \geq 0$ for $\lambda \in I$ and the result follows. □

Except when $a = 1$, the differential inequality (6.4) cannot be solved exactly. A roundabout is provided by the following lemma that compares the solutions of a possibly difficult differential inequality with solutions of a differential equation.

Lemma 6.26 *Let I be an interval containing 0 and let ρ be continuous on I. Let $a \geq 0$ and $v > 0$. Let $H : I \to \mathbb{R}$, be an infinitely many times differentiable function satisfying*

$$\lambda H'(\lambda) - H(\lambda) \leq \rho(\lambda)\left(aH'(\lambda) + v\right)$$

with

$$aH'(\lambda) + v > 0 \quad \text{for every} \quad \lambda \in I \text{ and } H'(0) = H(0) = 0.$$

Let $\rho_0 : I \to \mathbb{R}$ be a function. Assume that $G_0 : I \to \mathbb{R}$ is infinitely many times differentiable such that for every $\lambda \in I$,

$$aG_0'(\lambda) + 1 > 0 \quad \text{and } G_0'(0) = G_0(0) = 0 \text{ and } G_0''(0) = 1.$$

Assume also that G_0 solves the differential equation

$$\lambda G_0'(\lambda) - G_0(\lambda) = \rho_0(\lambda)\left(aG_0'(\lambda) + 1\right).$$

If $\rho(\lambda) \leq \rho_0(\lambda)$ for every $\lambda \in I$, then $H \leq vG_0$.

Proof Let $I, \rho, a, v, H, G_0, \rho_0$ be defined as in the statement of the lemma. Combining the assumptions on H, ρ_0, ρ and G_0,

$$\lambda H'(\lambda) - H(\lambda) \leq \frac{\left(\lambda G_0'(\lambda) - G_0(\lambda)\right)\left(aH'(\lambda) + v\right)}{aG_0'(\lambda) + 1}$$

for every $\lambda \in I$, or equivalently,

$$\left(\lambda + aG_0(\lambda)\right)H'(\lambda) - \left(1 + aG_0'(\lambda)\right)H(\lambda) \leq v\left(\lambda G_0'(\lambda) - G_0(\lambda)\right).$$

Setting $f(\lambda) = \lambda + aG_0(\lambda)$ for every $\lambda \in I$ and defining $g : I \to \mathbb{R}$ by

$$g(\lambda) = \frac{v\left(\lambda G_0'(\lambda) - G_0(\lambda)\right)}{\left(\lambda + aG_0(\lambda)\right)^2} \quad \text{if} \quad \lambda \neq 0 \text{ and } g(0) = \frac{v}{2},$$

our assumptions on G_0 imply that g is continuous on the whole interval I so that we may apply Lemma 6.25. Hence, for every $\lambda \in I$

$$H(\lambda) \leq f(\lambda)\int_0^\lambda g(x)dx = vf(\lambda)\int_0^\lambda \left(\frac{G_0(x)}{f(x)}\right)' dx$$

and the conclusion follows since $\lim_{x\to 0} G_0(x)/f(x) = 0$. □

Observe that the differential inequality in the statement of Lemma 6.22 has the same form as the inequalities considered in Lemma 6.26 where ϕ replaces ρ. Note also that for any $\gamma \geq 0$,

$$2G_\gamma(\lambda) = \frac{\lambda^2}{1 - \gamma\lambda}$$

solves the differential inequality

$$\lambda H'(\lambda) - H(\lambda) = \lambda^2(\gamma H'(\lambda) + 1). \tag{6.5}$$

So by choosing $\gamma = a$ and recalling that for $\lambda \geq 0$, $\phi(-\lambda) \leq \lambda^2/2$, it follows immediately from Lemma 6.26, that

$$G(\lambda) \leq \frac{\lambda^2 v}{2(1 - a\lambda)} \quad \text{for} \quad \lambda \in (0, 1/a).$$

As G is the logarithmic moment-generating function of $Z - \mathbf{E}Z$, this can be used to derive a Bernstein-type inequality for the left tail of Z. However, the obtained constants are not optimal, so proving that Lemma 6.22 requires some more care.

Proof of Lemma 6.22. The function $2G_\gamma$ may be the unique solution of equation (6.5) but this is not the only equation for which G_γ is the solution. Define

$$\rho_\gamma(\lambda) = \frac{\lambda G'_\gamma(\lambda) - G_\gamma(\lambda)}{1 + aG'_\gamma(\lambda)}.$$

Then, on some interval I, G_γ is the solution of the differential equation

$$\lambda H'(\lambda) - H(\lambda) = \rho_\gamma(\lambda)(1 + aH'(\lambda)),$$

provided $1 + aG'_\gamma$ remains positive on I.

Thus, we have to look for the smallest $\gamma \geq 0$ such that, on the relevant interval I (with $0 \in I$), we have both $\phi(-\lambda) \leq \rho_\gamma(\lambda)$ and $1 + aG'_\gamma(\lambda) > 0$ for $\lambda \in I$.

Introduce

$$\begin{aligned} D_\gamma(\lambda) &= (1 - \gamma\lambda)^2(1 + aG'_\gamma(\lambda)) = (1 - \gamma\lambda)^2 + a\lambda\left(1 - \frac{\gamma\lambda}{2}\right) \\ &= 1 + 2(a/2 - \gamma)\lambda - \gamma(a/2 - \gamma)\lambda^2. \end{aligned}$$

Observe that $\rho_\gamma(\lambda) = \lambda^2/(2D_\gamma(\lambda))$.

For any interval I, $1 + aG'_\gamma(\lambda) > 0$ for $\lambda \in I$ holds if and only if $D_\gamma(\lambda) > 0$ for $\lambda \in I$. Hence, if $D_\gamma(\lambda) > 0$ and $\phi(-\lambda) \leq \rho_\gamma(\lambda)$, then it follows from Lemma 6.26 that for every $\lambda \in I$, we have $G(\lambda) \leq vG_\gamma(\lambda)$.

We first deal with intervals of the form $I = [0, c_+^{-1})$ (with $c_+^{-1} = \infty$ when $c_+ = 0$). If $a \leq 1/3$, that is, $c_+ = 0$, $D_{c_+}(\lambda) = 1 + a\lambda > 0$ and $\rho_{c_+}(\lambda) \geq \lambda^2/(2(1+\lambda/3)) \geq \phi(-\lambda)$ for $\lambda \in I = [0, +\infty)$.

If $a > 1/3$, then $D_{c_+}(\lambda) = 1 + \lambda/3 - c_+\lambda^2/6$ satisfies $0 < 1 + \lambda/6 \leq D_{c_+}(\lambda) \leq 1 + \lambda/3$ on an interval I containing $[0, c_+^{-1})$, and therefore $\rho_{c_+}(\lambda) \geq \phi(-\lambda)$ on I.

Next we deal with intervals of the form $I = (-\theta, 0]$ where $\theta = a^{-1}$ if $c_- = 0$, and $\theta = c_-^{-1}(1 - \sqrt{1 - 6c_-})$ otherwise. Recall that for any $\lambda \in (-3, 0]$, $\phi(-\lambda) \leq \lambda^2/(2(1+\lambda/3))$.

If $a \geq 1/3$, that is, $c_- = 0$, $D_{-c_-}(\lambda) = 1 + a\lambda > 0$ for $\lambda \in (a^{-1}, 0]$,

$$\rho_{-c_-}(\lambda) = \frac{\lambda^2}{2(1+a\lambda)} \geq \frac{\lambda^2}{2(1+\lambda/3)}.$$

For $a \in (0, 1/3)$, note first that $0 < c_- \leq 1/6$, and that

$$0 < D_{-c_-}(\lambda) \leq 1 + \frac{\lambda}{3} + \frac{\lambda^2}{36} \leq \left(1 + \frac{\lambda}{6}\right)^2$$

for every $\lambda \in (-\theta, 0]$. This also entails that $\rho_{-c_-}(\lambda) \geq \phi(-\lambda)$ for $\lambda \in (-\theta, 0]$. □

6.13 Some Variations

Next we present a few inequalities that are based on slight variations of the entropy method. These versions differ in the assumptions on how V^+ or V^- are controlled by different functions of Z. These inequalities demonstrate the flexibility of the method, but our aim is not to give an exhaustive list of concentration inequalities that can be obtained this way. The message of this section is that by simple modifications of the main argument one may exploit many special properties of the function f.

We start with inequalities that use negative association between increasing and decreasing functions of Z.

Theorem 6.27 *Assume that for some nondecreasing function $g : \mathbb{R} \to \mathbb{R}$,*

$$V^- \leq g(Z).$$

Then for all $t > 0$,

$$\mathbf{P}\{Z < \mathbf{E}Z - t\} \leq e^{-t^2/(4\mathbf{E}g(Z))}.$$

Proof In order to prove lower-tail inequalities, it suffices to derive suitable upper bounds for the moment-generating function $F(\lambda) = \mathbf{E}e^{\lambda Z}$ for negative values of λ. By the third inequality of Theorem 6.15,

$$\begin{aligned}
&\lambda \mathbf{E}\left[Ze^{\lambda Z}\right] - \mathbf{E}\left[e^{\lambda Z}\right]\log \mathbf{E}\left[e^{\lambda Z}\right] \\
&\leq \sum_{i=1}^{n} \mathbf{E}\left[e^{\lambda Z}\tau(\lambda(Z_i' - Z)_+)\right] \\
&\leq \sum_{i=1}^{n} \mathbf{E}\left[e^{\lambda Z}\lambda^2(Z_i' - Z)_+^2\right] \\
&\quad (\text{using } \lambda < 0 \text{ and that } \tau(-x) \leq x^2 \text{ for } x > 0) \\
&= \lambda^2 \mathbf{E}\left[e^{\lambda Z}V^-\right] \\
&\leq \lambda^2 \mathbf{E}\left[e^{\lambda Z}g(Z)\right].
\end{aligned}$$

Since $g(Z)$ is a nondecreasing and $e^{\lambda Z}$ is a decreasing function of Z, Chebyshev's association inequality (Theorem 2.14) implies that

$$\mathbf{E}\left[e^{\lambda Z}g(Z)\right] \leq \mathbf{E}\left[e^{\lambda Z}\right]\mathbf{E}[g(Z)].$$

The inequality obtained has the same form as the differential inequality we saw in the proof of Theorem 6.2 (with $\mathbf{E}g(Z)$ in place of $v/2$) and it can be solved in an analogous way to obtain the announced lower-tail inequality. □

Often it is more natural to bound V^+ by an increasing function of Z than to bound V^-. In such situations one can still say something about lower tail probabilities of Z but we need the additional guarantee that $|Z - Z_i'|$ remains bounded and that the inequality only applies in a restricted range of the values of t.

Theorem 6.28 *Assume that there exists a nondecreasing function g such that $V^+ \leq g(Z)$ and for any value of $X = (X_1, \ldots, X_n)$ and X_i', $|Z - Z_i'| \leq 1$. Then for all $K > 0$, if $\lambda \in [0, 1/K)$, then*

$$\log \mathbf{E}e^{-\lambda(Z-\mathbf{E}Z)} \leq \lambda^2 \frac{\tau(K)}{K^2}\mathbf{E}g(Z).$$

Moreover, for all $0 < t \leq (e-1)\mathbf{E}g(Z)$, we have

$$\mathbf{P}\{Z < \mathbf{E}Z - t\} \leq \exp\left(-\frac{t^2}{4(e-1)\mathbf{E}g(Z)}\right).$$

Proof The key observation is that the function $\tau(x)/x^2 = (e^x - 1)/x$ is increasing if $x > 0$. Choose $K > 0$. Thus, for $\lambda \in (-1/K, 0)$, the second inequality of Theorem 6.15 implies that

$$\begin{aligned}\lambda \mathbf{E}\left[Ze^{\lambda Z}\right] - \mathbf{E}\left[e^{\lambda Z}\right]\log \mathbf{E}\left[e^{\lambda Z}\right] &\le \sum_{i=1}^{n} \mathbf{E}\left[e^{\lambda Z}\tau(-\lambda(Z - Z_i')_+)\right] \\ &\le \lambda^2 \frac{\tau(K)}{K^2}\mathbf{E}\left[e^{\lambda Z}V^+\right] \\ &\le \lambda^2 \frac{\tau(K)}{K^2}\mathbf{E}\left[g(Z)e^{\lambda Z}\right],\end{aligned}$$

where at the last step we used the assumption of the theorem.

As in the proof of Theorem 6.27, we bound $\mathbf{E}\left[g(Z)e^{\lambda Z}\right]$ by $\mathbf{E}[g(Z)]\mathbf{E}\left[e^{\lambda Z}\right]$. The rest of the proof is identical to that of Theorem 6.27. Here, we took $K = 1$. □

Our last general result deals with a frequently faced situation. In these cases V^+ may be bounded by the product of Z and another random variable W with well-behaved moment-generating function. The following theorem provides a way to deal with such functionals efficiently and painlessly.

Theorem 6.29 *Assume that f is nonnegative and that there exists a random variable W, such that*

$$V^+ \le WZ.$$

Then for all $\theta > 0$ and $\lambda \in (0, 1/\theta)$,

$$\log \mathbf{E}e^{\lambda\left(\sqrt{Z}-\mathbf{E}\sqrt{Z}\right)} \le \frac{\lambda\theta}{1-\lambda\theta}\log \mathbf{E}e^{\lambda W/\theta}.$$

Note that this theorem only bounds the moment-generating function of $\sqrt{Z}$. However, one may easily obtain bounds for the upper-tail probability of Z by observing that, since $\sqrt{\mathbf{E}Z} \ge \mathbf{E}\sqrt{Z}$, and by writing $x = \sqrt{\mathbf{E}Z + t} - \sqrt{\mathbf{E}Z}$, we have, for $\lambda > 0$,

$$\mathbf{P}\{Z > \mathbf{E}Z + t\} \le \mathbf{P}\left\{\sqrt{Z} > \mathbf{E}\sqrt{Z} + x\right\} \le \mathbf{E}e^{\lambda\left(\sqrt{Z}-\mathbf{E}\sqrt{Z}\right)}e^{-\lambda x}$$

by Markov's inequality.

Proof Introduce $Y = \sqrt{Z}$ and $Y^{(i)} = \sqrt{Z^{(i)}}$. Then

$$\begin{aligned}\mathbf{E}'\left[\sum_{i=1}^{n}(Y - Y^{(i)})_+^2\right] &= \mathbf{E}'\left[\sum_{i=1}^{n}\left(\sqrt{Z} - \sqrt{Z^{(i)}}\right)_+^2\right] \\ &\le \mathbf{E}'\left[\sum_{i=1}^{n}\left(\frac{(Z - Z^{(i)})_+}{\sqrt{Z}}\right)^2\right] \\ &\le \frac{1}{Z}\mathbf{E}'\left[\sum_{i=1}^{n}\left(Z - Z^{(i)}\right)_+^2\right] \\ &\le W.\end{aligned}$$

Thus, applying Theorem 6.16 for Y proves the statement. □

Example 6.30 (TRIANGLES IN A RANDOM GRAPH) Consider the Erdős–Rényi $G(n,p)$ model of a random graph. Recall that such a graph has n vertices and for each pair (u,v) of vertices an edge is inserted between u and v with probability p, independently. We write $m = \binom{n}{2}$, and denote the indicator variables of the m edges by $X_1, \ldots, X_m$ (i.e. $X_i = 1$ if edge $i = (u,v)$ is present in the random graph and $X_i = 0$ otherwise). Three edges form a *triangle* if there are vertices u, v, w such that the edges are of the form (u,v), (v,w), and (w,u). Concentration properties of the number of triangles in a random graph have received a great deal of attention and sharp bounds have been derived by various sophisticated methods for different ranges of the parameter p of the random graph (see the bibliographical remarks at the end of the chapter). Interestingly, the left tail is substantially easier to handle, as Janson's inequality, presented in the next section, offers sharp estimates. However, proving sharp inequalities for the upper tail was much more challenging. Here we only show some sub-optimal versions that are easy to obtain from the general results of this chapter.

Let $Z = f(X_1, \ldots, X_m)$ denote the number of triangles in a random graph. Note that

$$EZ = \frac{n(n-1)(n-2)}{6}p^3 \approx \frac{n^3p^3}{6}$$

and

$$\mathrm{Var}\,(Z) = \binom{n}{3}(p^3 - p^6) + \binom{n}{4}\binom{4}{2}(p^5 - p^6).$$

To obtain exponential upper-tail inequalities, we estimate the random variable

$$V^+ = \sum_{i=1}^{n} \mathbf{E}'(Z - Z_i')_+^2.$$

If v and u denote the extremities of edge i $(1 \le i \le m)$, then we denote by B_i the number of vertices w such that both edges (u,w) and (v,w) exist in the random graph. Then

$$V^+ = \sum_{i=1}^{m} X_i(1-p)B_i^2.$$

Since $\sum_{i=1}^{m} X_iB_i = 3Z$, we have

$$\begin{aligned} V^+ &\le (1-p)\sum_{i=1}^{m} X_i\left(\max_{j=1,\ldots,m} B_j\right)B_i \\ &= (1-p)\left(\max_{j=1,\ldots,m} B_j\right)\sum_{i=1}^{m} X_iB_i \\ &= 3(1-p)\left(\max_{j=1,\ldots,m} B_j\right)Z. \end{aligned}$$

By bounding $\max_{j=1,\ldots,m} B_j$ trivially by n, we have $V^+ \le 3(1-p)nZ$. Define $f_i(X^{(i)})$ as the number of triangles when we force the i-th edge to be absent in the graph. Then clearly $\sum_{i=1}^{n}(f(X) - f_i(X^{(i)}))^2 = V^+/(1-p)$ and therefore, using the terminology of Section 6.11, f is weakly $(3n, 0)$-self-bounding. Thus, by Theorem 6.19,

$$\boldsymbol{P}\{Z \geq \boldsymbol{E}Z + t\} \leq \exp\left(-\frac{t^2}{n^4p^3 + 3nt}\right).$$

It is clear that in the argument above a lot is lost by bounding $W \stackrel{\text{def}}{=} 3\max_{j=1,\dots,m} B_j$ by n. Indeed, one may achieve a significant improvement by using Theorem 6.29. In order to do so, we need to bound the moment-generating function of W. This may be done by another application of Theorem 6.19. Let $W^{(i)}$ denote the value of W when edge i is deleted from the random graph (if the graph contained that edge). Then $W^{(i)} \leq W$ and

$$\sum_{i=1}^{n}\left(W - W^{(i)}\right)^2 \leq 18W,$$

so W is weakly $(18, 0)$-self-bounding. Hence, by Theorem 6.19,

$$\log \boldsymbol{E}e^{\lambda(W-\boldsymbol{E}W)} \leq \frac{9\lambda^2\boldsymbol{E}W}{1-9\lambda}.$$

Denoting $Y = \sqrt{Z}$, Theorem 6.29 leads to

$$\log \boldsymbol{E}e^{\lambda(Y-\boldsymbol{E}Y)} \leq \frac{\lambda}{1-\lambda}\left(\frac{9\lambda^2\boldsymbol{E}W}{1-9\lambda} + \lambda\boldsymbol{E}W\right) \leq \frac{\lambda^2\boldsymbol{E}W}{1-10\lambda}.$$

This is a sub-gamma bound for the moment-generating function of Y, and the computations of Sections 2.4 and 2.8 imply

$$\boldsymbol{P}\{Y > \boldsymbol{E}Y + t\} \leq \exp\left(-\frac{t^2}{4\boldsymbol{E}W + 20t}\right).$$

Now it remains to bound the expected value of W. Note that $W/3$ is the maximum of $m = \binom{n}{2}$ binomial random variables with parameters (n, p^2). In order to obtain a quick upper bound for $\boldsymbol{E}W/3$, it is convenient to use the technique presented in Section 2.5 as follows: let S_i with $i \leq m$ denote a sequence of binomially distributed random variables with parameters n and p^2. By Jensen's inequality,

$$\begin{aligned}
\boldsymbol{E}W/3 &\leq \log\left(\boldsymbol{E}\max_{i=1,\dots,m} e^{S_i}\right) \\
&\leq \log\left(\boldsymbol{E}\left[me^{S_1}\right]\right) \\
&= \log m + \log\left(\boldsymbol{E}e^{S_1}\right) \\
&\leq \log m + (e-1)np^2 \\
&\leq 2\log n + 2np^2.
\end{aligned}$$

Arguably, the most interesting values for p are those when p is at most of the order of $n^{-1/2}$ and in this case, the dominating term in the above expression is $2\log n$. Hence, we obtain the following bound for the tail of $Y = \sqrt{Z}$

$$P\{Y \geq EY + t\} \leq \exp\left(-\frac{t^2}{24(np^2 + \log n) + 20t}\right).$$

It is now easy to get tail bounds for the number Z of triangles. We spare the reader from the straightforward details (see the exercises).

6.14 Janson's Inequality

As we saw in the examples of Section 6.13, in many cases the special structure of the function of independent random variables can be used to deduce concentration inequalities. In this section we present another general result, a celebrated exponential lower-tail inequality for Boolean polynomials.

More precisely, consider independent binary random variables $X_1, \ldots, X_n$ such that $P\{X_i = 1\} = 1 - P\{X_i = 0\} = p_i$ for some $p_1, \ldots, p_n \in [0, 1]$. To simplify notation, we identify every binary vector $\alpha \in \{0, 1\}^n$ with the subset of $\{1, \ldots, n\}$ defined by the non-zero components of α. For example, for $i \in \{1, \ldots, n\}$, we write $i \in \alpha$ to denote that the i-th component of α equals 1. Then for each $\alpha \in \{0, 1\}^n$, we introduce the binary random variable

$$Y_\alpha = \prod_{i \in \alpha} X_i.$$

Given a collection $\mathcal{I}$ of subsets of the binary hypercube $\{0, 1\}^n$, we may define

$$Z = \sum_{\alpha \in \mathcal{I}} Y_\alpha,$$

which is a polynomial of the binary vector $X = (X_1, \ldots, X_n)$.

Boolean polynomials of this type are common in many applications of the probabilistic method in discrete mathematics and also in the theory of random graphs, and their concentration properties have been the subject of intensive study. Note that for any $\alpha, \beta \in \mathcal{I}$ with $\alpha \cap \beta = \emptyset$ (i.e. if $\alpha_i \beta_i = 0$ for all $i = 1, \ldots, n$), $EY_\alpha Y_\beta = EY_\alpha EY_\beta$ and therefore the variance of Z equals

$$\begin{aligned}
\mathrm{Var}(Z) &= EZ^2 - (EZ)^2 = \sum_{\alpha,\beta \in \mathcal{I}} EY_\alpha Y_\beta - \sum_{\alpha,\beta \in \mathcal{I}} EY_\alpha EY_\beta \\
&= \sum_{\alpha,\beta \in \mathcal{I}: \alpha \cap \beta \neq \emptyset} (EY_\alpha Y_\beta - EY_\alpha EY_\beta) \\
&\leq \sum_{\alpha,\beta \in \mathcal{I}: \alpha \cap \beta \neq \emptyset} EY_\alpha Y_\beta \\
&\stackrel{\text{def}}{=} \Delta.
\end{aligned}$$

Thus, by Chebyshev's inequality,

$$P\{|Z - EZ| > t\} \leq \frac{\Delta}{t^2}.$$

The next theorem shows the surprising fact that, at least for the lower tail, there is always an exponential version of this inequality.

Theorem 6.31 (JANSON'S INEQUALITY) *Let $\mathcal{I}$ denote a collection of subsets of $\{0,1\}^n$ and define Z and Δ as above. Then for all $\lambda \leq 0$,*

$$\log \boldsymbol{E} e^{\lambda(Z-EZ)} \leq \phi\left(\frac{\lambda\Delta}{EZ}\right)\frac{(EZ)^2}{\Delta}$$

where $\phi(x) = e^x - x - 1$. In particular, for all $0 \leq t \leq EZ$,

$$P\{Z \leq EZ - t\} \leq e^{-t^2/(2\Delta)}.$$

The proof of Janson's inequality shown here shows certain similarities with the entropy method. In particular, the proof is based on bounding the derivative of the logarithmic moment-generating function of Z. However, sub-additivity inequalities can be avoided because of a positive association property that can be exploited by an appropriate use of Harris' inequality (Theorem 2.15).

Proof Denote the logarithmic moment generating function of $Z - EZ$ by $G(\lambda) = \log \boldsymbol{E} e^{\lambda(Z-EZ)}$. Then the derivative of G equals

$$G'(\lambda) = \frac{E[Ze^{\lambda Z}]}{Ee^{\lambda Z}} - EZ = \sum_{\alpha\in\mathcal{I}} \frac{E\left[Y_\alpha e^{\lambda Z}\right]}{Ee^{\lambda Z}} - EZ.$$

In the following, we derive an upper bound for each term $E\left[Y_\alpha e^{\lambda Z}\right]$ of the sum on the right-hand side.

Fix an $\alpha \in \mathcal{I}$ and introduce $U_\alpha = \sum_{\beta:\beta\cap\alpha\neq\emptyset} Y_\beta$ and $Z_\alpha = \sum_{\beta:\beta\cap\alpha=\emptyset} Y_\beta$. Clearly, regardless of what α is, $Z = U_\alpha + Z_\alpha$. Since

$$E\left[Y_\alpha e^{\lambda Z}\right] = E\left[e^{\lambda Z} \mid Y_\alpha = 1\right] EY_\alpha,$$

it suffices to bound the conditional expectation. The key observation is that since $\lambda \leq 0$, both $\exp(\lambda U_\alpha)$ and $\exp(\lambda Z_\alpha)$ are decreasing functions of $X_1, \ldots X_n$.

$$
\begin{aligned}
& E\left[e^{\lambda Z} \mid Y_\alpha = 1\right] \\
&= E\left[e^{\lambda U_\alpha} e^{\lambda Z_\alpha} \mid Y_\alpha = 1\right] \\
&\geq E\left[e^{\lambda U_\alpha} \mid Y_\alpha = 1\right] E\left[e^{\lambda Z_\alpha} \mid Y_\alpha = 1\right] \quad \text{(by Harris' inequality)} \\
&= E\left[e^{\lambda U_\alpha} \mid Y_\alpha = 1\right] Ee^{\lambda Z_\alpha} \quad \text{(since } Z_\alpha \text{ and } Y_\alpha \text{ are independent)} \\
&\geq E\left[e^{\lambda U_\alpha} \mid Y_\alpha = 1\right] Ee^{\lambda Z} \quad \text{(as } Z_\alpha \leq Z) \\
&\geq e^{\lambda E[U_\alpha \mid Y_\alpha = 1]} Ee^{\lambda Z} \quad \text{(by Jensen's inequality).}
\end{aligned}
$$

Note that we apply Harris' inequality above conditionally, given $Y_\alpha = 1$. This condition simply forces $X_i = 1$ for all $i \in \alpha$, so both U_α and Z_α are increasing functions of the independent random variables X_i, $i \notin \alpha$ and Harris' inequality is used legally. Thus, we obtain

$$
\begin{aligned}
& \frac{E\left[Ze^{\lambda Z}\right]}{EZ} \\
&\geq Ee^{\lambda Z} \sum_{\alpha \in \mathcal{I}} \frac{EY_\alpha}{EZ} e^{E[\lambda U_\alpha \mid Y_\alpha = 1]} \\
&\geq Ee^{\lambda Z} \exp\left(\sum_{\alpha \in \mathcal{I}} \frac{EY_\alpha}{EZ} E\left[\lambda U_\alpha \mid Y_\alpha = 1\right] \right) \quad \text{(by Jensen's inequality)} \\
&= Ee^{\lambda Z} \exp\left(\lambda \frac{\Delta}{EZ} \right)
\end{aligned}
$$

where we use the fact that

$$
\Delta = \sum_{\alpha \in \mathcal{I}} E\left[Y_\alpha U_\alpha\right].
$$

Summarizing, we have, for all $\lambda \leq 0$,

$$
G'(\lambda) \geq EZ\left(e^{\lambda \Delta / EZ} - 1\right).
$$

Thus, integrating this inequality between λ and 0 and using $G(0) = 0$, we find that for $\lambda \leq 0$,

$$
G(\lambda) \leq -EZ \int_\lambda^0 \left(e^{u \frac{\Delta}{EZ}} - 1\right) du = \phi\left(\frac{\lambda \Delta}{EZ}\right) \frac{(EZ)^2}{\Delta}
$$

as desired. The second inequality follows from the simple fact that for $x > 0$, $\phi(-x) \leq x^2/2$. □

Remark 6.6 (PROBABILITY OF NON-EXISTENCE) In many applications of Janson's inequality, one wishes to show that in a random draw of the vector $X = (X_1, \ldots, X_n)$, with high probability, there exists at least one element $\alpha \in \mathcal{I}$ for which $Y_\alpha = 1$. In other words, the goal is to show that $Z > 0$ with high probability. To this end, one may write

$$P\{Z = 0\} = P\{Z \leq EZ - EZ\} \leq \exp\left(-\frac{(EZ)^2}{2\Delta}\right),$$

which is guaranteed to be exponentially small whenever $\sqrt{\Delta}$ is small compared to EZ.

Example 6.32 (TRIANGLES IN A RANDOM GRAPH) A prototypical application of Janson's inequality is the case of the number of triangles in an Erdős–Rényi random graph $G(n,p)$, discussed in Example 6.30 in the previous section. If Z denotes the number of triangles in $G(n,p)$, then recall that

$$EZ = \binom{n}{3}p^3 \quad \text{and} \quad \mathrm{Var}(Z) = \binom{n}{3}p^3(1-p^3) + 2\binom{n}{4}\binom{4}{2}p^5(1-p).$$

The value of Δ may also be computed in a straightforward way. One obtains

$$\Delta = \binom{n}{3}p^3 + 2\binom{n}{4}\binom{4}{2}p^5$$

which is only slightly larger than $\mathrm{Var}(Z)$. For the probability that the random graph does not contain any triangle, we may use Janson's inequality with $t = EZ$:

$$P\{Z = 0\} \leq \exp\left(-\frac{\binom{n}{3}^2 p^6}{2\left(\binom{n}{3}p^3 + 2\binom{n}{4}\binom{4}{2}p^5\right)}\right) \leq \exp\left(-\frac{\binom{n}{3}p^2}{2\left(1 + 2np^2\right)}\right).$$

6.15 Bibliographical Remarks

The key principles of the entropy method rely on the ideas of proving Gaussian concentration inequalities based on logarithmic Sobolev inequalities. These are summarized in Chapter 5, where we also give some of the main references. It was Michel Ledoux (1997) who realized that these ideas may be used as an alternative route to some of Talagrand's exponential concentration inequalities for empirical processes and Rademacher chaos. Ledoux's ideas were taken further by Massart (2000*a*), Bousquet (2002*a*), Klein (2002), Rio (2001), and Klein and Rio (2005), while the core of the material of this chapter is based on Boucheron, Lugosi, and Massart (2000, 2003, 2009).

Different versions of the modified logarithmic Sobolev inequalities used in this chapter are due to Ledoux (1997, 1999, 2001) and Massart (2000*a*).

The bounded differences inequality is perhaps the simplest and most widely used exponential concentration inequality. The basic idea of writing a function of independent random variables as a sum of martingale differences, and using exponential inequalities for martingales, was first used in various applications by mathematicians including Yurinskii (1976), Maurey (1979), Milman and Schechtman (1986), and Shamir and

Spencer (1987). The inequality was first laid down explicitly and illustrated by a wide variety of applications in an excellent survey paper by McDiarmid (1989), and the result itself has often been referred to as McDiarmid's inequality. Martingale methods have served as a flexible and versatile tool for proving concentration inequalities (see the more recent surveys of McDiarmid (1998), Chung and Lu (2006*b*), and Dubhashi and Panconesi (2009)).

The exponential tail inequality for sums of independent Hilbert-space valued random variables derived in Example 6.3 is just a simple example. There is a vast literature dealing with tails of sums of vector-valued random variables. It is outside the scope of this book to derive the sharpest and most general results. Here we merely try to make the point that general concentration inequalities prove to be a versatile tool in such applications. In fact, applications of this type motivated some of the most significant advances in the theory of concentration inequalities. In Chapters 11, 12, and 13 we discuss many of the principal modern tools for analyzing the tails of sums of independent vector-valued random variables and empirical processes. For some of the classical references, the interested reader is referred to Yurinskii (1976, 1995), Ledoux and Talagrand (1991), and Pinelis (1995).

The inequality described in Exercise 6.4 was proved independently by Guntuboyina and Leeb (2009) and Bordenave, Caputo, and Chafaï (2011).

Theorem 6.5 is due to McDiarmid (1998) who proved it using martingale methods. The proof presented here is due to Andreas Maurer who kindly permitted us to reproduce his elegant work.

The exponential inequality for the largest eigenvalue of a random symmetric matrix described in Example 6.8 was proved by Alon, Krivelevich, and Vu (2002) who used Talagrand's convex distance inequality. Maurer (2006) obtained a better exponent with a more careful analysis. Alon, Krivelevich, and Vu (2002) show, with a simple extension of the argument, that for the k-th largest (or k-th smallest) eigenvalue the upper bounds become $e^{-t^2/(16k^2)}$, though it is not clear whether the factor k^{-2} in the exponent is necessary.

Theorem 6.9 appears in Maurer (2006). Theorem 6.10 was first established by Talagrand (1996*c*) who also proves a corresponding lower tail inequality which is presented in Section 7.5. The proof given here is due to Ledoux (1997).

Self-bounding functions were introduced by Boucheron, Lugosi, and Massart (2000) who prove Theorem 6.12 building on techniques developed by Massart (2000*a*). Various generalizations of the self-bounding property were considered by Boucheron, Lugosi, and Massart (2003, 2009), Boucheron *et al.* (2005*b*), Devroye (2002), Maurer (2006), and McDiarmid and Reed (2006). In particular, McDiarmid and Reed (2006) considered what we call strongly (a, b)-self-bounding functions and proved results that are only slightly weaker than those presented in Section 6.11. The weak self-bounding property was first considered by Maurer (2006), and Theorem 6.19 is due to him. Theorems 6.21 and 6.20 appear in Boucheron, Lugosi, Massart (2009).

We note here that the inequality linking the expected and annealed VC entropies answers, in a positive way, a question raised by Vapnik (1995, pp. 53–54): the empirical risk minimization procedure is *non-trivially consistent* and *rapidly convergent* if and only if the annealed entropy rate $(1/n)\log_2 ET(X)$ converges to zero. For the definitions and discussion we refer to Vapnik (1995).

The material of Sections 6.8, 6.9, and 6.13 is based on Boucheron, Lugosi, and Massart (2003).

Klaassen (1985) showed that Poisson distributions satisfy the "modified Poincaré" inequality

$$\operatorname{Var}(f(Z)) \leq \boldsymbol{E}Z \times \boldsymbol{E}[|Df(Z)|^2]$$

(see Exercise 3.21).

The search for modified logarithmic Sobolev inequalities, that is, functional inequalities which capture the tail behavior of distributions that are less concentrated than the Gaussian distribution, was initiated by Bobkov and Ledoux (1997). Their aim was to recover some results of Talagrand concerning concentration properties of the exponential distribution. Bobkov and Ledoux (1997, 1998) pointed out that the Poisson distribution cannot satisfy an analog of the Gaussian logarithmic Sobolev inequality. They establish the second inequality of Theorem 6.17. The first inequality of Theorem 6.17 is due to Wu (2000). Other modified logarithmic Sobolev inequalities have been investigated by Ané and Ledoux (2000), Chafaï (2006), Bobkov and Tetali (2006), and others.

Janson's inequality (Theorem 6.31) was first established by Janson (1990). This inequality has since become one of the basic standard tools of the probabilistic method of discrete mathematics and random graph theory, and many variations, refinements, and alternative proofs are now known. We refer the reader to the monographs of Alon and Spencer (1992), and Janson, Łuczak, and Ruciński (2000) for surveys and further references.

The number of triangles, and more generally, the number of copies of a fixed subgraph, in a random graph $G(n, p)$ has been a subject of intensive study. For the lower-tail probabilities, Janson's inequality, shown in Section 6.14, gives an essentially tight bound. However, obtaining sharp bounds for the upper tail has been an important non-trivial challenge. For such upper-tail inequalities we refer the interested reader to the papers Kim and Vu (2000, 2004), Vu (2000, 2001), Janson and Ruciński (2004, 2002), Janson, Oleszkiewicz, and Ruciński (2004), Bolthausen, Comets, and Dembo (2009), Döring and Eichelsbacher (2009), Chatterjee and Dey (2010), Chatterjee (2010), DeMarco and Kahn (2010), and Schudy and Sviridenko (2012).

The inequalities derived in Example 6.30 are not the best possible.

6.16 EXERCISES

6.1. Relax the condition of Theorem 6.7 in the following way. Show that if $X = (X_1, \ldots, X_n)$ and

$$\boldsymbol{E}\left[\sum_{i=1}^{n}(Z - Z_i')_+^2 \,\middle|\, X\right] \leq v$$

then for all $t > 0$,

$$P\{Z > EZ + t\} \leq e^{-t^2/(2v)}$$

and if

$$E\left[\sum_{i=1}^{n}(Z - Z_i')_-^2 \middle| X\right] \leq v,$$

then

$$P\{Z < EZ - t\} \leq e^{-t^2/(2v)}.$$

6.2. (THE CAUCHY INTERLACING THEOREM) Let A be an $n \times n$ Hermitian matrix with eigenvalues $\alpha_1 \leq \alpha_2 \leq \cdots \leq \alpha_n$. Denote by R_A the *Rayleigh quotient* defined, for every $x \in \mathbb{C}^n \setminus \{0\}$, by

$$R_A(x) = \frac{x^*Ax}{x^*x}.$$

Prove the min-max formulas

$$\alpha_k = \max\left\{\min\left\{R_A(x) : x \in U \text{ and } x \neq 0\right\} : \dim(U) = n - k + 1\right\}$$

and

$$\alpha_k = \min\left\{\max\left\{R_A(x) : x \in U \text{ and } x \neq 0\right\} : \dim(U) = k\right\}.$$

Let P be an orthogonal projection matrix with rank m and define the Hermitian matrix $B = PAP$. Denoting by $\beta_1 \leq \beta_2 \leq \cdots \leq \beta_m$ the eigenvalues of B, using the minmax formulas, show that the eigenvalues of A and B interlace, that is, for all $j \leq m$, $\alpha_j \leq \beta_j \leq \alpha_{n-m+j}$. (See Bai and Silverstein (2010).)

6.3. (RANK INEQUALITY FOR SPECTRAL MEASURES) Let A and B be $n \times n$ Hermitian matrices and denote by F_A and F_B the distribution functions related to the spectral measures L_A and L_B of A and B, respectively. Setting $k = \text{rank}(A - B)$, prove the rank inequality

$$\|F_A - F_B\|_\infty \leq \frac{k}{n}.$$

Hint: show that one can always assume that

$$A = \begin{bmatrix} A_{11} & A_{12} \\ A_{21} & A_{22} \end{bmatrix} \text{ and } B = \begin{bmatrix} B_{11} & A_{12} \\ A_{21} & A_{22} \end{bmatrix}$$

where the order of A_{22} is $n-k \times n-k$. Use the Cauchy interlacing theorem (see Exercise 6.2 above) for the pairs of Hermitian matrices A and A_{22} on the one hand and B and A_{22} on the other hand. (See Bai and Silverstein (2010).)

6.4. Show that the convexity assumption is essential in Theorem 6.10, by considering the following example: let n be an even positive integer and define $A = \{x \in [0,1]^n : \sum_{i=1}^n x_i \le n/2\}$. Let $f(x) = \inf_{y \in A} \|x - y\|$. Then clearly f is Lipschitz but not convex. Let the components of $X = (X_1, \ldots, X_n)$ be i.i.d. with $\mathbf{P}\{X_i = 0\} = \mathbf{P}\{X_i = 1\} = 1/2$. Show that there exists a constant $c > 0$ such that $\mathbf{P}\{f(X) > \mathbf{M}f(X) + cn^{1/4}\} \ge 1/4$ for all sufficiently large n. (This example is taken from Ledoux and Talagrand (1991, p. 17).)

6.5. Prove the following generalization of Theorem 6.10. Let $\mathcal{X} \subset \mathbb{R}^d$ be a convex compact set with diameter B. Let $X_1, \ldots, X_n$ be independent random variables taking values in $\mathcal{X}$ and assume that $f : \mathcal{X}^n \to \mathbb{R}$ is *separately convex* and Lipschitz, that is, $|f(x) - f(y)| \le \|x - y\|$ for all $x, y \in \mathcal{X}^n \subset \mathbb{R}^{dn}$. Then $Z = f(X_1, \ldots, X_n)$ satisfies, for all $t > 0$,

$$\mathbf{P}\{Z > \mathbf{E}Z + t\} \le e^{-t^2/(2B^2)}.$$

6.6. Let $X_1, \ldots, X_n$ be independent vector-valued random variables taking values in a compact convex set $\mathcal{X} \subset \mathbb{R}^d$ with diameter B. Let A denote the $d \times n$ matrix whose columns are $X_1, \ldots, X_n$ and let Z denote the largest singular value of A. Show that

$$\mathbf{P}\{Z > \mathbf{E}Z + t\} \le e^{-t^2/(2B^2)}.$$

Compare the result with Example 6.11.

6.7. Assume that $Z = f(X) = f(X_1, \ldots, X_n)$ where $X_1, \ldots, X_n$ are independent real-valued random variables and f is a nondecreasing function of each variable. Suppose that there exists another nondecreasing function $g : \mathbb{R}^n \to \mathbb{R}$ such that

$$\sum_{i=1}^n (Z - Z_i')_-^2 \le g(X).$$

Show that for all $t > 0$,

$$\mathbf{P}\{Z < \mathbf{E}Z - t\} \le e^{-t^2/(4\mathbf{E}g(X))}.$$

Hint: use Harris' inequality (Theorem 2.15).

6.8. (ALMOST BOUNDED DIFFERENCES) Assume that $Z = f(X) = f(X_1, \ldots, X_n)$ where $X_1, \ldots, X_n$ are independent real-valued random variables. Assume there exists a monotone set $A \subset \mathbb{R}^n$ and constants $v, C > 0$ such that for $x = (x_1, \ldots, x_n) \in A$, $\sum_{i=1}^n (f(x) - \inf_{x_i'} f(x_1, \ldots, x_i', \ldots, x_n))^2 \le v$ and for all $x \notin A$, $\sum_{i=1}^n (f(x) - \inf_{x_i'} f(x_1, \ldots, x_i', \ldots, x_n))^2 \le C$. (A monotone set is such that if $x \in A$ and $y \ge x$ (component-wise) then $y \in A$.) Show that for all $t > 0$,

$$P\{Z > EZ + t\} \leq \exp\left(\frac{-t^2}{2(v + CP\{X \notin A\})}\right).$$

Hint: use Harris' inequality (Theorem 2.15).

6.9. (RADEMACHER CHAOS OF ORDER TWO) Let $\mathcal{T}$ be a finite set of $n \times n$ symmetric matrices with zero diagonal entries. Let $\varepsilon = (\varepsilon_1, \ldots, \varepsilon_n)$ be a vector of independent Rademacher variables. Let

$$Z = \max_{M \in \mathcal{T}} \sum_{i=1}^{n} \sum_{j=1}^{n} M_{i,j} \varepsilon_i \varepsilon_j$$

and

$$Y = \max_{M \in \mathcal{T}} \left(\sum_{i=1}^{n} \left(\sum_{j=1}^{n} \varepsilon_j M_{i,j} \right)^2 \right)^{1/2}.$$

Let $B = \max_{M \in \mathcal{T}} \|M\|^2$ where $\|M\|$ denotes the (operator) norm of matrix M. Prove that

$$\begin{aligned}
\operatorname{Var}(Z) &\leq 8E\left[Y^2\right] \\
\operatorname{Var}(Y^2) &\leq 8BE\left[Y^2\right] \\
\log E e^{\lambda(Y^2 - EY^2)} &\leq \frac{\lambda^2}{(1 - 8B\lambda)} 8BE[Y^2] \\
\log E e^{\lambda(Z - EZ)} &\leq \frac{16\lambda^2}{2(1 - 64B\lambda)} E[Y^2],
\end{aligned}$$

where $\lambda \geq 0$. *Hint:* use Theorem 6.16 twice. Show that $8Y^2$ upper bounds an Efron–Stein estimate of the variance of Z. Then use the fact that Y may be represented as the supremum of a Rademacher process, and prove that Y^2 is $(16B, 0)$-weakly self-bounding. Note that

$$E[Y^2] = E\left[\sup_{M \in \mathcal{T}} \sum_{i,j=1}^{n} \varepsilon_i \varepsilon_j M_{i,j}^2\right].$$

See Talagrand (1996*b*), Ledoux (1997), and Boucheron, Lugosi, and Massart (2003).

6.10. Prove Theorem 6.17. *Hint:* use Lemma 6.18 and the so-called "law of rare events," that is, the convergence of the binomial distribution to a Poisson.

6.11. (A LOGARITHMIC SOBOLEV INEQUALITY FOR THE EXPONENTIAL DISTRIBUTION). Assume X is exponentially distributed, that is, it has density $\exp(-x)$for $x > 0$. Prove that if $f : [0, \infty) \to \mathbb{R}$ is differentiable, then

$$\text{Ent}\left((f(X))^2\right) \le 4E\left[X(f'(X))^2\right].$$

Hint: use the fact that if X_1 and X_2 are independent standard Gaussian random variables, $(X_1^2 + X_2^2)/2$ is exponentially distributed, and use the Gaussian logarithmic Sobolev inequality.

6.12. (SQUARE ROOT OF A POISSON RANDOM VARIABLE) Let X be a Poisson random variable. Prove that for $0 \le \lambda < 1/2$,

$$\log Ee^{\lambda(\sqrt{X}-E\sqrt{X})} \le \frac{\lambda^2}{1-2\lambda}.$$

Show that

$$\log Ee^{\lambda(\sqrt{X}-E\sqrt{X})} \le v\lambda(e^\lambda - 1)$$

where $v = (EX)E[1/(4X+1)]$. Use Markov's inequality to show that

$$P\left\{\sqrt{X} \ge E\sqrt{X} + t\right\} \le \exp\left(-\frac{t}{2}\log\left(1 + \frac{t}{2v}\right)\right).$$

Hint: the first inequality may be derived from Theorem 6.29. The second inequality may be derived from Theorem 6.17.

6.13. (ENTROPIC VERSION OF THE LAW OF RARE EVENTS) Let X be a random variable taking nonnegative integer values and define $p(k) = P\{X = k\}$ for $k = 0, 1, 2, \ldots$. The *scaled Fisher information* of X is defined by

$$K(X) = (EX)E\left[\left(\frac{(X+1)p(X+1)}{(EX)p(X)} - 1\right)^2\right].$$

Let $\mu = EX$. Use Theorem 6.17 to prove that the Kullback–Leibler divergence of X and a Poisson(μ) random variable is at most $K(X)$.

Let S be the sum of the independent integer-valued random variables $X_1, \ldots, X_n$ with $EX_i = p_i$. Let $\mu = \sum_{i=1}^n p_i$. Prove that

$$K(S) \le \sum_{i=1}^n \frac{p_i}{\mu} K(X_i).$$

From this sub-additivity property, prove that the Kullback–Leibler divergence of S and a Poisson(μ) random variable is at most $(1/\mu)\sum_{i=1}^n p_i^3/(1-p_i)$. (See Kontoyiannis, Harremoës, and Johnson (2005).)

6.14. Consider the maximal degree D of any vertex in a random $G(n,p)$ graph defined as in Example 6.13. Show that for any sequence $a_n \to \infty$, with probability tending to 1 as $n \to \infty$,

$$\left|D - np - \sqrt{2p(1-p)n\log n}\right| \leq a_n \sqrt{\frac{p(1-p)n}{\log n}}$$

(see Bollobás (2001, Corollary 3.14)). What do you obtain if you combine Lemma 2.4 with Theorem 6.12?

6.15. (LOWER BOUND FOR TRIANGLES) Let Z denote the number of triangles in a random graph $G(n, p)$ where $p \geq 1/n$. Show that for every $a > 0$ there exists a constant $c = c(a)$ such that

$$\mathbf{P}\left\{Z > \mathbf{E}Z + an^3p^3\right\} \geq e^{-cp^2n^2\log(1/p)}.$$

Hint: the lower bound is the probability that a fixed clique of size proportional to np exists in $G(n, p)$. (Vu (2001).)

6.16. Let Z be as in the previous exercise. Use the inequality for $\sqrt{Z}$ shown in the text to prove that for any $K > 1$, if $t \leq (K^2 - 1)\mathbf{E}Z$, then

$$\mathbf{P}\{Z > \mathbf{E}Z + t\} \leq \exp\left(-\frac{t^2}{(K+1)^2\mathbf{E}Z\left(24np^2 + 24\log n + \frac{20t}{(K+1)\sqrt{\mathbf{E}Z}}\right)}\right).$$

7

Concentration and Isoperimetry

The concentration inequalities discussed in this book are intimately related to isoperimetric problems. In this chapter we discuss some aspects of the rich relationship between isoperimetric problems and concentration inequalities. In Section 7.1 we start by establishing a connection between isoperimetric inequalities in general metric spaces and concentration of Lipschitz functions. We also give an equivalent formulation of the bounded differences inequality (Theorem 6.2) which shows that every not-too-small set in a product probability space has the property that the probability of those points whose Hamming distance from the set is much larger than $\sqrt{n}$ is exponentially small.

In Section 7.2 we show how the classical isoperimetric theorem follows from the Brunn–Minkowski theorem and discuss isoperimetric inequalities on the surface of the n-dimensional Euclidean ball and for the standard multivariate Gaussian measure.

In Section 7.3 we discuss the vertex isoperimetric theorem on the binary hypercube and its relationship to concentration inequalities such as the bounded differences inequality.

Then, in Section 7.4, we present a powerful concentration inequality, known as Talagrand's *convex distance inequality,* as a consequence of the concentration results for self-bounding functions from Section 6.11. In Sections 7.5 and 7.6 we describe its applications for convex Lipschitz functions and to a bin packing problem.

7.1 Lévy's Inequalities

The classical isoperimetric theorem (proved in Section 7.2 below) states that among all subsets of $\mathbb{R}^n$ of a given volume, Euclidean balls minimize their surface area. An equivalent formulation is that, for any $t > 0$, among all (measurable) sets $A \subset \mathbb{R}^n$ of a given volume, the ones for which the volume of the *blowup* of A, defined by

$$A_t = \{x \in \mathbb{R}^n : d(x, A) < t\},$$

have minimal volume are Euclidean balls. Here $d(x, A) = \inf_{y \in A} d(x, y)$ denotes the distance of x to the set A. The advantage of this formulation is that it avoids the notion of surface area and the problem can be generalized to arbitrary metric spaces. In fact, countless versions of the classical isoperimetric problem have been studied.

In particular, given a metric space $\mathcal{X}$ with corresponding distance d, consider the measure space formed by $\mathcal{X}$, the σ-algebra of all Borel sets of $\mathcal{X}$, and a probability measure $\mathbf{P}$. Let X be a random variable taking values in $\mathcal{X}$, distributed according to $\mathbf{P}$. The isoperimetric problem in this case is the following: given $p \in (0,1)$ and $t > 0$, determine the sets A with $\mathbf{P}\{X \in A\} \geq p$ for which the measure $\mathbf{P}\{d(X, A) \geq t\}$ is maximal. Even though the exact solution is only known in a few special cases, useful bounds for $\mathbf{P}\{d(X, A) \geq t\}$ can be derived under remarkably general circumstances. Such bounds are usually referred to as *isoperimetric inequalities.*

By introducing the so-called *concentration function*, defined, for all $t > 0$, by

$$\alpha(t) = \sup_{A \subset \mathcal{X}: \mathbf{P}\{A\} \geq 1/2} \mathbf{P}\{d(X,A) \geq t\} = \sup_{A \subset \mathcal{X}: \mathbf{P}\{A\} \geq 1/2} \mathbf{P}\{A_t^c\},$$

we see that isoperimetric inequalities may be formulated in terms of bounds for $\alpha(t)$. (Note that, generalizing the notion of a blowup of a set, we write $A_t = \{x \in \mathcal{X} : d(x, A) < t\}$ for any $A \subset \mathcal{X}$.) The next two simple theorems show that $\alpha(t)$ is intimately related to concentration of Lipschitz functions defined on $\mathcal{X}$. The first result points out that isoperimetric inequalities (more precisely, upper bounds for the concentration function) imply concentration of Lipschitz functions. Recall that a function $f : \mathcal{X} \to \mathbb{R}$ is Lipschitz if for all $x, y \in X$, $|f(x) - f(y)| \leq d(x, y)$. Recall also that $\mathbf{M}f(X)$ denotes a median of the random variable $f(X)$, that is, any number for which both $\mathbf{P}\{f(X) \leq \mathbf{M}f(X)\} \geq 1/2$ and $\mathbf{P}\{f(X) \geq \mathbf{M}f(X)\} \geq 1/2$ hold.

Theorem 7.1 (LÉVY'S INEQUALITIES) *For any Lipschitz function f,*

$$\mathbf{P}\{f(X) \geq \mathbf{M}f(X) + t\} \leq \alpha(t) \quad \textit{and} \quad \mathbf{P}\{f(X) \leq \mathbf{M}f(X) - t\} \leq \alpha(t).$$

Proof Consider the set $A = \{x : f(x) \leq \mathbf{M}f(X)\}$. By the definition of a median, $\mathbf{P}\{A\} \geq 1/2$. On the other hand, by the Lipschitz property of f,

$$A_t = \{x : d(x, A) < t\} \subseteq \{x : f(x) < \mathbf{M}f(X) + t\}.$$

The first inequality now follows from the definition of the concentration function. The second inequality follows from the first by considering $-f$. □

By an obvious modification of the proof one sees that if f is Lipschitz with constant C (i.e. $|f(x) - f(y)| \leq Cd(x, y)$ for all $x, y \in \mathcal{X}$), then

$$\mathbf{P}\{f(X) \geq \mathbf{M}f(X) + t\} \leq \alpha(t/C) \quad \text{and} \quad \mathbf{P}\{f(X) \leq \mathbf{M}f(X) - t\} \leq \alpha(t/C).$$

The next converse shows that concentration of Lipschitz functions implies an isoperimetric inequality.

Theorem 7.2 (CONVERSE) *If $\beta : \mathbb{R}_+ \to [0,1]$ is a function such that for every Lipschitz function $f : \mathcal{X} \to \mathbb{R}$*

$$\mathbf{P}\{f(X) \geq \mathbf{M}f(X) + t\} \leq \beta(t),$$

then $\beta(t) \geq \alpha(t)$.

Proof Simply observe that for any $A \subset \mathcal{X}$, the function f_A defined by $f_A(x) = d(x, A)$ is Lipschitz. Also, if $\mathbf{P}\{A\} \geq 1/2$, then 0 is a median of $f_A(X)$ and therefore

$$\alpha(t) = \sup_{A\subset\mathcal{X}:\mathbf{P}\{A\}\geq 1/2} \mathbf{P}\left\{f_A(X) \geq \mathbf{M}f(X) + t\right\} \leq \beta(t).$$ □

It is instructive to cast the bounded differences inequality (Theorem 6.2) in the framework described above.

Example 7.3 (BOUNDED DIFFERENCES INEQUALITY REVISITED) Consider independent random variables $X_1, \ldots, X_n$ taking their values in a (measurable) set $\mathcal{X}$ and denote the vector of these variables by $X = (X_1, \ldots, X_n)$ taking its value in $\mathcal{X}^n$. For an arbitrary (measurable) set $A \subset \mathcal{X}^n$ we write $\mathbf{P}\{A\} = \mathbf{P}\{X \in A\}$. The *Hamming distance* $d_H(x, y)$ between the vectors $x, y \in \mathcal{X}^n$ is defined as the number of coordinates in which x and y differ. With this distance the product space $\mathcal{X}^n$ becomes a metric space and Theorem 6.2 implies that if $f : \mathcal{X}^n \to \mathbb{R}$ is Lipschitz with respect to the Hamming distance, then

$$\mathbf{P}\{f(X) \geq \mathbf{E}f(X) + t\} \leq e^{-2t^2/n}.$$

The argument of Theorem 7.2 leads to the following.

Corollary 7.4 *For any $t > 0$,*

$$\mathbf{P}\left\{d_H(X, A) \geq t + \sqrt{\frac{n}{2}\log\frac{1}{\mathbf{P}\{A\}}}\right\} \leq e^{-2t^2/n}.$$

Proof Since the function $f(x) = d_H(x, A)$ is Lipschitz with respect to the Hamming distance, by the bounded differences inequality (Theorem 6.2),

$$\mathbf{P}\left\{\mathbf{E}d_H(X, A) - d_H(X, A) \geq t\right\} \leq e^{-2t^2/n}.$$

However, by taking $t = \mathbf{E}d_H(X, A)$, the left-hand side becomes $\mathbf{P}\{d_H(X, A) \leq 0\} = \mathbf{P}\{A\}$, so the above inequality implies

$$\mathbf{E}d_H(X, A) \leq \sqrt{\frac{n}{2}\log\frac{1}{\mathbf{P}\{A\}}}.$$

Then, by using the bounded differences inequality again, we obtain

$$\mathbf{P}\left\{d_H(X, A) \geq t + \sqrt{\frac{n}{2}\log\frac{1}{\mathbf{P}\{A\}}}\right\} \leq e^{-2t^2/n}$$

as desired. □

To interpret this corollary, observe that on the right-hand side we have the measure of the complement of the $t + \sqrt{(n/2)\log(1/P\{A\})}$-*blowup* of the set A, that is, the measure of the set of points whose Hamming distance from A is at least $t + \sqrt{(n/2)\log(1/P\{A\})}$. To appreciate the meaning of this fact, consider a set, say, with $P\{A\} = 1/10^6$. Then the measure of the set of points whose Hamming distance to A is more than $10\sqrt{n}$ is smaller than e^{-108}. In other words, product measures are concentrated on extremely small sets – hence the name "concentration of measure."

As in Theorem 7.1, the bounded differences inequality may also be derived from Corollary 7.4.

7.2 The Classical Isoperimetric Theorem

In this section we show how the classical isoperimetric theorem follows from a simple application of the Brunn–Minkowski inequality. The same inequality also yields interesting isoperimetric inequalities for the important case of the uniform distribution over the surface of the Euclidean unit ball in $\mathbb{R}^n$ and the canonical Gaussian distribution in $\mathbb{R}^n$.

The classical isoperimetric theorem in $\mathbb{R}^n$ states that, among all sets with a given volume, the Euclidean unit ball minimizes the surface area. More precisely, let $A \subset \mathbb{R}^n$ be a measurable set and denote by $\mathrm{Vol}(A)$ its Lebesgue measure. The *surface area* of A is defined by

$$\mathrm{Vol}(\partial A) = \lim_{t \to 0} \frac{\mathrm{Vol}(A_t) - \mathrm{Vol}(A)}{t},$$

provided that the limit exists. Here A_t denotes the t-blowup of A. Observe that if $B = \{x \in \mathbb{R}^n : \|x\| < 1\}$ denotes the unit open ball, then, recalling the notion of Minkowski sum from Section 4.14,

$$A_t = A + tB.$$

Theorem 7.5 (ISOPERIMETRIC THEOREM) *Let* $A \subset \mathbb{R}^n$ *be such that* $\mathrm{Vol}(A) = \mathrm{Vol}(B)$. *Then for any* $t > 0$, $\mathrm{Vol}(A_t) \geq \mathrm{Vol}(B_t)$. *Moreover, if* $\mathrm{Vol}(\partial A)$ *exists, then* $\mathrm{Vol}(\partial A) \geq \mathrm{Vol}(\partial B)$.

Proof By the Brunn–Minkowski inequality (Theorem 4.23),

$$\begin{aligned}
\mathrm{Vol}(A_t)^{1/n} &= \mathrm{Vol}(A + tB)^{1/n} \\
&\geq \mathrm{Vol}(A)^{1/n} + t\mathrm{Vol}(B)^{1/n} \\
&= \mathrm{Vol}(B)^{1/n}(1 + t) = \mathrm{Vol}(B_t)^{1/n},
\end{aligned}$$

establishing the first statement. The second follows simply because

$$\mathrm{Vol}(A_t) - \mathrm{Vol}(A) \geq \mathrm{Vol}(B)\left((1 + t)^n - 1\right) \geq nt\mathrm{Vol}(B)$$

where we used, for $a, b \geq 0$, $(a + b)^n \geq a^n + na^{n-1}b$. Thus, $\mathrm{Vol}(\partial A) \geq n\mathrm{Vol}(B)$. The isoperimetric theorem now follows from the fact (see Exercise 7.7) that $\mathrm{Vol}(\partial B) = n\mathrm{Vol}(B)$. □

There are few more examples of metric spaces and corresponding measures for which the exact solution of the isoperimetric problem is known. Two important and closely related cases are the surface of the unit Euclidean ball in $\mathbb{R}^n$ equipped with the uniform measure and $\mathbb{R}^n$ with the canonical Gaussian measure. Both of these isoperimetric theorems are significantly more intricate than Theorem 7.5 above but approximate isoperimetric inequalities are easy to derive, as is pointed out below.

Gaussian isoperimetric inequalities In the Gaussian isoperimetric problem one considers $\mathbb{R}^n$ equipped with the Euclidean metric and the canonical Gaussian measure P defined, for any measurable set $A \subset \mathbb{R}^n$, by $P(A) = \int_A \phi(x)dx$ where

$$\phi(x) = (2\pi)^{-n/2} e^{-\|x\|^2/2}.$$

The Gaussian isoperimetric theorem, proved in Chapter 10, states that among all measurable sets, half-spaces minimize the Gaussian surface area. More precisely, for any measurable set $A \subset \mathbb{R}^n$, the t-blowup of A satisfies

$$P(A_t) \geq \Phi\left(\Phi^{-1}(P(A) + t)\right)$$

where Φ denotes the standard Gaussian distribution function

$$\Phi(x) = \int_{-\infty}^{x} \frac{e^{-y^2/2}}{\sqrt{2\pi}} dy.$$

Equality holds if and only if A is a half-space (see Theorem 10.15). Equivalently, the concentration function, introduced in Section 7.1, equals

$$\alpha(t) = 1 - \Phi(t).$$

In other words, for any set $A \subset \mathbb{R}^n$ with $P(A) \geq 1/2$, $P(A_t^c) \leq 1 - \Phi(t)$. By well-known approximations of the standard Gaussian distribution function, $1 - \Phi(t) \approx 1/(t\sqrt{2\pi})e^{-t^2/2}$ (see Exercise 7.8). Observe that the Gaussian concentration inequality (Theorem 5.6) implies an isoperimetric inequality that already captures the essence of this. Indeed, if X is a standard Gaussian vector then for any Lipschitz function $f : \mathbb{R}^n \to \mathbb{R}$ and $t > 0$,

$$\begin{aligned} \mathbf{P}\left\{f(X) \geq \mathbf{M}f(X) + t\right\} &= \mathbf{P}\left\{f(X) \geq \mathbf{E}f(X) + (\mathbf{M}f(X) - \mathbf{E}f(X) + t)\right\} \\ &\leq e^{-(\mathbf{M}f(X) - \mathbf{E}f(X) + t)^2/2} \end{aligned}$$

by Theorem 5.6. Now the Gaussian Poincaré inequality (Theorem 3.20) and Exercise 2.1 imply that

$$\mathbf{M}f(X) - \mathbf{E}f(X) \leq \sqrt{\mathrm{Var}\,(f(X))} \leq 1$$

and therefore, by Theorem 7.2, $\alpha(t) \leq e^{-(t-1)^2/2}$, which is only slightly weaker than that derived from the Gaussian isoperimetric theorem.

Isoperimetric inequalities on the unit sphere A problem closely related to the Gaussian isoperimetric inequalities discussed above is the isoperimetric problem on the unit sphere. More precisely, let $S^{n-1} = \{x \in \mathbb{R}^n : \|x\| = 1\}$ denote the surface of the unit ball in $\mathbb{R}^n$. The isoperimetric problem in S^{n-1} (equipped with the Euclidean distance in $\mathbb{R}^n$) is of fundamental importance in many applications. To understand the relationship with the Gaussian isoperimetric problem, note that if X is a standard Gaussian vector in $\mathbb{R}^n$, then $X/\|X\|$ is uniformly distributed over S^{n-1}. Moreover, for large n, $\|X\|$ is concentrated around its expected value and therefore the canonical Gaussian measure in $\mathbb{R}^n$ resembles the uniform measure over S^{n-1}. Indeed, Lévy's isoperimetric theorem on S^{n-1} states that, as in the case of the Gaussian measure, the extremal sets in the isoperimetric problem on S^{n-1} are half-spaces of $\mathbb{R}^n$ as well. The intersection of a half-space and S^{n-1}, called a *spherical cap*, are just the balls in the metric space S^{n-1}. For $u \in S^{n-1}$ and $s \in [0, 1]$, let $C(u, s) = \{x \in S^{n-1} : x \cdot u \geq s\}$ denote a spherical cap of height $1 - s$ around u. According to Lévy's isoperimetric theorem, for any $t > 0$, for any measurable set $A \subset S^{n-1}$, if $C = C(u, s)$ is a spherical cap with $\mu(A) = \mu(C)$ (where μ denotes the uniform probability measure over S^{n-1}), then for any $t > 0$, $\mu(A_t) \geq \mu(C_t)$ where A_t and C_t denote the t-blowups of A and C. Note that C_t is a spherical cap as well. For simplicity of the discussion, consider the special case when $\mu(A) = 1/2$. In that case we may take $C = C(u, 0)$ to be any hemisphere for any $u \in S^{n-1}$ and the complement of C_t is a spherical cap of height $s = 1/\sqrt{1 + t^{-2}}$ (see Fig. 7.1).

By the isoperimetric theorem,

$$1 - \mu(A_t) \leq \mu\left(C\left(u, 1/\sqrt{1 + t^{-2}}\right)\right).$$

To better understand the implications of this bound, one may approximate the area of a spherical cap $C(u, s)$. An easy bound is obtained by observing that if $s \leq 1/\sqrt{2}$, the whole spherical cone defined as the convex hull of $C(u, s)$ and the origin is included in a ball of radius $\sqrt{1 - s^2}$ (see Fig. 7.2) and therefore the area of the cap is at most the proportion of

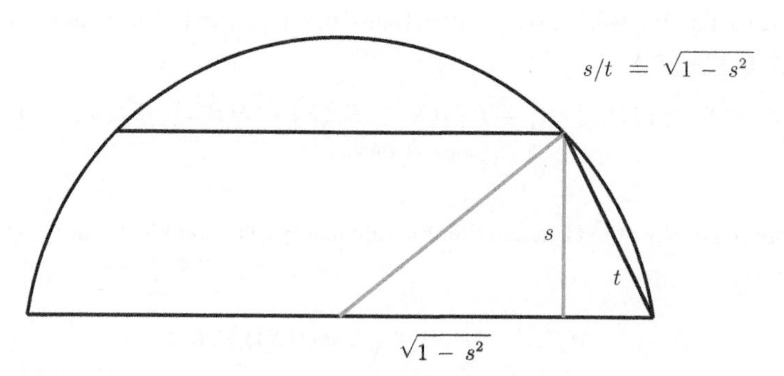

Figure 7.1 The height of the spherical cap defined as the complement of t-blowup of a hemisphere is $1 - s = 1 - 1/\sqrt{1 + t^{-2}}$

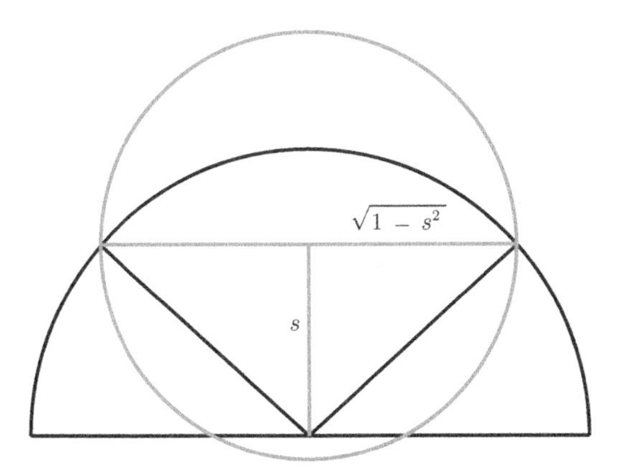

Figure 7.2 Bounding the area of a spherical cap

the unit ball $\{x : \|x\| \leq 1\}$ that falls in the spherical cone, which is at most $(1-s^2)^{n/2}$, and therefore, for any set with $\mu(A) \leq 1/2$,

$$1 - \mu(A_t) \leq (1+t^2)^{-n/2} \leq e^{-nt^2/2}.$$

By a more careful bounding of the integral that defines the area of the spherical cap, sharper bounds can be obtained. For example, for $\sqrt{2/n} \leq s \leq 1$, one has

$$\frac{1}{6s\sqrt{n}}(1-s^2)^{\frac{n-1}{2}} \leq \mu(C(u,s)) \leq \frac{1}{2s\sqrt{n}}(1-s^2)^{\frac{n-1}{2}}$$

(see Exercise 7.9) which gives slightly better bounds for $\mu(A_t)$.

As in the case of the Gaussian measure, it is not necessary to prove the full isoperimetric theorem to obtain inequalities of the type described above. A possible way to proceed is to use the fact that if X is a standard Gaussian vector in $\mathbb{R}^n$, then $X/\|X\|$ has the uniform distribution over S^{n-1} and use the Gaussian concentration inequality. Here we show how the Brunn–Minkowski inequality may be used.

To this end, consider an arbitrary subset $C \subset B = \{x \in \mathbb{R}^n : \|x\| \leq 1\}$ of the unit ball in $\mathbb{R}^n$. By the parallelogram rule, if $x \in C$ and $y \in C_t^c \cap B$, then

$$\|x+y\|^2 = 2\|x\|^2 + 2\|y^2\| - \|x-y\|^2 \leq 4 - t^2$$

and therefore $\|(x+y)/2\| \leq \sqrt{1-t^2/4} \leq 1 - t^2/8$. This implies that

$$\frac{1}{2}(C + C_t^c) \subset \left(1 - \frac{t^2}{8}\right)B$$

and therefore, by the "weaker form" of the Brunn–Minkowski inequality (Corollary 4.25), we have

$$\text{Vol}(C)\text{Vol}(C_t^c) \leq \mu\left(\frac{1}{2}(C + C_t^c)\right) \leq \left(1 - \frac{t^2}{8}\right)^{2n} \text{Vol}(B)^2.$$

Now let $A \subset S^{n-1}$ be a measurable subset of the surface of B and let $t > 0$. Defining $C = \cup_{a\in[1/2,1]} a \cdot A$, we find that $\cup_{a\in[1/2,1]} a \cdot A_t^c \subset C_{t/2}^c$. Moreover, $\mu(A) \le (1/2)\,\mathrm{Vol}(C)/\mathrm{Vol}(B)$ and $\mu(A_t^c) \le \mathrm{Vol}(\cup_{a\in[1/2,1]} a \cdot A_t^c)/\mathrm{Vol}(B)$, and therefore

$$\mu(A)\mu(A_t^c) \le 2\left(1 - \frac{t^2}{32}\right)^{2n}.$$

For example, for all sets $A \subset S^{n-1}$ with $\mu(A) \le 1/2$, we obtain

$$1 - \mu(A_t) \le 4e^{-nt^2/16}.$$

Even though the constants are worse than those obtained directly from the isoperimetric theorem, the inequality has a qualitatively similar form. Also, the proof is considerably simpler and can be generalized to other normed spaces (see Exercise 7.10).

7.3 Vertex Isoperimetric Inequality in the Hypercube

A thoroughly studied special case of isoperimetric problems is when the underlying metric space corresponds to the vertex set of a finite graph. Depending on how the size of the boundary of a subset of vertices is defined, we distinguish two variants, the *vertex isoperimetric problem* and the *edge isoperimetric problem*. In this section we discuss a basic but important special case of such discrete isoperimetric problems, when the graph is the binary hypercube.

Consider a graph G and let A be a set of its vertices. The *vertex boundary* of A is defined as the set of those vertices, not in A, which are connected to some vertex in A by an edge. We denote the vertex boundary of A by $\partial_V(A)$. The vertex isoperimetric problem in a graph G is to determine the sets A of a given cardinality whose vertex boundary contains a minimal number of vertices. In the edge isoperimetric problem one minimizes the number of edges between A and its complement.

The most classical and best understood special case is when the graph G is the binary hypercube $\{-1,1\}^n$ in which two vertices are connected by an edge if and only if their Hamming distance equals 1. We have already discussed briefly, in Chapter 4, the edge isoperimetric problem and we showed that when $|A| = 2^k$ for some integer $0 \le k \le n$, then k-dimensional sub-cubes minimize the size of the edge-boundary (see Theorem 4.3).

In order to describe the subsets of $\{-1,1\}^n$ with minimal vertex boundary, we define the so-called *simplicial order* of the elements of the binary hypercube. We say that $x = (x_1,\ldots,x_n) \in \{-1,1\}^n$ precedes $y = (y_1,\ldots,y_n) \in \{-1,1\}^n$ in the simplicial order if either $\|x\| < \|y\|$ (where $\|x\| = \sum_{i=1}^n \mathbb{1}_{\{x_i=1\}}$) or $\|x\| = \|y\|$ and $x_i = 1$ and $y_i = -1$ for the smallest i for which $x_i \ne y_i$. For example, if $n = 4$, $(1,-1,-1,1)$ precedes $(-1,1,1,1)$ and $(1,-1,1,-1)$ precedes $(1,-1,-1,1)$. Harper's classical vertex isoperimetric theorem states that initial segments of the simplicial ordering minimize the vertex boundary. More precisely, for $N = 1,\ldots,2^n$, let S_N denote the set of first N elements of $\{-1,1\}^n$ in the simplicial order. Then the following is true.

Theorem 7.6 (HARPER'S VERTEX ISOPERIMETRIC THEOREM) *For any subset $A \subset \{-1,1\}^n$,*

$$\partial_V(A) \geq \partial_V\left(S_{|A|}\right).$$

We leave the proof of this theorem to the reader (see Exercises 7.11–7.13 for detailed guidance).

Observe that if N has the form $N = \sum_{i=0}^{k}\binom{n}{i}$ for some $k = 0,\ldots,n$ then the initial segment S_N contains exactly those vectors x whose Hamming distance to $(-1,\ldots,-1)$ is at most k. In other words, S_N is a *Hamming ball* centered at the vector $(-1,\ldots,-1)$. The fact that among all sets with a given volume balls minimize the surface area is in close analogy with the classical isoperimetric theorem. Observe that if S_N is a Hamming ball with radius k $\left(\text{i.e. } N = \sum_{i=0}^{k}\binom{n}{i}\right)$ then $S_N \cup \partial_V(S_N)$ is the Hamming ball of radius $k+1$. This implies that for any set $A \subset \{-1,1\}^n$ with $|A| \geq \sum_{i=0}^{k}\binom{n}{i}$, we have $|A \cup \partial_V(A)| \geq \sum_{i=0}^{k+1}\binom{n}{i}$. By iterating this argument, we obtain the following simple consequence of Harper's theorem. For any $A \subset \{-1,1\}^n$ and $x \in \{-1,1\}^n$, let $d_H(x,A) = \min_{y\in A} d_H(x,y)$ be the Hamming distance of x to the set A. Also, denote by

$$A_t = \left\{x \in \{-1,1\}^n : d_H(x,A) < t\right\}$$

the t – *blowup* of the set A, that is, the set of points whose Hamming distance from A is at most t.

Corollary 7.7 *Let $A \subset \{-1,1\}^n$ such that $|A| \geq \sum_{i=0}^{k}\binom{n}{i}$. Then for any $t = 1,2,\ldots,$ $n-k+1$,*

$$|A_t| \geq \sum_{i=0}^{k+t-1}\binom{n}{i}.$$

In particular, if $|A|/2^n \geq 1/2$ then we may take $k = \lfloor n/2 \rfloor$ in the corollary above and

$$\frac{|A_t|}{2^n} \geq \mathbf{P}\{B(n,1/2) < \mathbf{E}B(n,1/2) + t\} \geq 1 - e^{-2t^2/n}$$

where $B(n,1/2)$ is a binomial random variable with parameters n and $1/2$. The last inequality follows from standard tail estimates of a symmetric binomial distribution, for example, from Hoeffding's inequality.

This simple fact reveals the concentration-of-measure phenomenon we have already encountered: consider any set A containing at least half of the points of $\{-1,1\}^n$. According to the corollary above, the fraction of those points which cannot be obtained by changing at most $c\sqrt{n}$ bits of some point in A is at most e^{-2c^2}. In other words, an immense majority of the points in $\{-1,1\}^n$ is within Hamming distance of the order of $\sqrt{n}$ of A.

We may also apply Lévy's inequality (Theorem 7.1) in this situation and observe that this simple consequence of Harper's vertex isoperimetric theorem implies a version of the bounded differences inequality for functions defined on $\{-1,1\}^n$ under the uniform measure. More precisely, let f be a function defined on $\{-1,1\}^n$ satisfying the bounded differences property such that

$$\max_{x\in\{-1,1\}^n,i} |f(x) - f(\tilde{x}^{(i)})| \le 1$$

where $\tilde{x}^{(i)} = (x_1, \ldots, x_{i-1}, -x_i, x_{i+1}, \ldots, x_n)$ is obtained by flipping the i-th bit of x. If $X = (X_1, \ldots, X_n)$ is uniformly distributed over $\{-1,1\}^n$, we may consider the random variable $Z = f(X)$. Then f is Lipschitz with respect to the Hamming distance and Theorem 7.1 implies that $\mathbf{P}\{Z > \mathbf{M}Z + t\} \le e^{-2t^2/n}$. This is just like in the bounded differences inequality, except that the expected value of Z is replaced by its median. However, by Exercise 2.1 and the Efron–Stein inequality,

$$\mathbf{M}f(X) - \mathbf{E}f(X) \le \sqrt{\mathrm{Var}\,(f(X))} \le \sqrt{n/4}.$$

Exact isoperimetric results like Harper's theorem are extremely valuable as they allow one to deduce the sharpest possible concentration inequalities for Lipschitz functions. Unfortunately, there are only special examples of exact isoperimetric results. Here we mention just one of them, without proof, and refer to the bibliographical remarks for further pointers in the literature.

Consider again the binary hypercube $\{-1,1\}^n$ but now equipped with the product measure of n i.i.d. Bernoulli random variables, that is, for any $x \in \{-1,1\}^n$, $\mathbf{P}\{x\} = p^{\|x\|}(1-p)^{n-\|x\|}$ where $p \in (0,1)$ is the parameter of the Bernoulli distribution. The isoperimetric problem now is to determine the sets A, with a given probability content, that minimize the probability $\mathbf{P}\{\partial_V(A)\}$ of the vertex boundary. This problem can be solved for the special case of monotone sets. Recall that a set $A \subset \{-1,1\}^n$ is monotone if $\mathbb{1}_{\{x\in A\}} \ge \mathbb{1}_{\{y\in A\}}$ for all $x = (x_1, \ldots, x_n)$ and $y = (y_1, \ldots, y_n)$ in $\{-1,1\}^n$ such that $x_i \ge y_i$ for all i. Surprisingly, Hamming balls are still isoperimetric sets in the sense of the following theorem whose proof we omit.

Theorem 7.8 *Let $k \in \{0, \ldots, n\}$ and let $S = \{x \in \{-1,1\}^n : \|x\| \le k\}$ be a Hamming ball of radius k. If $A \subset \{-1,1\}^n$ is a monotone set such that $\mathbf{P}\{A\} \ge \mathbf{P}\{S\}$ then $\mathbf{P}\{\partial_V(A)\} \ge \mathbf{P}\{\partial_V(S)\}$.*

7.4 Convex Distance Inequality

In a remarkable series of papers Talagrand developed an induction method to prove powerful concentration results in many cases when the bounded differences inequality fails. Perhaps the most widely used of these is the so-called "convex-distance inequality" which we present here as a consequence of the entropy method, namely Theorems 6.19 and 6.20.

To understand Talagrand's inequality, first observe that Corollary 7.4 may be easily generalized by allowing the distance of the point X from the set A to be measured by a *weighted Hamming distance*

$$d_\alpha(x, A) = \inf_{y\in A} d_\alpha(x, y) = \inf_{y\in A} \sum_{i:x_i\ne y_i} \alpha_i$$

where $\alpha = (\alpha_1, \ldots, \alpha_n)$ is a vector of nonnegative numbers. Repeating the argument of the proof of Corollary 7.4, we obtain, for all α,

$$\mathbf{P}\left\{d_\alpha(X,A) \geq t + \sqrt{\frac{\|\alpha\|^2}{2}\log\frac{1}{\mathbf{P}\{A\}}}\right\} \leq e^{-2t^2/\|\alpha\|^2},$$

where $\mathbf{P}\{A\} = \mathbf{P}\{X \in A\}$ and $\|\alpha\| = \sqrt{\sum_{i=1}^n \alpha_i^2}$ denotes the euclidean norm of α. Thus, for example, for all vectors α with unit norm $\|\alpha\| = 1$,

$$\mathbf{P}\left\{d_\alpha(X,A) \geq t + \sqrt{\frac{1}{2}\log\frac{1}{\mathbf{P}\{A\}}}\right\} \leq e^{-2t^2}.$$

Thus, denoting $u = \sqrt{\frac{1}{2}\log\frac{1}{\mathbf{P}\{A\}}}$, for any $t \geq u$,

$$\mathbf{P}\left\{d_\alpha(X,A) \geq t\right\} \leq e^{-2(t-u)^2}.$$

On the one hand, if $t \leq \sqrt{-2\log \mathbf{P}\{A\}}$, then $\mathbf{P}\{A\} \leq e^{-t^2/2}$. On the other hand, since $(t-u)^2 \geq t^2/4$ for $t \geq 2u$, for any $t \geq \sqrt{2\log\frac{1}{\mathbf{P}\{A\}}}$ the inequality above implies $\mathbf{P}\left\{d_\alpha(X,A) \geq t\right\} \leq e^{-t^2/2}$. Thus, for all $t > 0$, we have

$$\begin{aligned}\sup_{\alpha:\|\alpha\|=1} \mathbf{P}\{A\}\cdot\mathbf{P}\left\{d_\alpha(X,A) \geq t\right\} &\leq \sup_{\alpha:\|\alpha\|=1} \min\left(\mathbf{P}\{A\}, \mathbf{P}\left\{d_\alpha(X,A) \geq t\right\}\right) \\ &\leq e^{-t^2/2}.\end{aligned}$$

The main message of Talagrand's inequality is that the above inequality remains true even if the supremum is taken within the probability. To make this statement precise, introduce, for any $x = (x_1, \ldots, x_n) \in \mathcal{X}^n$, the *convex distance* of x from the set A by

$$d_T(x,A) = \sup_{\alpha\in[0,\infty)^n:\|\alpha\|=1} d_\alpha(x,A).$$

Theorem 7.9 (CONVEX DISTANCE INEQUALITY) *For any subset $A \subseteq \mathcal{X}^n$ and $t > 0$,*

$$\mathbf{P}\{A\}\mathbf{P}\left\{d_T(X,A) \geq t\right\} \leq e^{-t^2/4}.$$

The theorem is announced in the form by which Talagrand proved it, but the proof shown here gives a worse exponent in the upper bound ($e^{t^2/10}$ instead of $e^{t^2/4}$). We give a different proof of the form announced here in Section 8.4. The convex distance inequality is implied by Theorems 6.19 and 6.20 for weakly self-bounding functions. The key to the proof is establishing the following self-bounding property for the *square* of the convex distance. Recall from Section 6.11 the notion of weak self-bounding functions.

Lemma 7.10 *Let $A \in \mathcal{X}^n$ be a measurable set and define the function $f(x) = d_T(x,A)^2$. Introduce also*

$$f_i(x^{(i)}) = \inf_{x_i' \in \mathcal{X}} f(x_1, \ldots, x_{i-1}, x_i', x_{i+1}, \ldots, x_n).$$

Then for all $x \in \mathcal{X}^n$, $0 \leq f(x) - f_i(x^{(i)}) \leq 1$. Moreover, f is weakly $(4,0)$-self-bounding, that is,

$$\sum_{i=1}^{n} \left(f(x) - f_i(x^{(i)})\right)^2 \leq 4f(x).$$

Before proving the lemma, we show how it implies the convex distance inequality.

Proof of Theorem 7.9. By the definition of the convex distance, we have $A = \left\{x : d_T(x,A) = 0\right\}$. Observe now that thanks to Lemma 7.10, we may use Theorem 6.20 with $f(x) = d_T^2(x,A)$ and $t = \mathbf{E}d_T^2(X,A)$ to obtain

$$\mathbf{P}\{A\} = \mathbf{P}\left\{d_T^2(X,A) \leq \mathbf{E}d_T^2(X,A) - t\right\} \leq \exp\left(-\frac{\mathbf{E}d_T^2(X,A)}{8}\right)$$

or, equivalently,

$$\mathbf{P}\{A\} \exp\left(\frac{\mathbf{E}d_T^2(X,A)}{8}\right) \leq 1.$$

On the other hand, Theorem 6.19 implies that for $0 \leq \lambda \leq 1/2$,

$$\log \mathbf{E}\left[e^{\lambda(d_T^2(X,A) - \mathbf{E}d_T^2(X,A))}\right] \leq \frac{2\lambda^2 \mathbf{E}d_T^2(X,A)}{1 - 2\lambda}.$$

Choosing $\lambda = 1/10$, we have

$$\mathbf{E} \exp\left(\frac{d_T^2(X,A)}{10}\right) \leq \exp\left(\frac{\mathbf{E}d_T^2(X,A)}{8}\right)$$

which gives

$$\mathbf{P}\{X \in A\}\mathbf{E}e^{d_T^2(X,A)/10} \leq 1.$$

By Markov's inequality,

$$\mathbf{P}\left\{d_T(X,A) \geq t\right\} \leq \mathbf{E}e^{d_T^2(X,A)/10} e^{-t^2/10}$$

which completes the proof. □

It remains to prove the self-bounding property of the squared convex distance.

Proof of Lemma 7.10. The proof is based on different formulations of the convex distance. First we observe that $d_T(x, A)$ can be represented as a saddle point. Let $\mathcal{M}(A)$ denote the set of probability measures on A. Then

$$\begin{aligned} d_T(x, A) &= \sup_{\alpha:\|\alpha\|\leq 1} \inf_{\nu\in\mathcal{M}(A)} \sum_j \alpha_j \mathbf{E}_\nu \mathbb{1}_{\{x_j\neq Y_j\}} \\ &\quad \text{(where } Y = (Y_1, \ldots, Y_n) \text{ is distributed according to } \nu) \\ &= \inf_{\nu\in\mathcal{M}(A)} \sup_{\alpha:\|\alpha\|\leq 1} \sum_j \alpha_j \mathbf{E}_\nu \mathbb{1}_{\{x_j\neq Y_j\}} \end{aligned} \qquad (7.1)$$

where the saddle point is achieved. This follows from Sion's minmax theorem (1958) which states that if $f(x,y) : \mathcal{X} \times \mathcal{Y} \to \mathbb{R}$ is convex and lower-semi-continuous with respect to x, concave and upper-semi-continuous with respect to y, where $\mathcal{X}$ is convex and compact, then

$$\inf_x \sup_y f(x,y) = \sup_y \inf_x f(x,y).$$

We leave the details of checking the conditions of Sion's theorem to the reader (see Exercise 7.14).

By the Cauchy–Schwarz inequality,

$$d_T(x, A)^2 = \inf_{\nu\in\mathcal{M}(A)} \sum_{j=1}^n \left(\mathbf{E}_\nu \mathbb{1}_{\{x_j\neq Y_j\}}\right)^2.$$

Rather than minimizing in the large space $\mathcal{M}(A)$, we may perform minimization on the convex compact set of probability measures on $\{0,1\}^n$ by mapping $y \in A$ on $(\mathbb{1}_{\{y_j\neq X_j\}})_{1\leq j\leq n}$. Denote this mapping by χ. Note that the mapping depends on x but we omit this dependence to lighten notation. The set $\mathcal{M}(A)\circ\chi^{-1}$ of probability measures on $\{0,1\}^n$ coincides with $\mathcal{M}(\chi(A))$. It is convex and compact and therefore the infimum in the last display is achieved at some $\hat{\nu}$. Then $d_T(X, A)$ is just the Euclidean norm of the vector $\left(\mathbf{E}_{\hat{\nu}} \mathbb{1}_{\{x_j\neq Y_j\}}\right)_{j\leq n}$, and therefore the supremum in (7.1) is achieved by the vector $\hat{\alpha}$ of components

$$\hat{\alpha}_i = \frac{\mathbf{E}_{\hat{\nu}} \mathbb{1}_{\{x_i\neq Y_i\}}}{\sqrt{\sum_{j=1}^n \left(\mathbf{E}_{\hat{\nu}} \mathbb{1}_{\{x_j\neq Y_j\}}\right)^2}}.$$

For simplicity, assume that the infimum in the definition of $f_i(x^{(i)})$ is achieved (otherwise a standard approximation argument may be used).

Clearly, $f(x) - f_i(x^{(i)}) \geq 0$ for all i. On the other hand let $x_i^{(i)}$ and $\hat{\nu}_i$ denote the coordinate value and the probability distribution on A that witness the value of $f_i(x^{(i)})$, that is,

$$f_i(x^{(i)}) = \sum_{j\neq i} \left(\mathbf{E}_{\hat{\nu}_i} \mathbb{1}_{\{x_j\neq Y_j\}}\right)^2 + \left(\mathbf{E}_{\hat{\nu}_i} \mathbb{1}_{\{x_i^{(i)}\neq Y_i\}}\right)^2.$$

As $f(x) \le \sum_{j \ne i} \left(\mathbf{E}_{\hat{\nu}_i} \mathbb{1}_{\{x_j \ne Y_j\}}\right)^2 + \left(\mathbf{E}_{\hat{\nu}_i} \mathbb{1}_{\{x_i^{(i)} \ne Y_i\}}\right)^2$, we have

$$f(x) - f_i(x^{(i)}) \le \left(\mathbf{E}_{\hat{\nu}_i} \mathbb{1}_{\{x_i \ne Y_i\}}\right)^2 - \left(\mathbf{E}_{\hat{\nu}_i} \mathbb{1}_{\{x_i^{(i)} \ne Y_i\}}\right)^2 \le 1.$$

It remains to prove that f is weakly $(4, 0)$-self-bounding. To this end, we may use once again Sion's minmax theorem as in (7.1), to write the convex distance

$$\begin{aligned} d_T(x, A) &= \inf_{\nu \in \mathcal{M}(A)} \sup_{\alpha : \|\alpha\|_2 \le 1} \sum_{j=1}^n \alpha_j \mathbf{E}_\nu \mathbb{1}_{\{x_j \ne Y_j\}} \\ &= \sup_{\alpha : \|\alpha\|_2 \le 1} \inf_{\nu \in \mathcal{M}(A)} \sum_{j=1}^n \alpha_j \mathbf{E}_\nu \mathbb{1}_{\{x_j \ne Y_j\}}. \end{aligned}$$

Denote the pair (ν, α) at which the saddle point is achieved by $(\hat{\nu}, \hat{\alpha})$. Then

$$\sqrt{f_i(x^{(i)})} = \inf_{\nu \in \mathcal{M}(A)} \sup_{\alpha : \|\alpha\|_2 \le 1} \sum_{j=1}^n \alpha_j \mathbf{E}_\nu \mathbb{1}_{\{x_j^{(i)} \ne Y_j\}} \ge \inf_{\nu \in \mathcal{M}(A)} \sum_{j=1}^n \hat{\alpha}_j \mathbf{E}_\nu \mathbb{1}_{\{x_j^{(i)} \ne Y_j\}}.$$

Let $\tilde{\nu}$ denote the distribution on A that achieves the infimum in the latter expression. Then we have

$$\sqrt{f(x)} = \inf_\nu \sum_{j=1}^n \hat{\alpha}_j \mathbf{E}_\nu \mathbb{1}_{\{x_j \ne Y_j\}} \le \sum_{j=1}^n \hat{\alpha}_j \mathbf{E}_{\tilde{\nu}} \mathbb{1}_{\{x_j \ne Y_j\}}.$$

Hence,

$$\begin{aligned} \sqrt{f(x)} - \sqrt{f_i(x^{(i)})} &\le \sum_{j=1}^n \hat{\alpha}_j \mathbf{E}_{\tilde{\nu}} \left[\mathbb{1}_{\{x_j \ne Y_j\}} - \mathbb{1}_{\{x_j^{(i)} \ne Y_j\}}\right] \\ &= \hat{\alpha}_i \mathbf{E}_{\tilde{\nu}} \left[\mathbb{1}_{\{x_i \ne Y_i\}} - \mathbb{1}_{\{x_i^{(i)} \ne Y_i\}}\right] \le \hat{\alpha}_i, \end{aligned}$$

so

$$\left(\sqrt{f(x)} - \sqrt{f_i(x^{(i)})}\right)^2 \le \hat{\alpha}_i^2.$$

Finally, since $f_i(x^{(i)} \le f(x)$,

$$\begin{aligned} \sum_{i=1}^n \left(f(x) - f_i(x^{(i)})\right)^2 &= \sum_{i=1}^n \left(\sqrt{f(x)} - \sqrt{f_i(x^{(i)})}\right)^2 \left(\sqrt{f(x)} + \sqrt{f_i(x^{(i)})}\right)^2 \\ &\le \sum_{i=1}^n \hat{\alpha}_i^2 4f(x) \\ &\le 4f(x). \end{aligned}$$

□

Remark 7.2 Note that it follows from the proof above that

$$\sum_{i=1}^{n}\left(\sqrt{f(x)}-\sqrt{f_i(x^{(i)})}\right)^2 \leq 1,$$

which implies, by the Efron–Stein inequality, that $\mathrm{Var}\,(d_T(A,X)) \leq 1$. By Theorem 6.7, this property also implies that

$$\boldsymbol{P}\left\{d_T(A,X) - \boldsymbol{E}d_T(A,X) > t\right\} \leq e^{-t^2/2}.$$

This inequality is useful when $\boldsymbol{P}\{A\} \geq 1/2$ because in that case $\boldsymbol{M}\,d_T(A,X) = 0$, and by the bound for the variance, we have $\boldsymbol{E}d_T(A,X) \leq \sqrt{\mathrm{Var}\,(d_T(A,X))} \leq 1$. Thus, for all $A \subset \mathcal{X}^n$ with $\boldsymbol{P}\{A\} \geq 1/2$,

$$\boldsymbol{P}\{A\}\boldsymbol{P}\left\{d_T(A,X) > t\right\} \leq e^{-(t-1)^2/2},$$

which is just like the convex distance inequality. However, by this argument one cannot handle sets with small probability, which is important in many applications.

7.5 Convex Lipschitz Functions Revisited

Recall that in Section 6.6 we derived upper tail inequalities for convex Lipschitz functions (with respect to the Euclidean norm) of independent bounded random variables. The convex distance inequality may also be used to prove such a result. Moreover, we also obtain an analogous lower tail inequality.

The key is the following lemma which relates the Euclidean distance of a point to a convex subset of $[0,1]^n$ and the convex distance d_T.

For any $A \subset [0,1]^n$ and $x \in [0,1]^n$, define by

$$D(x,A) = \inf_{y\in A} \|x-y\|$$

the Euclidean distance of x and A (where $\|\cdot\|$ denotes the Euclidean norm).

Lemma 7.11 *Let $A \subset [0,1]^n$ be a convex set and let $x = (x_1,\ldots,x_n) \in [0,1]^n$. Then*

$$D(x,A) \leq d_T(x,A).$$

Proof The key to the proof is the saddle point representation (7.1) of the convex distance. Recall that $\mathcal{M}(A)$ is the set of all probability measures on A and $Y = (Y_1,\ldots,Y_n)$ denotes a random vector distributed according to a $\nu \in \mathcal{M}(A)$. Then

$$
\begin{aligned}
D(x,A) &= \inf_{\nu\in\mathcal{M}(A)} \|x - E_\nu Y\| \quad \text{(since } A \text{ is convex)} \\
&\le \inf_{\nu\in\mathcal{M}(A)} \sqrt{\sum_{j=1}^{n} \left(E_\nu \mathbb{1}_{\{x_j\neq Y_j\}}\right)^2} \quad \text{(since } x_j, Y_j \in [0,1]\text{)} \\
&= \inf_{\nu\in\mathcal{M}(A)} \sup_{\alpha:\|\alpha\|\le 1} \sum_{j=1}^{n} \alpha_j E_\nu \mathbb{1}_{\{x_j\neq Y_j\}} \quad \text{(by Cauchy–Schwarz)} \\
&= d_T(x,A).
\end{aligned}
$$

□

The following concentration inequality for convex Lipschitz functions now follows easily. In fact, it suffices to assume that f is *quasi-convex*, that is, $\{x : f(x) \le s\}$ is a convex set for all $s \in \mathbb{R}$.

Theorem 7.12 *Let $X = (X_1, \ldots, X_n)$ be a vector of independent random variables taking values in the interval $[0,1]$ and let $f : [0,1]^n \to \mathbb{R}$ be a quasi-convex function such that $|f(x) - f(y)| \le \|x - y\|$ for all $x, y \in [0,1]^n$. Then $f(X)$ satisfies, for all $t > 0$,*

$$
P\{f(X) > Mf(X) + t\} \le 2e^{-t^2/4}
$$

and

$$
P\{f(X) < Mf(X) - t\} \le 2e^{-t^2/4}.
$$

Proof For some $s \in \mathbb{R}$, define the set $A_s = \{x : f(x) \le s\} \subset [0,1]^n$. Because of quasi-convexity, A_s is convex. By the Lipschitz property and Lemma 7.11, for all $x \in [0,1]^n$,

$$
f(x) \le s + D(x, A_s) \le s + d_T(x, A_s),
$$

so the convex distance inequality implies

$$
P\{f(X) \ge s + t\} P\{f(X) \le s\} \le e^{-t^2/4}.
$$

Take $s = Mf(X)$ to get the upper tail inequality and $s = Mf(X) - t$ to get the lower tail inequality. □

7.6 Bin Packing

We now describe an application of the convex distance inequality for the bin packing discussed in Section 3.2.

Let $f(x)$ denote the minimum number of bins of size 1 into which the numbers $x_1, \ldots, x_n \in [0,1]$ can be packed. We consider the random variable $Z = f(X)$ where $X_1, \ldots, X_n$ are independent, taking values in $[0,1]$. The bounded differences inequality implies that

$$\mathbf{P}\{|Z - \mathbf{E}Z| \geq t\} \leq 2e^{-t^2/n}.$$

However, when the X_i are typically much smaller than 1, one expects that Z behaves similarly to $\sum_{i=1}^n X_i$. The following result shows that, in fact, the typical deviations of Z are of a much smaller order when $\mathbf{E}\sum_{i=1}^n X_i^2 \ll n$.

Corollary 7.13 *Denote* $\Sigma = \sqrt{\mathbf{E}\sum_{i=1}^n X_i^2}$. *Then for each* $t > 0$,

$$\mathbf{P}\left\{|Z - \mathbf{M}Z| \geq t + 1\right\} \leq 8e^{-t^2/(16(2\Sigma^2+t))}.$$

Proof First observe (and this is the only specific property of f we use in the proof) that for any $x, y \in [0,1]^n$,

$$f(x) \leq f(y) + 2\sum_{i:x_i \neq y_i} x_i + 1.$$

To see this it suffices to show that the x_i for which $x_i \neq y_i$ can be packed into at most $\lfloor 2\sum_{i:x_i\neq y_i} x_i \rfloor + 1$ bins. For this, it is enough to find a packing such that at most one bin is less than half full. Such a packing must exist because we can always pack the contents of two half-empty bins into one.

Denoting by $\alpha = \alpha(x) \in [0,\infty)^n$ the unit vector $x/\|x\|$, we clearly have

$$\sum_{i:x_i\neq y_i} x_i = \|x\| \sum_{i:x_i \neq y_i} \alpha_i = \|x\| d_\alpha(x,y).$$

Let a be a positive number and define the set $A_a = \{y : f(y) \leq a\}$. Then, by the argument above and by the definition of the convex distance, for each $x \in [0,1]^n$ there exists $y \in A_a$ such that

$$f(x) \leq f(y) + 2\sum_{i:x_i\neq y_i} x_i + 1 \leq a + 2\|x\| d_T(x, A_a) + 1,$$

from which we conclude that for each $a > 0$, $Z \leq a + 2\|X\| d_T(X, A_a) + 1$. Thus, writing $\Sigma = \sqrt{\mathbf{E}\sum_{i=1}^n X_i^2}$ for any $t \geq 0$,

$$\begin{aligned}
&\mathbf{P}\{Z \geq a + 1 + t\} \\
&\leq \mathbf{P}\left\{Z \geq a + 1 + t\frac{2\|X\|}{2\sqrt{2\Sigma^2 + t}}\right\} + \mathbf{P}\left\{\|X\| \geq \sqrt{2\Sigma^2 + t}\right\} \\
&\leq \mathbf{P}\left\{d_T(X, A_a) \geq \frac{t}{2\sqrt{2\Sigma^2 + t}}\right\} + e^{-(3/8)(\Sigma^2+t)}
\end{aligned}$$

where the bound on the second term follows by a simple application of Bernstein's inequality (see Exercise 7.17).

To obtain the desired inequality, we use the obtained bound with two different choices of a. To derive a bound for the upper tail of Z, we take $a = \mathbf{M}Z$. Then $\mathbf{P}\{A_a\} \geq 1/2$ and the convex distance inequality yields

$$P\{Z \geq MZ + 1 + t\} \leq 2\left(e^{-t^2/(16(2\Sigma^2+t))} + e^{-(3/8)(\Sigma^2+t)}\right) \leq 4e^{-t^2/(16(2\Sigma^2+t))}.$$

We obtain a similar inequality in the same way for $P\{Z \leq MZ - 1 - t\}$ by taking $a = MZ - t - 1$. □

7.7 Bibliographical Remarks

The connection between concentration inequalities and isoperimetric properties goes back to Lévy (1951) and has been an important research area in functional analysis and high-dimensional geometry. The importance of measure concentration in the asymptotic theory of Banach spaces is summarized in the now classical book of Milman and Schechtman (1986) where many of the early results are summarized. The more recent book of Ledoux (2001) is an excellent summary of measure concentration and its connections with isoperimetry and related geometric concepts. We recommend the surveys of Ball (1997), Schechtman (2003), and Gardner (2002) for the background, history, and many pointers to the related literature.

The inequalities of Theorem 7.1 were first pointed out by Lévy (1951) who also proved the isoperimetric theorem on the surface of the Euclidean ball in $\mathbb{R}^n$, along with Schmidt (1948). For extensions of Lévy's proof to Riemannian manifolds with positive curvature see Gromov (1980). The Gaussian isoperimetric theorem is due to Borell (1975) and Tsirelson and Sudakov (1974).

The vertex isoperimetric theorem for the discrete cube (Theorem 7.6) goes back to Harper (1966). Several simpler proofs have been published (see Katona (1975), Kleitman (1979), and Frankl and Füredi (1981)). A natural generalization of the discrete cube includes n-fold products of graphs. Given a graph with vertex set G and edge set E, the n-fold product of the graph has vertex set $G^n = G \times \cdots \times G$, and two vertices $g_1 = (g_{1,1}, \ldots, g_{1,n})$ and $g_2 = (g_{2,1}, \ldots, g_{2,n})$ are connected if and only if g_1 and g_2 agree in all but one component; if they differ in the i-th component then $g_{1,i}$ and $g_{2,i}$ are connected in the original graph. If G is K_2 (i.e. the complete graph on two vertices) then the product graph is just the binary hypercube. If G is a chain of length k then the product is a so-called *grid graph*. For grid graphs, Bollobás and Leader (1991*b*) established a vertex isoperimetric theorem, thus generalizing Harper's theorem. In general, there are very few examples of product graphs for which exact isoperimetric theorems have been established.

Bezrukov and Serra (2002) established a very general result which provides a powerful tool for finding further examples. On the other hand, isoperimetric *inequalities* have been established in great generality. For example, Alon and Milman (1985) show that if G is connected then powers of G form a *Lévy family*, that is, there exist constants C_1 and C_2 (depending on G) such that for any set $A \subset G^n$ whose cardinality is at least half of the cardinality of G^n, the complement of the t-blowup of A (defined in terms of the graph distance) is at most $C_1|G^n|e^{-C_2 t}$. Using martingale techniques, Bollobás and Leader (1991*a*) show a bound of the type $C_1|G^n|e^{-C_2 t^2}$.

Theorem 7.8 is due to Bollobás and Leader (1991*b*). For surveys on discrete isoperimetric inequalities we refer to Leader (1991) and Bezrukov (1994).

The "isoperimetric" approach to concentration inequalities was promoted and developed, in large part, in a remarkable series of papers by Talagrand (1995, 1996*b*, 1996*c*). The convex-distance inequality presented in Section 7.4 is perhaps the most useful representative of a family of inequalities established by Talagrand. The original proof (and its variants) is based on an induction argument, different from the one based on the entropy method presented here. We note that Talagrand's original proof gives a better constant in the exponent ($e^{-t^2/4}$ instead of $e^{-t^2/10}$). For several extensions and variations we refer to Talagrand (1995, 1996*b*, 1996*c*). Steele (1996), McDiarmid (1998), and Molloy and Reed (2002) survey a large variety of applications of the convex distance inequality. Pollard (2007) revisits Talagrand's original proof in order to make it more transparent. The proof presented here appears in Boucheron, Lugosi, and Massart (2009).

Theorem 7.12 is due to Talagrand (1996*c*).

The application of the convex distance inequality for the bin packing problem appears in Talagrand (1995).

7.8 EXERCISES

7.1. Let $\mathcal{X}$ be a metric space, let X be an $\mathcal{X}$-valued random variable distributed according to the probability measure $\mathbf{P}$, and let $\alpha(t)$ be the corresponding concentration function. Let $\varepsilon > 0$. Show that if $B \subset \mathcal{X}$ is such that $\mathbf{P}\{B\} \geq \varepsilon$ and t_0 is such that $\alpha(t_0) < \varepsilon$, then

$$\alpha(t) \geq \mathbf{P}\{d(X,B) \geq t_0 + t\}.$$

7.2. (A VARIANT OF THEOREM 7.2) Show that if $\beta : \mathbb{R}_+ \to [0,1]$ is a function such that for every Lipschitz function $f : \mathcal{X} \to \mathbb{R}$ with Lipschitz constant 1

$$\mathbf{P}\{f(X) \geq \mathbf{E}f(X) + t\} \leq \beta(t),$$

then $\beta(t) \geq \alpha(t/2)$. (See Ledoux (2001).)

7.3. (LAPLACE FUNCTIONAL) Define the *Laplace functional* by

$$L(\lambda) = \sup_f \mathbf{E}e^{\lambda f(X)}$$

where the supremum is taken over all Lipschitz functions $f : \mathcal{X} \to \mathbb{R}$ with Lipschitz constant 1 such that $\mathbf{E}f(X) = 0$. Show that for all $t > 0$,

$$\alpha(t) \leq \inf_{\lambda > 0} e^{-\lambda t/2} L(\lambda).$$

(See Ledoux (2001).)

7.4. (LAPLACE FUNCTIONAL OF A BOUNDED METRIC SPACE) Denote the diameter of $\mathcal{X}$ by $D = \sup_{x,y\in\mathcal{X}} d(x,y)$ and assume $D < \infty$. Show that

$$L(\lambda) \le e^{D^2\lambda^2/8}.$$

(See Ledoux (2001).)

7.5. (LAPLACE FUNCTIONAL OF A PRODUCT SPACE) Let $(\mathcal{X}_1, d_1), \ldots, (\mathcal{X}_n, d_n)$ be metric spaces with corresponding Borel σ-algebras and probability measures $P_1, \ldots, P_n$, respectively. Let $P = \otimes_{i=1}^n P_i$ be the product measure on the cartesian product space $\mathcal{X} = \mathcal{X}_1 \times \cdots \times \mathcal{X}_n$. $\mathcal{X}$ is a metric space with distance function $d(x,y) = \sum_{i=1}^n d(x_i, y_i)$. Show that if $L_i(\lambda)$ denotes the Laplace functional of $\mathcal{X}_i$ $(i = 1, \ldots, n)$ and $L(\lambda)$ is the Laplace functional of the product space $\mathcal{X}$, then

$$L(\lambda) \le \prod_{i=1}^n L(\lambda_i).$$

(See Ledoux (2001).)

7.6. (YET ANOTHER PROOF OF THE BOUNDED DIFFERENCES INEQUALITY) Combine the previous three exercises to get the following general version of the bounded differences inequality. Let $X_1, \ldots, X_n$ be independent random variables taking values in the metric spaces $(\mathcal{X}_1, d_1), \ldots, (\mathcal{X}_n, d_n)$, respectively. Let $f : \mathcal{X} \to \mathbb{R}$ be such that for all $x = (x_1, \ldots, x_n) \in \mathcal{X}$ and $y = (y_1, \ldots, y_n) \in \mathcal{X}$,

$$|f(x) - f(y)| \le \sum_{i=1}^n d_i(x_i, y_i).$$

Show that if D_i denotes the diameter of $\mathcal{X}_i$, then

$$\mathbf{P}\{|f(X_1, \ldots, X_n) - \mathbf{E}f(X_1, \ldots, X_n)| > t\} \le 2\exp\left(-\frac{t^2}{\sum_{i=1}^n D_i^2}\right).$$

(See Ledoux (2001).)

7.7. Show that if B denotes the Euclidean ball in $\mathbb{R}^n$ with radius 1, then $\mathrm{Vol}(\partial B) = n\mathrm{Vol}(B)$.

7.8. (GORDON'S INEQUALITY) Prove that if $\Phi(t) = (2\pi)^{-1/2}\int_{-\infty}^t e^{-x^2/2}dx$ denotes the standard normal distribution function and $\phi(t) = (2\pi)^{-1/2}e^{-t^2/2}$ is the standard normal density, then for all $t > 0$,

$$\frac{t}{t^2+1} \le \frac{1-\Phi(t)}{\phi(t)} \le \frac{1}{t}.$$

(See Gordon (1941), see also Birnbaum (1942).)

7.9. Show that for $\sqrt{2/n} \le s \le 1$, the normalized surface area of a spherical cap of height $1 - s$ in S^{n-1} satisfies

$$\frac{1}{6s\sqrt{n}}(1-s^2)^{\frac{n-1}{2}} \le \mu(C(u,s)) \le \frac{1}{2s\sqrt{n}}(1-s^2)^{\frac{n-1}{2}}.$$

(See e.g. Brieden *et al.* (2001).)

7.10. Extend the argument given in Section 7.2 for the proof of an isoperimetric inequality on the surface of the Euclidean ball to more general norms as follows. Consider a norm $\|\cdot\|$ on $\mathbb{R}^n$ and define its *modulus of convexity* by

$$\delta(\varepsilon) = \inf_{x,y \in \mathbb{R}: \|x\| \le 1, \|y\| \le 1, \|x-y\| \ge \varepsilon} \left(1 - \left\| \frac{x+y}{2} \right\|\right).$$

Let $S = \{x \in \mathbb{R}^n : \|x\| = 1\}$ be the "surface" of the unit ball in this norm and define the measure μ on S by

$$\mu(A) = \frac{\mathrm{Vol}(\cup_{a \in [0,1]} a \cdot A)}{\mathrm{Vol}(\{x : \|x\| \le 1\})}, \qquad A \subset S.$$

Show that for any measurable set $A \subset S$, the t-blowup A_t of A satisfies

$$\mu(A_t^c) \le \frac{2}{\mu(A)} e^{-2n\delta(t/2)}$$

(see Schechtman (2003) who also gives the history of this result).

The following few exercises ask the reader to prove Harper's vertex isoperimetric theorem (Theorem 7.6). Each exercise is a main step of the proof given by Kleitman (1979); see also Leader (1991).

7.11. (HARPER'S THEOREM: COMPRESSION) The proof of Harper's vertex isoperimetric theorem sketched here is based on the idea of *compression*. Let $A \subset \{-1, 1\}^n$ and define the i-sections of A by

$$A^{(i-)} = \left\{x^{(i,-1)} = (x_1, \dots, x_{i-1}, -1, x_{i+1}, \dots, x_n) : x^{(i,-1)} \in A\right\}$$

and

$$A^{(i+)} = \left\{x^{(i,1)} = (x_1, \dots, x_{i-1}, 1, x_{i+1}, \dots, x_n) : x^{(i,-1)} \in A\right\}.$$

Let $S_N^{(i,-)}$ (and $S_N^{(i,+)}$) denote the set of first N elements in the simplicial ordering of all vectors whose i-th component is 0 (and 1, respectively). Let $C_i(A)$ be the set whose i-sections are $S^{(i,-)}_{|A^{(i-)}|}$ and $S^{(i,+)}_{|A^{(i+)}|}$. Clearly, $|C_i(A)| = |A|$. Prove that $\partial_V(C_i(A)) \le \partial_V(A)$.

7.12. (HARPER'S THEOREM: ITERATION) Given $A \subset \{-1, 1\}^n$, define a sequence of sets recursively as follows: let $A_0 = A$. Having defined $A_0, A_1, \dots, A_k$, let $i \in \{1, \dots, n\}$ be any index such that $C_i(A_k) \ne A_k$ and define $A_{k+1} = C_i(A_k)$. If no such i exists, the process terminates. Show that the process terminates after a finite number of steps, that is, there exists a k such that $A_k = C_i(A_k)$ for every $i = 1, \dots, n$. Note that if B denotes the set obtained at the end of the process, then by the previous exercise, $|B| = |A|$ and $\partial_V(B) \le \partial_V(A)$.

7.13. (HARPER'S THEOREM: CONCLUSION) Let B be a set such that $C_i(B) = B$ for every $i = 1, \dots, n$. Show that either $B = S_{|B|}$ or, if n is even,

$$B = \left\{x \in \{-1, 1\}^n : \|x\| < \frac{n}{2}\right\} \cup \left\{x \in \{-1, 1\}^n : \|x\| = \frac{n}{2}, x_1 = 1\right\} - \{y\} \cup \{z\}$$

where $y = (y_1, \ldots, y_n)$ is such that $y_i = 1$ if and only if $i \in \{1, (n/2) + 2, (n/2) + 3, \ldots, n\}$, and $z = (z_1, \ldots, z_n)$ is such that $z_i = 1$ if and only if $i \in \{2, 3, \ldots, (n/2) + 1\}$ or, if n is odd,

$$B = \left\{x \in \{-1, 1\}^n : \|x\| < \frac{n}{2}\right\} - \{y\} \cup \{z\}$$

where $y = (y_1, \ldots, y_n)$ is such that $y_i = 1$ if and only if $i \in \{(n+3)/2, (n+5)/2, \ldots, n\}$ and $z = (z_1, \ldots, z_n)$ is such that $z_i = 1$ if and only if $i \in \{1, 2, \ldots, (n+1)/2\}$. Complete the proof of Harper's vertex isoperimetric theorem by noting that the two exceptional sets defined above have a larger boundary than the corresponding set $S_{|B|}$.

7.14. Check that the conditions of Sion's minmax theorem are satisfied in the representation of the convex distance as a saddle point in the proof of Lemma 7.10 (see Boucheron, Lugosi, and Massart (2003)).

7.15. (CONVEX DISTANCE AND CONFIGURATION FUNCTIONS) Recall the definition of a configuation function from Chapter 3. Assume $f : \mathcal{X}^n \to \mathbb{N}$ is a configuration function. Let $A_a = \{x : x \in \mathcal{X}^n, f(x) \leq a\}$. Check that for all $x \in \mathcal{X}^n$,

$$f(x) \leq a + \sqrt{f(x)} d_T(x, A_a).$$

Let $\boldsymbol{P}$ denote a product probability distribution over $\mathcal{X}^n$. Let $\boldsymbol{M}Z$ denote a median of $Z = f(X)$ under $\boldsymbol{P}$. Using Talagrand's convex distance inequality, show that

$$\boldsymbol{P}\{Z \geq \boldsymbol{M}Z + t\} \leq 2e^{-\frac{t^2}{4(\boldsymbol{M}Z+t)}},$$

and

$$\boldsymbol{P}\{Z \leq \boldsymbol{M}Z - t\} \leq 2e^{-\frac{t^2}{4\boldsymbol{M}Z}},$$

Hint: the function $t \mapsto (t-a)/\sqrt{t}$ is increasing for $t \geq a$ (see Talagrand (1995)).

7.16. Prove the following extension of Theorem 7.12. Let $X = (X_1, \ldots, X_n)$ be a vector of independent random variables taking values in the interval $[0, 1]$ and let $f : [0, 1]^n \to \mathbb{R}$ be a quasi-convex function. Suppose that there exists a convex set $S \subset [0, 1]^n$ such that $|f(x) - f(y)| \leq \|x - y\|$ for all $x, y \in S$ where $\boldsymbol{P}\{X \notin S\} < 1/2$. Then $f(X)$ satisfies, for all $t > 0$,

$$\boldsymbol{P}\{f(X) > \boldsymbol{M}f(X) + t\} \leq \boldsymbol{P}\{X \notin S\} + \frac{1}{1/2 - \boldsymbol{P}\{X \notin S\}} e^{-t^2/4}$$

and

$$\boldsymbol{P}\{f(X) < \boldsymbol{M}f(X) - t\} \leq 2\boldsymbol{P}\{X \notin S\} + 2e^{-t^2/4}$$

(Talagrand (1996c)).

7.17. Let $X_1, \ldots, X_n$ be independent random variables taking values is $[0, 1]$. Show that

$$\boldsymbol{P}\left\{\sqrt{\sum_{i=1}^n X_i^2} \geq \sqrt{2\boldsymbol{E}\sum_{i=1}^n X_i^2 + t}\right\} \leq e^{-(3/8)(\boldsymbol{E}\sum_{i=1}^n X_i^2 + t)}.$$

8

The Transportation Method

In this chapter we present the main ideas behind a different way of proving concentration inequalities that we call the *transportation method*. It is based on a beautiful idea of coupling and provides a simple and elegant approach that leads to concentration inequalities, some of which are difficult to prove with other general methods such as the ones described in Chapter 6. One of the strengths of the transportation method is that it is possible to extend it to weakly dependent random variables. However, we do not pursue this here.

The method is best described using the same basic framework defined in the previous chapters, that is, we let $X_1, \ldots, X_n$ be independent random variables taking values in a (measurable) set $\mathcal{X}$ and consider a measurable function $f : \mathcal{X}^n \to \mathbb{R}$ of n variables. As before, we define the real random variable $Z = f(X_1, \ldots, X_n)$. Once again, we need some general assumptions of regularity that can be formalized as follows. Let $d : \mathcal{X} \times \mathcal{X} \to [0, \infty)$ be a nonnegative function (typically a pseudo-metric) and let $c_1, \ldots, c_n \geq 0$ be constants. We assume that f satisfies the Lipschitz-type property

$$f(y) - f(x) \leq \sum_{i=1}^{n} c_i d\,(x_i, y_i) \tag{8.1}$$

for all $x, y \in \mathcal{X}^n$.

To demonstrate how transportation and concentration inequalities are connected, we recall the transportation lemma (Lemma 4.18) whose most basic special case states the following. Let Z be a real-valued random variable defined on a probability space $(\Omega, \mathcal{A}, P)$. The logarithm of the moment-generating function $\psi_{Z-\mathbf{E}_P Z}(\lambda) = \log \mathbf{E}_P \exp(\lambda(Z - \mathbf{E}_P Z))$ of the real-valued random variable satisfies

$$\psi_{Z-\mathbf{E}_P Z}(\lambda) \leq \frac{v\lambda^2}{2}$$

for every $\lambda > 0$ for some $v > 0$ if and only if for any probability measure Q absolutely continuous with respect to P and such that $D(Q\|P) < \infty$,

$$\mathbf{E}_Q f - \mathbf{E}_P f \leq \sqrt{2vD(Q\|P)}.$$

(Recall that $\mathbf{E}_P$ denotes integration with respect to the probability measure P.)

The lemma above suggests that one may prove sub-Gaussian concentration inequalities for $Z = f(X_1, \ldots, X_n)$ by proving a "transportation" inequality as above. The key to achieving this relies on *coupling*.

For $i = 1, \ldots, n$, denote by P_i the distribution of X_i, and let $P = P_1 \otimes \cdots \otimes P_n$ be the joint (product) distribution of $X_1, \ldots, X_n$ on $\mathcal{X}^n$. Consider a probability measure Q on $\mathcal{X}^n$, absolutely continuous with respect to P and let Y be a random variable (defined on the same probability space as X) such that Y has distribution Q. We say that the joint distribution $\mathbf{P}$ of the pair (X, Y) is a "coupling" of P and Q and we write $\mathcal{P}(P, Q)$ for the collection of all such probability distributions. Then, using the Lipschitz condition and the Cauchy–Schwarz inequality,

$$\begin{aligned} \mathbf{E}_Q f - \mathbf{E}_P f &= \mathbf{E}_{\mathbf{P}} \left[f(Y) - f(X) \right] \\ &\leq \sum_{i=1}^n c_i \mathbf{E}_{\mathbf{P}} d(X_i, Y_i) \\ &\leq \left(\sum_{i=1}^n c_i^2 \right)^{1/2} \left(\sum_{i=1}^n (\mathbf{E}_{\mathbf{P}} d(X_i, Y_i))^2 \right)^{1/2}. \end{aligned}$$

Thus, it suffices to upper bound

$$\sum_{i=1}^n (\mathbf{E}_{\mathbf{P}} d(X_i, Y_i))^2$$

by a constant multiple of $D(Q\|P)$. In particular, if one is able to prove that for some positive constant C

$$\min_{\mathbf{P} \in \mathcal{P}(P,Q)} \sum_{i=1}^n (\mathbf{E}_{\mathbf{P}} d(X_i, Y_i))^2 \leq 2C D(Q\|P), \tag{8.2}$$

then it follows from the argument described above that $\psi_{Z-\mathbf{E}_P Z}(\lambda) \leq v\lambda^2/2$ where $v = C \sum_{i=1}^n c_i^2$. This, of course, implies the sub-Gaussian concentration inequalities

$$\mathbf{P}\{Z \geq \mathbf{E}Z + t\} \leq e^{-t^2/(2v)} \quad \text{and} \quad \mathbf{P}\{Z \leq \mathbf{E}Z - t\} \leq e^{-t^2/(2v)}.$$

The bulk of the work therefore lies in proving the coupling inequality (8.2). By a general induction principle given in Lemma 8.13 at the end of this chapter, it suffices to prove the inequality for $n = 1$. Thus, the quantity of interest is

$$\min_{\mathbf{P} \in \mathcal{P}(P,Q)} \mathbf{E}_{\mathbf{P}} d(X, Y),$$

which quantifies the "effort" required to transport a mass distributed according to P into a mass distributed according to Q measured by the cost function d. This quantity is usually called the *transportation cost* from Q to P relatively to d. The transportation problem asks for constructing an optimal coupling $\mathbf{P} \in \mathcal{P}(P, Q)$, that is, a minimizer of the transportation cost $\mathbf{E}_{\mathbf{P}} d(X, Y)$. This explains the name *transportation method* for the technique of proving concentration inequalities discussed in this chapter.

The rest of the chapter is organized as follows. We first consider the case $d(x, y) = \mathbb{1}_{\{x \neq y\}}$. With this cost function, the Lipschitz condition (8.1) becomes just the bounded differences condition and, in fact, we recover an alternative proof of the bounded differences inequality (Theorem 6.2). Still dealing with the case $d(x, y) = \mathbb{1}_{\{x \neq y\}}$, in Section 8.2 we generalize the bounded differences condition by allowing the coefficients c_i in (8.1) to depend on the vector x. The resulting concentration inequalities do not have a proof based on the entropy method (especially the lower-tail bounds). In Section 8.5 we present a refinement of these ideas and use it to re-derive the Gaussian concentration inequality. The basic observation is that one may weaken condition (8.1) and assume instead only that

$$f(y) - f(x) \leq L \left(\sum_{i=1}^{n} d^2(x_i, y_i) \right)^{1/2}$$

if one is able to prove the stronger transportation inequality

$$\min_{\mathbf{P} \in \mathcal{P}(P,Q)} \sum_{i=1}^{n} \mathbf{E}_{\mathbf{P}} d^2(X_i, Y_i) \leq C D(Q \| P).$$

Indeed, in this case the Cauchy–Schwarz inequality implies that for every coupling of P and Q, one has

$$\mathbf{E}_Q f - \mathbf{E}_P f \leq L \left(\sum_{i=1}^{n} \mathbf{E}_{\mathbf{P}} d^2(X_i, Y_i) \right)^{1/2},$$

and therefore $\psi_{Z - \mathbf{E}_P Z}(\lambda) \leq v\lambda^2/2$ with $v = CL$. We show that the strengthened transportation inequality holds with $C = 1$ for the quadratic cost $d^2(x, y) = (x - y)^2$ in the case when P is the standard Gaussian distribution on $\mathbb{R}^n$. This provides an alternative proof of the Gaussian concentration inequality.

8.1 The Bounded Differences Inequality Revisited

Perhaps the simplest way to illustrate how the transportation method works is by re-proving the bounded differences inequality of Theorem 6.2. Recall that f satisfies the bounded differences condition if

$$|f(x_1, \ldots, x_i, \ldots, x_n) - f(x_1, \ldots, y_i, \ldots, x_n)| \leq c_i$$

for all $x, y \in \mathcal{X}^n$, and $i = 1, \ldots, n$. This implies that for all $x, y \in \mathcal{X}^n$,

$$|f(x) - f(y)| \leq \sum_{i=1}^{n} c_i \mathbb{1}_{\{x_i \neq y_i\}},$$

and therefore f satisfies condition (8.1) for the cost function $d(t, t') = \mathbb{1}_{\{t \neq t'\}}$. According to the argument described in the introduction of this chapter, we need to solve the transportation problem for this cost function. Clearly,

$$\min_{\mathbf{P} \in \mathcal{P}(P,Q)} \mathbf{E}_{\mathbf{P}} d(X, Y) = \min_{\mathbf{P} \in \mathcal{P}(P,Q)} \mathbf{P}\{X \neq Y\}.$$

The solution is given by the next lemma.

Lemma 8.1 *If P and Q are probability distributions on the same space $(\Omega, \mathcal{A})$, then*

$$\min_{\mathbf{P} \in \mathcal{P}(P,Q)} \mathbf{P}\{X \neq Y\} = V(P, Q)$$

where $V(P, Q)$ denotes the total variation distance

$$V(P, Q) = \sup_{A \in \mathcal{A}} |P(A) - Q(A)|.$$

Proof Note that if $\mathbf{P} \in \mathcal{P}(P, Q)$, then

$$\begin{aligned} |P(A) - Q(A)| &= \left|\mathbf{E}_{\mathbf{P}}\left[\mathbb{1}_{\{X \in A\}} - \mathbb{1}_{\{Y \in A\}}\right]\right| \\ &\leq \mathbf{E}_{\mathbf{P}}\left[\left|\mathbb{1}_{\{X \in A\}} - \mathbb{1}_{\{Y \in A\}}\right| \mathbb{1}_{\{X \neq Y\}}\right] \leq \mathbf{P}\{X \neq Y\}, \end{aligned}$$

which means that $V(P, Q) \leq \inf_{\mathbf{P} \in \mathcal{P}(P,Q)} \mathbf{P}\{X \neq Y\}$. Conversely, consider a probability measure μ which dominates P and Q and denote by p and q the corresponding densities of P and Q with respect to μ. Then

$$a \stackrel{\text{def}}{=} V(P, Q) = \int_{\Omega} (p - q)_+ d\mu = \int_{\Omega} (q - p)_+ d\mu = 1 - \int_{\Omega} \min(p, q) d\mu$$

and since we can assume that $a > 0$ (otherwise the result is trivial), we define the probability measure $\mathbf{P}$ as a mixture $\mathbf{P} = a\mathbf{P}_1 + (1 - a)\mathbf{P}_2$ where $\mathbf{P}_1$ and $\mathbf{P}_2$ are such that, for any measurable and bounded function Ψ,

$$a^2 \int_{\Omega \times \Omega} \Psi(x, y) d\mathbf{P}_1(x, y) = \int_{\Omega \times \Omega} (p(x) - q(x))_+ (q(y) - p(y))_+ \Psi(x, y) d\mu(x) d\mu(y)$$

and

$$(1 - a) \int_{\Omega \times \Omega} \Psi(x, y) d\mathbf{P}_2(x, y) = \int_{\Omega} \min(p(x), q(x)) \Psi(x, x) d\mu(x).$$

It is easy to check that $\mathbf{P} \in \mathcal{P}(P, Q)$. Moreover, since $\mathbf{P}_2$ is concentrated on the "diagonal" $\{(x, y) : x = y\}$, we have $\mathbf{P}\{X \neq Y\} = a\mathbf{P}_1\{X \neq Y\} \leq a$. □

It remains for us to prove a transportation inequality of the form (8.2). Given the interpretation of the variation distance as a transportation cost given by the previous lemma, it is now natural to use Pinsker's inequality (recall Theorem 4.19). This immediately gives us (8.2) for $n = 1$.

Theorem 8.2 (MARTON'S TRANSPORTATION INEQUALITY) *Let $P = P_1 \otimes \cdots \otimes P_n$ be a product probability measure on $\mathcal{X}^n$ and let Q be a probability measure absolutely continuous with respect to P. Then*

$$\min_{\mathbf{P} \in \mathcal{P}(P,Q)} \sum_{i=1}^{n} \mathbf{P}^2 \{X_i \neq Y_i\} \leq \frac{1}{2} D(Q\|P),$$

where $(X, Y) = (X_i, Y_i)_{i=1,\ldots,n}$ has distribution $\mathbf{P}$.

Proof We simply apply the general induction principle given in Lemma 8.13 at the end of this chapter, noticing that the basic induction assumption is satisfied by combining Pinsker's inequality (Theorem 4.19) and Lemma 8.1. □

Putting everything together, we see that (8.2) is satisfied with $C = 1/4$, which implies that if $f : \mathcal{X}^n \to \mathbb{R}$ satisfies the bounded differences condition and $X_1, \ldots, X_n$ are independent, then $Z = f(X_1, \ldots, X_n)$ is sub-Gaussian with variance factor $\sum_{i=1}^{n} c_i^2/4$, which implies the bounded differences inequality of Theorem 6.2.

8.2 Bounded Differences in Quadratic Mean

Next we take a step further and relax the bounded differences condition. We assume that $f : \mathcal{X}^n \to \mathbb{R}$ satisfies

$$f(y) - f(x) \leq \sum_{i=1}^{n} c_i(x) \mathbb{1}_{\{x_i \neq y_i\}}$$

for some functions $c_i : \mathcal{X}^n \to [0, \infty)$, $i = 1, \ldots, n$. Instead of forcing the c_i to be bounded we assume only that they are bounded in "quadratic mean" in the sense that

$$v \stackrel{\text{def}}{=} \mathbf{E} \sum_{i=1}^{n} c_i^2(X)$$

is finite. Under this assumption, the transportation method may be used as follows. Let Q be a probability distribution, absolutely continuous with respect to P, the distribution of X. Let $\mathbf{P}$ be a coupling of P and Q. Then

$$E_Q f - E_P f \leq \sum_{i=1}^{n} \mathbf{E}_P \left[c_i(X) \mathbf{P}\{X_i \neq Y_i \mid X\} \right]$$

which implies, by applying the Cauchy–Schwarz inequality twice,

$$\begin{aligned} E_Q f - E_P f &\leq \sum_{i=1}^{n} \left(\mathbf{E}_P c_i^2(X)\right)^{1/2} \left(\mathbf{E}_P \left[\mathbf{P}^2 \{X_i \neq Y_i \mid X\}\right]\right)^{1/2} \\ &\leq \left(\sum_{i=1}^{n} \mathbf{E}_P c_i^2(X)\right)^{1/2} \left(\sum_{i=1}^{n} \mathbf{E}_P \left[\mathbf{P}^2 \{X_i \neq Y_i \mid X\}\right]\right)^{1/2}. \end{aligned}$$

Using our assumption on f, this implies

$$E_Q f - E_P f \leq \sqrt{v} \left(\inf_{P \in \mathcal{P}(P,Q)} \sum_{i=1}^{n} \mathbf{E}_P \left[\mathbf{P}^2 \{X_i \neq Y_i \mid X\}\right] \right)^{1/2}.$$

Thus, by the "road map" laid down in the introduction to this chapter, if we can prove the inequality

$$\inf_{P \in \mathcal{P}(P,Q)} \sum_{i=1}^{n} \mathbf{E}_P \left[\mathbf{P}^2 \{X_i \neq Y_i \mid X\}\right] \leq 2D(Q \| P),$$

then Lemma 4.18 implies $\psi_{Z-\mathbf{E}Z}(\lambda) \leq v\lambda^2/2$ and the resulting sub-Gaussian tail inequality with variance factor v. Since Lemma 8.13 is applicable, it suffices to prove the transportation inequality above for $n = 1$. To this end, we first solve the corresponding transportation cost problem.

A conditional transportation cost problem First we need to introduce the analog of the total variation distance for our "conditional" transportation cost problem. Let P and Q be probability distributions and let μ be a measure dominating P and Q simultaneously. For concreteness, we may take $\mu = (P + Q)/2$. Then we may consider $p = dP/d\mu$ and $q = dQ/d\mu$ and define

$$d_2^2(Q, P) = \int \frac{(p-q)_+^2}{p} d\mu.$$

Observe that this definition does not depend on the dominating measure. Indeed, if ν is another measure dominating P and Q simultaneously, then μ is absolutely continuous with respect to ν and setting $g = d\mu/d\nu$ and we may write

$$\int \frac{(dP/d\nu - dQ/d\nu)_+^2}{dP/d\nu} d\nu = \int \frac{(gp - gq)_+^2}{gp} d\nu = \int \frac{(p-q)_+^2}{p} g d\nu = d_2^2(Q, P).$$

We are now ready to prove an analog of Lemma 8.1.

Lemma 8.3 *Let P and Q be probability distributions on a common measurable space $(\Omega, \mathcal{A})$. Then*

$$\min_{\mathbf{P}\in\mathcal{P}(P,Q)} \left(\mathbf{E}_{\mathbf{P}}\left[\mathbf{P}^2\{X \neq Y \mid X\}\right] + \mathbf{E}_{\mathbf{P}}\left[\mathbf{P}^2\{X \neq Y \mid Y\}\right]\right) = d_2^2(Q,P) + d_2^2(P,Q).$$

Proof Let $\mu = (P+Q)/2$ and denote by p and q the densities of P and Q with respect to μ. Introducing $\tilde{p}(x) = p(x)\mathbb{1}_{\{p(x)>0\}} + \mathbb{1}_{\{p(x)=0\}}$, notice that $\tilde{p}(X) = p(X)$ with probability one. Moreover, if $\mathbf{P} \in \mathcal{P}(P,Q)$, then

$$\mathbf{P}\{X = Y \mid X\} \leq \min\left(1, \frac{q(X)}{\tilde{p}(X)}\right)$$

with probability one. To see this, observe that for any nonnegative measurable function h,

$$\begin{aligned}
\mathbf{E}_{\mathbf{P}}\left[h(X)\mathbf{P}\{X = Y \mid X\}\right] &= \mathbf{E}_{\mathbf{P}}\left[h(X)\mathbb{1}_{\{X=Y\}}\right] \\
&\leq \mathbf{E}_{\mathbf{P}}\left[h(Y)\mathbb{1}_{\{p(Y)>0\}}\right] \\
&= \mathbf{E}_{\mathbf{P}}\left[h(X)\frac{q(X)}{\tilde{p}(X)}\right],
\end{aligned}$$

and therefore,

$$\mathbf{E}_{\mathbf{P}}\left[h(X)\left(\frac{q(X)}{\tilde{p}(X)} - \mathbf{P}\{X = Y \mid X\}\right)\right] \geq 0$$

from which the claim follows. This implies that

$$\mathbf{E}_{\mathbf{P}}\left[\mathbf{P}^2\{X \neq Y \mid X\}\right] \geq \mathbf{E}_{\mathbf{P}}\left[\left(1 - \frac{q(X)}{\tilde{p}(X)}\right)_+^2\right] = d_2^2(Q,P),$$

and therefore

$$d_2^2(Q,P) \leq \inf_{\mathbf{P}\in\mathcal{P}(P,Q)} \mathbf{E}_{\mathbf{P}}\left[\mathbf{P}^2\{X \neq Y \mid X\}\right].$$

Of course, symmetrically,

$$d_2^2(P,Q) \leq \inf_{\mathbf{P}\in\mathcal{P}(P,Q)} \mathbf{E}_{\mathbf{P}}\left[\mathbf{P}^2\{X \neq Y \mid Y\}\right],$$

which implies that

$$\begin{aligned}
&d_2^2(Q,P) + d_2^2(P,Q) \\
&\leq \inf_{\mathbf{P}\in\mathcal{P}(P,Q)} \left\{\mathbf{E}_{\mathbf{P}}\left[\mathbf{P}^2\{X \neq Y \mid X\}\right] + \mathbf{E}_{\mathbf{P}}\left[\mathbf{P}^2\{X \neq Y \mid Y\}\right]\right\}.
\end{aligned}$$

Conversely, if $a = V(P, Q) = 0$, there is nothing to prove. Otherwise we consider the same coupling $\boldsymbol{P} \in \mathcal{P}(P, Q)$ as in the proof of Lemma 8.1, that is, $\boldsymbol{P}$ is defined as a mixture $\boldsymbol{P} = a\boldsymbol{P}_1 + (1-a)\boldsymbol{P}_2$ where $\boldsymbol{P}_1$ and $\boldsymbol{P}_2$ are such that, for any measurable and bounded function Ψ,

$$a^2 \int_{\Omega\times\Omega} \Psi(x,y)d\boldsymbol{P}_1(x,y) = \int_{\Omega\times\Omega} (p(x)-q(x))_+(q(y)-p(y))_+\Psi(x,y)d\mu(x)d\mu(y)$$

and

$$(1-a)\int_{\Omega\times\Omega} \Psi(x,y)d\boldsymbol{P}_2(x,y) = \int_{\Omega} \min(p(x), q(x))\Psi(x,x)\, d\mu(x).$$

By construction of this coupling, we have, with probability one,

$$\begin{aligned}\boldsymbol{P}\{X \neq Y \mid X\} &= \left(\frac{\int_\Omega (q(y)-p(y))_+ d\mu(y)}{a}\right)\left(\frac{(p(X)-q(X))_+}{p(X)}\right) \\ &= \frac{(p(X)-q(X))_+}{p(X)},\end{aligned}$$

and therefore

$$\boldsymbol{E}_{\boldsymbol{P}}\left[\boldsymbol{P}^2\{X \neq Y \mid X\}\right] = \int_\Omega \frac{(p(x)-q(x))_+^2}{p^2(x)} p(x)d\mu(x) = d_2^2(Q,P).$$

Similarly we have

$$\boldsymbol{E}_{\boldsymbol{P}}\left[\boldsymbol{P}^2\{X \neq Y \mid Y\}\right] = d_2^2(P,Q),$$

concluding the proof of Lemma 8.3. □

The next step is an analog of Pinsker's inequality in which d_2 plays the role of the total variation distance.

Lemma 8.4 *Let P and Q be probability distributions on a common measurable space $(\Omega, \mathcal{A})$. If Q is absolutely continuous with respect to P, then*

$$d_2^2(Q,P) + d_2^2(P,Q) \leq 2D(Q\|P).$$

Proof Since $Q \ll P$, setting $q = dQ/dP$ we may write

$$d_2^2(Q,P) + d_2^2(P,Q) = \boldsymbol{E}_P\left[(1-q(X))_+^2\right] + \boldsymbol{E}_P\left[\frac{(q(X)-1)_+^2}{q(X)}\right].$$

Moreover, defining $h(t) = (1-t)\log(1-t) + t$ for $t < 1$ and $h(1) = 1$, we may write

$$D(Q\|P) = \boldsymbol{E}_P\left[h(1-q(X))\right] = \boldsymbol{E}_P\left[h\left((1-q(X))_+\right)\right] + \boldsymbol{E}_P\left[h\left(-(q(X)-1)_+\right)\right]$$

and the result follows by the inequalities

$$h(t) \geq \frac{t^2}{2} \quad \text{for } t \in [0,1] \quad \text{and} \quad h(-t) \geq \frac{t^2}{2(1+t)} \quad \text{for } t \geq 0$$

(recall Exercise 2.8). □

We are now ready to prove the main result of this section.

Theorem 8.5 (MARTON'S CONDITIONAL TRANSPORTATION INEQUALITY) *Let $P = P_1 \otimes \cdots \otimes P_n$ be a product probability measure on $\mathcal{X}^n$ and let Q be a probability measure absolutely continuous with respect to P. Then*

$$\min_{\mathbf{P} \in \mathcal{P}(P,Q)} \mathbf{E}_{\mathbf{P}} \sum_{i=1}^{n} \left(\mathbf{P}^2 \{X_i \neq Y_i \mid X_i\} + \mathbf{P}^2 \{X_i \neq Y_i \mid Y_i\} \right) \leq 2D(Q\|P),$$

where $(X, Y) = (X_i, Y_i)_{i=1,\ldots,n}$ has distribution $\mathbf{P}$.

Proof By Lemmas 8.3 and 8.4, for all $i = 1, \ldots, n$ and for every distribution ν which is absolutely continuous with respect to P_i,

$$\min_{\mathbf{P} \in \mathcal{P}(P_i,\nu)} \mathbf{E}_{\mathbf{P}} \left[\mathbf{P}^2 \{X_i \neq Y_i \mid X_i\} + \mathbf{P}^2 \{X_i \neq Y_i \mid Y_i\} \right] \leq 2D(\nu\|P_i).$$

The result follows by applying Lemma 8.13 with $\phi(x) = x^2/2$ and $w(x,y) = \mathbb{1}_{\{x \neq y\}}$. □

Marton's conditional transportation inequality implies the following powerful concentration inequality. It is, in essence, similar to some of the results of Chapter 6 but does not follow from any of them.

Theorem 8.6 *Let $f : \mathcal{X}^n \to \mathbb{R}$ be a measurable function and let $X_1, \ldots, X_n$ be independent random variables taking their values in $\mathcal{X}$. Define $Z = f(X_1, \ldots, X_n)$. Assume that there exist measurable functions $c_i : \mathcal{X}^n \to [0, \infty)$ such that for all $x, y \in \mathcal{X}^n$,*

$$f(y) - f(x) \leq \sum_{i=1}^{n} c_i(x) \mathbb{1}_{\{x_i \neq y_i\}}.$$

Setting

$$v = \mathbf{E} \sum_{i=1}^{n} c_i^2(X) \quad \text{and} \quad v_\infty = \sup_{x \in \mathcal{X}^n} \sum_{i=1}^{n} c_i^2(x),$$

for all $\lambda > 0$, we have

$$\psi_{Z-\mathbf{E}Z}(\lambda) \leq \frac{\lambda^2 v}{2} \quad \text{and} \quad \psi_{-Z+\mathbf{E}Z}(\lambda) \leq \frac{\lambda^2 v_\infty}{2}.$$

In particular, for all $t > 0$,

$$\mathbf{P}\{Z \geq \mathbf{E}Z + t\} \leq e^{-t^2/(2v)} \quad \text{and} \quad \mathbf{P}\{Z \leq \mathbf{E}Z - t\} \leq e^{-t^2/(2v_\infty)}.$$

Proof Let $P = P_1 \otimes \cdots \otimes P_n$ denote the distribution of the vector $X = (X_1, \ldots, X_n)$ and let Q be a probability distribution on $\mathcal{X}^n$ which is absolutely continuous with respect to P. If $\mathbf{P}$ is a coupling of P and Q, then, as we have seen at the beginning of this section,

$$\mathbf{E}_Q f - \mathbf{E}_P f \leq \sqrt{v} \left(\sum_{i=1}^n \mathbf{E}_P \left[\mathbf{P}^2 \{X_i \neq Y_i \mid X\} \right] \right)^{1/2},$$

and therefore

$$\mathbf{E}_Q f - \mathbf{E}_P f \leq \sqrt{v} \left(\inf_{\mathbf{P} \in \mathcal{P}(P,Q)} \sum_{i=1}^n \mathbf{E}_P \left[\mathbf{P}^2 \{X_i \neq Y_i \mid X\} \right] \right)^{1/2},$$

so by Theorem 8.5

$$\mathbf{E}_Q f - \mathbf{E}_P f \leq \sqrt{2vD(Q\|P)}.$$

Since this inequality holds for all $Q \ll P$, by Lemma 4.18, we have $\psi_{Z-\mathbf{E}Z}(\lambda) \leq \lambda^2 v/2$, proving the bound for the upper tail of Z.

To prove the inequalities for the lower tail of Z, introduce $g(x) = -f(x)$. Then the condition on f implies that for all $x, y \in \mathcal{X}^n$,

$$g(y) - g(x) \leq \sum_{i=1}^n c_i(y) \mathbb{1}_{\{x_i \neq y_i\}}.$$

Then, by repeating the argument at the beginning of the section, we get

$$\mathbf{E}_Q g - \mathbf{E}_P g \leq \sqrt{v_Q} \left(\inf_{\mathbf{P} \in \mathcal{P}(P,Q)} \sum_{i=1}^n \mathbf{E}_P \left[\mathbf{P}^2 \{X_i \neq Y_i \mid Y\} \right] \right)^{1/2},$$

where $v_Q = \sum_{i=1}^n \mathbf{E} c_i^2(Y)$. Unfortunately, v_Q depends on Q and it is therefore not a useful quantity. However, by bounding $v_Q \leq v_\infty$ and using Theorem 8.5, we get that, for all $Q \ll P$,

$$\mathbf{E}_Q g - \mathbf{E}_P g \leq \sqrt{2v_\infty D(Q\|P)}$$

and again we may conclude using Lemma 4.18. □

8.3 Applications of Marton's Conditional Transportation Inequality

Next we illustrate the use of Theorem 8.6 by revisiting some examples from earlier chapters such as the largest eigenvalue of a symmetric matrix with independent entries, configuration functions, and the bin packing problem.

Example 8.7 (THE LARGEST EIGENVALUE OF A RANDOM SYMMETRIC MATRIX) Consider again the example already investigated in Examples 3.14 and 6.8. Let A be a random symmetric real matrix with entries $X_{i,j}$, $1 \leq i \leq j \leq n$ where X the $X_{i,j}$ are independent random variables with $|X_{i,j}| \leq 1$. Let $Z = \lambda_1$ denote the largest eigenvalue of A. We already proved that $\mathrm{Var}(Z) \leq 16$ and that for all $t > 0$,

$$P\{Z > EZ + t\} \leq e^{-t^2/32}.$$

Here we show how Theorem 8.6 implies the same exponential bound and a similar lower tail inequality. If $x \in [-1,1]^{n(n+1)/2}$ is a vector with components $x_{i,j}$, $1 \leq i \leq j \leq n$, let $A(x) = \left((A(x))_{i,j}\right)_{n\times n}$ denote the corresponding symmetric matrix and $\lambda_1(x)$ its largest eigenvalue. Then for all $x, y \in [-1,1]^{n(n+1)/2}$,

$$\begin{aligned}
\lambda_1(x) - \lambda_1(y) &= \sup_{u\in\mathbb{R}^n:\|u\|=1} u^T A(x) u - \sup_{u\in\mathbb{R}^n:\|u\|=1} u^T A(y) u \\
&\leq v^T (A(x) - A(y)) v \\
&\qquad \text{(where } v = (v_1, \ldots, v_n) \text{ is a unit vector maximizing } u^T A(x) u\text{)} \\
&= \sum_{i=1}^n \sum_{j=1}^n v_i v_j \left(A(x)_{i,j} - A(y)_{i,j}\right) \\
&\leq 4 \sum_{1\leq i\leq j\leq n} \mathbb{1}_{\{x_{i,j}\neq y_{i,j}\}} |v_i v_j|.
\end{aligned}$$

Since v only depends on x, the function $f(x) = -\lambda_1(x)$ satisfies the condition of Theorem 8.6 with $c_{i,j}(x) = 4|v_i v_j|$. But $\sum_{1\leq i\leq j\leq n} c_{i,j}(x)^2 \leq 16$ for all x and therefore Theorem 8.6 implies the bounds

$$P\{Z > EZ + t\} \leq e^{-t^2/32} \quad \text{and} \quad P\{Z < EZ - t\} \leq e^{-t^2/32}$$

for all $t > 0$.

Example 8.8 (CONFIGURATION FUNCTIONS) Recall from Section 3.3 the definition of a configuration function $f : \mathcal{X}^n \to \{1, 2, \ldots, n\}$: a property Π is a sequence of sets $\Pi_1 \subset \mathcal{X}, \Pi_2 \subset \mathcal{X}^2, \ldots, \Pi_n \subset \mathcal{X}^n$. For $m \leq n$, a vector $(x_1, \ldots x_m) \in \mathcal{X}^m$ satisfies the property Π if $(x_1, \ldots x_m) \in \Pi_m$. Assume that Π is hereditary so that if $(x_1, \ldots x_m)$ satisfies Π then so does any sub-sequence $(x_{i_1}, \ldots x_{i_k})$ of $(x_1, \ldots x_m)$. The function f that maps any vector $x = (x_1, \ldots x_n)$ to the size of a largest sub-sequence satisfying Π is the configuration function associated with property Π.

If f is a configuration function and $X_1, \ldots, X_n$ are independent random variables taking values in $\mathcal{X}$, then define $Z = f(X_1, \ldots, X_n)$. Since configuration functions are self-bounding, Z satisfies the exponential inequalities of Theorem 6.12.

Let f be such a configuration function. For any $x \in \mathcal{X}^n$, fix a maximal sub-sequence $(x_{i_1}, \ldots, x_{i_m})$ satisfying property Π (so that $f(x) = m$). Let $c_i(x)$ denote the indicator that x_i belongs to the sub-sequence $(x_{i_1}, \ldots, x_{i_m})$. Thus, $\sum_{i=1}^n c_i(x)^2 = \sum_{i=1}^n c_i(x) = f(x)$. It follows from the definition of a configuration function that for all $x, y \in \mathcal{X}^n$,

$$f(y) \geq f(x) - \sum_{i=1}^{n} \mathbb{1}_{\{x_i \neq y_i\}} c_i(x).$$

This means that the function $g = -f$ satisfies the condition of Theorem 8.6 with $v = \mathbf{E}Z$. Thus, the first inequality of Theorem 8.6 implies that

$$\mathbf{P}\{Z \leq \mathbf{E}Z - t\} \leq e^{-t^2/(2\mathbf{E}Z)}.$$

Of course, we have already proved the same inequality as a consequence of Theorem 6.12.

To derive an exponential inequality for the upper tail of Z, we need to modify the proof of Theorem 8.6. Since for all $x, y \in \mathcal{X}^n$

$$f(y) - f(x) \leq \sum_{i=1}^{n} c_i(y) \mathbb{1}_{\{x_i \neq y_i\}},$$

it follows from Theorem 8.5 that for all $Q \ll P$,

$$\mathbf{E}_Q f - \mathbf{E}_P f \leq \sqrt{2D(Q\|P)\mathbf{E}_Q f},$$

where P denotes the distribution of the vector $X = (X_1, \ldots, X_n)$. But then

$$\mathbf{E}_Q f - \mathbf{E}_P f \leq \sqrt{2D(Q\|P)\mathbf{E}_P f + 2D(Q\|P)}$$

(see Exercise 8.3). By Lemma 4.18 this implies

$$\mathbf{P}\{Z \geq \mathbf{E}Z + t\} \leq \exp\left(-\mathbf{E}Z h_1\left(\frac{2t}{\mathbf{E}Z}\right)\right)$$

where $h_1(u) = 1 + u - \sqrt{1 + 2u}$, or, equivalently,

$$\mathbf{P}\left\{Z - \mathbf{E}Z \geq \sqrt{2t\mathbf{E}Z} + 2t\right\} \leq e^{-t}$$

(recall the calculations of Section 2.4). This inequality is similar, though not quite as sharp as the one that follows from Theorem 6.12.

Example 8.9 (BIN PACKING) Consider once again the random bin packing problem described in Example 3.3 and Section 7.6. Recall that $f(x)$ denotes the minimum number of bins of size 1 so that the numbers $x_1, \ldots, x_n \in [0,1]$ fit in $f(x)$ bins. We write $Z = f(X)$ when $X_1, \ldots, X_n$ are independent, taking values in $[0,1]$. In Section 7.6 we used the convex distance inequality to derive exponential tail inequalities for Z. The key property we used was that for all $x, y \in [0,1]^n$,

$$f(x) \leq f(y) + 2\sum_{i=1}^{n} \mathbb{1}_{\{x_i \neq y_i\}} x_i + 1,$$

so introducing $g(x) = -f(x)$, we have

$$g(y) \leq g(x) + 2\sum_{i=1}^{n} \mathbb{1}_{\{x_i \neq y_i\}} x_i + 1.$$

This looks very much like the condition of Theorem 8.6 except for the additional "+1" on the right-hand side. Thus, Theorem 8.6 is not directly applicable but Theorem 8.5 is still useful with a slight modification of the proof of Theorem 8.6. Indeed, it follows by Marton's conditional transportation inequality that, if P denotes the distribution of $X = (X_1, \ldots, X_n)$ and Q is absolutely continuous with respect to P, then

$$\mathbf{E}_Q g - \mathbf{E}_P g \leq \sqrt{2vD(Q\|P)} + 1$$

where $v = 4\sum_{i=1}^{n} \mathbf{E}X_i^2$. Then, by an easy application of Lemma 4.18, we have, for all $t > 0$,

$$\mathbf{P}\{Z < \mathbf{E}Z - t\} \leq \exp\left(\frac{-(t-1)^2}{8\sum_{i=1}^{n} \mathbf{E}X_i^2}\right).$$

We leave the details to the reader as an easy exercise. This bound for the lower tail of Z is slightly better than that which we obtained from the convex distance inequality. However, by a direct application of Theorem 8.6 we do not get an interesting bound for the upper tail because $v_\infty = \sup_x 4\sum_{i=1}^{n} x_i^2 = 4n$ leads to a bound that we could prove in a simpler way by the bounded differences inequality.

8.4 The Convex Distance Inequality Revisited

The power of Theorem 8.6 is best demonstrated by showing how easily it implies Talagrand's convex distance inequality which we proved by the entropy method (and with a suboptimal constant) in Section 7.4.

Recall that if $A \subset \mathcal{X}^n$ is a measurable set, then the convex distance of $x \in \mathcal{X}^n$ to the set A is defined as

$$d_T(x, A) = \sup_{\alpha \in [0,\infty)^n : \|\alpha\| \leq 1} \inf_{y \in A} \sum_{i=1}^{n} \alpha_i \mathbb{1}_{\{x_i \neq y_i\}}.$$

Denote by $c(x) = (c_1(x), \ldots, c_n(x))$ the vector of nonnegative components in the unit ball for which the supremum is achieved. Then

$$\begin{aligned} d_T(x, A) - d_T(y, A) &\leq \inf_{x' \in A} \sum_{i=1}^{n} c_i(x) \mathbb{1}_{\{x_i \neq x'_i\}} - \inf_{y' \in A} \sum_{i=1}^{n} c_i(x) \mathbb{1}_{\{y_i \neq y'_i\}} \\ &\leq \sum_{i=1}^{n} c_i(x) \mathbb{1}_{\{x_i \neq y_i\}}. \end{aligned}$$

This shows that $f(x) = -d_T(x, A)$ satisfies the condition of Theorem 8.6. Since $\sum_{i=1}^{n} c_i(x)^2 \leq 1$ for all x, Theorem 8.6 ensures that if X is a vector of independent random variables, then $d_T(X, A)$ is sub-Gaussian with variance factor 1. This property implies the convex distance inequality as follows. Let $Z = d_T(X, A)$. By Theorem 8.6, for all $t > 0$,

$$\mathbf{P}\{Z - \mathbf{E}Z \geq t\} \leq e^{-t^2/2}.$$

Since $t^2 \geq -(\mathbf{E}Z)^2 + (t + \mathbf{E}Z)^2/2$, this upper tail inequality implies

$$\mathbf{P}\{Z - \mathbf{E}Z \geq t\} \leq e^{(\mathbf{E}Z)^2/2} e^{-(t+\mathbf{E}Z)^2/4}.$$

Replacing t by $t - \mathbf{E}Z$, this inequality also implies that for $t > 0$,

$$\mathbf{P}\{Z \geq t\} \leq e^{(\mathbf{E}Z)^2/2} e^{-t^2/4}$$

(note that this bound is trivial whenever $t \leq \mathbf{E}Z$ and therefore we may always assume that $t > \mathbf{E}Z$). On the other hand, using the left-tail bound

$$\mathbf{P}\{\mathbf{E}Z - Z \geq t\} \leq e^{-t^2/2}$$

with $t = \mathbf{E}Z$, we get

$$\mathbf{P}\{X \in A\} = \mathbf{P}\{Z = 0\} \leq e^{-(\mathbf{E}Z)^2/2}.$$

Combining these bounds leads to

$$\mathbf{P}\{X \in A\}\mathbf{P}\{Z \geq t\} \leq e^{-t^2/4},$$

which is the convex distance inequality of Theorem 7.9.

8.5 Talagrand's Gaussian Transportation Inequality

The purpose of this section is to prove the following transportation inequality for the standard Gaussian measure.

Theorem 8.10 *Let P be the standard Gaussian probability measure on $\mathbb{R}^n$ and let Q be any probability measure which is absolutely continuous with respect to P. Then*

$$\min_{\mathbf{P}\in\mathcal{P}(P,Q)} \sum_{i=1}^{n} \mathbf{E}_{\mathbf{P}}(X_i - Y_i)^2 \leq 2D(Q\|P).$$

Before proving the theorem, we show how it implies the Tsirelson–Ibragimov–Sudakov inequality (Theorem 5.6), which we proved based on the Gaussian logarithmic Sobolev inequality and Herbst's argument.

Assume that $f : \mathbb{R}^n \to \mathbb{R}$ is a Lipschitz function, that is, for all $x, y \in \mathbb{R}^n$,

$$f(y) - f(x) \leq L\left(\sum_{i=1}^{n}(x_i - y_i)^2\right)^{1/2}.$$

Then, by Jensen's inequality, for every coupling $\mathbf{P}$ of P and $Q \ll P$, one has

$$\mathbf{E}_Q f - \mathbf{E}_P f = \mathbf{E}_{\mathbf{P}}\left[f(Y) - f(X)\right] \leq L\left(\sum_{i=1}^{n}\mathbf{E}_{\mathbf{P}}(X_i - Y_i)^2\right)^{1/2}.$$

Hence, Theorem 8.10 implies that

$$\mathbf{E}_Q f - \mathbf{E}_P f \leq \sqrt{2L^2 D(Q\|P)},$$

and it follows from Lemma 4.18 that $\psi_{Z-\mathbf{E}Z}(\lambda) \leq L^2\lambda^2/2$ for all $\lambda > 0$ where $Z = f(X)$. This implies the Gaussian concentration inequality.

Turning to the proof of Theorem 8.10, first note that the induction argument of Lemma 8.13 applies and therefore our main task is to deal with the one-dimensional case. Before proving the result, we describe a classical result which shows that the solution of the transportation cost problem for the quadratic loss is given by the so-called *quantile transform*, sometimes also called *monotone rearrangement*.

Lemma 8.11 *Let F and G be distribution functions on the real line. If X and Y are real-valued random variables with distribution functions F and G, respectively, $\mathbf{E}[(X - Y)^2]$ is minimal when X and Y are defined by the quantile transform of the same uniform random variable, that is, when $X = F^{-1}(U)$ and $Y = G^{-1}(U)$ where U is uniformly distributed on $[0, 1]$. The minimal value of $\mathbf{E}[(X - Y)^2]$ is therefore*

$$\int_0^1 \left(F^{-1}(t) - G^{-1}(t)\right)^2 dt.$$

Proof Since the marginal distributions of X and Y are given, minimizing $\mathbf{E}[(X-Y)^2]$ is equivalent to maximizing $\mathbf{E}[XY]$. We begin with the case when X and Y are nonnegative. Then, by Fubini's theorem,

$$\begin{aligned}\mathbf{E}[XY] &= \mathbf{E}\int_0^\infty\int_0^\infty \mathbb{1}_{\{x<X\}}\mathbb{1}_{\{y<Y\}}dxdy \\ &= \int_0^\infty\int_0^\infty \mathbf{P}\{X>x, Y>y\}\,dxdy.\end{aligned}$$

Applying this formula to the variables $F^{-1}(U)$ and $G^{-1}(U)$ yields

$$\begin{aligned}\mathbf{E}\left[F^{-1}(U)G^{-1}(U)\right] &= \int_0^\infty\int_0^\infty \mathbf{P}\left\{F^{-1}(U)>x, G^{-1}(U)>y\right\}dxdy \\ &= \int_0^\infty\int_0^\infty \mathbf{P}\left\{U>\max(F(x),G(y))\right\}dxdy,\end{aligned}$$

and therefore,

$$\begin{aligned}\mathbf{E}\left[F^{-1}(U)G^{-1}(U)\right] &= \int_0^\infty\int_0^\infty \min\left((1-F(x)),(1-G(y))\right)dxdy \\ &= \int_0^\infty\int_0^\infty \min\left(\mathbf{P}\{X>x\},\mathbf{P}\{Y>y\}\right)dxdy.\end{aligned}$$

Since $\mathbf{P}\{X>x, Y>y\} \le \min\left(\mathbf{P}\{X>x\},\mathbf{P}\{Y>y\}\right)$, we have shown that

$$\mathbf{E}[XY] \le \mathbf{E}\left[F^{-1}(U)G^{-1}(U)\right].$$

Dealing with the general case is more complicated but relies basically on the same arguments. Decomposing X and Y as $X = X^+ - X^-$ and $Y = Y^+ - Y^-$, we write $\mathbf{E}[XY]$ as the sum of four terms:

$$\mathbf{E}[XY] = \mathbf{E}[X^+Y^+] + \mathbf{E}[X^-Y^-] - \mathbf{E}[X^-Y^+] - \mathbf{E}[X^+Y^-]. \tag{8.3}$$

We now find that we can optimize each of these four terms individually. More precisely, the maximum of each term is achieved whenever $X = F^{-1}(U)$ and $Y = G^{-1}(U)$ which, of course, implies the desired result. For the first two terms, this is clear since the arguments above imply

$$\mathbf{E}[X^+Y^+] \le \mathbf{E}\left[\left(F^{-1}(U)\right)^+\left(G^{-1}(U)\right)^+\right],$$

and similarly,

$$\mathbf{E}[X^-Y^-] \le \mathbf{E}\left[\left(F^{-1}(U)\right)^-\left(G^{-1}(U)\right)^-\right].$$

Next we study the third term. Using Fubini's theorem once more, we may write

$$\mathbf{E}[X^-Y^+] = \int_0^\infty \int_0^\infty \mathbf{P}\{X \le -x, Y > y\}\, dxdy.$$

Note that

$$\mathbf{P}\{X \le -x, Y > y\} \ge \mathbf{P}\{X \le -x\} - \mathbf{P}\{Y \le y\},$$

which means that

$$\mathbf{P}\{X \le -x, Y > y\} \ge (F(-x) - G(y))^+ .$$

But the right-hand side of this inequality may be interpreted as

$$\begin{aligned}(F(-x) - G(y))^+ &= \mathbf{P}\left\{G(y) < U \le F(-x)\right\}\\ &= \mathbf{P}\left\{F^{-1}(U) \le -x, G^{-1}(U) > y\right\},\end{aligned}$$

and therefore,

$$\mathbf{P}\{X \le -x, Y > y\} \ge \mathbf{P}\left\{F^{-1}(U) \le -x, G^{-1}(U) > y\right\}.$$

It remains to integrate this inequality on $[0, \infty)^2$ to conclude that

$$\mathbf{E}[X^-Y^+] \ge \mathbf{E}\left[\left(F^{-1}(U)\right)^- \left(G^{-1}(U)\right)^+\right].$$

Exchanging the roles of X and Y yields the same result for the fourth term in (8.3), which finally leads to

$$\mathbf{E}[XY] \le \mathbf{E}\left[F^{-1}(U)G^{-1}(U)\right].$$

To finish the proof, it remains to compute the minimal value of $\mathbf{E}[(X - Y)^2]$ under the marginal constraints $X \sim F$ and $Y \sim G$. But we already know that the minimum is achieved whenever $X = F^{-1}(U)$ and $Y = G^{-1}(U)$ and in this case,

$$\mathbf{E}[(X - Y)^2] = \mathbf{E}\left(F^{-1}(U) - G^{-1}(U)\right)^2 = \int_0^1 \left(F^{-1}(t) - G^{-1}(t)\right)^2 dt. \qquad \square$$

Although it assists our understanding of the transportation approach to Gaussian concentration, we do not use Lemma 8.11 in the proof of Theorem 8.10 but, rather, prove directly the following inequality for the quantile transform.

Lemma 8.12 *Let γ be the standard normal distribution on the real line and ν be some probability distribution which is absolutely continuous with respect to γ. Denote by Φ and G the distribution functions of γ and ν and define the quantile transform*

$$T = G^{-1} \circ \Phi.$$

If X is a standard normal variable, then $Y = T(X)$ has distribution ν and

$$\mathbf{E}[(X - Y)^2] \leq 2D(\nu \| \gamma).$$

Proof Denote by g the density of ν with respect to γ. Assume first that g is bounded by a constant θ. We claim that this assumption implies that

$$|T(x)| \leq 2|x| \quad \text{when } |x| \text{ is large enough.} \tag{8.4}$$

Indeed, $g \leq \theta$ implies that for all x, $G(2x) \leq \theta\Phi(2x)$. Moreover, by Gordon's inequality for the tail behavior of Φ (see Exercise 7.8),

$$-\log \Phi(x) \sim \frac{x^2}{2} \quad \text{as } x \to -\infty,$$

and, in particular,

$$\lim_{x\to\infty} \frac{\Phi(2x)}{\Phi(x)} = 0.$$

Hence there exists $x_0 < 0$, such that $G(2x) \leq \Phi(x)$ or, equivalently, $2x \leq T(x)$ for all $x \leq x_0$. A similar argument for the right tail leads to (8.4).

The key observation for proving the lemma is that

$$T'(x) = \frac{\phi(x)}{g(T(x))\phi(T(x))},$$

where $\phi(t) = (2\pi)^{-1/2}e^{-t^2/2}$ denotes the standard normal density. Then we may write

$$\begin{aligned}
D(\nu\|\gamma) &= \mathbf{E}\log g(Y) \\
&= \mathbf{E}\left[\log\frac{\phi(X)}{\phi(Y)} - \log T'(X)\right] \\
&= \mathbf{E}\left[-\frac{X^2}{2} + \frac{Y^2}{2} - \log T'(X)\right] \\
&\geq \mathbf{E}\left[-\frac{X^2}{2} + \frac{Y^2}{2} + 1 - T'(X)\right]
\end{aligned}$$

where we use $-\log u \geq 1 - u$ for $u \geq 0$. From (8.4) we know, on the one hand, that Y has a finite second order moment and, on the other hand, that $\lim_{|x|\to\infty} T(x)\phi(x) = 0$. Hence, integrating by parts leads to

$$-\mathbf{E}T'(X) = -\int_{-\infty}^{+\infty} T'(x)\phi(x)dx = \int_{-\infty}^{+\infty} T(x)\phi'(x)dx$$
$$= -\int_{-\infty}^{+\infty} xT(x)\phi(x)dx = -\mathbf{E}[XY].$$

Then the inequality above becomes

$$D(\nu\|\gamma) \geq \mathbf{E}\left[-\frac{X^2}{2} + \frac{Y^2}{2}\right] + 1 - \mathbf{E}[XY] = \frac{\mathbf{E}[(X-Y)^2]}{2}$$

where we use $\mathbf{E}X^2 = 1$. This proves the lemma for the case when g is bounded.

The general case requires a truncation argument. We may assume $D(\nu\|\gamma) < \infty$ because otherwise there is nothing to prove. For any positive integer k, introduce the (bounded) density

$$g_k(x) = \frac{\min(g(x), k)}{c_k},$$

where $c_k = \int \min(g(x), k)\phi(x)dx$. By monotone convergence, the distribution function G_k of $\nu_k = g_k\phi$, converges pointwise to G, so $T_k = G_k^{-1} \circ \Phi$ converges pointwise to T. By Fatou's lemma and using the fact that the statement is true in the bounded case, we have

$$\mathbf{E}\left[(X - T(X))^2\right] \leq \liminf_{k\to\infty} \mathbf{E}\left[(X - T_k(X))^2\right] \leq 2 \liminf_{k\to\infty} D(\nu_k\|\gamma).$$

To complete the argument, it remains to prove that $\liminf_{k\to\infty} D(\nu_k\|\gamma) = D(\nu\|\gamma)$. Setting

$$H(u) = u \log u,$$

we may write

$$D(\nu_k\|\gamma) = \frac{1}{c_k}\int H(\min(g(x), k))\,\phi(x)dx - \log c_k.$$

By monotone convergence, $\lim_{k\to\infty} c_k = 1$. Moreover, since H increases on $[1, +\infty)$, the sequence of functions $H(\min(g(x), k))$ increases to $H(g(x))$ as $k \to \infty$. Furthermore, H is bounded from below by $-e^{-1}$ and Lebesgue's dominated convergence theorem allows us to conclude that

$$\lim_{k\to\infty} \int H(\min(g(x), k))\,\phi(x)dx = \int H(g(x))\,\phi(x)dx,$$

and therefore $\lim_{k\to\infty} D(\nu_k\|\gamma) = D(\nu\|\gamma)$, completing the proof of Lemma 8.12. □

Now that the one-dimensional transportation cost inequality is available, it is very easy to derive Talagrand's transportation cost inequality for the Gaussian measure via Lemma 8.13.

Proof of Theorem 8.10 Starting from Lemma 8.12 we may apply Lemma 8.13 with $\phi(x) = x/2$ and $w(x, y) = (x - y)^2$ to derive the theorem. □

8.6 Appendix: A General Induction Lemma

We close this chapter by a general induction principle that is an important part of the proofs of Theorems 8.2, 8.5, and 8.10. It allows us to extend the one-dimensional transportation inequalities to the multi-dimensional case.

Lemma 8.13 *Let $P = \bigotimes_{i=1}^n P_i$ be a product probability measure on a product measurable space $\mathcal{X}^n$ and let Q be a probability measure absolutely continuous with respect to P. Let $w : \mathcal{X} \times \mathcal{X} \to [0, \infty)$ be a measurable function and let $\phi : [0, \infty) \to [0, \infty)$ be a convex function. Suppose that for every $i = 1, \ldots, n$ and for every probability measure ν which is absolutely continuous with respect to P_i,*

$$\min_{\mathbf{P} \in \mathcal{P}(P_i, \nu)} \phi\left(\mathbf{E}_{\mathbf{P}} w(X_i, Y_i)\right) \leq D(\nu \| P_i). \tag{8.5}$$

Then

$$\min_{\mathbf{P} \in \mathcal{P}(P,Q)} \sum_{i=1}^n \phi\left(\mathbf{E}_{\mathbf{P}} w(X_i, Y_i)\right) \leq D(Q \| P).$$

Similarly, if for every $i = 1, \ldots, n$ and for every probability measure $\nu \ll P_i$

$$\min_{\mathbf{P} \in \mathcal{P}(P_i, \nu)} \mathbf{E}_{\mathbf{P}} \left[\phi\left(\mathbf{E}_{\mathbf{P}}\left[w(X_i, Y_i) \mid X_i\right]\right) + \phi\left(\mathbf{E}_{\mathbf{P}}\left[w(X_i, Y_i) \mid Y_i\right]\right)\right] \leq D(\nu \| P_i), \tag{8.6}$$

then

$$\min_{\mathbf{P} \in \mathcal{P}(P,Q)} \sum_{i=1}^n \mathbf{E}_{\mathbf{P}} \left[\phi\left(\mathbf{E}_{\mathbf{P}}\left[w(X_i, Y_i) \mid X_i\right]\right) + \phi\left(\mathbf{E}_{\mathbf{P}}\left[w(X_i, Y_i) \mid Y_i\right]\right)\right] \leq D(Q \| P)$$

and, a fortiori,

$$\min_{\mathbf{P} \in \mathcal{P}(P,Q)} \sum_{i=1}^n \mathbf{E}_{\mathbf{P}} \left[\phi\left(\mathbf{E}_{\mathbf{P}}\left[w(X_i, Y_i) \mid X\right]\right) + \phi\left(\mathbf{E}_{\mathbf{P}}\left[w(X_i, Y_i) \mid Y\right]\right)\right] \leq D(Q \| P).$$

Proof We start with the case when assumption (8.5) holds. We prove, by induction on $k \le n$, that for every Q absolutely continuous with respect to $P^k = \otimes_{i=1}^k P_i$,

$$\min_{P \in \mathcal{P}(P^k, Q)} \sum_{i=1}^{k} \phi\left(\mathbf{E}_{\mathbf{P}} w(X_i, Y_i)\right) \le D\left(Q \| P^k\right).$$

For $k = 1$, this is just assumption (8.5). Assume now that for any distribution Q', absolutely continuous with respect to P^{k-1}, the coupling inequality

$$\min_{P \in \mathcal{P}(P^{k-1}, Q')} \sum_{i=1}^{k-1} \phi\left(\mathbf{E}_{\mathbf{P}} w(X_i, Y_i)\right) \le D\left(Q' \| P^{k-1}\right) \tag{8.7}$$

holds. Now let $g = dQ/dP^k$ denote the density of Q with respect to P^k. Then, using the notation $H(u) = u \log u$,

$$D\left(Q \| P^k\right) = \int_{\mathcal{X}} \left[\int_{\mathcal{X}^{k-1}} H\left(g(x,t)\right) dP^{k-1}(x)\right] dP_k(t).$$

Denoting by g_k the marginal density $g_k(t) = \int_{\mathcal{X}^{k-1}} g(x,t) dP^{k-1}(x)$ and by q_k the corresponding marginal distribution of Q, $q_k = g_k P_k$, we may write $g(x,t) = g(x|t) g_k(t)$ and get, by Fubini's theorem,

$$D\left(Q \| P^k\right) = \int_{\mathcal{X}} g_k(t) \left[\int_{\mathcal{X}^{k-1}} H\left(g(x|t)\right) dP^{k-1}(x)\right] dP_k(t) + \int_{\mathcal{X}} H\left(g_k(t)\right) dP_k(t).$$

Introducing for any $t \in \mathcal{X}$, the conditional distribution

$$dQ(x|t) = g(x|t) dP^{k-1}(x),$$

the previous identity can be written as

$$D\left(Q \| P^k\right) = \int_{\mathcal{X}} D\left(Q\left(\cdot|t\right) \| P^{k-1}\right) dq_k(t) + D\left(q_k \| P_k\right),$$

which is known as the *chain rule* for relative entropy. Now (8.7) ensures that, for any $t \in \mathcal{X}$, there exists a probability distribution $\mathbf{P}_t$ on $\mathcal{X}^{k-1} \times \mathcal{X}^{k-1}$ belonging to $\mathcal{P}\left(P^{k-1}, Q\left(\cdot|t\right)\right)$ such that

$$\sum_{i=1}^{k-1} \phi\left(\mathbf{E}_{\mathbf{P}_t} w(X_i, Y_i)\right) \le D\left(Q\left(\cdot|t\right) \| P^{k-1}\right),$$

while (8.5) ensures that there exists a probability distribution Q_k on $\mathcal{X} \times \mathcal{X}$ belonging to $\mathcal{P}(P_k, q_k)$ such that

$$\phi\left(\mathbf{E}_{Q_k} w(X_k, Y_k)\right) \le D(\nu \| P_k).$$

Hence,

$$D\left(Q \| P^k\right) \ge \int_{\mathcal{X}} \sum_{i=1}^{k-1} \phi\left(\mathbf{E}_{\mathbf{P}_t} w(X_i, Y_i)\right) dq_k(t) + \phi\left(\mathbf{E}_{Q_k} w(X_k, Y_k)\right),$$

and by Jensen's inequality,

$$D\left(Q \| P^k\right) \ge \sum_{i=1}^{k-1} \phi\left[\int_{\mathcal{X}} \mathbf{E}_{\mathbf{P}_t} w(X_i, Y_i) dq_k(t)\right] + \phi\left(\mathbf{E}_{Q_k} w(X_k, Y_k)\right). \tag{8.8}$$

Now consider the probability distribution $\mathbf{P}$ on $\mathcal{X}^k \times \mathcal{X}^k$ with marginal distribution Q_k on $\mathcal{X} \times \mathcal{X}$ and such that the distribution of (X_i, Y_i) for $1 \le i \le k-1$, conditionally on (X_k, Y_k), is equal to $\mathbf{P}_{Y_k}$. More precisely, for any measurable and bounded function $\Psi : \mathcal{X}^k \times \mathcal{X}^k \to \mathbb{R}$, $\int_{\mathcal{X}^k \times \mathcal{X}^k} \Psi(x, y) d\mathbf{P}(x, y)$ is defined by

$$\int_{\mathcal{X} \times \mathcal{X}} \left[\int_{\mathcal{X}^{k-1} \times \mathcal{X}^{k-1}} \Psi\left[(x, x_k), (y, y_k)\right] d\mathbf{P}_{y_k}(x, y)\right] dQ_k(x_k, y_k).$$

Then, by construction, $\mathbf{P} \in \mathcal{P}\left(P^k, Q\right)$. Moreover,

$$\mathbf{E}_{\mathbf{P}} w(X_i, Y_i) = \int_{\mathcal{X}} \mathbf{E}_{\mathbf{P}_t} w(X_i, Y_i) dq_k(t) \quad \text{for all } i \le k-1$$

and

$$\mathbf{E}_{\mathbf{P}} w(X_k, Y_k) = \mathbf{E}_{Q_k} w(X_k, Y_k),$$

and therefore we obtain from (8.8) that

$$D\left(Q \| P^k\right) \ge \sum_{i=1}^{k-1} \phi\left(\mathbf{E}_{\mathbf{P}} w(X_i, Y_i)\right) + \phi\left(\mathbf{E}_{\mathbf{P}} w(X_k, Y_k)\right).$$

If we now consider assumption (8.6), the proof is very similar and we just sketch it. The induction argument ensures the existence of a coupling probability distribution $\mathbf{P}_t$ on $\mathcal{X}^{k-1} \times \mathcal{X}^{k-1}$ belonging to $\mathcal{P}\left(P^{k-1}, Q\left(.|t\right)\right)$ such that

$$\sum_{i=1}^{k-1} \int_{\mathcal{X}^{k-1}} \phi\left(\boldsymbol{E}_{P_t}\left[w(X_i,Y_i)|X_1 = x_1,\ldots,X_{k-1} = x_{k-1}\right]\right) dP^{k-1}(x_1,\ldots x_{k-1})$$
$$+ \int_{\mathcal{X}^{k-1}} \phi\left(\boldsymbol{E}_{P_t}\left[w(X_i,Y_i)|Y_1 = y_1,\ldots,Y_{k-1} = y_{k-1}\right]\right) dQ(y_1,\ldots,y_{k-1}|t)$$
$$\leq D\left(Q\left(.|t\right) \| P^{k-1}\right),$$

and one can define a coupling probability $Q_k \in \mathcal{P}(P_k, q_k)$ such that

$$\boldsymbol{E}_{Q_k}\left[\phi\left(\boldsymbol{E}_{Q_k}\left[w\left(X_k,Y_k\right)|X_k\right]\right) + \phi\left(\boldsymbol{E}_{Q_k}\left[w\left(X_k,Y_k\right)|Y_k\right]\right)\right] \leq D\left(\nu \| P_k\right).$$

We define the coupling probability $\boldsymbol{P}$ in exactly the same way as above. The proof can be completed in a similar way as before by using the chain rule, Fubini's Theorem, and Jensen's inequality. The last inequality of the theorem is easily obtained. □

8.7 Bibliographical Remarks

The transportation method for proving concentration inequalities was initiated by Marton (1986), building on earlier work on information theory by Ahlswede, Gács and Körner (1976) and Csiszár and Körner (1981). Marton first considered the case $d(x,y) = \mathbb{1}_{\{x \neq y\}}$, leading to the bounded differences inequality. Lemma 8.1 is due to Dobrushin (1970). Lemma 8.1 is a special instance of the *transportation cost* problem. The interested reader will find much more general results in Rachev (1991), including Kantorovich's theorem that relates the transportation cost to the bounded Lipschitz distance when the cost function is a distance and several analog coupling results for other types of distances between probability measures like the Prohorov distance (see also Strassen's theorem in Strassen (1965)).

Theorem 8.2 is a slightly stronger form of the original result of Marton (1986). By the Cauchy–Schwarz inequality, Theorem 8.2 implies that

$$\min_{P \in \mathcal{P}(P,Q)} \sum_{i=1}^{n} \boldsymbol{P}\{X_i \neq Y_i\} \leq \sqrt{\frac{n}{2} D(Q \| P)},$$

which is originally stated in Marton (1986).

The symmetric "Pinsker-type" inequality of Lemma 8.4 is due to Samson (2000).

The method is robust in the sense that it can be extended to functions of weakly dependent variables (see Marton (1996*b*, 2003, 2004), Rio (2000), and Samson (2000)).

The material in Section 8.2 is based on Marton (1996*a*) and Samson (2000). The results of Section 8.5 are due to Talagrand (1996*d*). Lemma 8.11 goes back to Fréchet (1957).

For more on the topic we refer to Dembo (1997), Ledoux (2001), and Samson (2003).

8.8 EXERCISES

8.1. Use Marton's transportation inequality (Theorem 8.2) to show that if P is a product probability measure on $\mathcal{X}^n$ then for any pair of measurable sets $A, B \subset \mathcal{X}^n$,

$$d_H(A,B) \le \sqrt{\frac{n}{2}\log\frac{1}{P(A)}} + \sqrt{\frac{n}{2}\log\frac{1}{P(B)}}$$

where $d_H(A,B) = \min_{x\in A, y\in B}\sum_{i=1}^n \mathbb{1}_{\{x_i\neq y_i\}}$ is the Hamming distance of A and B. What do you obtain if you take B to be the complement of the t-blowup of A?

8.2. Complete the details of the proof of the inequality in Example 8.9 for the left tail of the bin packing problem.

8.3. Let $a > 0$. Show that if $x, y > 0$ satisfy $y - a\sqrt{y} \le x$, then $y \le x + a\sqrt{x} + a^2$.

8.4. Let F and G be distribution functions on the real line. If X and Y are real-valued random variables with distribution functions F and G, show that $\mathbf{E}\,|X-Y|$ is minimal when X and Y are defined by the quantile transform of the same uniform random variable, that is, when $X = F^{-1}(U)$ and $Y = G^{-1}(U)$ where U is uniformly distributed on $[0,1]$. Conclude that the minimal value of $\mathbf{E}\,|X-Y|$ under the marginal constraints $X \sim F$ and $Y \sim G$ is

$$\int_0^1 \left|F^{-1}(t) - G^{-1}(t)\right| dt.$$

(*Hint:* use the formula $|X-Y| = X + Y - 2\max(X,Y)$ and begin with the case where X and Y are nonnegative).

8.5. Let F and G be distribution functions on the real line. Prove that

$$\int_0^1 \left|F^{-1}(t) - G^{-1}(t)\right| dt = \int_{-\infty}^{+\infty} |F(x) - G(x)|\, dx.$$

8.6. (RIO'S COVARIANCE INEQUALITY) Let X and Y be non-negative square-integrable random variables with distributions functions F and G, respectively. Prove the following bound, known as *Fréchet's inequality*:

$$Cov(X,Y) \le \int_0^1 F^{-1}(t)G^{-1}(t)dt - \int_0^1 F^{-1}(t)dt \int_0^1 G^{-1}(t)dt.$$

Let α be the (strong) mixing coefficient between X and Y defined as the supremum, over all Borels sets A and B, of $\left|Cov\left(\mathbb{1}_{\{X\in A\}}, \mathbb{1}_{\{Y\in B\}}\right)\right|$. Prove that

$$|Cov(X,Y)| \le \int_0^\infty \int_0^\infty \min\left(\alpha, \mathbf{P}\{X > u\}, \mathbf{P}\{Y > v\}\right) du dv$$

and derive that

$$|Cov(X,Y)| \leq \int_0^{\alpha} F^{-1}(1-t)G^{-1}(1-t)dt.$$

Let now X and Y be square integrable random variables, not necessarily non-negative. Denoting by F and G the distribution functions of $|X|$ and $|Y|$ respectively, prove the following covariance inequality:

$$|Cov(X,Y)| \leq 2\int_0^{2\alpha} F^{-1}(1-t)G^{-1}(1-t)dt.$$

(Rio (1993).)

9

Influences and Threshold Phenomena

This chapter is devoted to the study of functions defined on the n-dimensional binary hypercube $\{-1, 1\}^n$. The n-cube, with the uniform distribution, is the simplest product space and the tight connection between isoperimetric properties and concentration is revealed in the most transparent manner. Logarithmic Sobolev inequalities and hypercontractive estimates may be interpreted as generalized isoperimetric inequalities and have interesting consequences for the geometry of the hypercube. We are mostly interested in binary-valued (or *Boolean*) functions (or, equivalently, subsets of $\{-1, 1\}^n$) though in some cases it is convenient to deal with real-valued functions of the n binary variables.

An important notion that plays a crucial role in this chapter is the *influence* of a variable, already introduced in Chapter 4. We start by recalling some simple general isoperimetric inequalities for the hypercube, under the uniform distribution. In Section 9.2, using a logarithmic Sobolev inequality on the binary n-cube, we derive an improvement of the Efron–Stein inequality that implies some fundamental properties for influences of binary-valued functions. This inequality is used in Section 9.3 to derive "local" exponential concentration inequalities. In Section 9.4 another inequality for the variance, due to Talagrand, is proved.

Monotone sets play a central role in the study of influences, not only because their special properties make them an important object to study but also because one of the most important applications of the theory of influences, namely threshold phenomena, involves monotone sets. Section 9.5 is devoted to properties of influences of monotone sets, still under the uniform distribution.

Most results generalize easily to the case when the underlying measure is the product of n i.i.d. Bernoulli distributions with parameter $p \in (0, 1)$. The tools developed in this chapter allow one to study the evolution of the probability of monotone subsets of $\{-1, 1\}^n$ as p grows from 0 to 1. In particular, we establish general conditions under which an abrupt *phase transition* occurs around a certain critical value of p, that is, the probability of a monotone set jumps from values close to 0 to close to 1 in a narrow interval. Such effects are known as *threshold phenomena* and will be seen to occur for any monotone set that does not depend too much on any of the n variables.

9.1 Influences

Consider a subset A of the n-cube $\{-1,1\}^n$ and let P denote the uniform distribution on $\{-1,1\}^n$ so that $P(A) = 2^{-n}|A|$ where $|A|$ denotes the cardinality of the set A. We often find it convenient to work with Rademacher random variables $X_1, \ldots, X_n$ (i.e. the X_i are independent symmetric sign variables). Then the binary vector $X = (X_1, \ldots, X_n)$ is uniformly distributed in $\{-1,1\}^n$ and $P(A) = \mathbf{P}\{X \in A\}$.

Recall from Chapter 4 the definition of *influence* of a variable. We denote by $\overline{X}^{(i)} = (X_1, \ldots, X_{i-1}, -X_i, X_{i+1}, \ldots, X_n)$ the vector obtained by flipping the i-th component of the vector X and leaving the others intact. The influence of the i-th variable is

$$I_i(A) = \mathbf{P}\left\{\mathbb{1}_{\{X\in A\}} \neq \mathbb{1}_{\{\overline{X}^{(i)}\in A\}}\right\},$$

that is, the probability that changing the i-th variable changes the event $X \in A$. When this happens (i.e. when $\mathbb{1}_{\{X\in A\}} \neq \mathbb{1}_{\{\overline{X}^{(i)}\in A\}}$), we say that the i-th variable is *pivotal* for A.

The *total influence* is defined by the sum of individual influences

$$I(A) = \sum_{i=1}^{n} I_i(A).$$

Instead of subsets of $\{-1,1\}^n$, equivalently we may consider binary functions $f : \{-1,1\}^n \to \{0,1\}$. Such functions are sometimes called *Boolean*. If $f(x) = \mathbb{1}_{\{x\in A\}}$ then with some abuse of notation we can also write $I_i(f)$ for $I_i(A)$ and $I(f)$ for $I(A)$.

Example 9.1 (PARITY FUNCTION) Consider the *parity* function $f : \{-1,1\}^n \to \{0,1\}$ defined by $f(x) = 1$ if and only if the number of components of $x = (x_1, \ldots, x_n)$ equal to 1 is even. In this case, clearly for every $x \in \{-1,1\}^n$, every variable is pivotal and therefore $I_i(f) = 1$ for all $i = 1, \ldots, n$ and $I(f) = n$.

The parity function clearly maximizes the influence of all variables. The largest achievable total influence dramatically decreases if one considers *monotone* functions. Recall that a function $f : \{-1,1\}^n \to \{0,1\}$ is monotone if it is monotone in each of its variables, that is, $f(x) = 1$ implies $f(x_i^+) = 1$ where $x_i^+ = (x_1, \ldots, x_{i-1}, 1, x_{i+1}, \ldots, x_n)$ is obtained by fixing the i-th variable of x to be 1. If f is monotone, the corresponding set $A = \{x : f(x) = 1\}$ is called a *monotone set*. Monotone functions and sets play a central role in this chapter for many reasons, one of which is that they minimize total influence (see Theorem 9.10 below). One of the simplest monotone functions is the *majority* function that will be seen to maximize total influence among all monotone functions (see Theorem 9.11).

Example 9.2 (MAJORITY FUNCTION) Let n be odd and define $f(x) = 1$ if and only if $\sum_{i=1}^n x_i > 0$. f is obviously monotone. Since the function is symmetric, all influences $I_i(f)$ are equal. The first variable is pivotal if and only if $\sum_{i=2}^n x_i = 0$. Thus,

$I_1(A) = \mathbf{P}\{B = (n-1)/2\}$ where B is a binomial random variable with parameters $(n-1, 1/2)$. Therefore, for every $i = 1, \ldots, n$, by Stirling's formula,

$$I_i(f) = \binom{n-1}{(n-1)/2} 2^{-(n-1)} \sim \sqrt{\frac{2}{n\pi}}$$

and $I(f) \sim \sqrt{2n/\pi}$.

An interesting question we pursue in this chapter is how small the total influence of a function can be. A small total influence means that individual variables have little deciding power over the outcome of the function, a desirable property, for example, when the components represent votes of members of a society and the function represents a certain voting scheme.

The Efron–Stein inequality (Theorem 3.1) implies that

$$P(A)(1 - P(A)) = \mathrm{Var}\,(f(X)) \leq \frac{1}{4}\sum_{i=1}^{n} I_i(A) = \frac{1}{4}I(A).$$

In particular, if $P(A) = 1/2$, the total influence of A is at least 1. This bound is sharp when the value of the function is determined by only one variable, for example when $f(x) = (x_i + 1)/2$ for some $i \in \{1, \ldots, n\}$. Such a function is often called a *dictatorship*. Of course, in such a case, the influence $I_i(A)$ of the i-th variable equals one and the rest of the variables have zero influence. If a function f is such that there exists a small number of variables that determine the value of f, then f is called a *junta*. In this chapter we try to understand the behavior of functions of many variables, so we think about n as a large number and "small" in the previous definition means bounded, independently of n. Clearly, if f is a junta depending on k variables then $I(f) \leq k$. A fundamental result proved below is that any function with a small total influence is almost a junta in the sense that it can be tightly approximated by a junta. For the rigorous statement see Theorem 9.7 below.

A natural question is how small can the total influence be if the function f is symmetric in the sense that $I_1(f) = \cdots = I_n(f) = I(f)/n$. Below we reproduce a fundamental result of Kahn, Kalai, and Linial, implying that the total influence of a symmetric function is at least of the order of $\log n$, substantially larger than that of a dictatorship or a junta.

9.2 Some Fundamental Inequalities for Influences

If $P(A) < 1/2$, the bound obtained for the total influence from the Efron–Stein inequality is no longer sharp. One achieves a better bound by using the edge isoperimetric inequality of Theorem 4.3. Recall that this inequality states that for any $A \subset \{-1, 1\}^n$,

$$I(A) \geq 2P(A) \log_2 \frac{1}{P(A)}.$$

Observe that the latter inequality is a special case of the logarithmic Sobolev inequality Theorem 5.1 which states that for any real-valued function $f : \{-1, 1\}^n \to \mathbb{R}$,

$$\mathrm{Ent}(f^2) \leq 2\mathcal{E}(f)$$

where $\mathrm{Ent}(f^2) = E\left[f^2 \log(f^2)\right] - E\left[f^2\right] \log E\left[f^2\right]$ and

$$\mathcal{E}(f) = \frac{1}{4}\mathbf{E}\left[\sum_{i=1}^{n} \left(f(X) - f\left(\overline{X}^{(i)}\right)\right)^2\right].$$

Note that to lighten notation, we sometimes write $E[f]$ for $\mathbf{E}[f(X)]$ for any function $f : \{-1, 1\}^n \to \mathbb{R}$. Observe that the logarithmic Sobolev inequality applied for $f(x) = \mathbb{1}_{\{x \in A\}}$ recovers the edge isoperimetric inequality. The logarithmic Sobolev inequality of the n-cube also implies the following simple bound that we will find useful.

Lemma 9.3 *For any nonnegative function* $f : \{-1, 1\}^n \to [0, \infty)$,

$$E\left[f^2\right] \log \frac{E\left[f^2\right]}{E\left[f\right]^2} \leq 2\mathcal{E}(f).$$

Proof By Theorem 5.1 it suffices to prove that

$$\mathrm{Ent}(f^2) = E\left[f^2 \log(f^2)\right] - E\left[f^2\right] \log E\left[f^2\right] \geq E\left[f^2\right] \log \frac{E\left[f^2\right]}{(E\left[f\right])^2}.$$

This is trivial if $f \equiv 0$, otherwise, introducing $g(x) = f(x)/\sqrt{E\left[f^2\right]}$, it may be re-written as

$$E\left[g^2 \log(g^2)\right] \geq \log \frac{1}{(E\left[g\right])^2},$$

or, equivalently,

$$E\left[g^2 \log \frac{1}{gE[g]}\right] \leq 0.$$

This follows from the fact that $\log x \leq x - 1$ for $x > 0$ and that $E\left[g^2\right] = 1$:

$$E\left[g^2 \log \frac{1}{gE[g]}\right] \leq E\left[g^2 \left(\frac{1}{gE[g]} - 1\right)\right] = 0.$$

□

Next we prove an improvement of the Efron–Stein inequality that has various interesting consequences for the total influence of Boolean functions defined on the n-cube.

Consider a real-valued function $f : \{-1, 1\}^n \to \mathbb{R}$. As in Section 3.1, we express f as a sum of martingale differences for the natural filtration defined by the coordinate variables. More precisely, introduce

$$f_i(x) = 2^{i-n} \sum_{(x_{i+1},\dots,x_n)\in\{-1,1\}^{n-i}} f(x_1, \dots, x_n),$$

as the average of f over all binary vectors whose first i components agree with x, that is, $f_i(X) = \mathbf{E}[f(X)|X_1, \dots, X_i]$. Thus, $f_0(x) = E[f]$ and $f_n(x) = f(x)$. Define the martingale differences $\Delta_i : \{-1, 1\}^n \to \mathbb{R}$ by

$$\Delta_i(x) = f_i(x) - f_{i-1}(x), \quad i = 1, \dots, n.$$

Recall from Section 3.1 that $\operatorname{Var}(f) = \sum_{i=1}^n E[\Delta_i^2]$ where we use the shorthand notation $\operatorname{Var}(f) = \operatorname{Var}(f(X))$. We have the following general result.

Theorem 9.4 *For any* $f : \{-1, 1\}^n \to \mathbb{R}$,

$$\operatorname{Var}(f) \log \frac{\operatorname{Var}(f)}{\sum_{j=1}^n (E|\Delta_j|)^2} \le 2\mathcal{E}(f).$$

Recall that the Efron–Stein inequality implies $\operatorname{Var}(f) \le \mathcal{E}(f)$. The inequality of Theorem 9.4 presents an important improvement for functions defined on the binary n-cube whenever $\sum_{j=1}^n (E|\Delta_j|)^2 \ll \operatorname{Var}(f)$. We will see that this improvement has far-reaching consequences.

Proof The theorem follows easily from Lemma 9.3 and the decomposition $\mathcal{E}(f) = \sum_{i=1}^n \mathcal{E}(\Delta_i)$. To prove this decomposition, write, for any $j = 1, \dots, n$,

$$\begin{aligned}
&\mathcal{E}(\Delta_j)\\
&= \frac{1}{4}\sum_{i=1}^n \mathbf{E}\left[\left(\Delta_j(X) - \Delta_j\left(\overline{X}^{(i)}\right)\right)^2\right]\\
&= \frac{1}{4}\sum_{i=1}^n \mathbf{E}\left[\left(\Delta_j(X) - \Delta_j\left(\overline{X}^{(i)}\right)\right) \cdot \left(\left(f_j(X) - f_j\left(\overline{X}^{(i)}\right)\right) - \left(f_{j-1}(X) - f_{j-1}\left(\overline{X}^{(i)}\right)\right)\right)\right]\\
&= \frac{1}{4}\sum_{i=1}^n \mathbf{E}\left[\left(\Delta_j(X) - \Delta_j\left(\overline{X}^{(i)}\right)\right) \cdot \left(f_j(X) - f_j\left(\overline{X}^{(i)}\right)\right)\right]\\
&= \frac{1}{4}\sum_{i=1}^n \mathbf{E}\left[\left(\left(f_j(X) - f_j\left(\overline{X}^{(i)}\right)\right) - \left(f_{j-1}(X) - f_{j-1}\left(\overline{X}^{(i)}\right)\right)\right) \cdot \left(f_j(X) - f_j\left(\overline{X}^{(i)}\right)\right)\right]\\
&= \mathcal{E}(f_j) - \frac{1}{4}\sum_{i=1}^n \mathbf{E}\left[\left(\left(f_{j-1}(X) - f_{j-1}\left(\overline{X}^{(i)}\right)\right)\right) \cdot \left(f_j(X) - f_j\left(\overline{X}^{(i)}\right)\right)\right]\\
&= \mathcal{E}(f_j) - \mathcal{E}(f_{j-1})
\end{aligned}$$

where in the proof we used twice the fact that

$$\sum_{i=1}^{n} \mathbf{E}\left[\left(\Delta_j(X) - \Delta_j\left(\overline{X}^{(i)}\right)\right) \cdot \left(f_{j-1}(X) - f_{j-1}\left(\overline{X}^{(i)}\right)\right)\right] = 0.$$

Summing the obtained equation we have

$$\sum_{j=1}^{n} \mathcal{E}(\Delta_j) = \sum_{j=1}^{n} \left(\mathcal{E}(f_j) - \mathcal{E}(f_{j-1})\right) = \mathcal{E}(f).$$

This follows from $f_n = f$ and $f_0 = E[f]$. The theorem now follows easily by applying Lemma 9.3 to the absolute value of the martingale differences Δ_j:

$$\begin{aligned}
\mathcal{E}(f) &= \sum_{j=1}^{n} \mathcal{E}(\Delta_j) \\
&\geq \sum_{j=1}^{n} \mathcal{E}\left(|\Delta_j|\right) \\
&\geq \frac{1}{2} \sum_{j=1}^{n} E\left[\Delta_j^2\right] \log \frac{E\left[\Delta_j^2\right]}{\left(E|\Delta_j|\right)^2} \\
&= -\frac{1}{2} \mathrm{Var}(f) \sum_{j=1}^{n} \frac{E\left[\Delta_j^2\right]}{\mathrm{Var}(f)} \log \frac{\left(E|\Delta_j|\right)^2}{E\left[\Delta_j^2\right]} \\
&\geq -\frac{1}{2} \mathrm{Var}(f) \log \frac{\sum_{j=1}^{n} \left(E|\Delta_j|\right)^2}{\mathrm{Var}(f)} \\
&\quad \left(\text{by Jensen's inequality and } \textstyle\sum_j E\left[\Delta_j^2\right] = \mathrm{Var}(f)\right).
\end{aligned}$$

Rearranging, we obtain the stated inequality. □

To understand what Theorem 9.4 has to do with influences, consider a binary-valued function $f : \{-1, 1\}^n \to \{0, 1\}$ and recall from the proof of Theorem 3.1 that

$$\Delta_i = \mathbf{E}_i\left[f(X) - \mathbf{E}^{(i)} f(X)\right]$$

where $\mathbf{E}_i$ and $\mathbf{E}^{(i)}$ denote conditional expectation, conditioned on $X_1, \ldots, X_i$ and $X_1, \ldots, X_{i-1}, X_{i+1}, \ldots, X_n$, respectively. Thus, by Jensen's inequality,

$$\mathbf{E}|\Delta_i| \leq \mathbf{E}\left[\left|f(X) - \mathbf{E}^{(i)} f(X)\right|\right] = \frac{I_i(f)}{2}.$$

Since for binary-valued functions $\mathcal{E}(f) = I(f)/4$, it follows from Theorem 9.4 that

$$\sum_{i=1}^{n} I_i(f)^2 \geq 4\,\mathrm{Var}\,(f)\exp\left(-\frac{I(f)}{2\,\mathrm{Var}\,(f)}\right). \tag{9.1}$$

Recall from the previous section that the total influence of any function is at least a constant, namely $-2P(A)\log_2 P(A)$. This, of course, implies that the largest influence of any variable is at least of the order of $1/n$. Equation (9.1) implies a fundamental improvement of this: for every binary-valued function there exists a variable whose influence is at least of the order of $(\log n)/n$. In particular, the total influence of every symmetric function is at least of the order of $\log n$.

Theorem 9.5 *Let $f : \{-1, 1\}^n \to \{0, 1\}$ be a binary-valued function of n binary variables. Then*

$$\sum_{i=1}^{n} I_i(f)^2 \geq \frac{\mathrm{Var}\,(f)^2 \log^2 n}{n}.$$

In particular,

$$\max_{i=1,\dots,n} I_i(f) \geq \frac{\mathrm{Var}\,(f)\log n}{n}.$$

Proof Let $\varepsilon = (2\log(\mathrm{Var}\,(f)/4) + 4\log\log n)/\log n$. We consider two cases. If $I(f) \geq (2-\varepsilon)\,\mathrm{Var}\,(f)\log n$, then by the Cauchy–Schwarz inequality,

$$\sum_{i=1}^{n} I_i(f)^2 \geq \frac{1}{n}\left(\sum_{i=1}^{n} I_i(f)\right)^2 = \frac{I(f)^2}{n} \geq \frac{(2-\varepsilon)^2\,\mathrm{Var}\,(f)^2\log^2 n}{n}$$

and the stated bound holds since $\varepsilon < 1$. On the other hand, if $I(f) < (2-\varepsilon)\mathrm{Var}\,(f)\log n$, then by (9.1),

$$\sum_{i=1}^{n} I_i(f)^2 \geq 4\,\mathrm{Var}\,(f)\exp\left(-\frac{I(f)}{2\,\mathrm{Var}\,(f)}\right) \geq \frac{\mathrm{Var}\,(f)^2\log^2 n}{n}$$

as desired. □

Theorem 9.5 implies that if f is a symmetric function of its n variables, then the total influence is at least $\mathrm{Var}\,(f)\log n$, which is in sharp contrast with dictatorships and juntas that have a constant total influence. This is an essential improvement over the bound $2P(A)\log_2(1/P(A))$ that we derived from the edge isoperimetric inequality for an arbitrary function. The following example shows that the obtained bound cannot be improved essentially.

Example 9.6 (TRIBES) This example shows that there exist functions of n binary variables whose largest influence is as small as $O(n^{-1}\log n)$. To construct such an example, let $\ell = \lfloor \log_2 n - \log_2 \log_2 n \rfloor$ and assume, for simplicity, that n is an integer multiple of ℓ. Divide the n variables $x_1, \ldots, x_n$ into n/ℓ blocks of length ℓ (the so-called "tribes") and define $f(x) = 1$ if there exists a block such that all variables are equal to 1 in that block and let $f(x) = 0$ otherwise. First note that

$$P\{f(X) = 1\} = 1 - \left(1 - 2^{-\ell}\right)^{n/\ell} \to \frac{1}{e}$$

as $n \to \infty$. The variable x_1 is pivotal if and only if $x_2 = \cdots = x_\ell = 1$ and no other block has all variables equal to 1. The probability of this event is

$$\begin{aligned} I_1(f) &= 2^{-(\ell-1)}\left(1 - 2^{-\ell}\right)^{(n/\ell)-1} \\ &\le 4 \cdot 2^{-\log_2 n + \log_2 \log_2 n} \exp\left(-\left(\frac{n}{\ell 2^\ell} - \frac{1}{2^\ell}\right)\right) \qquad (\text{using } 1 - x \le e^{-x}) \\ &\le \frac{4\log_2 n}{n} e^{-1/2}. \end{aligned}$$

Since all variables of f have the same influence, the total influence is at most $I(f) \le 4e^{-1/2}\log_2 n$ and $\sum_{i=1}^n I_i(f)^2 \le (16e\log_2^2 n)/n$, showing the tightness of Theorem 9.5 up to constant factors.

Interestingly, one may use Theorem 9.4 to derive another fundamental property of influences of a binary-valued function, namely that any function with a small (i.e. constant) total influence must almost be determined by a small number of variables, in the sense that there exists a junta that closely approximates the function. This is made precise in the next theorem.

Theorem 9.7 *Let $f : \{-1,1\}^n \to \{0,1\}$ be a binary-valued function with total influence $I(f)$ and let $\varepsilon \in (0,1)$ be arbitrary. Let $m = \lfloor I(f)/\varepsilon \rfloor$. Then there exists a subset of m variables and a real-valued function $g : \{-1,1\}^n \to \mathbb{R}$ depending on these m variables only such that*

$$\mathbf{E}\left[(f-g)^2\right] \le \frac{I(f)}{\max(1, \log(2/\varepsilon))}.$$

Note that if $I(f)$ is bounded (i.e. does not grow with n) and ε is a constant, the function g is clearly a junta as it depends on a bounded number of variables. The error of approximation may be made arbitrarily small by choosing ε sufficiently small. The construction of g is simple and intuitive: one identifies m variables with largest influence (these are the variables g depends on) and takes averages with respect to all other variables. The key for the proof below is Theorem 9.4.

Proof Without loss of generality we may assume that the variables are ordered by decreasing influences, that is, $I_1(f) \geq \cdots \geq I_n(f)$. Clearly, $I_i(f) < \varepsilon$ for all $i > m$ by the definition of m, and therefore

$$\sum_{i=m+1}^{n} I_i(f)^2 \leq I(f) \max_{i=m+1,\ldots,n} I_i(f) \leq I(f)\varepsilon.$$

Recall the martingale decomposition $f(x) = \sum_{i=1}^{n} \Delta_i(x) = \sum_{i=1}^{n}(f_i(x) - f_{i-1}(x))$ introduced earlier in the proof of Theorem 9.4 and define $g = f_m$. Clearly, g depends on m variables only. In the rest of the proof we show that g approximates f as stated.

Recall from the proof of Theorem 9.4 that $\mathcal{E}(f) = \sum_{i=1}^{n} \mathcal{E}(\Delta_i)$. Applying this identity to $f(x) - g(x) = \sum_{i=m+1}^{n} \Delta_i(x)$, we have

$$\mathcal{E}(f) = \sum_{i=1}^{n} \mathcal{E}(\Delta_i) \geq \sum_{i=m+1}^{n} \mathcal{E}(\Delta_i) \geq \mathcal{E}(f - g).$$

Next we apply Theorem 9.4 for $f - g$ to get

$$\begin{aligned} I(f) &= 4\mathcal{E}(f) \\ &\geq 4\mathcal{E}(f-g) \\ &\geq 2\,\mathrm{Var}(f-g) \log \frac{\mathrm{Var}(f-g)}{\sum_{i=m+1}^{n} (E|\Delta_i|)^2} \\ &\geq 2\,\mathrm{Var}(f-g) \log \frac{4\,\mathrm{Var}(f-g)}{\sum_{i=m+1}^{n} I_i(f)^2} \qquad (\text{since } E|\Delta_i| \leq I_i(f)/2) \\ &\geq 2\,\mathrm{Var}(f-g) \log \frac{4\,\mathrm{Var}(f-g)}{I(f)\varepsilon}. \end{aligned}$$

Rearranging, we have

$$\frac{4\,\mathrm{Var}(f-g)}{I(f)\varepsilon} \log \frac{4\,\mathrm{Var}(f-g)}{I(f)\varepsilon} \leq \frac{2}{\varepsilon}.$$

To solve this inequality for $\mathrm{Var}(f-g)$, note that $x \log x \leq y$ implies $x \leq 2y/\log y$ if $y \geq e$ and $x > 0$. Therefore, when $2/\varepsilon > e$, we have

$$\frac{4\,\mathrm{Var}(f-g)}{I(f)\varepsilon} \leq \frac{4/\varepsilon}{\log(2/\varepsilon)}, \quad \text{that is,} \quad \mathrm{Var}(f-g) \leq \frac{I(f)}{\log(2/\varepsilon)}.$$

To finish the proof note that $E[f-g]=0$ and therefore $E\left[(f-g)^2\right] = \mathrm{Var}(f-g)$. □

The previous theorem guarantees the existence of a real-valued function g that closely approximates, in the L_2 sense, the binary-valued function f. It is now easy to construct a binary-valued junta that also approximates tightly f, see Exercise 9.1.

9.3 Local Concentration

In this section we apply Theorem 9.4 to derive local exponential concentration inequalities for functions defined on the binary hypercube. We use the argument already shown in Section 3.6, the only difference being that the Efron–Stein inequality is replaced by the improved variance inequality of Theorem 9.4. The improved bounds imply local sub-Gaussian tail bounds (as opposed to the sub-exponential estimates obtained in Section 3.6).

Consider a function $f : \{-1, 1\}^n \to \mathbb{R}$ such that there exists a constant $v > 0$ such that for all $x \in \{-1, 1\}^n$,

$$\sum_{i=1}^{n} \left(f(x) - f\left(\overline{x}^{(i)}\right)\right)_+^2 \le v.$$

Recall that the quantiles of f are defined, for any $\alpha \in (0, 1)$, by

$$Q_\alpha = \inf\{z : \mathbf{P}\{f(X) \le z\} \ge \alpha\}.$$

As in Section 3.6, for any $b \ge a \ge \mathbf{M}f = Q_{1/2}$, we introduce the function $g_{a,b} : \mathcal{X}^n \to \mathbb{R}$ by

$$g_{a,b}(x) = \begin{cases} b & \text{if } f(x) \ge b \\ f(x) & \text{if } a < f(x) < b \\ a & \text{if } f(x) \le a \end{cases}$$

and observe that

$$\operatorname{Var}(g_{a,b}) \ge \frac{\mathbf{P}\{g_{a,b}(X) = b\}}{4}(b-a)^2 = \frac{\mathbf{P}\{f(X) \ge b\}}{4}(b-a)^2.$$

Now, instead of the Efron–Stein inequality, we use Theorem 9.4 for the variance of $g_{a,b}$. Recall that this inequality implies

$$\operatorname{Var}(g_{a,b}) \log \frac{\operatorname{Var}(g_{a,b})}{\sum_{j=1}^{n} \left(\mathbf{E}|\Delta_j|\right)^2} \le 2\mathcal{E}(g_{a,b})$$

where $\Delta_i(x) = g^i_{a,b}(x) - g^{i-1}_{a,b}(x)$ and $g^i_{a,b}(X) = \mathbf{E}[g_{a,b}(X)|X_1, \dots, X_i]$. Since the function $x \log x$ is monotone whenever it is positive, the previous two inequalities for $\operatorname{Var}(g_{a,b})$ may be combined to get

$$\frac{\mathbf{P}\{f(X) \ge b\}}{4}(b-a)^2 \log \frac{\mathbf{P}\{f(X) \ge b\}(b-a)^2}{4\sum_{j=1}^{n} \left(\mathbf{E}|\Delta_j|\right)^2} \le 2\mathcal{E}(g_{a,b}). \tag{9.2}$$

Next, we derive suitable upper bounds for the quantities $\sum_{j=1}^{n} \left(E|\Delta_j|\right)^2$ and $\mathcal{E}(g_{a,b})$. First observe that

$$\begin{aligned}
\mathcal{E}(g_{a,b}) &= \frac{1}{4}\sum_{i=1}^{n} \mathbf{E}\left[\left(g_{a,b}(X) - g_{a,b}\left(\overline{X}^{(i)}\right)\right)^2\right] \\
&= \frac{1}{2}\sum_{i=1}^{n} \mathbf{E}\left[\left(g_{a,b}(X) - g_{a,b}\left(\overline{X}^{(i)}\right)\right)_+^2\right] \\
&= \frac{1}{2}\mathbf{E}\left[\mathbb{1}_{\{f(X)>a\}}\sum_{i=1}^{n}\left(g_{a,b}(X) - g_{a,b}\left(\overline{X}^{(i)}\right)\right)_+^2\right] \\
&\leq v\mathbf{P}\{f(X) > a\}/2.
\end{aligned}$$

On the other hand,

$$\begin{aligned}
E|\Delta_j| &\leq \mathbf{E}\left|g_{a,b}(X) - g_{a,b}\left(\overline{X}^{(i)}\right)\right| \\
&= 2\mathbf{E}\left[\left(g_{a,b}(X) - g_{a,b}\left(\overline{X}^{(i)}\right)\right)_+\right] \\
&= 2\mathbf{E}\left[\left(g_{a,b}(X) - g_{a,b}\left(\overline{X}^{(i)}\right)\right)_+ \mathbb{1}_{\{f(X)>a\}}\right] \\
&\quad \text{(by the definition of } g_{a,b}\text{)} \\
&\leq 2\sqrt{\mathbf{E}\left[\left(g_{a,b}(X) - g_{a,b}\left(\overline{X}^{(i)}\right)\right)_+^2\right]}\sqrt{\mathbf{P}\{f(X) > a\}} \\
&\quad \text{(by the Cauchy–Schwarz inequality)} \\
&= \sqrt{2\mathbf{E}\left[\left(g_{a,b}(X) - g_{a,b}\left(\overline{X}^{(i)}\right)\right)^2\right]}\sqrt{\mathbf{P}\{f(X) > a\}}.
\end{aligned}$$

Thus,

$$\sum_{j=1}^{n} \left(E|\Delta_j|\right)^2 \leq 8\mathbf{P}\{f(X) > a\}\mathcal{E}(g_{a,b}) \leq 4v\mathbf{P}\{f(X) > a\}^2.$$

Plugging these estimates into (9.2), we obtain

$$A \log \frac{A}{2\mathbf{P}\{f(X) > a\}} \leq 1$$

where we introduced $A = \mathbf{P}\{f(X) \geq b\}(b-a)^2/(4v\mathbf{P}\{f(X) > a\})$. The meaning of this inequality can be seen in the most transparent manner by taking $a = Q_{1-2^{-k}} \stackrel{\text{def}}{=} a_k$ and $b = Q_{1-2^{-(k+1)}} = a_{k+1}$ for some integer $k \geq 1$. Then $\mathbf{P}\{f(X) > a\} \leq 2^{-k}$, $\mathbf{P}\{f(X) \geq b\} \geq 2^{-(k+1)}$, and the inequality above implies

$$A \log(2^{k-1}A) \leq 1$$

or, equivalently, $y \log y \leq 2^{k-1}$ where $y = 2^{k-1}A$. It is easy to see that this implies $y \leq 2^k/k$, that is, $A \leq 2/k$. Since $A \geq (a_{k+1} - a_k)^2/(8v)$, we have derived the following theorem.

Theorem 9.8 *Let $f : \{-1, 1\}^n \to \mathbb{R}$ satisfy $\sum_{i=1}^n (f(x) - f(\overline{x}^{(i)}))_+^2 \leq v$ and let $a_k = Q_{1-2^{-k}}$. Then for all integers $k \geq 1$,*

$$a_{k+1} - a_k \leq 4\sqrt{\frac{v}{k}}.$$

This is an essential improvement over the bound $4\sqrt{v}$ obtained in Section 3.6 using the Efron–Stein inequality. Note that if $f(X)$ was a normal random variable with variance v, then one would have $a_k \sim \sqrt{2vk \log 2}$ and $a_{k+1} - a_k \sim \sqrt{v \log 2/k}$. The bound of the theorem has the same form, apart from a constant factor. This shows that functions satisfying the conditions of Theorem 9.8 not only have sub-Gaussian tail probabilities (as implied by Theorem 6.7) but the differences between quantiles of the distribution of f are dominated by corresponding differences of a normally distributed random variable. In this sense, Theorem 9.8 may be considered as a "local" concentration inequality.

Recall from earlier chapters that examples of functions satisfying the conditions of Theorem 9.8 include suprema of Rademacher averages, Talagrand's convex distance, the largest eigenvalue of a symmetric random matrix, etc. An important restriction in Theorem 9.8 is that it only holds for functions defined on the binary hypercube (as opposed to more general concentration inequalities as, for example Theorem 6.7).

With similar arguments one may also derive local concentration inequalities for self-bounding functions. We leave the details to the reader (see Exercise 9.5).

9.4 Discrete Fourier Analysis and a Variance Inequality

In the previous sections we saw how Theorem 9.4, an improvement of the Efron–Stein inequality, implies various interesting results about influences of a binary-valued function defined on the binary n-cube. In this section we present a closely related inequality for the variance of a real-valued function defined on the binary n-cube.

The proof of this inequality is based on Fourier analysis on the hypercube $\{-1, 1\}^n$, a technique that has proved powerful in a variety of problems. Discrete Fourier analysis is an elegant and intuitive tool in the study of functions of several binary variables. In this context the Bonami–Beckner hypercontractive inequality (Theorem 5.18) turns out to be a powerful tool.

We start by recalling some basic notions of Fourier analysis on the discrete n-cube $\{-1, 1\}^n$, introduced in Section 5.8.

We treat the set $\mathcal{F}$ of real-valued functions $f : \{-1, 1\}^n \to \mathbb{R}$ as a 2^n-dimensional Euclidean space with inner product

$$\langle f, g \rangle = E[fg] = \mathbf{E}[f(X)g(X)] = 2^{-n} \sum_{x \in \{-1,1\}^n} f(x)g(x), \quad f, g \in \mathcal{F}$$

and corresponding norm $\|f\|_2 = \sqrt{\langle f,f\rangle}$. To any of the 2^n subsets $S \subset \{1,\ldots,n\}$, we assign the function

$$u_S(x) = \prod_{i\in S} x_i.$$

(If $S = \emptyset$, we define $u_S \equiv 1$.) It can be seen immediately that the u_S form an orthonormal basis of $\mathcal{F}$ and therefore every $f \in \mathcal{F}$ may be expressed, in a unique way, as the *Fourier–Walsh expansion*

$$f(x) = \sum_{S\subset\{1,\ldots,n\}} \hat{f}(S)u_S(x)$$

where, for all $S \subset \{1,\ldots,n\}$, $\hat{f}(S) = \langle f, u_S\rangle$. The $\hat{f}(S)$ are called the *Fourier coefficients* of f. Using these definitions, we obtain *Parseval's identity*:

$$\|f\|_2^2 = \left\langle f, \sum_{S\subset\{1,\ldots,n\}} \hat{f}(S)u_S \right\rangle = \sum_{S\subset\{1,\ldots,n\}} \hat{f}(S)\langle f, u_S\rangle = \sum_{S\subset\{1,\ldots,n\}} \hat{f}(S)^2.$$

Since $\hat{f}(\emptyset) = E[f]$,

$$\mathrm{Var}(f) = \|f\|_2^2 - (E[f])^2 = \sum_{S\neq\emptyset} \hat{f}(S)^2.$$

In order to make the connection to influences, introduce the function

$$g_i(x) = \frac{f(x) - f(\overline{x}^{(i)})}{2}, \quad i = 1,\ldots,n$$

and denote the Fourier coefficients of g_i by $\hat{g}_i(S)$, $S \subset \{1,\ldots,n\}$. The key observation is that for every $i = 1,\ldots,n$ and $S \subset \{1,\ldots,n\}$,

$$\hat{g}_i(S) = \langle g_i, u_S\rangle = \frac{1}{2}\mathbf{E}\left[\left(f(X) - f\left(\overline{X}^{(i)}\right)\right)\prod_{j\in S} X_j\right] = \begin{cases} 0 & \text{if } i \notin S \\ \hat{f}(S) & \text{if } i \in S. \end{cases} \tag{9.3}$$

If $f : \{-1,1\}^n \to \{0,1\}$ is binary-valued, then $I_i(f) = E[g_i^2]/4$, and we may apply Parseval's identity to obtain

$$I_i(f) = 4\|g_i\|_2^2 = 4\sum_{S\subset\{1,\ldots,n\}} \hat{g}_i(S)^2 = 4\sum_{S:i\in S} \hat{f}(S)^2$$

and therefore the total influence may be written as

$$I(f) = 4 \sum_{S \subset \{1,\ldots,n\}} |S| \hat{f}(S)^2.$$

Equation (9.3) also implies that

$$\operatorname{Var}(f) = \sum_{S \neq \emptyset} \hat{f}(S)^2 = \sum_{S \neq \emptyset} \sum_{i=1}^{n} \frac{\hat{g}_i(S)^2}{|S|}.$$

Note that the last two identities immediately imply $\operatorname{Var}(f) \leq I(f)/4$, a special case of the Efron–Stein inequality.

The main result of this section is the following inequality for the variance.

Theorem 9.9 *Let $f : \{-1,1\}^n \to \mathbb{R}$ be a real-valued function. Then*

$$\operatorname{Var}(f) \leq C \sum_{i=1}^{n} \frac{\mathbf{E}\left[\left(f(X) - f\left(\overline{X}^{(i)}\right)\right)^2\right]}{1 + \log \frac{\sqrt{\mathbf{E}\left[\left(f(X) - f\left(\overline{X}^{(i)}\right)\right)^2\right]}}{\mathbf{E}\left|f(X) - f\left(\overline{X}^{(i)}\right)\right|}}$$

where $C \leq 3(6 \cdot e^{1/3} + 1)(\log 2)/8 \approx 3.297589$ is a universal constant.

The proof of Theorem 9.9 requires one more tool, namely the Bonami–Beckner inequality (Corollary 5.16) which we now recall. For every $f \in \mathcal{F}$ and for any $q \geq 2$ and $k = 1, \ldots, n$,

$$\left\| \sum_{S:|S|=k} \hat{f}(S) u_S \right\|_q \leq (q-1)^{k/2} \left\| \sum_{S:|S|=k} \hat{f}(S) u_S \right\|_2,$$

where $\|f\|_p$ is defined as $\left(E[f^p]\right)^{1/p}$ for any $p > 0$.

Proof of Theorem 9.9. Recalling the formula for the variance

$$\operatorname{Var}(f) = \sum_{S \neq \emptyset} \sum_{i=1}^{n} \frac{\hat{g}_i(S)^2}{|S|},$$

we see that in order to prove the theorem, it suffices to show that for any $f : \{-1,1\}^n \to \mathbb{R}$,

$$\sum_{S \neq \emptyset} \frac{\hat{f}(S)^2}{|S|} \leq 4C \frac{\|f\|_2^2}{1 + \log \frac{\|f\|_2}{\|f\|_1}},$$

which is what we do in the remaining part of the proof. Fix $k \leq n$ and observe that

$$\sum_{S:|S|=k} \hat{f}(S)^2 = \left\langle \sum_{S:|S|=k} \hat{f}(S) u_S, f \right\rangle$$

$$\leq \left\| \sum_{S:|S|=k} \hat{f}(S) u_S \right\|_3 \cdot \|f\|_{3/2} \quad \text{(by Hölder's inequality)}$$

$$\leq 2^{k/2} \left(\sum_{S:|S|=k} \hat{f}(S)^2 \right)^{1/2} \cdot \|f\|_{3/2}$$

(by the Bonami–Beckner inequality, used with $q = 3$).

This implies that, for all $k = 1, \ldots, n$,

$$\sum_{S:|S|=k} \hat{f}(S)^2 \leq 2^k \|f\|_{3/2}^2$$

and we have, for all positive integers m,

$$\sum_{S:1\leq|S|\leq m} \frac{\hat{f}(S)^2}{|S|} \leq \|f\|_{3/2}^2 \sum_{k=1}^{m} \frac{2^k}{k} \leq 3\frac{2^m}{m} \|f\|_{3/2}^2.$$

At the last step we used the fact that for $k \geq 3$, $2^{k+1}/(k+1) \geq (3/2)2^k/k$. Now we may write

$$\sum_{S\neq\emptyset} \frac{\hat{f}(S)^2}{|S|} = \sum_{S:1\leq|S|\leq m} \frac{\hat{f}(S)^2}{|S|} + \sum_{S:|S|>m} \frac{\hat{f}(S)^2}{|S|}$$

$$\leq 3\frac{2^m}{m} \|f\|_{3/2}^2 + \frac{1}{m+1} \sum_{S:|S|>m} \hat{f}(S)^2$$

$$\leq \frac{1}{m+1} \left(6 \cdot 2^m \|f\|_{3/2}^2 + \|f\|_2^2\right).$$

Now we choose m as the largest integer such that $2^m \|f\|_{3/2}^2 \leq e^{2/3} \|f\|_2^2$ so that

$$m + 1 \geq \frac{2}{\log 2} \log\left(e^{1/3} \|f\|_2 / \|f\|_{3/2}\right)$$

and

$$\sum_{S\neq\emptyset} \frac{\hat{f}(S)^2}{|S|} \leq \frac{(6 \cdot e^{2/3}) + 1)}{m+1} \|f\|_2^2 \leq 4C \cdot \frac{\|f\|_2^2}{\log\left(e^{1/3} \|f\|_2 / \|f\|_{3/2}\right)},$$

where $C = (6 \cdot e^{2/3}) + 1)(\log 2)/8$. The proof is concluded by observing that, by the Cauchy–Schwarz inequality,

$$E\left[|f|^{3/2}\right] \leq \|f\|_1^{1/2} \cdot \|f\|_2$$

and therefore

$$\frac{\|f\|_2}{\|f\|_1} \leq \left(\frac{\|f\|_2}{\|f\|_{3/2}}\right)^3 . \qquad \square$$

Remark 9.4 The constant C in Theorem 9.9 is not optimal and can easily be improved by a more careful analysis. In Exercise 9.3 we sketch a different proof yielding the improved constant $C = 9/10$. By considering $f(x) = \sum_{i=1}^{n} x_i$, we see that the best possible value of C is at least $1/4$.

9.5 Monotone Sets

Monotone subsets of the binary n-cube have a central importance in the study of influences for various reasons. First, their special form makes them crucial in understanding influences of general sets. Second, monotone sets appear naturally in the study of threshold phenomena and social choice theory, some of the most important applications of the theory of influences (see Section 9.6 below).

Recall that a function $f : \{-1, 1\}^n \to \{0, 1\}$ is termed monotone if it is non-decreasing in all of its components, that is, $f(x_1, \ldots, x_{i-1}, -1, x_{i+1}, \ldots, x_n) \leq f(x_1, \ldots, x_{i-1}, 1, x_{i+1}, \ldots, x_n)$ for all $x = (x_1, \ldots, x_n) \in \{-1, 1\}^n$ and $i \in \{1, \ldots, n\}$.

To present the ideas in the simplest possible setting, we still assume the uniform distribution over $\{-1, 1\}^n$, that is, in this section $X = (X_1, \ldots, X_n)$ is a vector of independent symmetric sign variables. However, most results extend, in a straightforward way, to the case when the components of X are i.i.d. with $\mathbf{P}\{X_i = 1\} = 1 - \mathbf{P}\{X_i = -1\} = p$ and with p possibly different from $1/2$ (see the exercises).

We begin by proving that monotone functions minimize the total influence.

Theorem 9.10 *For any function $f : \{-1, 1\}^n \to \{0, 1\}$ there exists a monotone function $g : \{-1, 1\}^n \to \{0, 1\}$ such that $E[g] = E[f]$ and $I(g) \leq I(f)$.*

Proof The proof is based on a simple "shifting" technique. By a sequence of transformations we replace $A = \{x : f(x) = 1\}$ by a monotone set of the same size as A with total influence not exceeding that of A. If A is not monotone, then there exists a variable $i \in \{1, \ldots, n\}$ such that for some x, $(x_1, \ldots, x_{i-1}, -1, x_{i+1}, \ldots, x_n) \in A$ and $(x_1, \ldots, x_{i-1}, 1, x_{i+1}, \ldots, x_n) \notin A$. Fix such a variable i and define the set $A^{(i)}$ by switching all such pairs of points (see Fig. 9.1), that is,

$$x \in A^{(i)} \quad \text{if and only if} \quad \begin{cases} \text{either } x \in A & \text{and } x_i = 1 \\ \text{or } x \in A & \text{and } \overline{x}^{(i)} \in A \\ \text{or } x \notin A & \text{and } \overline{x}^{(i)} \in A \text{ and } x_i = 1. \end{cases}$$

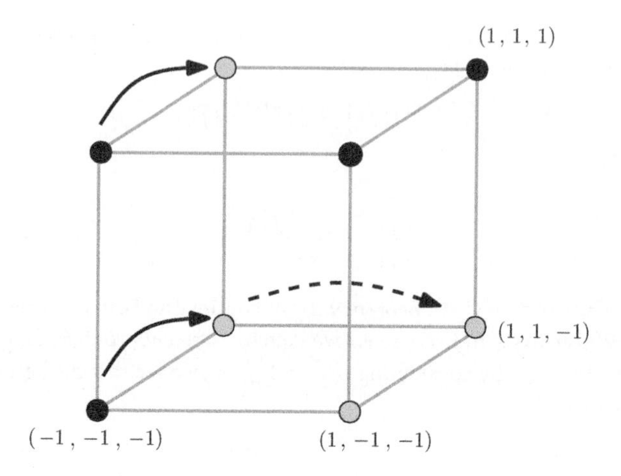

Figure 9.1 Shifting the non-monotone set A along the second variable to obtain $A^{(2)}$. In the next step $A^{(2)}$ is shifted along the first variable to obtain a monotone set of same size and decreased total influence

Clearly, $P(A^{(i)}) = P(A)$ and it is easy to see that $I(A^{(i)}) \leq I(A)$. If $A^{(i)}$ is not monotone, this transformation can be repeated (with a variable different from i). If, to each set A, we assign the "progress measure" $\phi(A) = \sum_{x\in A} \|x\|$ then we see that at each transformation step, the value of ϕ strictly increases by at least 1 and therefore the transformation process terminates after a finite number of steps. (Recall the notation $\|x\| = \sum_{i=1}^{n} \mathbb{1}_{\{x_i=1\}}$ for all $x = (x_1, \ldots, x_n) \in \{-1, 1\}^n$.) The obtained set must be monotone, has the same cardinality as A, and has a total influence not larger than $I(A)$. □

The next result shows that among all monotone functions, simple majority maximizes the total influence. Equivalently, Hamming balls centered at the vector $(1, 1, \ldots, 1)$ have a *maximal* edge boundary among all monotone sets. This is interesting in view of Harper's theorem (Theorem 7.6) which states that Hamming balls *minimize* the vertex boundary. Recall from Section 4.4 that without the restriction of monotonicity, the sub-cubes of $\{-1, 1\}^n$ minimize the edge-boundary (i.e. total influence).

Theorem 9.11 *Let $B = \{x : \|x\| > n/2\}$ be the Hamming ball of radius $n/2$ centered at the all-1 vector. Then for any monotone set $A \subset \{-1, 1\}^n$, $I(A) \leq I(B)$. If A is a monotone set with cardinality $|A| = \sum_{i=0}^{k} \binom{n}{i}$ for some $k \in \{0, 1, \ldots, n\}$ then $I(A) \leq I(B_k)$ where $B_k = \{x : \|x\| > n - k - 1\}$.*

Proof For binary vectors $x, y \in \{-1, 1\}^n$, we write $x \prec y$ if $x_i \leq y_i$ for all $i = 1, \ldots, n$ and $\|y\| = \|x\| + 1$. Using the monotonicity of A, we may write

$$I(A) = \mathbf{E} \sum_{i=1}^{n} \mathbb{1}_{\{X_i \text{ is pivotal}\}}$$

$$= 2^{-n} \sum_{x\in\{-1,1\}^n} \sum_{y:x\prec y} \left(\mathbb{1}_{\{y\in A\}} - \mathbb{1}_{\{x\in A\}}\right).$$

Observe that

$$\sum_{x\in\{-1,1\}^n} \sum_{y:x\prec y} \mathbb{1}_{\{y\in A\}} = \sum_{y\in A} \|y\|,$$

since every $y \in A$ is counted $\|y\|$ times in the double sum on the left-hand side. On the other hand,

$$\sum_{y:x\prec y} \mathbb{1}_{\{x\in A\}} = (n - \|x\|)\mathbb{1}_{\{x\in A\}},$$

and therefore

$$I(A) = 2^{-n} \sum_{x\in A} (2\|x\| - n).$$

This expression is clearly maximized if $A = \{x : \|x\| > n/2\}$. The second statement follows similarly. □

Note that if n is even, the "closed" Hamming ball $\{x : \|x\| \geq n/2\}$ has the same total influence as $B = \{x : \|x\| > n/2\}$ and therefore both sets have maximal influence among all monotone sets. For odd n, the set B is the unique maximizer. Now it follows immediately that for symmetric monotone functions all individual influences must go to zero at a rate of $O(n^{-1/2})$. In particular, for monotone symmetric functions, all individual influences converge to zero. More precisely, we have the following.

Corollary 9.12 *If A is a monotone set such that all individual influences $I_i(A)$ are equal then*

$$I_i(A) \leq I_i(B) = \binom{n-1}{\lfloor (n-1)/2 \rfloor} 2^{-(n-1)} \sim \sqrt{\frac{2}{n\pi}}.$$

For monotone sets one also has

$$\sum_{i=1}^{n} I_i(A)^2 \leq 4P(A)(1 - P(A)).$$

To see this, observe that monotonicity of A implies that for $f(x) = \mathbb{1}_{\{x\in A\}}$, the influence of the i-th variable equals twice the Fourier coefficient corresponding to the singleton $\{i\}$, that is, $I_i(f) = 2\hat{f}(\{i\})$ (see Section 9.4 for the definitions). Since $\mathrm{Var}(f) = \sum_{S\neq\emptyset} \hat{f}(S)^2$, we immediately have $\sum_{i=1}^{n} I_i(A)^2 \leq 4\,\mathrm{Var}(f) = 4P(A)(1 - P((A))$.

Equality is achieved, for example, if A is a dictatorship of the form $A = \{x : x_i = 1\}$. On the other hand, for the simple majority function $\sum_{i=1}^{n} I_i(A)^2 \sim (2/\pi)^2$ is also bounded away from zero.

9.6 Threshold Phenomena

One of the most beautiful applications of the theory of influences is in the study of phase transitions and threshold phenomena. In this section we give a brief overview of some of the basic results in this fascinating area.

Consider a monotone binary-valued function defined on the binary cube: $f : \{-1,1\}^n \to \{0,1\}$. In contrast to earlier sections in this chapter, now $\{-1,1\}^n$ is equipped with the product of Bernoulli(p) measures. In other words, the distribution of the random binary vector $X = (X_1, \dots, X_n)$ is such that the components X_i are independent with distribution $\mathbf{P}\{X_i = 1\} = 1 - \mathbf{P}\{X_i = -1\} = p$ for all $i = 1, \dots, n$, where $p \in [0,1]$. We denote the measure induced by X on $\{-1,1\}^n$ by P_p so that the notation makes explicit the dependence on the parameter p. We denote $A = \{x : f(x) = 1\}$. Since f is monotone, A is a monotone set. The main object of our study is the evolution of

$$P_p(A) \stackrel{\text{def}}{=} \mathbf{P}\{X \in A\} = \sum_{x \in A} p^{\|x\|}(1-p)^{n-\|x\|}$$

as p varies in $[0,1]$. (Recall that $\|x\| = \sum_{i=1}^n \mathbb{1}_{\{x_i=1\}}$.) If $A \neq \emptyset$ and $A \neq \{-1,1\}^n$, then monotonicity of A implies that $P_0(A) = 0$, $P_1(A) = 1$, and $P_p(A)$ is a strictly increasing differentiable function of p in $[0,1]$. The unique value $p_{1/2}$ for which $P_{p_{1/2}}(A) = 1/2$ is called the *critical* value of the parameter p.

The main message of this section is that if the function f does not depend too much on any of its variables then there is a sharp transition around $p_{1/2}$. In a narrow interval the value of $P_p(A)$ increases from near-zero values to near one.

To fix ideas, let $\varepsilon \in (0,1)$ and define p_ε such that $P_{p_\varepsilon}(A) = \varepsilon$. If $\varepsilon < 1/2$ is small, the difference $p_{1-\varepsilon} - p_\varepsilon$ indicates how quickly the probability of A grows close to the critical probability. If this difference is small, then a "phase transition" occurs around the critical value $p_{1/2}$ (see Fig. 9.2).

To gain some insight, consider first the simple examples of a dictatorship and simple majority.

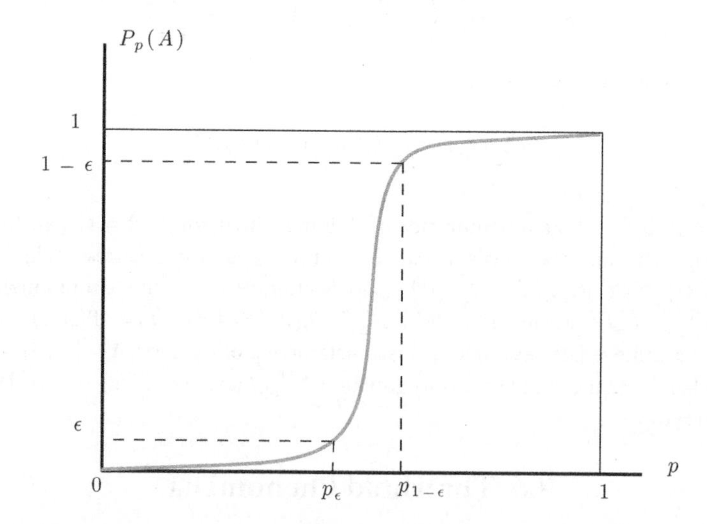

Figure 9.2 $p_{1-\varepsilon} - p_\varepsilon$ is the "threshold width", the length of the interval in which the probability of a monotone set A grows from ε to $1-\varepsilon$

Example 9.13 (DICTATORSHIP) Suppose that $f(x) = (x_1 + 1)/2$ is a monotone dictatorship function, that is, f is determined by just one of the n variables. Then clearly $P_p(A) = p$ and $p_{1-\varepsilon} - p_\varepsilon = 1 - 2\varepsilon$. This means that for the transition from small values of $P_p(A)$ to large ones, one needs to drastically change the value of p. In other words, no phase transition occurs in this example. We will see soon that this property is shared by any function with a small total influence.

Example 9.14 (SIMPLE MAJORITY: CONDORCET'S JURY THEOREM) One observes a qualitatively different behavior by considering the example of a simple majority function defined by $A = \{x : \sum_{i=1}^{n} x_i > 0\}$. For simplicity, assume that the number n of variables is odd. Then $p_{1/2} = 1/2$. To estimate the length of the threshold interval $(p_\varepsilon, 1 - p_\varepsilon)$, note that by Hoeffding's inequality, if $p < 1/2$,

$$
\begin{aligned}
P_p(A) &= P_p\left(\sum_{i=1}^{n} x_i > 0\right) \\
&= P_p\left(\sum_{i=1}^{n} x_i - (2p-1)n > (1-2p)n\right) \leq e^{-2n(1-2p)^2},
\end{aligned}
$$

and therefore $P_p(A) \leq \varepsilon$ whenever $p \leq 1/2 - \sqrt{\log(1/\varepsilon)/8n}$. By a symmetric argument $P_p(A) \geq 1 - \varepsilon$ for all $p \geq 1/2 + \sqrt{\log(1/\varepsilon)/8n}$, so the value of $P_p(A)$ jumps from ε to $1 - \varepsilon$ in an interval of length not more than $\sqrt{\log(1/\varepsilon)/2n}$. In other words, if the number of variables is large, one witnesses a sharp threshold around the critical parameter value $p = 1/2$. What we have just derived is a quantitative version of a classical result of social choice theory, known as *Condorcet's jury theorem.*

As shown in the sequel, the threshold phenomenon exhibited by the last example extends to a wide class of monotone functions. Apart from monotonicity, the only required property for such phase transitions is that the function should not depend too much on each variable. In other words, if all individual influences are small, there is a quick transition from very small to very large probabilities. The key tool for making the connection to the world of influences is a simple result known as Russo's lemma. Russo's lemma, stated and proved below, asserts that the derivative of the measure of A, with respect to the parameter p, is just the total influence. Recall the definition of the influence of the i-th variable:

$$
I_i^p(A) = I_i^p(f) = P_p\left(\left\{x : f(x) \neq f\left(\overline{x}^{(i)}\right)\right\}\right).
$$

The total influence is just $I^p(A) = \sum_{i=1}^{n} I_i^p(A)$. Note that we make the dependence on p explicit in the notation.

Theorem 9.15 (RUSSO'S LEMMA) *Let A be a monotone subset of $\{-1, 1\}^n$. Then for any $p \in (0, 1)$,*

$$
\frac{dP_p(A)}{dp} = I^p(A).
$$

Proof Let $\mathbf{p} = (p_1, \ldots, p_n)$ be a vector of n components $p_i \in (0,1)$ and define the probability measure $Q_{\mathbf{p}}$ on $\{-1,1\}^n$ as the product of n independent Bernoulli measures with parameters p_i. Thus, $P_p = Q_{\mathbf{p}}$ with $\mathbf{p} = (p, p, \ldots, p)$. Let $U_1, \ldots, U_n$ be independent uniformly distributed random variables in $[0,1]$. If we define $X_i = 2\mathbb{1}_{\{U_i \le p\}} - 1$ for $i = 1, \ldots, n$, then the joint distribution of $X = (X_1, \ldots, X_n)$ is just P_p. Let $p' \in (0,1)$ and for some fixed i, define $X_i' = \mathbb{1}_{\{U_i \le p'\}}$. Let $\hat{X}_i = (X_1, \ldots, X_{i-1}, X_i', X_{i+1}, \ldots, X_n)$ be obtained by replacing the i-th component of X by X_i' and keeping all other variables fixed. If we denote by $\mathbf{p}_i' = (p, \ldots, p, p', p, \ldots, p)$ the vector whose i-th component equals p' while all others are p, then the distribution of $\hat{X}_i$ is just $Q_{\mathbf{p}_i'}$.

Assume first that $p' \ge p$. Then by the monotone property of A,

$$\begin{aligned} Q_{\mathbf{p}_i'}(A) - Q_{\mathbf{p}}(A) &= \mathbf{P}\{\hat{X}_i \in A, X \notin A\} \\ &= \mathbf{P}\{U_i \in (p, p'] \text{ and } X_i \text{ is pivotal for } A\} \\ &= (p' - p)\mathbf{P}\{X_i \text{ is pivotal for } A\} \\ &= (p' - p) I_i^p(A). \end{aligned}$$

A similar argument shows that if $p' < p$, one also has $Q_{\mathbf{p}_i'}(A) - Q_{\mathbf{p}}(A) = (p' - p) I_i^p(A)$. By dividing both sides by $p' - p$ and letting $p' \to p$, we get

$$\frac{\partial Q_{\mathbf{p}}(A)}{\partial p_i} = I_i^p(A).$$

Russo's lemma now follows from a simple application of the chain rule:

$$\frac{dP_p(A)}{dp} = \sum_{i=1}^{n} \frac{\partial Q_{\mathbf{p}}(A)}{\partial p_i} = I^p(A). \qquad \square$$

As an immediate consequence, we obtain the following generalization of Theorem 9.11.

Corollary 9.16 *Let A be a monotone subset of $\{-1,1\}^n$. Then*

$$I^p(A) = \frac{1}{p(1-p)} \mathbf{E}\left[(\|X\| - np)\mathbb{1}_{\{X \in A\}}\right].$$

In particular,

$$I^p(A) \le \sqrt{\frac{nP_p(A)}{p(1-p)}}.$$

Proof By Russo's lemma,

$$\begin{aligned} I^p(A) &= \frac{dP_p(A)}{dp} \\ &= \frac{d}{dp} \sum_{x \in \{-1,1\}^n} p^{\|x\|}(1-p)^{n-\|x\|} \mathbb{1}_{\{x \in A\}} \end{aligned}$$

$$= \sum_{x \in \{-1,1\}^n} \left(\frac{\|x\|}{p} - \frac{n - \|x\|}{1-p} \right) p^{\|x\|}(1-p)^{n-\|x\|} \mathbb{1}_{\{x \in A\}}$$

$$= \frac{1}{p(1-p)} \sum_{x \in \{-1,1\}^n} (\|x\| - np) p^{\|x\|}(1-p)^{n-\|x\|} \mathbb{1}_{\{x \in A\}}$$

$$= \frac{1}{p(1-p)} \mathbf{E}\left[(\|X\| - np) \mathbb{1}_{\{X \in A\}} \right].$$

The second statement follows from the Cauchy–Schwarz inequality by noting that the distribution of $\|X\|$ is binomial with parameters n and p. □

Russo's lemma provides a convenient tool for studying the speed of growth of $P_p(A)$. It shows that $P_p(A)$ grows rapidly whenever the total influence is large. Thus, the inequalities for influences established in Section 9.2 carry important information about the length of the "threshold" interval in which $P_p(A)$ changes from, say, a small value ε to $1 - \varepsilon$. In order to make these inequalities useful, they need to be extended to the case when X is distributed according to P_p. However, this is immediate if one replaces the logarithmic Sobolev inequality used in the proof of Theorem 9.4 by an appropriate version for the measure P_p. For example, the generalization of Theorem 9.5 to non-symmetric distributions implies that if the set A is symmetric in the sense that all individual influences are equal, then the total influence (and hence the derivative of $P_p(A)$) is at least of the order of $\log n$ whenever $P_p(A)$ is neither too small nor too large (see Exercise 9.8 for the details). We also derive this result as a corollary of a more general principle (see Theorem 9.17 and Corollary 9.18).

On the other hand, Theorem 9.7 may also be easily generalized to the measure P_p. This theorem implies that if A is such that the total influence is small, then A is close to being determined by a small number of variables. Translated to the language of threshold phenomena, this means that if A has a coarse threshold (i.e. if $P_p(A)$ takes a long time to grow from small values to large), A must be almost a junta, provided that the critical probability is neither too close to 0 nor to 1 (see Exercise 9.9 for details).

In order to derive the main result of this section, we need to generalize Theorem 9.4 to the case when $X = (X_1, \ldots, X_n)$ is distributed according to P_p for values of p different from 1/2. This is immediate if we apply an appropriate logarithmic Sobolev inequality for the measure P_p, generalizing Theorem 5.1. Such an inequality is, of course, available. Simply recall that Theorem 5.2 states that for any real-valued function $f : \{-1, 1\}^n \to \mathbb{R}$,

$$\mathrm{Ent}(f^2) \leq c(p)\mathcal{E}(f)$$

where, denoting by E_p expectation with respect to the measure P_p, $\mathrm{Ent}(f^2) = E_p[f^2 \log(f^2)] - E_p[f^2] \log E_p[f^2]$,

$$\mathcal{E}(f) = p(1-p)E_p \left[\sum_{i=1}^{n} \left(f(x) - f(\overline{x}^{(i)}) \right)^2 \right],$$

and

$$c(p) = \frac{1}{1-2p} \log \frac{1-p}{p}.$$

With the help of this inequality, the proof of Theorem 9.4 may be repeated to obtain that for any $f : \{-1, 1\}^n \to \mathbb{R}$,

$$\operatorname{Var}(f) \log \frac{\operatorname{Var}(f)}{\sum_{j=1}^n \left(E_p|\Delta_j|\right)^2} \le c(p)\mathcal{E}(f).$$

Specializing to binary-valued functions $f : \{-1, 1\}^n \to \{0, 1\}$, and using the facts that $\mathcal{E}(f) = p(1-p)I^p(f)$ and

$$E_p|\Delta_i| \le 2p(1-p)I_i^p(f),$$

we obtain

$$\begin{aligned} c(p)p(1-p)I^p(f) &\ge \operatorname{Var}(f) \log \frac{\operatorname{Var}(f)}{(2p(1-p))^2 \sum_{i=1}^n \left(I_i^p(f)\right)^2} \\ &\ge \operatorname{Var}(f) \log \frac{\operatorname{Var}(f)}{(2p(1-p))^2 I^p(f)\delta_p} \end{aligned}$$

where $\delta_p = \max_{i=1,\dots,n} I_i^p(f)$ denotes the maximal influence. Introducing the notation

$$A = \frac{\operatorname{Var}(f)}{(2p(1-p))^2 I^p(f)\delta_p} \quad \text{and} \quad B = \frac{c(p)}{4p(1-p)\delta_p},$$

the above inequality may be written as $A \log A \le B$, which implies $A \le 2B/\log B$, that is,

$$I^p(f) \ge \frac{\operatorname{Var}(f) \log \frac{c(p)}{4p(1-p)\delta_p}}{2c(p)p(1-p)}.$$

In combination with Russo's lemma, we have obtained the following.

Theorem 9.17 *For any monotone set $A \subset \{-1, 1\}^n$, we have*

$$\frac{dP_p(A)}{dp} \ge \frac{P_p(A)(1-P_p(A))}{c(p)p(1-p)} \log \frac{c(p)}{4p(1-p)\delta_p}$$

where $\delta_p = \max_{i=1,\dots,n} I_i^p(A)$ and $c(p) = \frac{1}{1-2p} \log \frac{1-p}{p}$.

The theorem shows that if each variable has a small influence (i.e. if δ_p is small) then the derivative of $P_p(A)$ is large whenever $P_p(A)(1 - P_p(A))$ is large, that is, when $P_p(A)$ is close to $1/2$. This means that $P_p(A)$ grows rapidly in the vicinity of the critical parameter $p_{1/2}$, resulting in a quick transition from very small to very large values of the probability $P_p(A)$ of the monotone set A. We know from Corollary 9.16 that $I^p(A) \le \sqrt{\frac{nP_p(A)}{p(1-p)}}$ and therefore, if A is symmetric in the sense that all variables have the same influence, then $\delta_p \le \sqrt{\frac{P_p(A)}{np(1-p)}}$. We may use this estimate to derive the following quantitative result bounding the length of the "threshold" interval in which the $P_p(A)$ grows from ε to $1 - \varepsilon$.

Corollary 9.18 *Let $A \subset \{-1, 1\}^n$ be a monotone set such that for all $p \in (0, 1)$, all variables have the same influence, that is, $I_1^p(A) = \cdots = I_n^p(A)$. Then for any $\varepsilon \in (0, 1/2)$,*

$$p_{1-\varepsilon} - p_\varepsilon \le \frac{8 \log \frac{1}{2\varepsilon}}{\log \frac{n}{16}}.$$

The proof below handles some constants relatively generously, and it is not difficult to improve the constants in the corollary (see, e.g., Exercise 9.8 where a direct simple proof of Corollary 9.18 is suggested). The main message of this result is that regardless of what the monotone set is, if all variables have equal influence, then there is a sharp phase transition within an interval of length $O(1/\log n)$ around the critical value of p. This may be regarded as a powerful generalization of Condorcet's jury theorem. Of course, this statement is most interesting if the critical value $p_{1/2}$ is not too close to either 0 or 1. Indeed, it is common to define the monotone set A as having a *sharp threshold* if for all $\varepsilon \in (0, 1)$,

$$\frac{p_{1-\varepsilon} - p_\varepsilon}{\min(p_{1/2}, 1 - p_{1/2})} = o(1).$$

Corollary 9.18 states that A experiences a sharp threshold whenever $\min(p_{1/2}, 1 - p_{1/2}) \log n \to \infty$. Determining the location of the critical value is often a very challenging problem whose solution requires problem-specific tools. The study of such techniques goes beyond the scope of this book.

Proof As mentioned above, the symmetry assumption and Corollary 9.16 imply $\delta_p \le \sqrt{\frac{P_p(A)}{np(1-p)}} \le \sqrt{\frac{1}{np(1-p)}}$. Plugging this estimate into the lower bound of Theorem 9.17, we have

$$\begin{aligned}
\frac{dP_p(A)}{dp} &\ge \frac{P_p(A)(1-P_p(A))}{2c(p)p(1-p)} \log \frac{c(p)}{4p(1-p)\sqrt{\frac{1}{np(1-p)}}} \\
&= \frac{P_p(A)(1-P_p(A))}{2c(p)p(1-p)} \log\left(\sqrt{\frac{n}{16}} \cdot \frac{\log \frac{1-p}{p}}{\sqrt{p(1-p)}(1-2p)}\right) \\
&\quad \left(\text{using } c(p) = \tfrac{1}{1-2p} \log \tfrac{1-p}{p}\right) \\
&= \frac{P_p(A)(1-P_p(A))}{4c(p)p(1-p)} \log \frac{n}{16} \\
&\quad + \frac{P_p(A)(1-P_p(A))}{2} \cdot \frac{1-2p}{p(1-p)\log \frac{1-p}{p}} \log \frac{\log \frac{1-p}{p}}{\sqrt{p(1-p)}(1-2p)} \\
&\ge \frac{P_p(A)(1-P_p(A))}{2} \log \frac{n}{16}
\end{aligned}$$

where at the last step we use the fact that $c(p)p(1-p) \leq 1/2$ and that, since $p(1-p) \leq 1$,

$$\frac{1-2p}{p(1-p)\log\frac{1-p}{p}}\log\frac{\log\frac{1-p}{p}}{\sqrt{p(1-p)}(1-2p)} \geq \frac{1-2p}{\log\frac{1-p}{p}}\log\frac{\log\frac{1-p}{p}}{1-2p} \geq 0$$

simply because $x\log(1/x) \geq 0$ for all $x \in [0,1]$ and $(1-2p)/\log((1-p)/p) \in [0,1/2]$.

Therefore, for any $p \leq p_{1/2}$, since $1 - P_p(A) \geq 1/2$, we have

$$\frac{dP_p(A)}{dp} \geq \frac{P_p(A)}{4}\log\frac{n}{16}$$

or, equivalently,

$$\frac{d(\log P_p(A))}{dp} \geq \frac{1}{4}\log\frac{n}{16}.$$

Using this estimate in the interval $[p_\varepsilon, p_{1/2}]$, we obtain

$$\log\frac{1}{2} - \log\varepsilon \leq (p_{1/2} - p_\varepsilon)\frac{1}{4}\log\frac{n}{16},$$

that is, $p_{1/2} - p_\varepsilon \geq 4\log(1/(2\varepsilon))/\log(n/16)$. Since the same upper bound holds for $p_{1-\varepsilon} - p_{1/2}$, the proof is complete. □

9.7 Bibliographical Remarks

The study of influences was initiated by Ben-Or and Linial (1990) and was made popular by the influential paper of Kahn, Kalai, and Linial (1988). Since then it has become a rich area of research of which we merely offer some highlights. For an excellent survey of influences, threshold phenomena, and many related topics, we refer to Kalai and Safra (2006).

Our treatment of the Kahn–Kalai–Linial theorem (Theorem 9.5) is based on the elegant proofs of Falik and Samorodnitsky (2007) and Rossignol (2006). In particular, Lemma 9.3 and Theorem 9.4 appear in Falik and Samorodnitsky (2007). Rossignol (2006) uses essentially the same arguments. Theorem 9.5 was first proved by Kahn, Kalai, and Linial (1988). The "tribes" example (Example 9.6) was constructed by Ben-Or and Linial (1990). Theorem 9.7 is due to Friedgut (1998).

Theorem 9.9 was first proved by Talagrand (1994*a*). The original proofs of Theorems 9.5 and 9.7 both use the Fourier analysis techniques shown in Section 9.4.

Several attempts have been made to obtain analogs of Theorems 9.4 and 9.9 beyond the binary hypercube with a product measure. Indeed, O'Donnell and Wimmer (2009), Keller, Mossel, and Sen (2012*a*, 2012*b*), and Cordero-Erausquin and Ledoux (2011) obtain extensions in various directions.

Benjamini, Kalai, and Schramm (2003) apply Talagrand's inequality (Theorem 9.9) to prove a sub-linear bound for the variance of first passage percolation, improving the argument of Example 3.13. Benaïm and Rossignol (2006) use Theorem 9.4 to derive exponential concentration inequalities for first passage percolation.

The material of Section 9.3 comes from Devroye and Lugosi (2008). They extend Theorem 9.8 to real-valued functions defined on the r-ary cube $\{0, 1, \ldots, r-1\}^n$ for integers $r > 2$.

The proof of Theorem 9.11 appears in Friedgut and Kalai (1996).

For threshold phenomena, which have been studied extensively in the context of random graphs, see Erdős and Rényi (1960), Bollobás (2001), and Janson, Łuczak, and Ruciński (2000). The first general results concerning threshold phenomena are due to Margulis (1974), Russo (1982), and Bollobás and Thomason (1987). Russo's lemma appears in Margulis (1974) and Russo (1982), but see also Grimmett (1989). Corollary 9.18 is from Talagrand (1994*a*). The best known constants were proved by Rossignol (2006).

The results presented here only give satisfactory conditions for the existence of sharp thresholds if the critical value is bounded away from zero and one. Also, we only treat product distributions on the binary hypercube while more general product, and even some non-product, spaces are of great interest. Indeed, generalizations in these directions have been the focus of intensive research. A small sample of the literature that the interested reader may consult includes Bollobás and Riordan (2006*b*, 2006*a*), Bourgain and Kalai (1997), Bourgain *et al.* (1992), Friedgut (1999, 2005), Hatami (2012), Kalai (2004), Mossel, O'Donnell and Oleszkiewicz (2010), Talagrand (1993, 1997, 1999), and van den Berg (2008).

9.8 EXERCISES

9.1. Let $f : \{-1,1\}^n \to \{0,1\}$ be binary-valued and let $g : \{-1,1\}^n \to \mathbb{R}$ be a real-valued function. Write $A = \{x : f(x) = 1\}$ and define the set $B = \{x : g(x) \geq 1/2\}$. Show that

$$P(A \triangle B) \leq 4E\left[(f-g)^2\right]$$

where $A \triangle B = \{x : f(x) \neq \mathbb{1}_{\{g(x)\geq 1/2\}}$ denotes the symmetric difference of A and B.

9.2. Give a proof of Theorem 9.5 based on Theorem 9.9 (with possibly different constants).

9.3. Show that Theorem 9.9 holds with $C = 9/10$. *Hint:* show that for any f,

$$\sum_{S\neq\emptyset} \frac{\hat{f}(S)^2}{3|S|} \leq \sum_{S\subset\{1,\ldots,n\}} \frac{\hat{f}(S)^2}{2|S|+1} = \int_0^1 \sum_{S\subset\{1,\ldots,n\}} \hat{f}(S)^2 \gamma^{2|S|} d\gamma$$

and use the Bonami–Beckner inequality (Corollary 5.17) to show

$$\mathrm{Var}(f) \leq 3\sum_{i=1}^n \int_0^1 \sum_{S\subset\{1,\ldots,n\}} \hat{g}_i(S)^2 \gamma^{2|S|} d\gamma \leq 3\sum_{i=1}^n \int_0^1 \|g_i\|_{1+\gamma^2}^2 d\gamma.$$

Use Hölder's inequality and some calculus to bound

$$\int_0^1 \|g_i\|_{1+\gamma^2}^2 d\gamma \le \|g_i\|_2^2 \int_0^1 \left(\frac{\|g_i\|_1}{\|g_i\|_2}\right)^{2(1-\gamma^2)/(1+\gamma^2)} d\gamma \le \frac{6}{5}\frac{\|g_i\|_2^2}{1+\log\frac{\|g_i\|_1}{\|g_i\|_2}}.$$

(This simple proof of Talagrand's inequality was suggested by Benjamini, Kalai, and Schramm (2003).)

9.4. Prove the following version of Theorem 9.8. Assume $f : \{-1,1\}^n \to \mathbb{R}$ satisfies $\sum_{i=1}^n (f(x) - f(x^{(i)}))_-^2 \le v$ for some $v > 0$ and let $B = \max_{x,i} |f(x) - f(x^{(i)})|$. Show that there is a constant K such that for all $\delta < \gamma \le 1/2$, by taking $a = Q_{1-\gamma} + B$ and $b = Q_{1-\delta}$,

$$Q_{1-\delta} - Q_{1-\gamma} \le B + K\sqrt{\frac{v\gamma}{\delta \log \frac{e^2}{2\gamma}}}.$$

9.5. Let $f : \{-1,1\}^n \to \mathbb{R}$ be such that

$$|f(x) - f(x^{(i)})| \le B \quad \text{for all } x \text{ and } i \text{ and} \quad \sum_{i=1}^n \left(f(x) - f\left(x^{(i)}\right)\right)_+^2 \le \phi(f(x))$$

where ϕ is a nonnegative nondecreasing function. Show that there exists a constant K such that for all $\delta < \gamma \le 1/2$,

$$Q_{1-\delta} - Q_{1-\gamma} \le K\sqrt{\frac{\phi(Q_{1-\delta} + B)\gamma}{\delta \log \frac{e^2}{2\gamma}}}.$$

In particular, recalling the notation $a_k = Q_{1-2^{-k}}$,

$$a_{k+1} - a_k \le K\sqrt{\frac{\phi(a_{k+1} + B)}{k}}$$

(see Devroye and Lugosi (2008)).

9.6. Let $A \subset \{-1,1\}^n$ be a monotone set with critical parameter $p_{1/2}$. Prove that for every $0 < \varepsilon < 1/2$ there exists a constant c such that $p_{1-\varepsilon} - p_\varepsilon \le c\min(p_{1/2}, 1 - p_{1/2})$ see Bollobás and Thomason (1987).

9.7. Let $k \in \{1,\ldots,n\}$ and let $B = \{x : \|x\| \ge k\}$ be a Hamming ball in $\{-1,1\}^n$ and let $A \subset \{-1,1\}^n$ be any monotone set whose critical parameter $p_{1/2}$ is at least that of B. Show that for any $p \ge p_{1/2}$, $P_p(A) \le P_p(B)$.

9.8. Generalize Theorem 9.5 to the case when the distribution over $\{-1,1\}$ is P_p, the product of n independent Bernoulli(p) measures. More precisely, show that for any set $A \subset \{-1,1\}^n$,

$$\sum_{i=1}^n I_i^p(A)^2 \ge \frac{(2-\varepsilon)^2(P_p(A)(1-P_p(A)))^2 \log^2 n}{n}$$

and

$$\max_{i=1,\ldots,n} I_i^p(A) \geq \frac{(2-\varepsilon)P_p(A)(1-P_p(A))\log n}{n}$$

where $\varepsilon = \log(P_p(A)(1-P_p(A))p(1-p)\log^2 n)/(c(p)\log n)$. Use the second inequality, together with Russo's lemma, to prove Corollary 9.18.

9.9. Prove the following generalization of Theorem 9.7 to the distribution P_p: let $\varepsilon \in (0,1)$. There exists a subset of $m = \lfloor I(f)/\varepsilon \rfloor$ variables and a real-valued function $g : \{-1,1\}^n \to \mathbb{R}$ depending on these m variables only such that

$$E\left[(f-g)^2\right] \leq \frac{2I(f)c(p)}{p(1-p)\log(1/4\varepsilon) + \log(c(p)/(p(1-p)))}$$

where $c(p) = (1/(1-2p))\log((1-p)/p)$. Conclude, using Russo's lemma, that if $A \subset \{-1,1\}^n$ is a monotone set such that there exist absolute constants $K_1, K_2 \in (0,1)$ such that $p_{3/4} - p_{1/4} \geq K_1$ and $\min(p_{1/2}, 1-p_{1/2}) \geq K_2$ then A may be approximated by a junta (i.e. by a function depending on a bounded number of variables).

9.10. Let $A \subset \{-1,1\}^n$ be a monotone set such that $p_{1/2} = 1/2$. Show that there exists a universal constant $c > 0$ such that $p_{3/4} - p_{1/4} \geq c/\sqrt{n}$.

9.11. Prove the following generalization of Russo's lemma. Let $f : \{-1,1\}^n \to \mathbb{R}$ be a real-valued function and let $X = (X_1, \ldots, X_n)$ be a vector of independent, identically distributed components with $\mathbf{P}\{X_i = 1\} = 1 - \mathbf{P}\{X_i = -1\} = p$. Then

$$\frac{d\mathbf{E}f(X)}{dp} = \sum_{i=1}^{n} \mathbf{E}\left(f(X_1,\ldots,X_{i-1},1,X_{i+1},\ldots,X_n) - f(X_1,\ldots,X_{i-1},-1,X_{i+1},\ldots,X_n)\right).$$

(Rossignol, 2006.)

9.12. (CONCENTRATION AND INFLUENCE) Let $A \subset \{-1,1\}^n$ and let X be uniformly distributed on $\{-1,1\}^n$. Let $d(X,A) = \min_{y\in A}\sum_{i=1}^n \mathbb{1}_{\{X_i \neq y_i\}}$ denote the Hamming distance of X to the set A. Prove that

$$\mathbf{E}d(X,A) \leq \frac{I(A)}{2P(A)}.$$

(Talagrand, 1999.)

10

Isoperimetry on the Hypercube and Gaussian Spaces

The purpose of this chapter is to explore further the rich connection between concentration and isoperimetry on the n-dimensional binary cube and also on $\mathbb{R}^n$, equipped with the canonical Gaussian measure.

The close relationship between concentration inequalities and isoperimetry is a recurring theme of this book. Since our focus is on functions of independent random variables, the associated measure spaces are product spaces. The simplest product space is the binary hypercube, which deserves special attention not only because it is a canonical example but also because the complex isoperimetric behavior of subsets of the hypercube have much to teach us about more general product spaces. Also, isoperimetric results for the binary hypercube often lead naturally to their analogs for the canonical Gaussian measure via the central limit theorem.

As a first example, in Section 4.4, we have seen that among all sets $A \subset \{-1, 1\}^n$ of a given number of points (say, of size $|A| = 2^{n-1}$), sub-cubes have the smallest *edge boundary*. We proved this as an easy consequence of Han's inequality, which is also at the basis of logarithmic Sobolev inequalities, used in Chapter 5 to prove concentration via the entropy method. This is in a sharp contrast with the fact that the *vertex boundary* is minimized by Hamming balls (see Section 7.3). The contrast is sharp because, among all monotone sets, Hamming balls *maximize* the size of the edge boundary (see Theorem 9.11) and at the same time it is obvious that a sub-cube of size 2^{n-1} maximizes the size of the vertex boundary among all monotone sets. Recall also from Chapter 7 that the vertex isoperimetric theorem leads immediately to a sub-Gaussian concentration inequality for Lipschitz functions of n independent symmetric Bernoulli variables.

In this chapter we introduce a third alternative for measuring the size of the boundary of a subset of the binary hypercube and prove a corresponding isoperimetric inequality (see Corollary 10.7 below). This inequality shows that edge and vertex boundaries cannot be small at the same time, which explains, intuitively, the conflict between edge and vertex isoperimetric problems mentioned above. This isoperimetric result is the consequence of Bobkov's inequality (Theorem 10.2), a powerful functional inequality which may be regarded as a sharpening of the logarithmic Sobolev inequality of Theorem 5.1.

One of the most important corollaries of Bobkov's inequality is the Gaussian isoperimetric theorem. To describe this beautiful result, recall from Section 7.2 that the classical isoperimetric problem asks which subsets of $\mathbb{R}^n$ have minimal surface area among those with a given volume (where volume and surface area are measured with respect to the n-dimensional Lebesgue measure). According to the classical isoperimetric theorem (Theorem 7.5), Euclidean balls are these extremal sets. As is emphasized in Chapter 7, an equivalent formulation of the classical isoperimetric problem asks one to determine the sets of a given volume such that the set of points within a certain distance to the set has minimal volume. This second formulation avoids handling the notion of a surface area and allows one to ask the same question in any metric measurable space.

A case of fundamental importance is that of $\mathbb{R}^n$ equipped with the canonical Gaussian measure (i.e. with the standard normal distribution with mean vector $(0, \dots, 0)$ and identity covariance matrix). For this case, the (Gaussian) isoperimetric problem is formulated as follows: among all measurable sets in $\mathbb{R}^n$ with a given probability under the canonical Gaussian distribution, for which ones does the set of points within a certain Euclidean distance have minimal Gaussian probability?

In Section 7.2 we already pointed out that the Tsirelson–Ibragimov–Sudakov inequality may be used to obtain Gaussian isoperimetric inequalities. However, in this chapter we show that the Gaussian isoperimetric problem may be solved exactly.

The Gaussian isoperimetric theorem (see Theorems 10.15 and 10.14 below) states the beautiful fact that half-spaces are the solution of the Gaussian isoperimetric theorem. Following Bobkov, we prove this theorem starting from Bobkov's inequality on the hypercube and then applying the central limit theorem. This strategy is similar to that which we used in Section 3.7 to derive the Gaussian Poincaré inequality from the Efron–Stein inequality, and in Section 5.3 to prove the Gaussian logarithmic Sobolev inequality from its analog on the hypercube.

We also extend Bobkov's inequality on the hypercube to asymmetric Bernoulli distributions. This simple extension allows us to derive some further results on threshold widths for certain monotone sets. (Recall Chapter 9 for the basic results.) In particular, in Section 10.3 we provide a simple proof for some deep results pioneered by Margulis.

The main work of this chapter appears in Section 10.1. Once we prove Bobkov's inequality, the rest of the results follow easily.

10.1 Bobkov's Inequality for Functions on the Hypercube

The purpose of this section is to prove an inequality for functions $f : \{-1, 1\}^n \to \mathbb{R}$ defined on the binary hypercube. We restrict our attention to the case when the hypercube is equipped with the uniform distribution, that is, we let $X = (X_1, \dots, X_n) \in \{-1, 1\}^n$ be a vector of independent Rademacher random variables. We may think about Bobkov's inequality as another member of the family of inequalities to which the Efron–Stein inequality and the logarithmic Sobolev inequality belong. Denote the i-th component of the discrete gradient vector $\nabla f(x) = (\nabla_1 f(x), \dots, \nabla_n f(x))$ of f by $\nabla_i f(x) = \left(f(x) - f(\overline{x}^{(i)})\right)/2$, where $\overline{x}^{(i)} = (x_1, \dots, x_{i-1}, -x_i, x_{i+1}, \dots, x_n)$. Then the

Efron–Stein inequality, specialized in this case, states that $\mathrm{Var}(f(X)) \le \mathbf{E}\|\nabla f(X)\|^2$, while by the logarithmic Sobolev inequality of Theorem 5.1, $\mathrm{Ent}(f^2) \le 2\mathbf{E}\|\nabla f(X)\|^2$.

To state Bobkov's inequality, we need to introduce the function

$$\gamma(x) = \varphi(\Phi^{-1}(x)) \quad \text{for } x \in (0,1),$$

where $\varphi(x) = (1/\sqrt{2\pi})e^{-x^2/2}$ is the standard Gaussian density, and $\Phi(x) = \int_{-\infty}^{x} \varphi(y)\,dy$ is the Gaussian distribution function. We also define $\gamma(0) = \gamma(1) = 0$. We call γ *the Gaussian isoperimetric function.* (In statistics $1/\gamma = (\Phi^{-1})'$ is known as the quantile-density function of the normal distribution.) The Gaussian isoperimetric function γ is concave and symmetric around $1/2$. This is a consequence of the following lemma that summarizes some of the basic properties of γ. We leave the proof as an easy exercise.

Lemma 10.1 *The Gaussian isoperimetric function γ satisfies*

1. $\gamma'(x) = -\Phi^{-1}(x)$ *for all* $x \in (0,1)$,
2. $\gamma(x)\gamma''(x) = -1$ *for all* $x \in (0,1)$,
3. $(\gamma')^2$ *is convex over* $(0,1)$.

We are now ready to state the key result of this chapter.

Theorem 10.2 (BOBKOV'S INEQUALITY) *Suppose X is uniformly distributed over $\{-1,1\}^n$. Then for all $n \ge 1$ and for all functions $f : \{-1,1\}^n \to [0,1]$,*

$$\gamma(\mathbf{E}f(X)) \le \mathbf{E}\sqrt{\gamma(f(X))^2 + \|\nabla f(X)\|^2}.$$

The next lemma, which describes the behavior of γ, helps us interpret Bobkov's inequality.

Lemma 10.3 *For all $x \in [0,1/2]$,*

$$x\sqrt{\frac{1}{2}\log\frac{1}{x}} \le \gamma(x) \le x\sqrt{2\log\frac{1}{x}}.$$

Moreover,

$$\lim_{x\to 0} \frac{\gamma(x)}{x\sqrt{2\log\frac{1}{x}}} = 1.$$

This lemma implies that $\gamma(x)/(x\sqrt{\log(1/x)})$ remains bounded as x tends to 0 (see Fig. 10.1). The proof is left as an exercise (see Exercise 10.4).

Next we turn to the proof of Theorem 10.2. As in the logarithmic Sobolev inequalities of Chapter 5 and the Bonami–Beckner inequality (see Theorem 5.18), Bobkov's inequality is also proved by induction over dimension. First we prove the theorem for $n = 1$ and then use Minkowski's inequality in the induction argument to extend the result to all dimensions $n > 1$.

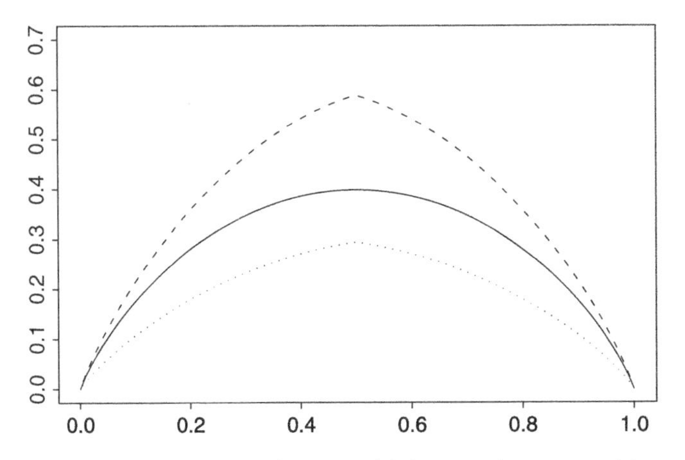

Figure 10.1 The Gaussian isoperimetric function $\gamma(x)$ between the upper and lower bounds of Lemma 10.3

Lemma 10.4 (THE CASE $n = 1$) *Let X be a Rademacher random variable (i.e. $\mathbf{P}\{X = 1\} = \mathbf{P}\{X = -1\} = 1/2$). For all functions $f : \{-1, 1\} \to [0, 1]$,*

$$\gamma(\mathbf{E}f(X)) \le \mathbf{E}\sqrt{\gamma(f(X))^2 + \frac{(f(X) - f(-X))^2}{4}}.$$

The proof is based on elementary algebraic manipulations based on the following technical lemma.

Lemma 10.5 *For any $c \in (0, 1/2]$ and $x \in [0, c]$, we have*

$$\gamma(c + x)^2 + \gamma(c - x)^2 + 2x^2 - 2\gamma(c)^2 \ge 2(\gamma'(c))^2 x^2$$

and

$$\frac{\gamma(c + x)^2 - \gamma(c - x)^2}{x} \le 4\gamma(c)\gamma'(c).$$

Proof The first inequality follows by observing that both the function $\gamma(c + x)^2 + \gamma(c - x)^2 + 2x^2 - 2\gamma(c)^2 - 2(\gamma'(c))^2 x^2$ and its derivative are zero at $x = 0$, and it is convex on $[0, c)$. (This follows by the second and third properties of γ in Lemma 10.1.)

To prove the second inequality, first note that

$$\frac{\gamma(c + x)^2 - \gamma(c - x)^2}{x} = \frac{\int_0^x ((\gamma^2(c + s))' - (\gamma^2(c - s))')ds}{x}.$$

Since for $s = 0$ the integrand on the right-hand side is $4\gamma(c)\gamma'(c)$, it suffices to prove that the integrand is a nonincreasing function of s. To prove this, observe that since $\gamma\,\gamma'' = -1$, (by Lemma 10.1), the derivative of the integrand with respect to s is

$$(\gamma^2(c+s))'' - (\gamma^2(c-s))'' = 2\left((\gamma'(c+s))^2 - (\gamma'(c-s))^2\right).$$

Since $(\gamma'(x))^2$, is convex and symmetric around $1/2$, it is non-increasing on $(0, 1/2]$. Thus, if $c + s \leq 1/2$ then $(\gamma'(c+s))^2 - (\gamma'(c-s))^2 \leq 0$ while if $c + s \geq 1/2$ then $(\gamma'(c+s))^2 = (\gamma'(1-c-s))^2 \leq (\gamma'(c-s))^2$ as $1 - c - s \geq c - s$. In all cases, $(\gamma'(c+s))^2 - (\gamma'(c-s))^2 \leq 0$. This concludes the proof. □

Proof of Lemma 10.4. Introducing $c = (f(1) + f(-1))/2$ and $x = |f(1) - f(-1)|/2$, the statement of the lemma may be written, equivalently, as

$$\gamma(c) \leq \frac{1}{2}\sqrt{\gamma(c+x)^2 + x^2} + \frac{1}{2}\sqrt{\gamma(c-x)^2 + x^2}. \tag{10.1}$$

where $c \in (0, 1)$ and $x \in [0, \min(c, 1-c)]$. As γ is symmetric around $1/2$, we may assume, without loss of generality, that $0 \leq x \leq c \leq 1/2$. It is convenient to introduce the notation

$$h(x) = \gamma(c+x)^2 + x^2 - \gamma(c)^2.$$

Squaring both sides of (10.1) and rearranging, (10.1) becomes

$$(\gamma(c)^2 - h(x)) + (\gamma(c)^2 - h(-x)) \leq 2\sqrt{(\gamma(c)^2 + h(x))(\gamma(c)^2 + h(-x))}.$$

We may assume that the left-hand side is positive, otherwise there is nothing to prove. Then, squaring both sides of the last inequality and rearranging, (10.1) is found to be equivalent to

$$(h(x) - h(-x))^2 \leq 8\gamma(c)^2(h(x) + h(-x)),$$

which may be rewritten as

$$\left(\gamma(c+x)^2 - \gamma(c-x)^2\right)^2 \leq 8\gamma(c)^2\left(\gamma(c+x)^2 + \gamma(c-x)^2 + 2x^2 - 2\gamma(c)^2\right).$$

By the first inequality of Lemma 10.5, for all $c \in (0, 1/2]$ for all $x \in [0, c]$,

$$8\gamma(c)^2\left(\gamma(c+x)^2 + \gamma(c-x)^2 + 2x^2 - 2\gamma(c)^2\right) \geq 16\gamma(c)^2(\gamma'(c))^2x^2.$$

So, in order to prove (10.1) under the assumption that $(\gamma(c)^2 - h(x)) + (\gamma(c)^2 - h(-x)) > 0$, it suffices to check that

$$(\gamma(c+x)^2 - \gamma(c-x)^2)^2 \leq 16\gamma(c)^2(\gamma'(c))^2x^2.$$

But as $0 \leq x \leq c \leq 1/2$, $\gamma(c+x) \geq \gamma(c-x)$, and $\gamma'(c) \geq 0$, this last inequality follows from the second inequality of Lemma 10.5. □

With the case of $n = 1$ proved, we now turn to the induction step to complete the proof of Bobkov's inequality. We state and prove this step in a somewhat more general scenario than what is needed for the proof of Bobkov's theorem. This generality will be helpful when we extend Bobkov's inequality to the product of not necessarily symmetric Bernoulli distributions on the hypercube. We show that if, for some function $\alpha : (0,1) \to \mathbb{R}$, an inequality like (10.1) holds for a distribution P on a set $\mathcal{X}$, then it also holds for the n-fold product of P on the product space $\mathcal{X}^n$.

Consider a vector $X = (X_1, \ldots, X_n)$ of independent random variables, whose components take their values in a measurable set $\mathcal{X}$. Assume that for each $i = 1, \ldots, n$, we have an operator $\tilde{\nabla}_i$ that assigns a real-valued function $\mathcal{X} \to \mathbb{R}$ to a real-valued function $\mathcal{X} \to \mathbb{R}$. When we write $\tilde{\nabla}_i f(x_1, \ldots, x_n)$, it is implicitely understood that $\tilde{\nabla}_i$ acts on the function $f(x_1, \ldots, x_{i-1}, \cdot, x_{i+1}, \ldots, x_n)$ with $x_1, \ldots, x_{i-1}, x_{i+1}, \ldots, x_n$ fixed. Thus, in this case we consider f as a function of its i-th variable only. The only requirement for $\tilde{\nabla}_i$ is that it should be measurable in the sense that if X is a random variable taking values in $\mathcal{X}^n$ and f is measurable, then $\tilde{\nabla}_i f(X)$ should be a random variable. Moreover, we assume that for any $f : \mathcal{X}^n \to \mathbb{R}$,

$$\left|\tilde{\nabla}_i \mathbf{E}\left[f(X) \mid X_i\right]\right| \leq \mathbf{E}\left[\left|\tilde{\nabla}_i f(X)\right| \mid X_i\right]. \tag{10.2}$$

Note that if $\mathcal{X} = \{-1, 1\}$, the components of the discrete gradient vector $\tilde{\nabla}_i f(x) = \nabla_i f(x) = \frac{f(x) - f(\overline{x}^{(i)})}{2}$ appearing in Bobkov's inequality (Theorem 10.2) satisfy this requirement. The main induction argument is summarized in the following lemma.

Lemma 10.6 (INDUCTION LEMMA) *Let $X = (X_1, \ldots, X_n)$ be a vector of independent random variables taking values in the set $\mathcal{X}^n$. Assume that the operators $\tilde{\nabla}_i$, $i = 1, \ldots, n$ satisfy condition (10.2). Assume that the function $\alpha : [0,1] \to [0, \infty)$ is such that for all $i \leq n$ and for all functions $g : \mathcal{X} \to [0,1]$,*

$$\alpha(\mathbf{E}g(X_i)) \leq \mathbf{E}\sqrt{\alpha(g(X_i))^2 + |\tilde{\nabla}_i g(X_i)|^2}.$$

Then for all functions $f : \mathcal{X}^n \to [0,1]$,

$$\alpha(\mathbf{E}f(X)) \leq \mathbf{E}\sqrt{\alpha(f(X))^2 + \|\tilde{\nabla} f(X)\|^2}$$

where $\|\tilde{\nabla} f(X)\|^2 = \sum_{i=1}^n \left(\tilde{\nabla}_i f(X)\right)^2$.

Proof The lemma is proved by induction over n. For $n = 1$ there is nothing to prove, so let $n \geq 2$. The induction hypothesis is that the lemma holds for $1, \ldots, n-1$.

Recall that $\overline{\mathbf{E}}_{n-1}$ stands for the conditional expectation operator conditioned on X_n (i.e. integration with respect to $X_1, \ldots, X_{n-1}$) and $\mathbf{E}^{(n)}$ stands for expectation with respect to the variable X_n only (i.e. conditional on $X_1, \ldots, X_{n-1}$). For each $y \in \mathcal{X}$ and $x^{(n)} \in \mathcal{X}^{n-1}$, denote $f_y(x^{(n)}) = f(x^{(n)}, y)$. As usual, we denote $X^{(n)} = (X_1, \ldots, X_{n-1})$.

Fix $x_n \in \mathcal{X}$ for now. Then

$$\overline{\mathbf{E}}_{n-1}\sqrt{\frac{1}{2}\left(\alpha\left(f\left(X^{(n)}, x_n\right)\right)^2 + \|\tilde{\nabla} f\left(X^{(n)}, x_n\right)\|^2\right)}$$

$$= \overline{\mathbf{E}}_{n-1}\sqrt{\frac{\alpha\left(f_{x_n}(X^{(n)})\right)^2 + \|\tilde{\nabla} f_{x_n}(X^{(n)})\|^2}{2} + \frac{\left(\tilde{\nabla}_n f(X^{(n)}, x_n)\right)^2}{2}}$$

$$\geq \left(\frac{\left(\overline{\mathbf{E}}_{n-1}\sqrt{\alpha\left(f_{x_n}\left(X^{(n)}\right)\right)^2 + \|\tilde{\nabla} f_{x_n}\left(X^{(n)}\right)\|^2}\right)^2}{2} + \frac{\left(\overline{\mathbf{E}}_{n-1}\left|\tilde{\nabla}_n f\left(X^{(n)}, x_n\right|\right)\right)^2}{2}\right)^{1/2}$$

where we used Minkowski's inequality by taking, in Theorem 2.16, $q = 2$, the X-variable to be uniform on $\{1, 2\}$, and the Y-variable to be $X^{(n)}$, and

$$Z = \sqrt{\alpha\left(f_{x_n}\left(X^{(n)}\right)\right)^2 + \|\tilde{\nabla} f_{x_n}\left(X^{(n)}\right)\|^2}$$

if the X-variable equals 1 and $\tilde{\nabla}_n f\left(X^{(n)}, x_n\right)$ if it equals 2.

As for each fixed x_n, f_{x_n} is a function of $n - 1$ identically distributed independent random variables, we may apply the induction hypothesis to the first term on the right-hand side of the inequality above and get

$$\overline{\mathbf{E}}_{n-1}\sqrt{\frac{1}{2}\left(\alpha\left(f\left(X^{(n)}, x_n\right)\right)^2 + \|\tilde{\nabla} f\left(X^{(n)}, x_n\right)\|^2\right)}$$

$$\geq \sqrt{\frac{\alpha\left(\overline{\mathbf{E}}_{n-1} f_{x_n}\left(X^{(n)}\right)\right)^2}{2} + \frac{\left(\overline{\mathbf{E}}_{n-1}\left|\tilde{\nabla}_n f\left(X^{(n)}, x_n\right)\right|\right)^2}{2}}$$

$$\geq \sqrt{\frac{\alpha\left(\overline{\mathbf{E}}_{n-1} f_{x_n}\left(X^{(n)}\right)\right)^2}{2} + \frac{\left(\tilde{\nabla}_n \overline{\mathbf{E}}_{n-1} f\left(X^{(n)}, x_n\right)\right)^2}{2}},$$

where the last line follows from our assumption on the operator $\tilde{\nabla}_n$.

Taking now expectation with respect to the distribution of X_n, we obtain

$$\mathbf{E}\sqrt{\alpha(f(X))^2 + \|\tilde{\nabla} f(X)\|^2}$$

$$= \mathbf{E}^{(n)}\left[\overline{\mathbf{E}}_{n-1}\sqrt{\alpha\left(f\left(X^{(n)}, X_n\right)\right)^2 + \|\tilde{\nabla} f\left(X^{(n)}, X_n\right)\|^2}\right]$$

$$\geq \mathbf{E}^{(n)}\left[\sqrt{\alpha\left(\overline{\mathbf{E}}_{n-1} f_{X_n}\left(X^{(n)}\right)\right)^2 + \left(\tilde{\nabla}_n \overline{\mathbf{E}}_{n-1} f_{X_n}\left(X^{(n)}\right)\right)^2}\right].$$

Now let the function $g : \mathcal{X} \to [0, 1]$ be defined by $g(x) = \overline{E}_{n-1} f_x\left(X^{(n)}\right)$. Then the right-hand side of the last inequality is just

$$E^{(n)} \sqrt{\alpha\left(g(X_n)\right)^2 + \left(\tilde{\nabla} g(X_n)\right)^2}$$

which can be lower bounded by invoking the induction hypothesis once more. This finally leads to

$$\begin{aligned} E\sqrt{\alpha(f(X))^2 + \|\tilde{\nabla} f(X)\|^2} &\geq \alpha\left(E^{(n)} g(X_n)\right) \\ &= \alpha\left(Ef(X)\right). \end{aligned}$$

□

Combining Lemmas 10.4 and 10.6, we obtain Theorem 10.2.

10.2 An Isoperimetric Inequality on the Binary Hypercube

While the original purpose of Bobkov's inequality was to provide a functional inequality that implies the Gaussian isoperimetric theorem (see Theorem 10.14 below), it has proved to be a powerful tool in understanding the isoperimetric structure of the binary hypercube. Indeed, a trivial application of Theorem 10.2, described in this section, provides an interesting interpolation between the edge and vertex isoperimetric theorems on the hypecube.

As explained in the introduction of this chapter, the vertex isoperimetric theorem states that, among all sets $A \subset \{-1, 1\}^d$ of a given size, Hamming balls minimize the size of the vertex boundary $\partial_V(A)$ (defined as the set of vertices of $\{-1, 1\}^n$ that are outside A but are connected with at least one vertex that belongs to A), while the size of the edge boundary $\partial_E(A)$ (i.e. the set of edges between A and A^c) is minimized by sub-cubes. At the same time, among all monotone sets, Hamming balls maximize the size of the edge boundary and sub-cubes maximize the size of the vertex boundary. This apparent conflict between the sizes of edge and vertex boundaries suggests that no monotone set can have a simultaneously small edge and vertex boundary. This is indeed the case as we show it below in Corollary 10.10.

As a simple result of the same flavor, consider the following corollary of Bobkov's inequality.

Let $A \in \{-1, 1\}^n$ be a (not necessarily monotone) set and consider the indicator function $f(x) = \mathbb{1}_{\{x \in A\}}$. Since $\gamma(0) = \gamma(1) = 0$, Bobkov's inequality, applied to f, implies

$$\gamma(P(A)) \leq E\|\nabla \mathbb{1}_{\{x \in A\}}\| = \frac{1}{2} E \sqrt{\sum_{i=1}^{n} \left(\mathbb{1}_{\{x \in A\}} - \mathbb{1}_{\{\overline{x}^{(i)} \in A\}}\right)^2}$$

where $P(A) = |A|2^{-n}$ is the probability of A under the uniform distribution. Defining the *symmetric vertex boundary* of A by $\overline{\partial}_V(A) = \partial_V(A) \cup \partial_V(A^c)$, we clearly have

$$\gamma(P(A)) \le \mathbf{E}\|\nabla \mathbb{1}_{\{x\in A\}}\| = \mathbf{E}\mathbb{1}_{\{X\in\overline{\partial}_V(A)\}}\|\nabla \mathbb{1}_{\{x\in A\}}\|,$$

so by the Cauchy–Schwarz inequality, we obtain

$$\gamma(P(A)) \le \sqrt{P(\overline{\partial}_V(A))}\sqrt{\mathbf{E}\|\nabla \mathbb{1}_{\{X\in A\}}\|^2}.$$

Now clearly, $4\|\nabla \mathbb{1}_{\{x\in A\}}\|^2 = \sum_{i=1}^n (\mathbb{1}_{\{x\in A\}} - \mathbb{1}_{\{\overline{x}^{(i)}\in A\}})^2$ is the number of edges between A to A^c incident to x if $x \in \overline{\partial}_V(A)$. Thus, $\mathbf{E}\|\nabla \mathbb{1}_{\{X\in A\}}\|^2 = |\partial_E(A)|2^{-(n+1)} = I(A)/2$ where $I(A)$ is the total influence of A. Thus, we have the following result.

Corollary 10.7 (AN ISOPERIMETRIC INEQUALITY ON THE CUBE) *For any subset A of $\{-1,1\}^n$,*

$$P\left(\overline{\partial}_V(A)\right) \cdot I(A) \ge 2\gamma\left(P(A)\right)^2.$$

The lemma asserts that the size of the edge boundary (total influence) and the vertex perimeter (in the sense of $|\overline{\partial}_V(A)|$) cannot be simultaneously small if $|A|$ is large. For example, for any set with cardinality $|A| = 2^{n-1}$, we see that $P(\overline{\partial}_V(A))I(A) \ge 1/\pi$. Note that for the sub-cube $A = \{x : x_1 = 1\}$, $P(\overline{\partial}_V(A))I(A) = 1$ and for the Hamming ball $A = \{x : \sum_i x_i > 0\}$ (for n even), $P(\overline{\partial}_V(A))I(A) \sim 4/\pi$ is also bounded by a constant so the inequality above is essentially saturated by both extremes.

In the next section we show that for monotone sets, Corollary 10.7 may be refined by replacing the symmetric vertex boundary $\overline{\partial}_V(A)$ by the "real" vertex boundary $\partial_V(A)$.

10.3 Asymmetric Bernoulli Distributions and Threshold Phenomena

In this section we present a variant of Bobkov's inequality (Theorem 10.2). The result shown here differs from that of Bobkov in three aspects. First, it holds not only for uniformly distributed vectors over the discrete hypercube $\{-1,1\}^n$, but also for products of asymmetric Bernoulli distributions. That is, $X = (X_1,\ldots,X_n)$ is supposed to have independent, identically distributed components with $\mathbf{P}\{X_i = 1\} = 1 - \mathbf{P}\{X_i = -1\} = p$ for some $p \in (0,1)$. Second, we restrict out attention to *monotone* functions $f : \{-1,1\}^n \to [0,1]$, that is, we assume that for all $i = 1,\ldots,n$ and $x_1,\ldots,x_{i-1},x_{i+1},\ldots,x_n \in \{-1,1\}$,

$$f(x_1,\ldots,x_{i-1},-1,x_{i+1},\ldots,x_n) \le f(x_1,\ldots,x_{i-1},1,x_{i+1},\ldots,x_n).$$

Finally, the discrete gradient ∇f is replaced by its "positive part" $\nabla^+ f = (\nabla_1^+ f, \ldots, \nabla_n^+ f)$ whose components are defined by

$$\nabla_i^+ f = \left(f(x) - f(\overline{x}^{(i)})\right)_+ .$$

Theorem 10.8 (TILLICH-ZÉMOR INEQUALITY) *Let $p \in (0,1)$, let $f : \{-1,1\}^n \to [0,1]$ be a monotone function and let $X = (X_1, \ldots, X_n)$ be a vector of independent Bernoulli random variables with $\mathbf{P}\{X_i = 1\} = 1 - \mathbf{P}\{X_i = -1\} = p$. Then*

$$\gamma(\mathbf{E}f(X)) \leq \mathbf{E}\sqrt{\gamma(f(X))^2 + 2\log(1/p)\|\nabla^+ f(X)\|^2}.$$

The proof of the theorem is similar to that of Bobkov's inequality. First we prove it for $n = 1$ and then use the induction lemma (Lemma 10.6) to extend it to higher dimensions.

Lemma 10.9 *Let X be a random sign, with $\mathbf{P}\{X = 1\} = p$. For all monotone functions $f : \{-1,1\} \to [0,1]$,*

$$\gamma(\mathbf{E}f(X)) \leq \mathbf{E}\sqrt{\gamma(f(X))^2 + 2\log(1/p)(f(x) - f(-x))_+^2}.$$

Proof Let $f(-1) = c$ and $f(1) = c + x$, with $x \in [0, 1-c]$. Let $q = 1 - p$. The inequality in the lemma is equivalent to

$$\gamma(c + px) - q\gamma(c) \leq p\sqrt{\gamma(c+x)^2 + 2x^2\log(1/p)}.$$

The left-hand side is nonnegative since γ is concave and nonnegative. Thus, equivalently, we need to prove

$$\left(\gamma(c + px) - q\gamma(c)\right)^2 - p^2\gamma(c+x)^2 - 2p^2x^2\log(1/p) \leq 0.$$

For a fixed value of c, let us denote the expression on the left-hand side by $F(x)$. Whatever the value of c, $F(0) = 0$ and $F'(0) = 0$, so it suffices to prove that F is concave on $[0, 1-c)$. The first and second derivatives may be computed and simplified using Lemma 10.1, and we obtain

$$\frac{F'(x)}{2} = p\gamma'(c+px)\left(\gamma(c+cx) - q\gamma(c)\right) - 2p^2x\log(1/p) - p^2\gamma'(c+x)\gamma(c+x)$$

and

$$\frac{F''(x)}{2} = p^2\left((\gamma'(c+px))^2 - (\gamma'(c+x))^2\right) - 2p^2\log(1/p) - p^2q\gamma(c)\gamma''(c+px).$$

Note that

$$\begin{aligned}(\gamma'(c+px))^2-(\gamma'(c+x))^2 &= 2\int_{c+x}^{c+px}\gamma'(t)\gamma''(t)dt\\ &= -2\int_{c+x}^{c+px}\frac{\gamma'(t)}{\gamma(t)}dt \quad \text{(by Lemma 10.1)}\\ &= 2\log\frac{\gamma(c+x)}{\gamma(c+px)}.\end{aligned}$$

This allows us to further simplify the expression of $F''(x)$:

$$\begin{aligned}\frac{F''(x)}{2p^2} &= 2\log\frac{p\gamma(c+x)}{\gamma(c+px)}+\frac{q\gamma(c)}{\gamma(c+px)}\\ &= 2\log\left(\frac{p\gamma(c+x)+q\gamma(c)}{\gamma(c+px)}-\frac{q\gamma(c)}{\gamma(c+px)}\right)+\frac{q\gamma(c)}{\gamma(c+px)}\\ &\le 2\log\left(1-\frac{q\gamma(c)}{\gamma(c+px)}\right)+\frac{q\gamma(c)}{\gamma(c+px)},\end{aligned}$$

where the last inequality follows from the fact that $\gamma(c+x)$ is a concave and nonnegative function of x on $[0,1-c)$ while log is increasing.

The lemma then follows by observing that $2\log(1-u)+u$ is zero at $u=0$ and nonincreasing on $[0,1)$. □

Proof of Theorem 10.8. The proof is an almost immediate consequence of Lemma 10.9 and the general induction argument. The only issue is raised by the fact that Lemma 10.6 formally does not handle the restriction to monotone functions. However, if f is monotone, so are $\overline{E}_{n-1}f_x(X^{(n)})$ and f_{X_n} and the proof of Lemma 10.6 goes through. □

We may now apply Theorem 10.8 for the indicator function $f(x)=\mathbb{1}_{\{x\in A\}}$ of any monotone subset of the binary hypercube.

For any set $A\subset\{-1,1\}^n$, define the function

$$h_A(x)=\|\nabla^+\mathbb{1}_{\{x\in A\}}\|^2=\sum_{i=1}^n\left(\mathbb{1}_{\{x\in A\}}-\mathbb{1}_{\{\overline{x}^{(i)}\in A\}}\right)_+.$$

Clearly, $h_A(x)=0$ if $x\notin A$ and otherwise $h_A(x)$ is the number of edges leaving A from x. $E\sqrt{h_A(X)}$ may be interpreted as some kind of a "surface area" of the set A and the following corollary is an isoperimetric inequality based on this notion. It sharpens and generalizes Corollary 10.7 for monotone sets.

Observe that $\left(\mathbb{1}_{\{x\in A\}}-\mathbb{1}_{\{\overline{x}^{(i)}\in A\}}\right)_+=1$ if and only if the i-th variable is pivotal for A and $x_i=1$. Since these two events are independent, we conclude that $Eh_A(X)=p\cdot I^p(A)$ is p

times the total influence of the set A (recall Section 9.5). Then by the Cauchy–Schwarz inequality,

$$\mathbf{E}\sqrt{h_A(X)} = \mathbf{E}\sqrt{h_A(X)}\mathbb{1}_{\{X\in\partial_V A\}} \leq \sqrt{pI^p(A)P_p(\partial_V(A)}$$

(recall that $P_p(A) = \sum_{x\in A} p^{\|x\|}(1-p)^{n-\|x\|}$).

Corollary 10.10 (AN ISOPERIMETRIC INEQUALITY FOR MONOTONE SETS) *Let $p \in (0,1)$, and let $X = (X_1,\dots,X_n)$ be a vector of independent Bernoulli random variables with $\mathbf{P}\{X_i = 1\} = 1 - \mathbf{P}\{X_i = -1\} = p$. Then for any monotone set $A \in \{-1,1\}^n$,*

$$\mathbf{E}\sqrt{h_A(X)} \geq \frac{1}{\sqrt{2\log(1/p)}}\gamma\left(P_p(A)\right).$$

In particular,

$$2P_p(\partial_V(A)) \cdot I^p(A)p\log\frac{1}{p} \geq \gamma(P(A))^2.$$

Part of the beauty of these inequalities lies in their dimension-free nature. Indeed, n does not appear anywhere in the expressions. To appreciate the sharpness of this result, it is instructive to check the cases of the Hamming ball $A = \left\{x : \sum_{i=1}^n x_i \geq n/2\right\}$, the sub-cube $\{x : x_1 = 1\}$, and the singleton $\{(1,\dots,1)\}$.

As we saw in Section 9.6, inequalities for the total influence may be used to study the evolution of the probability $P_p(A)$ of a monotone set $A \in \{-1,1\}^n$. Indeed, by Russo's lemma, $dP_p(A)/dp = I^p(A)$. For example, Corollary 10.10 immediately implies that for sets with a small vertex boundary, $P_p(A)$ experiences a sharp transition around the critical probability $p_{1/2}$ (defined as the value for which $P_{p_{1/2}}(A) = 1/2$ (see Exercise 10.1). Another application of Corollary 10.10 is shown by the next result, which interestingly gives sufficient conditions for a monotone set to guarantee narrow thresholds. This corollary of the Tillich–Zémor inequality sharpens a classical result defined by Margulis. Recall that Φ denotes the Gaussian distribution function.

Theorem 10.11 (MARGULIS'S GRAPH CONNECTIVITY THEOREM) *Let $k > 0$ and let $A \subset \{-1,1\}^n$ be a monotone set such that for all $x \in \partial_V(A)$, $h_A(x) \geq k$. Then*

$$P_p(A) \leq \Phi\left(\sqrt{2k}\left(\sqrt{-\log p_{1/2}} - \sqrt{-\log p}\right)\right) \quad \text{for } 0 < p < p_{1/2}$$

$$P_p(A) \geq \Phi\left(\sqrt{2k}\left(\sqrt{-\log p_{1/2}} - \sqrt{-\log p}\right)\right) \quad \text{for } p_{1/2} < p < 1.$$

Proof Since $h_A(x) \geq k\mathbb{1}_{\{x\in\partial_V(A)\}}$, the total influence may be bounded from below as

$$pI^p(A) = \mathbf{E}h_A(X) \geq \mathbf{E}\sqrt{kh_A(X)} \geq \sqrt{\frac{k}{2\log(1/p)}}\gamma\left(P_p(A)\right)$$

by Corollary 10.10. Thus, by Russo's lemma, we have

$$\frac{dP_p(A)}{dp} \geq \sqrt{\frac{k}{2p^2 \log(1/p)}} \gamma\left(P_p(A)\right).$$

Since $\gamma(s)\gamma''(s) = -1$ (by Lemma 10.1), this may be re-written as

$$\gamma''(P_p(A)) \frac{dP_p(A)}{dp} \leq -\sqrt{\frac{k}{2p^2 \log(1/p)}}.$$

Suppose $p < p_{1/2}$. Integrating this inequality between p and $p_{1/2}$, we obtain

$$\gamma'(P_{p_{1/2}}(A)) - \gamma'(P_p(A)) \leq \sqrt{2k}\left(\sqrt{-\log p_{1/2}} - \sqrt{-\log p}\right).$$

The proof of the first inequality is completed by noting that $\gamma'(P_{p_{1/2}}(A)) = \gamma'(1/2) = 0$ and $\gamma'(P_p(A))) = -\Phi^{-1}(P_p(A))$, by Lemma 10.1. The second inequality follows similarly. □

The theorem above implies that if k is large, that is, if every vertex on the boundary of A has many edges connecting it to the complement of A, then $P_p(A)$ experiences a sharp transition around the critical value $p_{1/2}$. The following example explains the name of the theorem.

Example 10.12 (CONNECTIVITY OF RANDOM SUBGRAPHS OF A FINITE GRAPH) Consider the following random graph model. Let $G = (V, E)$ be a finite connected graph with vertex set V and edge set E. Now remove every edge of G at random, independently, with probability p. Let A be the event that the remaining graph is disconnected. We may represent A as a subset of $\{-1, 1\}^{|E|}$ as follows: every binary vector $x \in \{-1, 1\}^{|E|}$ represents a graph such that a component 1 represents a removed edge of G and -1 a remaining edge. Then the set A representing all connected graphs is clearly monotone. We are interested in the evolution of $P_p(A)$, the probability that the remaining graph is disconnected. Suppose that the graph G is $k + 1$-edge-connected, that is, the graph remains connected after the removal of any set of k edges. In this case, for any disconnected graph $x \in \partial_V(A)$ on the boundary of A, $h_A(x) \geq k$ and therefore Theorem 10.11 is applicable. It shows that as p grows from 0 to 1, the probability $P_p(A)$ jumps from values close to 0 to values close to 1 in an interval of length $O(1/\sqrt{k})$.

Remark 10.3 The "surface area" $\mathbf{E}\sqrt{h_A(X)}$ may also be related to the sum of the squared influences studied in Chapter 9. In fact,

$$\sqrt{\sum_{i=1}^n I_i^p(A)^2} = \left(\frac{1}{p^2}\sum_{i=1}^n \left(\mathbf{E}\left(\mathbb{1}_{\{X\in A\}} - \mathbb{1}_{\{\overline{X}^{(i)}\in A\}}\right)_+\right)^2\right)^{1/2}$$
$$\leq \frac{1}{p}\mathbf{E}\sqrt{\sum_{i=1}^n \left(\mathbb{1}_{\{X\in A\}} - \mathbb{1}_{\{\overline{X}^{(i)}\in A\}}\right)_+^2}$$
$$= \frac{1}{p}\mathbf{E}\sqrt{h_A(X)},$$

where the inequality follows from the fact that if $Y_1, \ldots, Y_n$ are random variables with a finite second moment, then $\left(\sum_{i=1}^n (\mathbf{E}Y_i)^2\right)^{1/2} \leq \mathbf{E}\left(\sum_{i=1}^n Y_i^2\right)^{1/2}$ by the convexity of the Euclidean norm and Jensen's inequality.

10.4 The Gaussian Isoperimetric Theorem

In this section we prove the celebrated Gaussian isoperimetric theorem, which states that, among all sets of a given Gaussian measure, the Gaussian surface area (defined below) is minimized by half-spaces. The proof presented here is based on Bobkov's inequality (Theorem 10.2). The key ingredient is a functional inequality which extends Theorem 10.2 from the uniform distribution over the hypercube to the Gaussian distribution using the central limit theorem. The argument is similar to that by which we obtained the Gaussian Poincaré inequality or the Gaussian logarithmic Sobolev inequality from their discrete analogues.

Theorem 10.13 (BOBKOV'S GAUSSIAN INEQUALITY) *Let $X = (X_1, \ldots, X_n)$ be a vector of independent standard Gaussian random variables. Let $f : \mathbb{R}^n \to [0, 1]$ be a differentiable function with gradient ∇f. Then*

$$\gamma(\mathbf{E}f(X)) \leq \mathbf{E}\sqrt{(\gamma(f(X))^2) + \|\nabla f(X)\|^2},$$

where $\gamma = \varphi \circ \Phi^{-1}$ is the Gaussian isoperimetric function.

Proof It suffices to prove that the theorem holds for all $f : \mathbb{R}^n \to [0, 1]$ that are twice differentiable and have a compact support because the extension to all differentiable f may be achieved through a routine density argument.

Let k be a positive integer, and let $\varepsilon = (\varepsilon_{i,j})_{i=1,\ldots,n,j=1,\ldots,k}$ be a vector of independent Rademacher random variables. Define the function $f_k : \{-1, 1\}^{nk} \to [0, 1]$ by

$$f_k(\varepsilon) = f\left(\sum_{j=1}^k \frac{\varepsilon_{1,j}}{\sqrt{k}}, \ldots, \sum_{j=1}^k \frac{\varepsilon_{n,j}}{\sqrt{k}}\right).$$

Now we may apply Theorem 10.2 for the function f_k and obtain

$$\gamma\left(\mathbf{E}f_k(\varepsilon)\right) \leq \mathbf{E}\sqrt{\gamma^2(f_k(\varepsilon)) + \|\nabla f_k(\varepsilon)\|^2}.$$

(Note that, with an abuse of notation, here ∇ stands for the discrete gradient, introduced in Section 10.1.)

Now all we have to do is to let k go to infinity on both sides of the inequality. Indeed, by the central limit theorem, $\lim_{k\to\infty} \gamma(\mathbf{E}f_k(\varepsilon)) = \gamma(\mathbf{E}f(X))$, where X is a vector of n independent standard Gaussian random variables. On the other hand, proceeding exactly the same way as in the proofs of the Gaussian Poincaré and Gaussian logarithmic Sobolev inequalities, by the central limit theorem, we also have

$$\lim_{k\to\infty} \mathbf{E}\sqrt{\gamma^2(f_k(\varepsilon)) + \|\nabla f_k(\varepsilon)\|^2} = \mathbf{E}\sqrt{\gamma(f(X))^2 + \|\nabla f(X)\|^2}. \qquad \square$$

In the rest of this section we show how Bobkov's Gaussian inequality may be used to derive the Gaussian isoperimetric theorem.

Recall that the t-blowup of a set $A \subset \mathbb{R}^n$ is defined by

$$A_t = \{x : d(A;x) < t\}$$

where $d(A,x) = \inf_{y\in A} \|x-y\|$ is the Euclidean distance of x to the set A.

In analogy with the definition of surface area used in Section 7.2, me may define the Gaussian boundary measure of a Borel set A by

$$\lim_{t\searrow 0} \frac{P(A_t \setminus A)}{t}$$

whenever the limit exists, where P is the canonical Gaussian measure on $\mathbb{R}^n$.

The Gaussian isoperimetric problem is to determine which (Borel) sets A have minimal Gaussian boundary measure among all sets in $\mathbb{R}^n$ with a given probability p. The Gaussian isoperimetric theorem states the beautiful fact that the extremal sets are linear half-spaces in all dimensions and for all p:

Theorem 10.14 (GAUSSIAN ISOPERIMETRIC THEOREM) *Let P be the canonical Gaussian distribution on $\mathbb{R}^n$ and let $A \in \mathbb{R}^n$ be a Borel set. Then*

$$\liminf_{t\to 0} \frac{P(A_t \setminus A)}{t} \geq \gamma(P(A)).$$

Moreover if A is a half-space defined by $A = \{x : x \in \mathbb{R}^n, x_1 \leq z\}$, then

$$\lim_{t\to 0} \frac{P(A_t \setminus A)}{t} = \gamma(P(A)) = \varphi(z).$$

(Recall that $\gamma(x) = \varphi(\Phi^{-1}(x))$ denotes the Gaussian isoperimetric function.)

Proof The theorem is an almost immediate consequence of Theorem 10.13. However, the Gaussian isoperimetric theorem is concerned with characteristic functions of sets while

Bobkov's Gaussian inequality deals with differentiable functions. We apply Bobkov's inequality to smooth approximations of indicator functions.

First note that if $P(\overline{A} \setminus A) > 0$ there is nothing to prove where $\overline{A}$ denotes the closure of A. Hence we may assume that $P(\overline{A} \setminus A) = 0$ and indeed, without loss of generality, we may even assume that A is open.

For each $t > 0$, define $f_t : \mathbb{R}^n \to [0,1]$ by

$$f_t(x) = \left(1 - \frac{d(A,x)}{t}\right)_+ .$$

Clearly, $f_t(x) = 1$ for all $x \in A$, $f_t(x) = 0$ for $x \notin A_t$, and f_t is $1/t$-Lipschitz. However, f_t is not differentiable. We may further smooth it by convolution with a Gaussian kernel. Thus, for $\sigma > 0$, we define

$$f_{t,\sigma}(x) = \int_{\mathbb{R}^n} f_t(y) \frac{1}{(\sqrt{2\pi}\sigma)^n} \exp\left(-\frac{\|x-y\|^2}{2\sigma^2}\right) dy.$$

For each $\sigma > 0$, the function $f_{t,\sigma}$ still maps $\mathbb{R}^n$ to $[0,1]$, it is infinitely many times differentiable, and remains $1/t$-Lipschitz (Exercise 10.11).

Now we may apply Bobkov's Gaussian inequality. Indeed, if X denotes a standard Gaussian vector, then for each $t, \sigma > 0$, by Theorem 10.13,

$$\gamma\left(\mathbf{E}f_{t,\sigma}(X)\right) \le \mathbf{E}\gamma\left(f_{t,\sigma}(X)\right) + \mathbf{E}\|\nabla f_{t,\sigma}(X)\|.$$

If we let σ tend to 0, by the dominated convergence theorem, we have

$$\lim_{\sigma\to 0} \gamma\left(\mathbf{E}f_{t,\sigma}(X)\right) = \gamma\left(\mathbf{E}f_t(X)\right), \quad \text{and} \quad \lim_{\sigma\to 0} \mathbf{E}\gamma\left(f_{t,\sigma}(X)\right) = \mathbf{E}\gamma\left(f_t(X)\right).$$

Since $f_t(x) = 1$ for $x \in A$ and $f_t(x) = 0$ for $x \notin A_t$, we have $\gamma(f_t(x)) \le (1/\sqrt{2\pi})\mathbb{1}_{\{A_t \setminus A\}}$ and therefore $\mathbf{E}\gamma(f_t(X)) \le (1/\sqrt{2\pi})\mathbf{P}\{A_t \setminus A\}$. Now by the Lipschitz property of $f_{t,\sigma}$, we have $\|\nabla f_{t,\sigma}(X)\| \le 1/t$. Also, if x is in the interior of A or outside the closure of A_t, $\|\nabla f_{t,\sigma}(x)\| \to 0$ as $\sigma \to 0$. If x is in the interior of $A_t \setminus A$, then $\|\nabla f_{t,\sigma}(x)\| \to 1/t$. Hence, by dominated convergence,

$$\lim_{\sigma\to 0} \mathbf{E}\|\nabla f_{t,\sigma}\| = \frac{P(A_t \setminus A)}{t}.$$

Combining all the above,

$$\gamma(\mathbf{E}f_t(x)) \le \left(1/\sqrt{2\pi}\right) P(A_t \setminus A) + \frac{P(A_t \setminus A)}{t}.$$

Letting finally $t \to 0$,

$$\gamma(P(A)) \leq \liminf_{t \to 0} \frac{P(A_t \setminus A)}{t}$$

which is just the lower bound in the Gaussian isoperimetric theorem.

The fact that half-spaces achieve equality is obvious. □

Next we describe an equivalent version of the Gaussian isoperimetric theorem in the manner of measure concentration described in the introduction of Chapter 7. It gives a sharp lower bound for the Gaussian measure of the blowup of any set in terms of the measure of the set.

Theorem 10.15 (GAUSSIAN CONCENTRATION THEOREM) *Let P be the canonical Gaussian distribution on $\mathbb{R}^n$ and let $A \subset \mathbb{R}^n$ be a Borel set. Then for all $t \geq 0$,*

$$P(A_t) \geq \Phi\left(\Phi^{-1}(P(A)) + t\right).$$

Equality holds if A is a half-space.

In the proof we need the following simple technical observation whose proof is left as an exercise.

Proposition 10.16 *If A is a finite union of open balls in $\mathbb{R}^n$, then $P(A_t)$ is a differentiable function of $t > 0$.*

Proof of Theorem 10.15. We call a Borel set $A \subset \mathbb{R}^n$ *smooth* if $P(A_t)$ is a differentiable function of t on $(0, \infty)$.

Observe that if A is smooth, then

$$\frac{d\Phi^{-1}(P(A_t))}{dt} = \frac{dP(A_t)}{dt} \times \frac{1}{\gamma\left(P(A_t)\right)}.$$

It follows from the Gaussian isoperimetric theorem that the right-hand side is at least 1 since $\frac{dP(A_t)}{dt} = \lim_{s \to 0} \frac{P(A_{t+s} \setminus A_t)}{s}$. By integrating this inequality,

$$\begin{aligned}\Phi^{-1}\left(P\left(A_t\right)\right) &= \Phi^{-1}\left(P(A)\right) + \int_0^r \frac{d\Phi^{-1}(P(A_s))}{ds} ds \\ &\geq \Phi^{-1}\left(P(A)\right) + t.\end{aligned}$$

Hence, the theorem holds for all smooth sets. The remaining work is to extend this to all Borel sets.

Note first that if $P(A) = 0$, the theorem is automatically satisfied and therefore we may focus on Borel sets A with positive probability. By Proposition 10.16, the concentration property holds for any finite union of open balls.

Now let A be any Borel set with $P(A) > 0$. Let $0 < \varepsilon < t$. Then by Vitali's covering theorem, there exists a countable collection of disjoint open balls $\{B_1, B_2, \ldots\}$, all intersecting A and diameter at most ε, such that $P(A \setminus \cup_{n=1}^{\infty} B_n) = 0$. But then

$$\begin{aligned} P(A_t) &\geq P(\cup_{n=1}^{\infty}(B_n)_{t-\varepsilon}) \\ &= \lim_{n\to\infty} P\left(\cup_{i=1}^{n}(B_i)_{t-\varepsilon}\right) \\ &\geq \lim_{n\to\infty} \Phi\left(\Phi^{-1}\left(P(\cup_{i=1}^{n} B_i)\right) + t - \varepsilon\right) \\ &= \Phi\left(\Phi^{-1}\left(P(\cup_{n=1}^{\infty} B_n)\right) + t - \varepsilon\right) \\ &\geq \Phi\left(\Phi^{-1}\left(P(A)\right) + t - \varepsilon\right). \end{aligned}$$

The argument is completed by taking ε to 0. □

Note that the Gaussian concentration theorem and the Gaussian isoperimetric theorem are, in fact, equivalent in the sense that the Gaussian concentration theorem implies that for every Borel set $A \subset \mathbb{R}^n$,

$$\begin{aligned} \liminf_{t\to 0} \frac{P(A_t \setminus A)}{t} &\geq \liminf_{t\to 0} \frac{\Phi\left(\Phi^{-1}(P(A)) + t\right) - \Phi\left(\Phi^{-1}(P(A))\right)}{t} \\ &= \gamma(P(A)) \end{aligned}$$

which is just the statement of Theorem 10.14. See Exercise 10.7 for a more general argument.

10.5 Lipschitz Functions of Gaussian Random Variables

Recall that by the Gaussian concentration inequality (Theorem 5.6), any Lipschitz function of independent Gaussian random variables has sub-Gaussian tails. This result may be sharpened by combining the Gaussian concentration theorem (Theorem 10.15) with Lévy's inequality (Theorem 7.1). We obtain the following.

Theorem 10.17 *Let $X = (X_1, \ldots, X_n)$ be a vector of n independent standard normal random variables. Let $f : \mathbb{R}^n \to \mathbb{R}$ denote a Lipschitz function with Lipschitz constant L and let $\mathbf{M}f(X)$ denote a median of $f(X)$. Then, for all $t > 0$,*

$$\mathbf{P}\left\{f(X) - \mathbf{M}f(X) \geq t\right\} \leq 1 - \Phi(t/L).$$

Recall that, according to Gordon's inequality (Exercise 7.8), $1 - \Phi(t) \leq (1/t\sqrt{2\pi})e^{-t^2/2}$. The Gaussian concentration inequality fails to capture the corrective factor t^{-1}. The inequality of Theorem 10.17 cannot be improved in general as for $f(x) = n^{-1/2}\sum_{i=1}^{n} x_i$, equality is achieved for all $t > 0$. Note, however, that the refinement above bounds the probability of deviations around the median rather than around the mean.

10.6 Bibliographical Remarks

Bobkov's inequality (Theorem 10.2) on the hypercube was first established by Bobkov (1997) as a first step in his elementary proof of the Gaussian isoperimetric inequality. It is Bobkov's argument that we follow in this chapter. Bobkov's induction lemma (Lemma 10.6) was further generalized by Bobkov and Götze (1999, Lemma 2.1).

Theorem 10.8 is due to Tillich and Zémor (2001), as is Corollary 10.10. The history of this inequality goes back to Margulis (1974) and was subsequently improved by Talagrand (1993), and Bobkov and Götze (1999). It was Talagrand (1993) who promoted the use of $\boldsymbol{E}\sqrt{h_A(X)}$ as an adequate measure of surface area for monotone subsets of the hypercube. Linial and Rozenman (2002) characterized the monotone sets of a given volume that minimize the surface area $\boldsymbol{E}\sqrt{h_A(X)}$ under the uniform distribution.

Talagrand (1997) proved that under the uniform probability on $\{-1,1\}^n$, $\boldsymbol{E}\sqrt{h_A(X)}$ can be $O(1)$ only if $\sum_{i=1}^n I_i(A)^2$ is $\Omega(1)$. Talagrand (1997) also proves that there exist $b > 0$ and $a \in (0, 1/2)$ such that for every monotone set $A \subset \{-1,1\}^n$,

$$\boldsymbol{E}\|\nabla \mathbb{1}_{\{A\}}\| \geq bP(A)(1-P(A))\left(\log\frac{e}{P(A)(1-P(A))}\right)^{1/2-a}\left(\log\frac{e}{\sum_{i\leq n} I_i^2(A)}\right)^{a},$$

where P is the uniform distribution over $\{-1,1\}^n$.

The Gaussian isoperimetric theorem was first established independently by Borell (1975) and Tsirelson and Sudakov (1974). Borell's proof relied on Lévy's isoperimetric theorem for Euclidean spheres.

In a series of papers Ehrhard developed a proof of the Gaussian isoperimetric theorem based on Gaussian rearrangment techniques (see Ehrhard (1982, 1983*b*, 1983*a*, 1984, 1986), and also Borell (1986).

Ledoux (1996) proved a version of Theorem 10.14 with sub-optimal constants using semigroup techniques.

The equivalence between the Gaussian isoperimetric theorem (Theorem 10.15) and Bobkov's Gaussian inequality (Theorem 10.13) was pointed out by Ehrhard (1984), but see also Bakry and Ledoux (1996), Capitaine, Hsu, and Ledoux (1997), and Barthe and Maurey (2000).

The proof of Theorem 10.13 given here is from Bobkov (1997). The theorem is extended in Barthe and Maurey (2000) who also provide a proof for Theorem 10.13 based on stochastic calculus.

Borell and Ehrhard (see Ehrhard (1986)) obtained the following Gaussian version of the Brunn–Minkowski inequality: for all convex sets A, B in $\mathbb{R}^n$,

$$\Phi^{-1}\left(P(\lambda A + (1-\lambda)B)\right) \geq \lambda\Phi^{-1}\left(P(A)\right) + (1-\lambda)\Phi^{-1}\left(P(B)\right) \tag{10.4}$$

where P is the canonical Gaussian measure on $\mathbb{R}^n$.

10.7 EXERCISES

10.1. (SHARP THRESHOLD FOR SETS WITH SMALL VERTEX BOUNDARY) Let $A \in \{-1,1\}^n$ be a nonempty monotone set and let $P_p(A)$ be its probability under the product of Bernoulli(p) measures. For $a \in [0,1]$, let p_a be the unique value such that $P_{p_a}(A) = a$. Show that for any $\varepsilon \in (0, 1/2)$,

$$p_{1-\varepsilon} - p_{1/2} \leq \frac{(\log 2)(1/2 - \varepsilon)}{\varepsilon^2 \log(1/\varepsilon)} \inf_{q \in (1/2, 1-\varepsilon)} P_q(\partial_V(A)).$$

Derive a similar upper bound for $p_{1/2} - p_\varepsilon$. *Hint:* use Corollary 10.10.

10.2. (CHEEGER CONSTANT FOR UNIVARIATE DISTRIBUTIONS) Let P denote a probability distribution on $\mathbb{R}^n$. Let α_P denote the associated isoperimetric function defined, for $p \in (0,1)$,

$$\alpha_P(p) = \inf_{A \subset \mathbb{R}^n : P(A) = p} \liminf_{t \to 0} \frac{P(A_t \setminus A)}{t}$$

where the infimum is taken over all measurable sets with probability p. The *Cheeger constant* of P is defined by

$$\kappa(P) = \inf_{p \in (0,1)} \frac{\alpha_P(p)}{\min(p, 1-p))}.$$

When $n = 1$, let F denote the distribution function of P. Assume that P is absolutely continuous and let f denote its density. Prove that

$$\kappa(P) = \operatorname*{ess\,inf}_{a < x < b} \frac{f(x)}{\min(F(x), 1 - F(x))},$$

where $a = \inf\{x : F(x) > 0\}$ and $b = \sup\{x : F(x) < 1\}$.

Compute $\kappa(P)$ for the standard Gaussian distribution and for the Laplace distribution (whose density is $\frac{1}{2}\exp(-|x|)$). (See Bobkov and Houdré (1997).)

10.3. (CHEEGER CONSTANT IN PRODUCT SPACES) With the notation of the previous exercise, let P^n denote the n-fold product of the measure P on the real line. Prove that

$$\kappa(P^n) \geq \frac{1}{2\sqrt{6}} \kappa(P).$$

(Bobkov and Houdré (1997, Theorem 1.1)).

10.4. (APPROXIMATION OF THE GAUSSIAN ISOPERIMETRIC FUNCTION) Prove Lemma 10.3. *Hint:* the lower bound on γ follows easily from Gordon's inequality (Exercise 7.8). The upper bound follows from Lemma 10.9.

10.5. (GAUSSIAN MEASURE) Let $A = \{x : x \in \mathbb{R}^n, \langle x, u\rangle < \lambda\}$ be a half-space in $\mathbb{R}^n$ for some $u \in \mathbb{R}^n$ and $\lambda \in \mathbb{R}$. Let P denote the canonical Gaussian distribution on $\mathbb{R}^n$. Show that for any $t > 0$,

$$\Phi^{-1}(P(A_t)) = \Phi^{-1}(P(A)) + t.$$

10.6. (PROOF OF PROPOSITION 10.16) Let P be an absolutely continuous probability distribution on $\mathbb{R}^n$ and let A be a finite union of open balls. Prove that $P(A_t)$ is a differentiable function of $t > 0$.

10.7. (FROM ISOPERIMETRY TO CONCENTRATION) Assume that a probability distribution P on $\mathbb{R}^n$ satisfies, for all Borel sets $A \subset \mathbb{R}^n$,

$$\liminf_{t \searrow 0} \frac{P(A_t \setminus A)}{t} \geq cf(F^{-1}(P(A))),$$

where $c \in (0, 1]$ is a constant and F is a continuously differentiable distribution function over $\mathbb{R}$ and f its derivative. Prove that for all Borel sets A and all $t \geq 0$,

$$P(A_t) \geq F\left(F^{-1}(P(A)) + ct\right).$$

10.8. (FUNCTIONAL INEQUALITIES INVOLVING $\alpha(x) = x(1-x)$) Consider the discrete gradient ∇_i on the cube $\{-1, 1\}^n$ whose components are defined by $\nabla_i^+ f(x) = (f(x) - f(\overline{x}^{(i)}))_+$. Let $X = (X_1, \ldots, X_n)$ be a vector of independent random signs such that $\mathbf{P}\{X = 1\} = 1 - \mathbf{P}\{X = -1\} = p$. Denote $\kappa(p) = (1-p)/(1+p)$. Prove that for all functions $f : \{-1, 1\}^n \to [0, 1]$,

$$\alpha(\mathbf{E}f(X)) \leq \mathbf{E}\sqrt{\alpha(f(X))^2 + \max(\kappa(p), \kappa(1-p))\|\nabla^+ f(X)\|^2}.$$

Show that if f is monotone increasing,

$$\alpha(\mathbf{E}f(X)) \leq \mathbf{E}\sqrt{\alpha(f(X))^2 + \kappa(p)\|\nabla^+ f(X)\|^2}.$$

Show that for any monotone set $A \subset \{-1, 1\}^n$,

$$\mathbf{E}\|\nabla^+ \mathbb{1}_{\{A\}}\| \geq \sqrt{\frac{1+p}{1-p}} P(A)(1 - P(A)).$$

Is this inequality tight for the majority function, dictatorships, singletons? See Bobkov and Götze (1999, page 254, Proposition 2.3).

10.9. (SHARP THRESHOLD IN CHANNEL CODING) A *binary linear block code* C of length n is a subset of $\{0,1\}^n$ such that for any $u, v \in C$, $u \oplus v \in C$ where $\oplus$ denotes coordinate-wise addition modulo 2. Let C be a binary linear block code such that the Hamming distance of any two distinct elements of C is at least 2Δ. Suppose an element (a *codeword*) $v \in C$ of the code is transmitted over a *binary symmetric channel* with crossover probability $p \in (0,1)$. This means that the received message is $v \oplus X$ where X is a vector of i.i.d. Bernoulli (p) random variables. If maximum likelihood decoding is used, then the decoder picks an element of C that is closest (in Hamming distance) to $v \oplus X$. Let $A_v \subset \{0,1\}^n$ be the set such that for all $x \in A_v$, $v \oplus x$ is decoded back to v. The *decoding error* associated with the codeword v is defined by $\mathrm{err}_v(p) = 1 - \mathbf{P}\{X \in A_v\}$. Denote by p_c by $\mathrm{err}_v(p_c) = 1/2$. Use Theorem 10.11 to show that for all $v \in C$,

$$\mathrm{err}_v(p) \leq \Phi\left(\sqrt{2\Delta}\left(\sqrt{-\log p_c} - \sqrt{-\log p}\right)\right) \quad \text{for } 0 < p < p_c$$

$$\mathrm{err}_v(p) \geq \Phi\left(\sqrt{2\Delta}\left(\sqrt{-\log p_c} - \sqrt{-\log p}\right)\right) \quad \text{for } p_c < p < 1.$$

(Tillich and Zémor (2001, Theorem 3)).

10.10. (BOBKOV'S GAUSSIAN INEQUALITY IMPLIES THE GAUSSIAN LOGARITHMIC SOBOLEV INEQUALITY) Derive the Gaussian logarithmic Sobolev inequality (Theorem 5.4) from Bobkov's inequality (Theorem 10.13). Is it possible to derive Theorem 5.1 from Theorem 10.2 in the same way? *Hint:* consider

$$\frac{\sqrt{\log \frac{1}{\varepsilon}}}{\varepsilon}\left(\mathbf{E}\sqrt{\gamma^2(\varepsilon f^2) + \|\varepsilon \nabla f^2\|^2} - \gamma(\varepsilon \mathbf{E} f^2)\right),$$

let $\varepsilon \to 0$, and use $\lim_{\varepsilon \to 0} \gamma(\varepsilon) / \left(\varepsilon\sqrt{2\log(1/\varepsilon)}\right) = 1$. (Ledoux (2000) notes that the fact that the logarithmic Sobolev inequality is a consequence of Bobkov's inequality was pointed out by W. Beckner.)

10.11. (REGULARIZATION ARGUMENT IN THE PROOF OF THE GAUSSIAN ISOPERIMETRIC THEOREM) Show that the function $f_{t,\sigma} : \mathbb{R}^n \to [0,1]$, defined in the proof of the Gaussian isoperimetric theorem is infinitely many times differentiable, and $1/t$-Lipschitz for all $\sigma > 0$.

Check furthermore that when $\sigma \to 0$, $\|\nabla f_{t,\sigma}\| \to 0$ outside $A_t \setminus A$, and tends to $1/t$ in $A_t \setminus A$.

10.12. (GAUSSIAN ISOPERIMETRY AND THE BRUNN–MINKOWSKI INEQUALITY) Derive the Gaussian isoperimetric theorem for convex sets from the Gaussian Brunn–Minkowski inequality (see inequality (10.4)).

11

The Variance of Suprema of Empirical Processes

One of the principal driving forces behind the development of concentration inequalities has been the interest in understanding the magnitude of stochastic fluctuations of a norm of a sum of independent vector-valued random variables, or, equivalently, the supremum of an empirical process. This, and the next two chapters are dedicated to this subject. We are now prepared to apply the machinery developed in the previous chapters to this particular case. Concentration inequalities for the suprema of empirical processes have countless applications in probability, statistics, machine learning, harmonic analysis, and high-dimensional geometry, to name but a few principal areas.

In the first chapter devoted to this topic, we focus our attention on the *variance* of the supremum of an empirical process. In this relatively simple problem, we gain insight into some of the principal phenomena in a transparent way. In the subsequent two chapters, technically more challenging exponential concentration inequalities are developed and some tools for bounding the expected value are surveyed.

We start by defining what we mean by an empirical process. To avoid complications arising from measurability problems, we only consider processes indexed by countable index sets. In fact, the reader will not lose the essence of any argument by considering finite index sets only.

Let $\mathcal{T}$ denote a countable index set. Suppose that we are given, for each $i = 1, \ldots, n$, a collection $X_i = (X_{i,s})_{s\in\mathcal{T}}$ of real-valued random variables and assume that $X_1, \ldots, X_n$ are independent but not necessarily identically distributed. The *empirical process* indexed by $\mathcal{T}$ is the collection of random variables $\sum_{i=1}^n X_{i,s}$, $s \in \mathcal{T}$. The supremum of this empirical process is simply

$$Z = \sup_{s\in\mathcal{T}} \sum_{i=1}^n X_{i,s}.$$

If $\mathcal{T}$ contains only one element, then Z is a sum of real-valued random variables. However, if $\mathcal{T}$ has more elements, then understanding the behavior of the random variable Z is more complicated.

One may attempt to analyze suprema of empirical processes by decomposing the problem in two parts. One part is understanding the behavior of the expected value $\mathbf{E}Z$ of the supremum and the other is determining, or at least bounding, the random fluctuations of Z around its expectation. It is possible to make meaningful statements about the fluctuations of suprema of empirical processes without understanding much of their expected value. The concentration inequalities discussed in this book prove to be useful tools, as illustrated in this and the following chapter. Interestingly, concentration inequalities even prove helpful in investigating the behavior of the expected value $\mathbf{E}Z$. This is explored in Chapter 13.

We have already encountered suprema of some special empirical processes. The first example we discussed were Rademacher averages (see Section 3.2). Indeed, let $(\alpha_{i,s})$ be a collection of real numbers indexed by $i = 1, \ldots, n$ and $s \in \mathcal{T}$ and let $\varepsilon_1, \ldots, \varepsilon_n$ be independent Rademacher variables (that is, $\mathbf{P}\{\varepsilon_i = -1\} = \mathbf{P}\{\varepsilon_i = 1\} = 1/2$). Then $X_{i,s} = \alpha_{i,s}\varepsilon_i$, and

$$Z = \sup_{s \in \mathcal{T}} \sum_{i=1}^{n} X_{i,s} = \sup_{s \in \mathcal{T}} \sum_{i=1}^{n} \alpha_{i,s}\varepsilon_i$$

is the Rademacher average already mentioned in Chapter 3.

Another important example of a supremum of an empirical process is the *norm* of a sum of random vectors. To see the connection, consider first the ℓ_p norm of a vector $y = (y_1, \ldots, y_d) \in \mathbb{R}^d$ defined as $\|y\|_p = (\sum_{i=1}^d |y_i|^p)^{1/p}$ where $p \in (1, \infty)$. Let $q = p/(p-1)$ be the conjugate of p. The space $\mathbb{R}^d$ with the ℓ_q norm is called the dual space of $\mathbb{R}^d$ endowed with ℓ_p. Now let $\mathcal{T} = \{\alpha : \alpha \in \mathbb{Q}^d, \|\alpha\|_q \leq 1\}$ be the countable set of vectors in $\mathbb{R}^d$ with rational coordinates whose ℓ_q norm is at most 1. Then each α in $\mathcal{T}$ defines the linear functional on $\mathbb{R}^d$ by $\alpha(y) = \sum_{i=1}^d y_i\alpha_i$. Then $\sup_{\alpha \in \mathcal{T}} \alpha(y) = \|y\|_p$ and therefore if $Y_1, \ldots, Y_n$ are independent random vectors taking values in $\mathbb{R}^d$, then

$$Z = \left\| \sum_{i=1}^{n} Y_i \right\|_p = \sup_{\alpha \in \mathcal{T}} \sum_{i=1}^{n} \sum_{s=1}^{d} Y_{i,s}\alpha_s$$

is the supremum of an empirical process. Indeed, the same argument applies not only for the ℓ_p norm but also in any separable Banach space $\mathbb{B}$ with norm $\|\cdot\|_{\mathbb{B}}$. If $\mathcal{T}$ denotes a dense countable subset of the unit ball of the dual $\mathbb{B}'$ of $\mathbb{B}$, then

$$\|y\|_{\mathbb{B}} = \sup_{\alpha \in \mathcal{T}} \alpha(y).$$

Defining the random variable $X_{i,\alpha} = \alpha(Y_i)$, the supremum of the empirical process indexed by $\mathcal{T}$ is

$$\left\| \sum_{i=1}^{n} Y_i \right\|_{\mathbb{B}} = \sup_{\alpha \in \mathcal{T}} \sum_{i=1}^{n} \alpha(Y_i) = \sup_{\alpha \in \mathcal{T}} \sum_{i=1}^{n} X_{i,\alpha}.$$

We have already considered some examples of concentration inequalities for norms of sums of independent random vectors. Norms of Gaussian random vectors are discussed in Section 5.4 and the Bonami–Beckner inequalities of Section 5.8 translate into concentration inequalities for suprema of Rademacher processes. Note also that the largest eigenvalue of a random symmetric matrix discussed in Examples 3.14 and 6.8 also fits within this framework.

Often the indices $s \in \mathcal{T}$ may be associated with measurable functions $f_s : \mathcal{X} \to \mathbb{R}$, $s \in \mathcal{T}$ defined on some set $\mathcal{X}$. If $Y_1, \ldots, Y_n$ are independent random variables taking values in $\mathcal{X}$, then by defining $X_{i,s} = f_s(Y_i)$, the supremum of the empirical process equals

$$\sup_{s\in\mathcal{T}} \sum_{i=1}^{n} X_{i,s} = \sup_{s\in\mathcal{T}} \sum_{i=1}^{n} f_s(Y_i).$$

In most applications of empirical processes appearing in statistics and machine learning, this is the most frequent notation. A classical example is the Kolmogorov–Smirnov statistics. The Y_i's are assumed to be independently and uniformly distributed over $[0, 1]$ and for each rational $s \in [0, 1]$, a function f_s is defined by $f_s(x) = \mathbb{1}_{\{x\le s\}} - s$. Then the (one-sided) Kolmogorov–Smirnov statistics is the supremum

$$\sup_{s\in[0,1]} \sum_{i=1}^{n} (\mathbb{1}_{\{Y_i\le s\}} - s) = \sup_{s\in[0,1]\cap\mathbb{Q}} \sum_{i=1}^{n} (\mathbb{1}_{\{Y_i\le s\}} - s).$$

This chapter is mostly devoted to upper bounding the variance of suprema of empirical processes, but we also mention upper bounds for the second moment $\mathbf{E}Z^2$. Our main tool is, once again, the Efron–Stein inequality. Various estimates of the variance are derived and the bounds often involve one of the following three quantities:

$$V = \sum_{i=1}^{n} \mathbf{E} \sup_{s\in\mathcal{T}} X_{i,s}^2$$

$$\Sigma^2 = \mathbf{E} \sup_{s\in\mathcal{T}} \sum_{i=1}^{n} X_{i,s}^2$$

$$\sigma^2 = \sup_{s\in\mathcal{T}} \sum_{i=1}^{n} \mathbf{E} X_{i,s}^2.$$

Clearly, $\sigma^2 \le \Sigma^2 \le V$. Given the lack of a standard terminology, we will refer to V, Σ^2, and σ^2 as the *strong variance, weak variance,* and *wimpy variance,* respectively. In general, there may be significant gaps between any two of these quantities. A notable difference is the case of Rademacher averages when $\sigma^2 = \Sigma^2$.

In this and the next two chapters, results are stated and proved for empirical processes indexed by finite or countable sets. These results can often be easily extended to suprema of empirical processes indexed by uncountable sets. The empirical process is said to be *separable* if there exists a countable subset $\mathcal{S} \subset \mathcal{T}$ such that, almost surely, for all $i = 1, \ldots, n$ and for all $t \in \mathcal{T}$, there exists a sequence $\{t_n\}$ of elements of $\mathcal{S}$ such that X_{i,t_n} converges to $X_{i,t}$, that is, $\mathcal{T}$ contains a dense countable subset $\mathcal{S}$ with respect to the topology of pointwise convergence. The subset $\mathcal{S}$ is sometimes called the separant.

If $\mathcal{T}$ admits a countable separant $\mathcal{S}$, then

$$Z = \sup_{t \in \mathcal{T}} \sum_{i=1}^{n} X_{i,t} = \sup_{s \in \mathcal{S}} \sum_{i=1}^{n} X_{i,s} \quad \text{almost surely.}$$

This shows that the supremum of a separable empirical process is measurable. Let $Y = \sup_{s \in \mathcal{S}} \sum_{i=1}^{n} X_{i,s}$ be the supremum of the empirical process indexed by the countable separant. Note that Y is the monotone limit of maxima computed over finite sets. One can easily check that $\mathrm{Var}(Z) = \mathrm{Var}(Y)$, whether the quantities are finite or not. Moreover, almost surely, we also have

$$\sup_{t \in \mathcal{T}} \sum_{i=1}^{n} X_{i,t}^2 = \sup_{s \in \mathcal{S}} \sum_{i=1}^{n} X_{i,s}^2.$$

All results stated in this and the next two chapters for suprema of empirical processes indexed by finite of countable sets extend to suprema of separable processes.

In Section 11.1 we begin by showing how the Efron–Stein inequality implies that $\mathrm{Var}(Z) \le V$ and $\mathrm{Var}(Z) \le \Sigma^2 + \sigma^2$. A natural question is then whether $\mathrm{Var}(Z) \le \sigma^2$. However, this is easily shown to be false by a counter-example (see Exercise 11.1).

In Section 11.2 we proceed with a discussion of Nemirovski's inequality. Nemirovski's original inequality relates $\mathbf{E}Z^2$ to V. We argue that it makes sense to bound $\mathbf{E}Z^2$ in terms of Σ^2 and that this may significantly improve the original inequality. We point out that the difference between the weak and strong variances can be quite substantial.

Even though σ^2 may be smaller than $\mathrm{Var}(Z)$, Σ^2 (and therefore also $\mathrm{Var}(Z)$) can be upper bounded by a linear combination of $\mathbf{E}Z$ and σ^2. This result follows from some basic symmetrization and contraction inequalities from empirical process theory presented in Section 11.3. The connexion between wimpy and weak variances $\Sigma^2 \le 2\mathbf{E}Z + \sigma^2$ can be established in a simple way for centered empirical processes uniformly bounded by 1 and with identically distributed summands (see Theorem 11.10).

Connecting the wimpy and weak variances without the uniform boundedness assumption requires more effort, namely an appropriate truncation argument. This is the subject of the Hoffmann–Jørgensen inequalities described in Section 11.5.

11.1 General Upper Bounds for the Variance

In Section 3.2 we saw how the Efron–Stein inequality allows one to derive sharp upper bounds for the variance of Rademacher averages. Here we show that they prove equally useful when dealing with suprema of general empirical processes. The following proposition describes easy and general upper bounds for the variance. Despite their simplicity, these bounds can be sharp, for example when the index set $\mathcal{T}$ contains only one element.

Theorem 11.1 *Let $Z = \sup_{s \in \mathcal{T}} \sum_{i=1}^n X_{i,s}$ be the supremum of an empirical process as defined above. Then*

$$\mathrm{Var}(Z) \leq V.$$

If $\mathbf{E}X_{i,s} = 0$ for all $i = 1, \ldots, n$ and for all $s \in \mathcal{T}$, then

$$\mathrm{Var}(Z) \leq \Sigma^2 + \sigma^2.$$

Proof To prove the first inequality, introduce $Z_i = \sup_{s \in \mathcal{T}} \sum_{j: j \neq i} X_{j,s}$. Let $\hat{s} \in \mathcal{T}$ be such that $Z = \sum_{i=1}^n X_{i,\hat{s}}$ and let $\hat{s}_i$ be such that $Z_i = \sum_{j \neq i} X_{j,\hat{s}_i}$. (We implicitly assume here that the suprema in the definition of Z and Z_i are achieved. This is not necessarily the case if $\mathcal{T}$ is not a finite set. In that case one can define $\hat{s}$ and $\hat{s}_i$ as appropriate approximate minimizers and the argument carries over.) Then

$$(Z - Z_i)_+ \leq (X_{i,\hat{s}})_+ \leq \sup_{s \in \mathcal{T}} |X_{i,s}|$$

and

$$(Z - Z_i)_- \leq (X_{i,\hat{s}_i})_- \leq \sup_{s \in \mathcal{T}} |X_{i,s}|,$$

so

$$\sum_{i=1}^n (Z - Z_i)^2 \leq \sum_{i=1}^n \sup_{s \in \mathcal{T}} X_{i,s}^2.$$

The first inequality follows from Efron–Stein inequality.

To prove the second, for each $i = 1, \ldots, n$, let $Z_i' = \sup_{s \in \mathcal{T}} \left(\sum_{j \neq i} X_{j,s} + X_{i,s}' \right)$ where X_i' is an independent copy of X_i. Note that

$$(Z - Z_i')_+^2 \leq \left(X_{i,\hat{s}} - X_{i,\hat{s}}' \right)^2.$$

Denoting by $\mathbf{E}'$ the expectation with respect to the random variables $X_1', \ldots, X_n'$, by the Efron–Stein inequality,

$$
\begin{aligned}
\operatorname{Var}(Z) &\leq E \sum_{i=1}^{n} (Z - Z_i')_+^2 \\
&\leq E \sum_{i=1}^{n} E' \left[\left(X_{i,\hat{s}} - X_{i,\hat{s}}' \right)^2 \right] \\
&\leq E \sum_{i=1}^{n} \left(X_{i,\hat{s}}^2 + E' \left[X_{i,\hat{s}}'^2 \right] \right) \\
&\quad \left(\text{as } X_{i,\hat{s}}' \text{ is independent of } \hat{s} \text{ and } E' X_{i,\hat{s}}' = 0 \right) \\
&\leq E \sup_{s \in \mathcal{T}} \sum_{i=1}^{n} X_{i,s}^2 + \sup_{s \in \mathcal{T}} \sum_{i=1}^{n} E X_{i,s}^2. \qquad \square
\end{aligned}
$$

Note that Theorem 11.1 still holds if the process is assumed to be separable (see Exercise 11.3).

11.2 Nemirovski's Inequality

Next we show how the upper bounds for the variance described in the previous section can be used to obtain bounds for EZ^2 where Z is the norm of a sum of independent vector-valued random variables. Let $X_1, \ldots, X_n$ be independent random variables with values in a complete separable normed space $\mathbb{B}$ satisfying $EX_i = 0$ and let

$$S_n = \sum_{i=1}^{n} X_i.$$

In this section, we assume that all $E\left[\|X_i\|_{\mathbb{B}}^2 \right]$ are finite. Our purpose is to relate $E\left[\|S_n\|_{\mathbb{B}}^2 \right]$ to the values $E\left[\|X_i\|_{\mathbb{B}}^2 \right]$, $i = 1, \ldots, n$. The original question raised by Nemirovski was whether there exists a constant $\kappa = \kappa(\mathbb{B})$ such that

$$E\left[\|S_n\|_{\mathbb{B}}^2 \right] \leq \kappa(\mathbb{B}) \sum_{i=1}^{n} E\left[\|X_i\|_{\mathbb{B}}^2 \right].$$

If the normed space $\mathbb{B}$ is a Euclidean space (or more generally a Hilbert space), then $E\left[\|S_n\|_{\mathbb{B}}^2 \right] = \sum_{i=1}^{n} E\left[\|X_i\|_{\mathbb{B}}^2 \right]$ and we may take $\kappa(\mathbb{B}) = 1$. When considering other norms, this simple connection breaks down. If $\mathbb{B}$ is finite-dimensional (e.g. $\mathbb{B} = \mathbb{R}^d$ endowed with the ℓ_p norm with $p \in [1, \infty]$), then $\kappa(\mathbb{B}) \leq \dim(\mathbb{B})$ (see Section 13.5 and the bibliographical remarks for references). However, in some interesting cases, much better bounds are possible. In this section, we focus on $\mathbb{R}^d$ equipped with the ℓ_∞ norm $\|y\|_\infty = \max_{i=1,\ldots,d} |y_i|$. In Section 13.4, we point out that the results presented here extend to Gaussian and Rademacher sums of symmetric matrices equipped with an operator norm. In Section 13.5, we discuss the case of $\mathbb{R}^d$ under the ℓ_p norm for $1 \leq p < \infty$.

Theorem 11.2 *Let $X_1, \ldots, X_n$ be independent random variables taking their values in $\mathbb{R}^d$ such that they are symmetric (i.e. $-X_i$ has the same distribution as X_i). Let $S_n = \sum_{i=1}^n X_i$. If $\Sigma^2 = \mathbf{E} \max_{j=1,\ldots,d} \sum_{i=1}^n X_{i,j}^2$, define*

$$C(n, d) = \sup \frac{\mathbf{E} \|S_n\|_\infty^2}{\Sigma^2},$$

where the supremum is taken over all distributions of independent, $\mathbb{R}^d$-valued symmetric random variables $X_1, \ldots, X_n$ with finite Σ^2.

Then, $C(n, d)$ is a non-decreasing function of both d and n, and

$$C(n, d) \leq 2(1 + \log(2d)).$$

Moreover, letting $C(\infty, d) = \lim_{n\to\infty} C(n, d)$, we have, for $d \geq 2$,

$$C(\infty, d) \geq \left(\Phi^{-1}\left(1 - \frac{1}{2(d+1)}\right)\right)^2$$

and

$$\lim_{d\to\infty} \frac{C(\infty, d)}{2\log(d)} = 1.$$

The proof of Theorem 11.2, uses the next technical lemma.

Lemma 11.3 *Let $X_1, \ldots, X_n$ be independent standard Gaussian random variables. Let Φ denote the distribution function of the standard Gaussian distribution. Then*

$$\Phi^{-1}\left(1 - 1/(2(n+1))\right) \leq \mathbf{E} \max_{i=1,\ldots,n} |X_i| \leq \sqrt{2\log(2n)}$$

and

$$\mathbf{E} \max_{i=1,\ldots,n} X_i^2 \leq 1 + 2\log(2n).$$

Moreover,

$$\lim_{n\to\infty} \frac{\mathbf{E} \max_{i=1,\ldots,n} |X_i|}{\sqrt{2\log n}} = \lim_{n\to\infty} \frac{\mathbf{E} \max_{i=1,\ldots,n} X_i^2}{2\log n} = 1.$$

Proof As

$$\mathbf{E} \max_{i=1,\ldots,n} |X_i| = \mathbf{E} \max_{i=1,\ldots,n} \max(-X_i, X_i),$$

Theorem 2.5 implies that $\mathbf{E} \max_{i=1,\ldots,n} |X_i| \leq \sqrt{2\log(2n)}$. The random variable $\max_{i=1,\ldots,n} |X_i|$ is distributed like $\Phi^{-1}\left((1 + \max(U_1, \ldots, U_n))/2\right)$, where $U_1, \ldots, U_n$

are independent uniformly distributed over $[0, 1]$. By convexity of Φ^{-1} over $[1/2, 1)$, Jensen's inequality implies

$$E \max_{i=1,\ldots,n} |X_i| \geq \Phi^{-1}\left(\frac{1 + E\max(U_1, \ldots, U_n)}{2}\right) = \Phi^{-1}\left(1 - \frac{1}{2(n+1)}\right).$$

By the Gaussian Poincaré inequality (Theorem 3.20), as the maximum of the absolute values is a 1-Lipschitz function,

$$E \max_{i=1,\ldots,n} X_i^2 \leq 1 + \left(E \max_{i=1,\ldots,n} |X_i|\right)^2 .$$

The last statement follows from the fact that $\lim_{t\to\infty} \Phi^{-1}(1 - 1/t)/\sqrt{2\log t} = 1$ (see Exercise 11.7). □

Proof Throughout this proof $\|\cdot\|$ stands for $\|\cdot\|_\infty$. The fact that $C(n, d)$ is non-decreasing in both n and d is obvious from the definition.

By definition of the variance,

$$E\left[\|S_n\|^2\right] = \mathrm{Var}\left(\|S_n\|\right) + \left(E\|S_n\|\right)^2.$$

Since

$$\|S_n\| = \max_{\substack{j=1,\ldots,d \\ b\in\{-1,1\}}} \sum_{i=1}^n b X_{i,j},$$

we may write $\|S_n\|$ as the supremum of an empirical process with a finite index set. By Theorem 11.1,

$$\mathrm{Var}\left(\|S_n\|\right) \leq 2\Sigma^2.$$

As the random variables X_i are assumed to be symmetric, $(X_1, \ldots, X_n)$ is distributed as $\varepsilon_1 X_1, \ldots, \varepsilon_n X_n$ where the ε_i are independent Rademacher variables. Then

$$E\|S_n\| = E \max_{j=1,\ldots,d} \left|\sum_{i=1}^n \varepsilon_i X_{i,j}\right|,$$

and fixing the variables X_i's and bounding the expectation with respect to the ε_i, it follows from Theorem 2.5 that

$$E\left[\max_{j=1,\ldots,d} \left|\sum_{i=1}^n \varepsilon_i X_{i,j}\right| \mid X_1, \ldots, X_n\right] \leq \sqrt{2\log(2d) \max_{j=1,\ldots,d} \sum_{i=1}^n X_{i,j}^2}.$$

Taking expectation on both sides of this inequality and using Jensen's inequality to bring the expectation under the square root sign, we obtain

$$\mathbf{E}\|S_n\| = \mathbf{E} \max_{j=1,\ldots,d} \left| \sum_{i=1}^n \varepsilon_i X_{i,j} \right| \le \sqrt{2\log(2d)\, \mathbf{E} \max_{j=1,\ldots,d} \sum_{i=1}^n X_{i,j}^2}.$$

Combining this inequality with the bound $\mathrm{Var}(\|S_n\|) \le 2\Sigma^2$ and the definition of the variance, we finally get

$$\mathbf{E}\,\|S_n\|^2 \le 2(1 + \log(2d))\Sigma^2.$$

This proves that $C(n,d) \le 2(1 + \log(2d))$ for all n and d.

Let χ_d^2 denote the chi-square distribution with d degrees of freedom. (This is the distribution of the sum of the squares of d independent standard Gaussian random variables.)

The lower bound for $C(\infty, d)$ follows simply by considering the case when each X_i is a d-dimensional standard Gaussian vector. Then for each $j = 1, \ldots, d$, $(1/n)(\sum_{i=1}^n X_{i,j})^2$ is distributed according to χ_1^2. Hence $\|S_n\|/\sqrt{n}$ is distributed as the maximum of the absolute values of d independent standard Gaussian random variables. By Jensen's inequality, Lemma 11.3 implies

$$\frac{\mathbf{E}\|S_n\|^2}{n} \ge \left(\Phi^{-1}\left(1 - \frac{1}{2(d+1)}\right)\right)^2.$$

On the other hand, Σ^2 is the expected value of the maximum of d independent χ_n^2-distributed random variables. By Theorem 2.7, we have

$$\Sigma^2 \le n + 2\sqrt{n \log d} + 2\log d.$$

Putting these two bounds together,

$$C(n,d) \ge \frac{\left(\Phi^{-1}\left(1 - \frac{1}{2(d+1)}\right)\right)^2}{1 + 2\sqrt{\log d/n} + 2\log d/n}.$$

The last statement follows from the fact that $\lim_{t\to\infty} \Phi^{-1}(1 - 1/t)/\sqrt{2\log t} = 1$. □

In the rest of this section we compare the strong and weak variances

$$V = \sum_{i=1}^n \mathbf{E} \max_{j=1,\ldots,d} X_{i,j}^2 \quad \text{and} \quad \Sigma^2 = \mathbf{E} \max_{j=1,\ldots,d} \sum_{i=1}^n X_{i,j}^2.$$

Since

$$\Sigma^2 \geq \mathbf{E} \max_{i=1,\dots,n} \max_{j=1,\dots,d} X_{i,j}^2,$$

we clearly have

$$\Sigma^2 \leq V \leq n\Sigma^2.$$

On the other hand,

$$V \leq \mathbf{E} \sum_{j=1}^{d} \sum_{i=1}^{n} X_{i,j}^2 \leq d\Sigma^2,$$

and therefore

$$\Sigma^2 \leq V \leq \min(n, d)\,\Sigma^2.$$

Next we illustrate in two different examples that the ratio V/Σ^2 can indeed be of the order of $\min(n, d)$.

Consider first the case when $X_{i,j} = \varepsilon_i a_{i,j}$ where $\varepsilon_1, \dots, \varepsilon_n$ are i.i.d. Rademacher random variables. Then

$$\Sigma^2 = \max_{j=1,\dots,d} \sum_{i=1}^{n} a_{i,j}^2 \qquad \text{while} \qquad V = \sum_{i=1}^{n} \max_{j=1,\dots,d} a_{i,j}^2.$$

Choosing $a_{j,j} = 1$ for every $j \leq \min(n, d)$ and $a_{i,j} = 0$ otherwise, we see that $\Sigma^2 = 1$ while $V = \min(n, d)$. This shows that the ratio V/Σ^2 can indeed achieve the maximal possible value $\min(n, d)$.

It is even more interesting to notice that the ratio V/Σ^2 can be large even when the variables $X_{i,j}$ are i.i.d. standard normal random variables. In this case

$$\mathbf{E}\, \|S_n\|_\infty^2 = V = n\mathbf{E} \max_{j=1,\dots,d} Y_j^2$$

where $Y_1, \dots, Y_d$ are independent standard normal random variables. On the other hand, as we have seen in the proof of Theorem 11.2,

$$\Sigma^2 \leq n + 2\left(\sqrt{n \log d} + \log d\right).$$

Since $\Sigma^2 \geq n$, we obtain

$$\frac{\mathbf{E} \max_{j=1,\dots,d} Y_j^2}{1 + 2\sqrt{(\log d)/n} + 2(\log d)/n} \leq \frac{V}{\Sigma^2} \leq \mathbf{E} \max_{j=1,\dots,d} Y_j^2.$$

To interpret this bound, we distinguish three different asymptotic regimes of dependence of $d = d_n$ on n as $n \to \infty$.

Case 1. Assume first $\lim_{n\to\infty}(\log d_n)/n = 0$. By Lemma 11.3, we have $V/\Sigma^2 \sim 2\log(d_n)$ as $n \to \infty$. Since the convergence of $(\log d_n)/n$ to 0 can be arbitrarily slow, the ratio V/Σ^2 is close to its maximal value n.

Case 2. If $\lim_{n\to\infty}(\log d_n)/n = \alpha$ for some $\alpha > 0$, then V/Σ^2 is of the order of n in the sense that

$$\liminf_{n\to\infty} \frac{V}{n\Sigma^2} \geq \frac{2\alpha}{1+\sqrt{2\alpha}+2\alpha}.$$

Case 3. Finally, if $\lim_{n\to\infty}(\log d_n)/n = \infty$, then $\lim_{n\to\infty} V/(n\Sigma^2) = 1$.

11.3 The Symmetrization and Contraction Principles

The bound $\mathrm{Var}(Z) \leq \Sigma^2 + \sigma^2$ of Theorem 11.1 shows that by understanding Σ^2, one has a good grasp on the size of the random fluctuations of the supremum of the empirical process. However, often the wimpy variance σ^2 is easier to interpret than Σ^2. Luckily, if an upper bound for $\sup_{s\in\mathcal{T}} |X_{i,s}|$ is available, σ^2 and Σ^2 may be related by a simple inequality which we present next. This inequality is based on symmetrization inequalities and contraction principles, which are useful and frequently used tools in the theory of empirical processes. We start with a simple symmetrization inequality.

Lemma 11.4 (SYMMETRIZATION INEQUALITIES) *Let $X_1,\ldots,X_n$ be independent random vectors where $X_i = (X_{i,s})_{s\in\mathcal{T}}$. Assume that the process is centered, that is, for each $i = 1,\ldots,n$ and $s \in \mathcal{T}$, $\mathbf{E}X_{i,s} = 0$. Let $\varepsilon_1,\ldots,\varepsilon_n$ be a sequence of independent Rademacher variables independent of $X_1,\ldots,X_n$. Then*

$$\frac{1}{2}\mathbf{E}\sup_{s\in\mathcal{T}}\left|\sum_{i=1}^n \varepsilon_i X_{i,s}\right| \leq \mathbf{E}\sup_{s\in\mathcal{T}}\left|\sum_{i=1}^n X_{i,s}\right| \leq 2\mathbf{E}\sup_{s\in\mathcal{T}}\left|\sum_{i=1}^n \varepsilon_i X_{i,s}\right|$$

and

$$\mathbf{E}\sup_{s\in\mathcal{T}}\sum_{i=1}^n X_{i,s} \leq 2\mathbf{E}\sup_{s\in\mathcal{T}}\sum_{i=1}^n \varepsilon_i X_{i,s}.$$

Proof We start with the second inequality. The proof of the last inequality is similar. Let $X'_1,\ldots,X'_n$ be distributed as $X_1,\ldots,X_n$ but independent of them. This means that the random vectors $X_i - X'_i$ are independent and symmetric, distributed as the $\varepsilon_i(X_i - X'_i)$. Thus,

$$
\begin{aligned}
\mathbf{E} \sup_{s \in \mathcal{T}} \left| \sum_{i=1}^{n} X_{i,s} \right| &= \mathbf{E} \sup_{s \in \mathcal{T}} \left| \sum_{i=1}^{n} \left(X_{i,s} - \mathbf{E} X'_{i,s} \right) \right| \\
&\leq \mathbf{E} \sup_{s \in \mathcal{T}} \left| \sum_{i=1}^{n} \left(X_{i,s} - X'_{i,s} \right) \right| \qquad \text{(by Jensen's inequality)} \\
&= \mathbf{E} \sup_{s \in \mathcal{T}} \left| \sum_{i=1}^{n} \varepsilon_i \left(X_{i,s} - X'_{i,s} \right) \right| \\
&\leq 2\mathbf{E} \sup_{s \in \mathcal{T}} \left| \sum_{i=1}^{n} \varepsilon_i X_{i,s} \right| .
\end{aligned}
$$

The first inequality follows by a similar argument:

$$
\begin{aligned}
\frac{1}{2} \mathbf{E} \sup_{s \in \mathcal{T}} \left| \sum_{i=1}^{n} \varepsilon_i X_{i,s} \right| &= \frac{1}{2} \mathbf{E} \sup_{s \in \mathcal{T}} \left| \sum_{i=1}^{n} \varepsilon_i (X_{i,s} - \mathbf{E} X'_{i,s}) \right| \\
&\leq \frac{1}{2} \mathbf{E} \sup_{s \in \mathcal{T}} \left| \sum_{i=1}^{n} \varepsilon_i (X_{i,s} - X'_{i,s}) \right| \qquad \text{(by Jensen's inequality)} \\
&= \frac{1}{2} \mathbf{E} \sup_{s \in \mathcal{T}} \left| \sum_{i=1}^{n} (X_{i,s} - X'_{i,s}) \right| \\
&\leq \mathbf{E} \sup_{s \in \mathcal{T}} \left| \sum_{i=1}^{n} X_{i,s} \right| . \qquad \square
\end{aligned}
$$

Symmetrization inequalities motivate the use of conditional Rademacher averages in empirical process theory. The conditional Rademacher average associated with the empirical process $\sum_{i=1}^{n} X_i$ is the conditional expectation of the supremum of the symmetrized empirical process $\sum_{i=1}^{n} \varepsilon_i X_i$, given $X_1, \ldots, X_n$, defined as

$$
\mathbf{E} \left[\sup_{s \in \mathcal{T}} \sum_{i=1}^{n} \varepsilon_i X_{i,s} \middle| X_1, \ldots, X_n \right] .
$$

We have already encountered conditional Rademacher averages in Section 3.3 where we showed that if $\sup_{s \in \mathcal{T}, i \leq n} |X_{i,s}| \leq 1$ almost surely, then the conditional Rademacher average is a self-bounding function. This implies that conditional Rademacher averages are relatively stable (their variance is not larger than their expected value) and have sub-Poissonian tails (see Section 6.7). Lemma 11.4 complements this observation. It shows that, up to a constant factor, conditional Rademacher averages estimate the expected value of the supremum of the underlying empirical process.

Another simple and useful tool in empirical process theory is the so-called contraction principle. We start by an easy version followed by a more general formulation.

Theorem 11.5 *Let $x_1, \dots, x_n$ be vectors whose real-valued components are indexed by $\mathcal{T}$, that is, $x_i = (x_{i,s})_{s\in\mathcal{T}}$. Let $\alpha_i \in [0,1]$ for $i = 1, \dots, n$. Let $\varepsilon_1, \dots, \varepsilon_n$ be independent Rademacher random variables. Then*

$$\mathbf{E} \sup_{s\in\mathcal{T}} \sum_{i=1}^{n} \varepsilon_i \alpha_i x_{i,s} \le \mathbf{E} \sup_{s\in\mathcal{T}} \sum_{i=1}^{n} \varepsilon_i x_{i,s}.$$

Proof Let $\Psi : (\mathbb{R}^{\mathcal{T}})^n \to \mathbb{R}$ be defined by

$$\Psi(x_1, \dots, x_n) = \mathbf{E} \sup_{s\in\mathcal{T}} \sum_{i=1}^{n} \varepsilon_i x_{i,s}.$$

The function Ψ is convex since it is a linear combination of suprema of linear functions. It is also invariant under sign change in the sense that for all $(\eta_1, \dots, \eta_n) \in \{-1,1\}^n$,

$$\Psi(x_1, \dots, x_n) = \Psi(\eta_1 x_1, \dots, \eta_n x_n).$$

Fix $(x_1, \dots, x_n) \in (\mathbb{R}^{\mathcal{T}})^n$. Consider the restriction of Ψ to the convex hull of the 2^n points of the form $(\eta_1 x_1, \dots, \eta_n x_n)$, with $(\eta_1, \dots, \eta_n) \in \{-1,1\}^n$. The supremum of Ψ is achieved at one of the vertices $(\eta_1 x_1, \dots, \eta_n x_n)$. The sequence of vectors $(\alpha_1 x_1, \dots, \alpha_n x_n)$ lies inside the convex hull of $(\eta_1 x_1, \dots, \eta_n x_n)$, $(\eta_1, \dots, \eta_n) \in \{-1,1\}^n$ and therefore

$$\begin{aligned}\mathbf{E} \sup_{s\in\mathcal{T}} \sum_{i=1}^{n} \varepsilon_i \alpha_i x_{i,s} &= \Psi(\alpha_1 x_1, \dots, \alpha_n x_n) \\ &\le \Psi(x_1, \dots, x_n) \\ &= \mathbf{E} \sup_{s\in\mathcal{T}} \sum_{i=1}^{n} \varepsilon_i x_{i,s}.\end{aligned}$$ □

The next theorem generalizes Theorem 11.5. It serves not only for comparing expectations but also higher moments, moment-generating functions, and tail probabilities.

Theorem 11.6 (CONTRACTION PRINCIPLE) *Let $x_1, \dots, x_n$ be vectors whose real-valued components are indexed by $\mathcal{T}$, that is, $x_i = (x_{i,s})_{s\in\mathcal{T}}$. For each $i = 1, \dots, n$ let $\varphi_i : \mathbb{R} \to \mathbb{R}$ be a 1-Lipschitz function such that $\varphi_i(0) = 0$. Let $\varepsilon_1, \dots, \varepsilon_n$ be independent Rademacher random variables, and let $\Psi : [0, \infty) \to \mathbb{R}$ be a non-decreasing convex function. Then*

$$\mathbf{E}\left[\Psi\left(\sup_{s\in\mathcal{T}} \sum_{i=1}^{n} \varepsilon_i \varphi_i(x_{i,s})\right)\right] \le \mathbf{E}\left[\Psi\left(\sup_{s\in\mathcal{T}} \sum_{i=1}^{n} \varepsilon_i x_{i,s}\right)\right]$$

and

$$\mathbf{E}\left[\Psi\left(\frac{1}{2}\sup_{s\in\mathcal{T}}\left|\sum_{i=1}^{n}\varepsilon_i\varphi_i(x_{i,s})\right|\right)\right] \leq \mathbf{E}\left[\Psi\left(\sup_{s\in\mathcal{T}}\left|\sum_{i=1}^{n}\varepsilon_i x_{i,s}\right|\right)\right].$$

The proof is based on the following technical lemma.

Lemma 11.7 *Let $\Psi : \mathbb{R} \to \mathbb{R}$ denote a convex nondecreasing function. Let $\varphi : \mathbb{R} \to \mathbb{R}$ be a 1-Lipschitz function such that $\varphi(0) = 0$. Let $\mathcal{T} \subset \mathbb{R}^2$. Then*

$$\Psi\left(\sup_{s\in\mathcal{T}}(s_1 + \varphi(s_2))\right) + \Psi\left(\sup_{s\in\mathcal{T}}(s_1 - \varphi(s_2))\right)$$

$$\leq \Psi\left(\sup_{s\in\mathcal{T}}(s_1 + s_2)\right) + \Psi\left(\sup_{s\in\mathcal{T}}(s_1 - s_2)\right).$$

Proof Since Ψ is convex and nondecreasing, if a, b, c, d are such that $0 \leq d - c \leq b - a$ and $c \leq a$, then

$$\Psi(d) - \Psi(c) \leq \Psi(b) - \Psi(a). \tag{11.1}$$

Denote by $\hat{s} = (\hat{s}_1, \hat{s}_2)$ and $\hat{t} = (\hat{t}_1, \hat{t}_2)$ the elements of $\mathcal{T}$ that achieve the suprema on the left-hand side. It suffices to show that

$$\Psi(\hat{s}_1 + \varphi(\hat{s}_2)) + \Psi(\hat{t}_1 - \varphi(\hat{t}_2)) \leq \Psi(\hat{s}_1 + \hat{s}_2) + \Psi(\hat{t}_1 - \hat{t}_2),$$

or, equivalently,

$$\Psi(\hat{t}_1 - \varphi(\hat{t}_2)) - \Psi(\hat{t}_1 - \hat{t}_2) \leq \Psi(\hat{s}_1 + \hat{s}_2) - \Psi(\hat{s}_1 + \varphi(\hat{s}_2)).$$

As Ψ is nondecreasing we have both

$$\hat{s}_1 + \varphi(\hat{s}_2) \geq \hat{t}_1 + \varphi(\hat{t}_2)$$

and

$$\hat{s}_1 - \varphi(\hat{s}_2) \leq \hat{t}_1 - \varphi(\hat{t}_2),$$

which implies

$$\varphi(\hat{t}_2) - \varphi(\hat{s}_2) \leq \hat{s}_1 - \hat{t}_1 \leq \varphi(\hat{s}_2) - \varphi(\hat{t}_2),$$

and therefore

$$|\hat{s}_1 - \hat{t}_1| \le \varphi(\hat{s}_2) - \varphi(\hat{t}_2) \le |\hat{s}_2 - \hat{t}_2|,$$

where the last inequality follows from the fact that φ is 1-Lipschitz.

First consider the case when $\hat{s}_2$ and $\hat{t}_2$ are both positive. We may assume that $\hat{s}_2 \ge \hat{t}_2 \ge 0$ because otherwise we may exchange the roles of $\hat{s}$ and $\hat{t}$ and change the sign of φ. This implies that $\hat{s}_2 - \varphi(\hat{s}_2) \ge \hat{t}_2 - \varphi(\hat{t}_2) \ge 0$. Moreover, as $\hat{s}_1 + \varphi(\hat{s}_2) \ge \hat{t}_1 + \varphi(\hat{t}_2) \ge \hat{t}_1 - \hat{t}_2$, (11.1) allows us to conclude.

Consider now the case where $\hat{s}_2$ and $\hat{t}_2$ are both negative. Similarly to the previous case, we may assume that $\hat{t}_2 \le \hat{s}_2 \le 0$. Now we have $0 \le \varphi(\hat{s}_2) - \hat{s}_2 \le \varphi(\hat{t}_2) - \hat{t}_2$ and $\hat{s}_1 + \hat{s}_2 \le \hat{s}_1 - \varphi(\hat{s}_2) \le \hat{t}_1 - \varphi(\hat{t}_2)$. Once again, (11.1) allows us to conclude.

To end the proof, consider the situation when $\hat{t}_2 \le 0 \le \hat{s}_2$. Then $\Psi(\hat{t}_1 - \varphi(\hat{t}_2)) - \Psi(\hat{t}_1 - \hat{t}_2) \le 0$ and $0 \le \Psi(\hat{s}_1 + \hat{s}_2) - \Psi(\hat{s}_1 + \varphi(\hat{s}_2))$ as $-\hat{t}_2 \le -\varphi(\hat{t}_2)$ and $\varphi(\hat{s}_2) \le \hat{s}_2$. This is enough to conclude. The last case can be handled by changing the sign of φ and permuting $\hat{s}$ and $\hat{t}$. □

Proof of Theorem 11.6. We begin by proving the first inequality. It suffices to prove that, if $\mathcal{T} \subset \mathbb{R}^n$ is a finite set of vectors $s = (s_1, \ldots, s_n)$, then

$$\mathbf{E}\left[\Psi\left(\sup_{s\in\mathcal{T}} \sum_{i=1}^{n} \varepsilon_i \varphi_i(s_i)\right)\right] \le \mathbf{E}\left[\Psi\left(\sup_{s\in\mathcal{T}} \sum_{i=1}^{n} \varepsilon_i s_i\right)\right].$$

The key step is that for an arbitrary function $A : \mathcal{T} \to \mathbb{R}$,

$$\mathbf{E}\left[\Psi\left(\sup_{s\in\mathcal{T}} A(s) + \sum_{i=1}^{n} \varepsilon_i \varphi_i(s_i)\right)\right] \le \mathbf{E}\left[\Psi\left(\sup_{s\in\mathcal{T}} A(s) + \sum_{i=1}^{n} \varepsilon_i s_i\right)\right]. \tag{11.2}$$

The base case $n = 1$ is handled by Lemma 11.7. In this case (11.2) is equivalent to

$$\mathbf{E}\left[\Psi\left(\sup_{u\in\mathcal{U}}(u_1 + \varepsilon\varphi(u_2))\right)\right] \le \mathbf{E}\left[\Psi\left(\sup_{u\in\mathcal{U}}(u_1 + \varepsilon u_2)\right)\right],$$

where $\mathcal{U} = \{(A(s), s) : s \in \mathcal{T}\}$.

The proof of (11.2) goes by induction on n:

$$
\begin{aligned}
&\boldsymbol{E}\left[\Psi\left(\sup_{s\in\mathcal{T}} A(s)+\sum_{i=1}^{n}\varepsilon_i\varphi_i(s_i)\right)\right] \\
&= \boldsymbol{E}\left[\boldsymbol{E}\left[\Psi\left(\sup_{s\in\mathcal{T}} A(s)+\sum_{i=1}^{n-1}\varepsilon_i\varphi_i(s_i)+\varepsilon_n\varphi_n(s_n)\right)\middle|\varepsilon_1,\ldots,\varepsilon_{n-1}\right]\right] \\
&\leq \boldsymbol{E}\left[\boldsymbol{E}\left[\Psi\left(\sup_{s\in\mathcal{T}} A(s)+\varepsilon_n s_n+\sum_{i=1}^{n-1}\varepsilon_i\varphi_i(s_i)\right)\middle|\varepsilon_1,\ldots,\varepsilon_{n-1}\right]\right] \\
&= \boldsymbol{E}\left[\boldsymbol{E}\left[\Psi\left(\sup_{s\in\mathcal{T}} A(s)+\varepsilon_n s_n+\sum_{i=1}^{n-1}\varepsilon_i\varphi_i(s_i)\right)\middle|\varepsilon_n\right]\right] \\
&\leq \boldsymbol{E}\left[\boldsymbol{E}\left[\Psi\left(\sup_{s\in\mathcal{T}} A(s)+\varepsilon_n s_n+\sum_{i=1}^{n-1}\varepsilon_i s_i\right)\middle|\varepsilon_n\right]\right] \\
&= \boldsymbol{E}\left[\Psi\left(\sup_{s\in\mathcal{T}} A(s)+\varepsilon_n s_n+\sum_{i=1}^{n-1}\varepsilon_i s_i\right)\right]
\end{aligned}
$$

where the first inequality follows from the base case, and the second by assuming that (11.2) holds for $n-1$ Rademacher variables.

We turn to the proof of the second inequality in the theorem. By Jensen's inequality,

$$
\begin{aligned}
&\boldsymbol{E}\left[\Psi\left(\frac{1}{2}\sup_{s\in\mathcal{T}}\left|\sum_{i=1}^{n}\varepsilon_i\varphi_i(s_i)\right|\right)\right] \\
&= \boldsymbol{E}\left[\Psi\left(\frac{1}{2}\sup_{s\in\mathcal{T}}\left(\sum_{i=1}^{n}\varepsilon_i\varphi_i(s_i)\right)_+ +\frac{1}{2}\sup_{s\in\mathcal{T}}\left(\sum_{i=1}^{n}-\varepsilon_i\varphi_i(s_i)\right)_+\right)\right] \\
&\leq \frac{1}{2}\boldsymbol{E}\left[\Psi\left(\sup_{s\in\mathcal{T}}\left(\sum_{i=1}^{n}\varepsilon_i\varphi_i(s_i)\right)_+\right)\right]+\frac{1}{2}\boldsymbol{E}\left[\Psi\left(\sup_{s\in\mathcal{T}}\left(\sum_{i=1}^{n}-\varepsilon_i\varphi_i(s_i)\right)_+\right)\right].
\end{aligned}
$$

The second inequality in the theorem now follows by invoking twice the first inequality and noting that the function $\Psi((x)_+)$ is convex and nondecreasing. □

11.4 Weak and Wimpy Variances

In this section, we bound the weak variance Σ^2 by its wimpy counterpart for empirical processes with uniformly bounded random summands. More precisely, we show how symmetrization (Lemma 11.4) and contraction (Theorem 11.6) allow us to upper bound Σ^2 using $\boldsymbol{E}Z$ and σ^2.

Theorem 11.8 *Define* $Z = \sup_{s\in\mathcal{T}} \sum_{i=1}^{n} X_{i,s}$ *where* $EX_{i,s} = 0$ *and* $|X_{i,s}| \le 1$ *for all* $i = 1, \ldots, n$ *and* $s \in \mathcal{T}$. *Then*

$$\operatorname{Var}(Z) \le \Sigma^2 + \sigma^2 \le 8E \sup_{s\in\mathcal{T}} \left| \sum_{i=1}^{n} X_{i,s} \right| + 2\sigma^2.$$

The key to the proof of Theorem 11.8 is the following simple lemma.

Lemma 11.9 *Under the conditions of Theorem 11.8,*

$$\Sigma^2 \le \sigma^2 + 2E \sup_{s\in\mathcal{T}} \sum_{i=1}^{n} \varepsilon_i X_{i,s}^2,$$

where $\varepsilon_1, \ldots, \varepsilon_n$ *are independent Rademacher variables,*

Proof Clearly,

$$\begin{aligned}\Sigma^2 &= E \sup_{s\in\mathcal{T}} \sum_{i=1}^{n} \left(\left(X_{i,s}^2 - EX_{i,s}^2 \right) + EX_{i,s}^2 \right) \\ &\le E \sup_{s\in\mathcal{T}} \sum_{i=1}^{n} \left(X_{i,s}^2 - EX_{i,s}^2 \right) + \sigma^2.\end{aligned}$$

On the other hand, by Lemma 11.4,

$$E \sup_{s\in\mathcal{T}} \sum_{i=1}^{n} \left(X_{i,s}^2 - EX_{i,s}^2 \right) \le 2E \sup_{s\in\mathcal{T}} \sum_{i=1}^{n} \varepsilon_i X_{i,s}^2.$$

□

Proof of Theorem 11.8. By Theorem 11.1, it suffices to prove that $\Sigma^2 \le 8E \sup_{s\in\mathcal{T}} \left| \sum_{i=1}^{n} X_{i,s} \right| + \sigma^2$. But by Lemma 11.9, this amounts to showing that $2E \sup_{s\in\mathcal{T}} \sum_{i=1}^{n} \varepsilon_i X_{i,s}^2 \le 4E \sup_{s\in\mathcal{T}} \left| \sum_{i=1}^{n} X_{i,s} \right|$. As $\varphi(x) = x^2$ is 2-Lipschitz on $[-1, 1]$, by Theorem 11.6,

$$E \sup_{s\in\mathcal{T}} \sum_{i=1}^{n} \varepsilon_i X_{i,s}^2 \le 2E \sup_{s\in\mathcal{T}} \sum_{i=1}^{n} \varepsilon_i X_{i,s}.$$

Finally, as each $X_{i,s}$ is centered, by the symmetrization inequalities,

$$E \sup_{s\in\mathcal{T}} \sum_{i=1}^{n} \varepsilon_i X_{i,s}^2 \le 4E \sup_{s\in\mathcal{T}} \left| \sum_{i=1}^{n} X_{i,s} \right|.$$

□

When the random vectors X_i are identically distributed and uniformly bounded, the bound of Theorem 11.8 can be improved as is shown next.

Theorem 11.10 *Let $Z = \sup_{s\in\mathcal{T}} \sum_{i=1}^{n} X_{i,s}$ be the supremum of an empirical process such that $X_1, \ldots, X_n$ are independent and identically distributed and for all $i = 1, \ldots, n$ and $s \in \mathcal{T}$, $|X_{i,s}| \le 1$ with probability 1 and $\mathbf{E}X_{i,s} = 0$. Then*

$$\mathrm{Var}(Z) \le 2\mathbf{E}Z + \sigma^2.$$

We prove that even if we do not assume that the summands are identically distributed,

$$\mathrm{Var}(Z) \le 2\mathbf{E}Z + \sum_{i=1}^{n} \sup_{s\in\mathcal{T}} \mathbf{E}X_{i,s}^2.$$

Of course, if the random vectors X_i are identically distributed, the second expression on the right-hand side equals σ^2.

The theorem follows from careful usage of the Efron–Stein inequality. The key observation is that the supremum of the empirical process satisfies a certain self-bounding property. Recall that various versions of self-bounding functions are investigated in Chapters 3 and 6. Here we need a slightly different notion. In order to show the essence of the argument, we generalize the statement to such self-bounding random variables. To this end, consider a random variable Z that is a function of independent random variables $X_1, \ldots, X_n$ for which the following assumptions hold: for every $i = 1, \ldots, n$, there exists a measurable function Z_i of $X^{(i)} = (X_1, \ldots, X_{i-1}, X_{i+1}, \ldots, X_n)$ and a random variable Y_i such that for some constant $a \in [0, 1]$,

$$Y_i \le Z - Z_i \le 1, \quad \mathbf{E}^{(i)}Y_i \ge 0, \quad \text{and} \quad Y_i \le a, \tag{11.3}$$

where $\mathbf{E}^{(i)}$ denotes the conditional expectation given $X^{(i)}$, and

$$\sum_{i=1}^{n}(Z - Z_i) \le Z. \tag{11.4}$$

Note that if these assumptions are satisfied, then $Z_i \le \mathbf{E}^{(i)}Z$ as $\mathbf{E}^{(i)}Z - Z_i = \mathbf{E}^{(i)}[Z - Z_i] \ge \mathbf{E}^{(i)}Y_i \ge 0$. Also observe that if $Y_i \equiv 0$, then the condition simplifies to the self-bounding property introduced in Chapter 3.

Lemma 11.11 *Let Z be a real-valued function of the independent random variables $X_1, \ldots, X_n$ satisfying assumptions (11.3) and (11.4). Then for every $i = 1, \ldots, n$,*

$$\mathbf{E}^{(i)}\left(Z - \mathbf{E}^{(i)}Z\right)^2 \le \mathbf{E}^{(i)}(Z - Z_i)^2 \le (1 + a)\,\mathbf{E}^{(i)}[Z - Z_i] + \mathbf{E}^{(i)}Y_i^2.$$

Proof The first inequality is obvious. To prove the second, set $\varphi(x) = x^2 - (1 + a)x$. Then, since $(Z - Z_i) - Y_i \ge 0$ and $((Z - Z_i) - 1) + (Y_i - a) \le 0$, we have

$$\varphi(Z - Z_i) - \varphi(Y_i) = [(Z - Z_i) - Y_i]\,[((Z - Z_i) - 1) + (Y_i - a)] \le 0.$$

Hence,

$$\mathbf{E}^{(i)}\varphi(Z - Z_i) \le \mathbf{E}^{(i)}\varphi(Y_i),$$

and therefore

$$E^{(i)}(Z - Z_i)^2 \leq (1+a)\,E^{(i)}[Z - Z_i] + E^{(i)}Y_i^2 - (1+a)\,E^{(i)}Y_i$$

which leads to the desired result thanks to the assumption $E^{(i)}Y_i \geq 0$. □

Proof of Theorem 11.10. The assumptions of Lemma 11.11 are satisfied by suprema of centered empirical processes with $\sup_{s\in\mathcal{T}} |X_{i,s}|$ upper bounded by 1 if we choose $a = 1$. Indeed, let $\hat{s}$ denote an element of $\mathcal{T}$ that achieves the supremum in the definition of Z. For each $i = 1, \ldots, n$, let $Z_i = \sup_{s\in\mathcal{T}} \sum_{j\neq i} X_{j,s}$. Let $\hat{s}_i$ denote an element of $\mathcal{T}$ that achieves the supremum in Z_i. Then

$$X_{i,\hat{s}_i} = \sum_{j=1}^{n} X_{j,\hat{s}_i} - \sum_{j\neq i} X_{j,\hat{s}_i} \leq Z - Z_i \leq \sum_{j=1}^{n} X_{j,\hat{s}} - \sum_{j\neq i} X_{j,\hat{s}} = X_{i,\hat{s}}.$$

Summing over i in the inequalities on the right-hand side, we get the self-bounding condition

$$\sum_{i=1}^{n} (Z - Z_i) \leq Z.$$

We assume furthermore that for every $i = 1, \ldots, n$ and $s \in \mathcal{T}$, $EX_{i,s} = 0$ and $|X_{i,s}| \leq 1$ almost surely. Then defining Y_i by $Y_i = X_{i,\hat{s}_i}$, we get

$$E^{(i)}Y_i = 0 \text{ and } Y_i \leq 1.$$

Now Theorem 11.10 follows as an immediate consequence of Lemma 11.11 and the Efron–Stein inequality (Theorem 3.1). □

11.5 Unbounded Summands

The bounds presented in the previous section are only useful when the random variable $\max_{i=1,\ldots,n} \sup_{s\in\mathcal{T}} |X_{i,s}|$ is uniformly bounded. In other cases, one way to bound the variance of suprema of empirical process proceeds by complementing the contraction principle with some kind of truncation. A convenient device is the so-called Hoffmann–Jørgensen inequality. Before describing this device, we establish the following "maximal" inequality.

Lemma 11.12 (LÉVY'S MAXIMAL INEQUALITY) *Let $X_1, \ldots, X_n$ be independent (not necessarily identically distributed) symmetric random variables where $X_i = (X_{i,s})_{s\in\mathcal{T}}$. Define $S_k = \sum_{i\leq k} X_i$ and $S_{k,s} = \sum_{i=1}^{k} X_{i,s}$ for $k = 1, \ldots, n$. Let $Z_k = \sup_{s\in\mathcal{T}} |S_{k,s}|$. Then, for $t \geq 0$,*

$$P\left\{\max_{k\leq n} Z_k \geq t\right\} \leq 2P\{Z_n \geq t\}.$$

Proof Let E denote the event $\{Z_n \geq t\}$ and for each $k = 1, \ldots, n$, let A_k denote the event $\{\max_{j<k} Z_j < t \text{ and } Z_k \geq t\}$. The collection of events A_k forms a partition of the event $\{\max_{k\leq n} Z_k \geq t\}$. Note that for each k, the random vectors $S_n - S_k$ and $-(S_n - S_k)$ are identically distributed and independent of S_k. Observe that

$$\begin{aligned} 2Z_k &= \sup_{s\in\mathcal{T}} \left|S_{k,s} + (S_{n,s} - S_{k,s}) + S_{k,s} - (S_{n,s} - S_{k,s})\right| \\ &\leq \sup_{s\in\mathcal{T}} \left|S_{k,s} + (S_{n,s} - S_{k,s})\right| + \sup_{s\in\mathcal{T}} \left|S_{k,s} - (S_{n,s} - S_{k,s})\right| \\ &= Z_n + \sup_{s\in\mathcal{T}} \left|S_{k,s} - (S_{n,s} - S_{k,s})\right|. \end{aligned}$$

The two expressions on the right-hand side of the last display are identically distributed thanks to the symmetry assumption. On A_k, as $Z_k \geq t$, we have either $Z_n \geq t$ or $\sup_{s\in\mathcal{T}} \left|S_{k,s} - (S_{n,s} - S_{k,s})\right| \geq t$. Thus,

$$\begin{aligned} &2\boldsymbol{P}\{Z_n \geq t \text{ and } A_k\} \\ &= \boldsymbol{P}\{Z_n \geq t \text{ and } A_k\} \\ &\quad + \boldsymbol{P}\left\{\sup_{s\in\mathcal{T}} \left|S_{k,s} - (S_{n,s} - S_{k,s})\right| \geq t \text{ and } A_k\right\} \\ &\geq \boldsymbol{P}\left\{\left(Z_n \geq t \text{ or } \sup_{s\in\mathcal{T}} \left|S_{k,s} - (S_{n,s} - S_{k,s})\right| \geq t\right) \text{ and } A_k\right\} \\ &= \boldsymbol{P}\{A_k\}. \end{aligned}$$

Summing over all $k = 1, \ldots, n$,

$$2\boldsymbol{P}\{Z_n \geq t\} = \sum_{k=1}^{n} 2\boldsymbol{P}\{Z_n \geq t \text{ and } A_k\} \geq \sum_{k=1}^{n} \boldsymbol{P}\{A_k\} = \boldsymbol{P}\left\{\max_{k\leq n} Z_k \geq t\right\}. \qquad \square$$

The next lemma is the simplest representative of a family of results known as the Hoffmann–Jørgensen inequalities.

Lemma 11.13 (HOFFMANN–JØRGENSEN INEQUALITY) *Let $X_i = (X_{i,s})_{s\in\mathcal{T}}$, $i = 1, \ldots, n$ be independent (not necessarily identically distributed) random variables. For $k = 1, \ldots, n$, let $S_k = \sum_{i=1}^k X_i$ and $S_{k,s} = \sum_{i=1}^k X_{i,s}$. Let $Z_k = \sup_{s\in\mathcal{T}} \left|S_{k,s}\right|$ and $M = \sup_{i\leq n, s\in\mathcal{T}} \left|X_{i,s}\right|$. Then for all $t, u, v > 0$,*

$$\begin{aligned} &\boldsymbol{P}\left\{\max_{k\leq n} Z_k \geq t + u + v\right\} \\ &\leq \boldsymbol{P}\{M \geq v\} + \boldsymbol{P}\left\{\max_{k\leq n} Z_k \geq t\right\} \boldsymbol{P}\left\{\sup_{\substack{1\leq j\leq k\leq n,\\ s\in\mathcal{T}}} \left|S_{k,s} - S_{j,s}\right| \geq u\right\}. \end{aligned}$$

Proof Let the event A_k be defined by $Z_j < t$ for all $j < k$ and $Z_k \geq t$. The event $E = \{\max_{k \leq n} Z_k \geq t + u + v\}$ can be partitioned as $E = \cup_{k=1}^{n} E \cap A_k$.

On $A_k \cap E$, for some $k \leq m \leq n$, $Z_m \geq t + u + v$. But

$$Z_m \leq Z_{k-1} + \sup_{s \in T} |X_{k,s}| + \sup_{s \in T} |S_{m,s} - S_{k,s}|,$$

so

$$\begin{aligned} A_k \cap E &\subseteq \left(A_k \cap \left\{ \sup_{s \in T} |X_{k,s}| \geq v \right\} \right) \cup \left(A_k \cap \left\{ \sup_{\substack{m \geq k, \\ s \in T}} |S_{m,s} - S_{k,s}| \geq u \right\} \right) \\ &\subseteq (A_k \cap \{M \geq v\}) \cup \left(A_k \cap \left\{ \sup_{\substack{m \geq k, \\ s \in T}} |S_{m,s} - S_{k,s}| \geq u \right\} \right). \end{aligned}$$

As A_k and $(S_m - S_k)_{m \geq k}$ are independent,

$$\begin{aligned} \boldsymbol{P}\{A_k \cap E\} &\leq \boldsymbol{P}\{A_k \text{ and } M \geq v\} + \boldsymbol{P}\{A_k\} \boldsymbol{P} \left\{ \sup_{\substack{m \geq k, \\ s \in T}} |S_{m,s} - S_{k,s}| \geq u \right\} \\ &\leq \boldsymbol{P}\{A_k \text{ and } M \geq v\} + \boldsymbol{P}\{A_k\} \boldsymbol{P} \left\{ \sup_{\substack{0 \leq j \leq m, \\ s \in T}} |S_{m,s} - S_{j,s}| \geq u \right\}. \end{aligned}$$

Summing over all k leads to the desired result. □

If we assume that the random vectors are symmetrically distributed, combining the Hoffmann–Jørgensen inequality and Lévy's maximal inequality, we obtain the following corollary.

Corollary 11.14 *Consider the conditions and notation of Lemma 11.13 and assume that each $X_{i,s}$ has a symmetric distribution. Then for all $t, v > 0$,*

$$\begin{aligned} \boldsymbol{P}\{Z_n \geq 2t + v\} &\leq \boldsymbol{P} \left\{ \max_{k \leq n} Z_k \geq 2t + v \right\} \\ &\leq \boldsymbol{P}\{M \geq v\} + 4 \left(\boldsymbol{P}\{Z_n \geq t\} \right)^2. \end{aligned}$$

This result may be used to relate the expectation of Z_n, the tail probability of Z_n and the expectation of M as follows.

Corollary 11.15 *Under the conditions of Corollary 11.14, let $t > 0$ be such that $P\{Z_n > t\} < 1/4$. Then*

$$EZ_n \le \left(\frac{\sqrt{4t} + \sqrt{EM}}{1 - (4P\{Z_n > t\})^{1/2}} \right)^2 .$$

Proof Let α, β, γ be positive and such that $\alpha + \beta + \gamma = 1$. Then, by Corollary 11.14,

$$\begin{aligned} EZ_n &\le \int_0^\infty P\left\{ \max_{k \le n} Z_k > x \right\} dx \\ &\le \int_0^\infty P\{M > \alpha x\}\, dx + \int_0^\infty 4P\{Z_n > \beta x\} P\{Z_n > \gamma x\}\, dx \\ &= \frac{EM}{\alpha} + \frac{4}{\beta} \int_0^\infty P\{Z_n > x\} P\left\{ Z_n > \frac{\gamma x}{\beta} \right\} dx \\ &\le \frac{EM}{\alpha} + \frac{4t}{\beta} + \frac{4}{\gamma} P\{Z_n > t\}\, EZ_n. \end{aligned}$$

Now letting $\delta = \sqrt{EM} + \sqrt{4t} + \sqrt{4P\{Z_n > t\}\, EZ_n}$ and choosing $\alpha = \sqrt{EM}/\delta$, $\beta = \sqrt{4t}/\delta$ and $\gamma = \sqrt{4P\{Z_n > t\}\, EZ_n}/\delta$, we obtain

$$EZ_n \le \left(\sqrt{EM} + \sqrt{4t} + \sqrt{4P\{Z_n > t\}\, EZ_n} \right)^2 . \qquad \square$$

Next we relate the expected value of the Rademacher process generated by the large values of the $\sup_{s \in \mathcal{T}} |X_{i,s}|$ with the expected value of M.

Lemma 11.16 *Under the conditions of Corollary 11.14, let $\varepsilon_1, \ldots, \varepsilon_n$ denote independent Rademacher variables. Let $\lambda > 4$ and define $t_0 = \lambda EM$. Then*

$$E \max_{s \in \mathcal{T}} \left| \sum_{i=1}^n \varepsilon_i X_{i,s} \mathbb{1}_{\{\sup_{s \in \mathcal{T}} |X_{i,s}| > t_0\}} \right| \le \left(\frac{1 + 2\sqrt{\lambda}}{1 - 2/\sqrt{\lambda}} \right)^2 EM.$$

Proof We use Corollary 11.15 with $Z = \max_{s \in \mathcal{T}} \left| \sum_{i=1}^n \varepsilon_i X_{i,s} \mathbb{1}_{\{\sup_{s \in \mathcal{T}} |X_{i,s}| > t_0\}} \right|$ and $M = \max_{i \le n, s \in \mathcal{T}} |X_{i,s}|$. We obtain

$$EZ \le \left(\frac{\sqrt{4t_0} + \sqrt{E[M]}}{1 - (4P\{Z > t_0\})^{1/2}} \right)^2 .$$

The right-hand side may be bounded further by observing that, by Markov's inequality,

$$P\{Z > t_0\} \le P\{M > t_0\} \le \frac{EM}{t_0} = \frac{1}{\lambda}. \qquad \square$$

Even though the statement and derivation of Lemma 11.16 resort to heavy notation and sophisticated arguments, the statement lends itself to a simple interpretation. Given the choice of t_0, with high probability there is at most one index $1 \le i \le n$ such that $X_{i,s}\mathbb{1}_{\{\sup_{s\in\mathcal{T}}|X_{i,s}|>t_0\}} \neq 0$. The sum then reduces to a single summand, and thus with high probability, Z is distributed like M.

Now that we are prepared to establish a connection between the wimpy and the weak variances of the supremum of an empirical process, it generalizes Theorem 11.8.

Theorem 11.17 *Let $Z = \sup_{s\in\mathcal{T}} \left|\sum_{i=1}^n X_{i,s}\right|$ denote the supremum of an empirical process. Assume that the random variables X_i are symmetric for $i = 1, \ldots, n$. Let $M = \sup_{i=1,\ldots,n,s\in\mathcal{T}} X_{i,s}^2$. Then*

$$\Sigma^2 \le \sigma^2 + 16\sqrt{EM}\,EZ + 2 \times 18^2 EM.$$

Proof Let $\varepsilon_1, \ldots, \varepsilon_n$ denote independent Rademacher random variables, and let $t_0 = \lambda EM$ with $\lambda > 4$. By Lemma 11.9, we have

$$\Sigma^2 \le \sigma^2 + 2E \sup_{s\in\mathcal{T}} \sum_{i=1}^n \varepsilon_i X_{i,s}^2.$$

The last expression can be split into two parts:

$$\begin{aligned} E \sup_{s\in\mathcal{T}} \sum_{i=1}^n \varepsilon_i X_{i,s}^2 &\le E \sup_{s\in\mathcal{T}} \sum_{i=1}^n \varepsilon_i X_{i,s}^2 \mathbb{1}_{\{\sup_{s\in\mathcal{T}} X_{i,s}^2 \le t_0\}} \\ &\quad + E \sup_{s\in\mathcal{T}} \sum_{i=1}^n \varepsilon_i X_{i,s}^2 \mathbb{1}_{\{\sup_{s\in\mathcal{T}} X_{i,s}^2 > t_0\}} \\ &\le E \sup_{s\in\mathcal{T}} \sum_{i=1}^n \varepsilon_i X_{i,s}^2 \mathbb{1}_{\{\sup_{s\in\mathcal{T}} X_{i,s}^2 \le t_0\}} \\ &\quad + E \sup_{s\in\mathcal{T}} \left| \sum_{i=1}^n \varepsilon_i X_{i,s}^2 \mathbb{1}_{\{\sup_{s\in\mathcal{T}} X_{i,s}^2 > t_0\}} \right|. \end{aligned}$$

The first term on the right-hand side is bounded by $2\sqrt{t_0}E \sup_{s\in\mathcal{T}} \left|\sum_{i=1}^n \varepsilon_i X_{i,s}\right|$ thanks to the contraction principle (Theorem 11.6). The second term may be handled using Lemma 11.16:

$$E \sup_{s\in\mathcal{T}} \left| \sum_{i=1}^n \varepsilon_i X_{i,s}^2 \mathbb{1}_{\{\sup_{s\in\mathcal{T}} |X_{i,s}^2| > t_0\}} \right| \le \left(\frac{1 + 2\lambda^{1/2}}{1 - 2/\lambda^{1/2}} \right)^2 EM.$$

The proof is completed by taking $\lambda = 16$. □

A shallow comparison between Theorem 11.17 and Theorem 11.10 might suggest that the last lemma is completely satisfactory. Considering a Gaussian setting, where $(X_{i,j})_{1\le j\le d}$ is a standard Gaussian vector for each $i = 1, \ldots, n$, shows that is not the case. The upper bound on $\Sigma^2 - \sigma^2$ derived from Theorem 11.17 is approximately $\sqrt{4n\log(nd)\log(d)}$, while a better upper bound is $\sqrt{4n\log(d)} + 2\log d$ as $\Sigma^2 - \sigma^2$ is the expected value of the maximum of d independent sub-gamma random variables with variance factor $2n$ and scale factor 2.

11.6 Bibliographical Remarks

Suprema of empirical processes play a fundamental role in statistics and machine learning (see, for example, the monographs of van der Vaart and Wellner (1996), van de Geer (2000), and Massart (2006) for surveys).

Nemirovski's inequality was first stated in Nemirovski (2000). It has been used in high-dimensional statistics by Greenshtein and Ritov (2004). Nemirovski showed that if $\mathbb{B}$ is $\mathbb{R}^d$ endowed with the ℓ_p norm where $2 \le p \le \infty$, then there exists a constant $K(p, d)$ such that

$$\boldsymbol{E}\|S_n\|_p^2 \le K(p, d)V$$

where $V = \boldsymbol{E}\sum_{i=1}^n \|X_i\|_p^2$. Duembgen *et al.* (2010) re-examined Nemirovski's results and established, using a variety of methods from linear analysis, convex geometry, and high-dimensional probability, that for $d \ge 3$,

$$K(p, d) \le \min(d, 2e\log d, p - 1).$$

Morover they proved that for $p = \infty$, $\liminf_{d\to\infty} K(\infty, d)/(2\log d) \ge 1$. Note that, for the example used in the derivation of the lower bound for $C(\infty, d)$ in Theorem 11.2, we had $\boldsymbol{E}\|S_n\|_p^2 = V$.

Symmetrization techniques were popularized by Paul Lévy. Ledoux and Talagrand (1991) provide a thorough description of the impact of symmetrization on the analysis of sums of independent random vectors. In the field of empirical process theory, Lemma 11.4 was advocated by Giné and Zinn (1984). Symmetrization had been used in different ways by Vapnik and Chervonenkis (1971, 1974, 1981) in their influential papers (see also Vapnik 1982,1998) in order to develop deviation inequalities for suprema of empirical processes (see Exercises 12.1 and 12.3). Symmetrization still plays an important role in empirical process theory (see Panchenko 2003).

The contraction principle for Rademacher sums (Theorem 11.6) is due to Ledoux and Talagrand (1991, Chapter 4). Theorem 11.6 is part of a collection of related results also called contraction principles (see Exercises 11.12, 11.13, and 11.14 for some related results). Note that while all these exercises can be solved by invoking Theorem 11.6, simpler proofs exist, for example Ledoux and Talagrand (1991, Chapter 4). The proof of Theorem 11.8 can be found in Massart (2000*a*).

Theorem 11.10 is described by Rio (2001). Variants can be found in Bousquet (2002*b*) (see Exercise 11.16).

Lévy's inequalities were derived by Paul Lévy in a general investigation of sums of independent random vectors; see again Ledoux and Talagrand (1991) for an excellent exposition. The Hoffmann-Jørgensen inequality appears in Hoffmann-Jørgensen (1974) as an extension of Kolmogorov's converse maximal inequality. The latter relates moments of a sum of centered independent real-valued symmetric random variables with its quantiles and with the moments of the maxima of the summands (see Exercise 11.18). A more general version of Lemma 11.16, as well as a thorough discussion of the topic and its implications, can be found in de la Peña and Giné (1999, Section 1.2). Lemma 11.16 is due to Giné, Latała, and Zinn (2000).

11.7 EXERCISES

11.1. Show that it is not necessarily true that the variance of the supremum of an empirical process is upper bounded by the wimpy variance. *Hint:* consider $n = 1, \mathcal{T} = \{1, 2\}$, and binary-valued random variables.

11.2. (A BAD EXAMPLE FOR VARIANCE BOUNDS) Consider $\mathcal{T} = \{1, \ldots, n\}$ and assume that the $X_{i,s}$ are i.i.d. exponential random variables with mean 1. Then $Z = \sup_{s \in \mathcal{T}} \sum_{i=1}^n X_{i,s}$ is the supremum of n i.i.d. random variables. Compute the expectation and the variance of Z. Compute the variance upper bound provided by Theorem 11.1. Compare. Letting $Z_i = \sup_{s \in \mathcal{T}} \sum_{j \neq i}^n X_{j,s}$, compute the upper bound $\sum_{i=1}^n \mathbf{E}\left[(Z - Z_i)^2\right]$ and compare the result with the true value of Var (Z). *Hint:* the expectation and variance of Z are respectively H_n the n-th harmonic number, and H_n^2 the n-th harmonic number of the second kind ($H_n^2 \leq \pi^2/6$).

11.3. Prove that Theorem 11.1 still holds if the index set is not assumed to be countable but the process is separable. *Hint:* let $\mathcal{S} \subseteq \mathcal{T}$ be a separant. Let $s_1, s_2, \ldots$ be an enumeration of the elements of $\mathcal{S}$. Apply Theorem 11.1 to the empirical process indexed by $s_1, \ldots, s_n$ and use the monotone convergence theorem.

11.4. (SYMMETRIZATION AND ASYMMETRIC PROCESSES) For $i = 1, \ldots, n$ and $s = 1, \ldots, n$, let $X_{i,s}$ be independent random variables with $\mathbf{P}\{X_{i,s} = n/(n-1)\} = (n-1)/n$ and $\mathbf{P}\{X_{i,s} = -n\} = 1/n$. Let $\varepsilon_1, \ldots, \varepsilon_n$ be independent random Rademacher variables that are independent of $(X_{i,s})$, $1 \leq i, s \leq n$. Let $\mathcal{T} = \{1, \ldots, n\}$. Prove that for sufficiently large n,

$$\mathbf{E} \sup_{s \in \mathcal{T}} \sum_{i=1}^n \varepsilon_i X_{i,s} \geq \frac{n}{2} \frac{\log \frac{n}{4}}{\log \log n}.$$

Deduce from this observation that

$$\frac{1}{2} \mathbf{E} \sup_{s \in \mathcal{T}} \sum_{i=1}^n \varepsilon_i X_{i,s} > \mathbf{E} \sup_{s \in \mathcal{T}} \sum_{i=1}^n X_{i,s}.$$

Compare with Theorem 11.4. *Hint:* note that with high probability, at least $n/4$ Rademacher variables are negative. Use and prove the fact that the maximum of n independent binomial random variables with parameters $n/4$ and $1/n$ is at least $\log(n/4)/\log\log n$.

11.5. (IMPROVED SYMMETRIZATION INEQUALITIES) Let $X_1, \ldots, X_n$ be independent random vectors $X_i = (X_{i,s})_{s\in\mathcal{T}}$. Let Ψ denote a convex increasing function. Assume that for each $i = 1, \ldots, n$ and $s \in \mathcal{T}$, $X_{i,s}$ is integrable and centered. Let $\varepsilon_1, \ldots, \varepsilon_n$ be independent of Rademacher random variables. Prove that

$$\mathbf{E}\left[\Psi\left(\frac{1}{2}\sup_{s\in\mathcal{T}}\left|\sum_{i=1}^n \varepsilon_i X_{i,s}\right|\right)\right] \le \mathbf{E}\left[\Psi\left(\sup_{s\in\mathcal{T}}\left|\sum_{i=1}^n X_{i,s}\right|\right)\right]$$

$$\le \mathbf{E}\left[\Psi\left(2\sup_{s\in\mathcal{T}}\left|\sum_{i=1}^n \varepsilon_i X_{i,s}\right|\right)\right].$$

11.6. Prove that Theorem 11.4 still holds if the index set is assumed to be separable. *Hint:* proceed as in Exercise 11.3.

11.7. (BOUNDING THE GAUSSIAN QUANTILE FUNCTION) Let Φ be the standard Gaussian distribution function. Prove that for $t \ge 5$,

$$\Phi^{-1}(1 - 1/t) \ge \sqrt{2\log t - \log\log t - \log(4\pi)}$$

and that for $t \ge 2$

$$\Phi^{-1}(1 - 1/t) \le \sqrt{2\log t - \log\log t - \log(\pi)}.$$

Hint: use Lemma 10.1.

11.8. (NEMIROVSKI'S INEQUALITY IN THE NON-SYMMETRIC CASE) Using the notation of Theorem 11.2, prove that even if the X_i are not assumed to be symmetric but centered,

$$\mathbf{E}\,\|S_n\|_\infty^2 \le 2\,(1 + 4\log(2d))\,\Sigma^2.$$

11.9. (OPTIMALITY OF THE CONSTANT IN THE CONTRACTION PRINCIPLE) Prove the optimality of the constant $1/2$ on the left-hand side of the contraction principle (Theorem 11.6). *Hint:* let $\mathcal{T} = \{1, 2\}$, X deterministic, and $X_{1,1} = X_{2,1} = 1$ while $X_{1,2} = X_{2,2} = -1$. Let $\varphi_1(x) = x$, $\varphi_2(x) = -|x|$ and Ψ the identity. (See the remark following statement of Theorem 4.12 in Ledoux and Talagrand (1991).)

11.10. (CONTRACTION PRINCIPLE AND TAIL BOUNDS) Let $\mathbb{B}$ denote a separable Banach space with norm $\|\cdot\|_{\mathbb{B}}$. Let $\varepsilon_1, \ldots, \varepsilon_n$ be independent Rademacher random variables. Let $1 \ge \lambda_1 \ge \cdots \ge \lambda_n \ge 0$. Let $v_1, \ldots, v_n \in \mathbb{B}$. Prove that for all $t > 0$,

$$P\left\{\left\|\sum_{i=1}^{n}\lambda_i\varepsilon_i v_i\right\|_{\mathbb{B}} > t\right\} \leq 2P\left\{\left\|\sum_{i=1}^{n}\varepsilon_i v_i\right\|_{\mathbb{B}} > t\right\}.$$

(See Theorem 4.4 in Ledoux and Talagrand (1991).)

11.11. (CONTRACTION PRINCIPLE FOR GAUSSIAN SUMS) Let $\Psi : \mathbb{R}_+ \to \mathbb{R}$ be a nondecreasing convex function. Let $X_1, \ldots, X_n$ be independent random vectors $X_i = (X_{i,s})_{s\in\mathcal{T}}$. Assume that for each i, s, $X_{i,s}$ is integrable and centered. For each $i = 1, \ldots, n$, let $\varphi_i : \mathbb{R} \to \mathbb{R}$ denote a 1-Lipschitz function such that $\varphi_i(0) = 0$. Let $Y_1, \ldots, Y_n$ be independent standard Gaussian random variables. Prove that

$$\boldsymbol{E}\left[\Psi\left(\frac{1}{2}\sup_{s\in\mathcal{T}}\left|\sum_{i=1}^{n}Y_i\varphi_i(X_{i,s})\right|\right)\right] \leq \boldsymbol{E}\left[\Psi\left(2\sup_{s\in\mathcal{T}}\left|\sum_{i=1}^{n}Y_iX_{i,s}\right|\right)\right].$$

(See Corollary 3.17 in Ledoux and Talagrand (1991).)

11.12. (COROLLARY OF CONTRACTION PRINCIPLE I) Let $\mathbb{B}$ denote a separable Banach space with norm $\|\cdot\|_{\mathbb{B}}$. Let $X_1, \ldots, X_n$ be independent $\mathbb{B}$-valued symmetric random variables. Let $\lambda_1, \ldots, \lambda_n$ be real numbers with $\|\lambda\|_\infty = \sup_{i=1,\ldots,n}|\lambda_i|$. Prove that for all $p \geq 1$,

$$\boldsymbol{E}\left[\left\|\sum_{i=1}^{n}\lambda_iX_i\right\|_{\mathbb{B}}^{p}\right] \leq \|\lambda\|_\infty^p\boldsymbol{E}\left[\left\|\sum_{i=1}^{n}X_i\right\|_{\mathbb{B}}^{p}\right].$$

(See Garling (2007, p. 188).)

11.13. (COROLLARY OF CONTRACTION PRINCIPLE II) Let $\mathbb{B}$ denote a separable Banach space with norm $\|\cdot\|_{\mathbb{B}}$. Let $X_1, \ldots, X_n$ and $Y_1, \ldots, Y_n$ denote independent real-valued symmetric random variables with $|X_n| \leq |Y_n|$ for all n almost surely. Let $v_1, \ldots, v_n \in \mathbb{B}$. Prove that for all $p \geq 1$,

$$\boldsymbol{E}\left[\left\|\sum_{i=1}^{n}X_iv_i\right\|_{\mathbb{B}}^{p}\right] \leq \boldsymbol{E}\left[\left\|\sum_{i=1}^{n}Y_iv_i\right\|_{\mathbb{B}}^{p}\right].$$

(See Garling (2007, p. 188).)

11.14. (COROLLARY OF CONTRACTION PRINCIPLE III) Let $\mathbb{B}$ denote a separable Banach space with norm $\|\cdot\|_{\mathbb{B}}$. Let $X_1, \ldots, X_n$ and $Y_1, \ldots, Y_n$ denote independent real-valued symmetric random variables with $\boldsymbol{E}|Y_i| \geq 1/C$ for all i and $X_n = \mathrm{sign}(Y_n)$. Let $v_1, \ldots, v_n \in \mathbb{B}$. Prove that for all $p \geq 1$,

$$\boldsymbol{E}\left[\left\|\sum_{i=1}^{n}X_iv_i\right\|_{\mathbb{B}}^{p}\right] \leq C^p\boldsymbol{E}\left[\left\|\sum_{i=1}^{n}Y_iv_i\right\|_{\mathbb{B}}^{p}\right].$$

(See Garling (2007, p. 188).)

11.15. (COMPARISON OF GAUSSIAN AND RADEMACHER SUMS) Let $\mathbb{B}$ denote a separable Banach space with norm $\|\cdot\|_{\mathbb{B}}$. Let $X_1, \ldots, X_n$ be independent Rademacher variables and let $Y_1, \ldots, Y_n$ be independent standard Gaussian variables. Let $v_1, \ldots, v_n \in \mathbb{B}$. Prove that

$$\mathbf{E}\left[\left\|\sum_{i=1}^{n} Y_i v_i\right\|_{\mathbb{B}}\right] \leq \sqrt{2\log n}\, \mathbf{E}\left[\left\|\sum_{i=1}^{n} X_i v_i\right\|_{\mathbb{B}}\right].$$

Hint: see Inequality (4.9) in Ledoux and Talagrand (1991).

11.16. (ANOTHER VARIANCE BOUND) Let $|\alpha_{i,s}| \leq 1$ for all $i = 1, \ldots, n$ and $s \in \mathcal{T}$. Let $Y_1, \ldots, Y_n$ be independent centered random variables such that for all integers $q \geq 2$,

$$\mathbf{E}|Y_i|^q \leq q!\frac{c^{q-2}\sigma^2}{2}$$

for some constants c and σ. Let $Z = \sup_{s\in\mathcal{T}} \sum_{i=1}^{n} \alpha_{i,s} Y_i$. Show that

$$\operatorname{Var}(Z) \leq n\sigma^2 + 2\mathbf{E}Z.$$

Hint: this result is a by-product of Bousquet (2002*b*, Theorem 2.12).

11.17. Using the notation of Theorem 11.10, letting $E = \mathbf{E}Z/n$, prove that

$$\operatorname{Var}(Z) \leq n\sigma^2 + (2 - E)\mathbf{E}Z.$$

Hint: use the same pattern of proof, but replace Z_i by $Z_i - E$. See Rio (2012).

11.18. (HOFFMANN–JØRGENSEN INEQUALITY FOR HIGHER MOMENTS) Let $X_i = (X_{i,s})_{s\in\mathcal{T}}$ for $i = 1, \ldots, n$ be independent symmetric (not necessarily identically distributed) random vectors. Let $Z = \sup_{s\in\mathcal{T}} \left|\sum_{i=1}^{n} X_{i,s}\right|$ and $M = \max_{i\leq n, s\in\mathcal{T}} |X_{i,s}|$. Prove that for any $p \geq 2$, there exists a constant κ_p such that

$$\mathbf{E}\left[Z^p\right]^{1/p} \leq \kappa_p \left(\mathbf{E}Z + \mathbf{E}[M^p]^{1/p}\right).$$

Show that as $p \to \infty$, κ_p/p remains bounded. *Hint:* follow the pattern of proof of Corollary 11.15. (Note that it is possible to choose κ_p so that $(\log p)\kappa_p/p$ remains bounded as $p \to \infty$; see Latała (1997) and de la Peña and Giné (1999, Theorem 1.5.11 and Example 1.5.12).)

11.19. (SUB-GAMMA SUMMANDS) Let $(\alpha_{i,s})$ be a collection of real numbers indexed by $i = 1, \ldots, n$ and $s \in \mathcal{T}$ such that $|\alpha_{i,s}| \leq 1$ for all i and s. Let $X_1, \ldots, X_n$ be independent centered random variables such that for all integers $q \geq 2$,

$$\mathbf{E}|X_i|^q \leq q!\frac{c^{q-2}\sigma^2/n}{2}$$

for some constants c and σ. Let $Z = \sup_{s\in\mathcal{T}} \sum_{i=1,\ldots,n} \alpha_{i,s}X_i$. Check that $\mathrm{Var}(Z) \le 2\sigma^2$ and that

$$\mathrm{Var}(Z) \le \sigma^2 \sup_{s\in\mathcal{T}} \sum_{i=1}^{n} \alpha_{i,s}^2/n + \mathbf{E} \sup_{s\in\mathcal{T}} \sum_{i=1}^{n} \alpha_{i,s}^2 X_i^2$$

and

$$\mathrm{Var}(Z) \le 2\sigma^2 \sup_{s\in\mathcal{T}} \sum_{i=1}^{n} \alpha_{i,s}^2/n + 2\mathbf{E} \sup_{s\in\mathcal{T}} \sum_{i=1}^{n} \varepsilon_i \alpha_{i,s}^2 X_i^2 .$$

Hint: use Theorem 11.1, then use Theorem 11.17 to upper bound the last summand.

11.20. Let $X = (X_1, \ldots, X_n)$ be uniformly distributed over $[-1, 1]^n$. Let $Z = \sqrt{\sum_{i=1}^{n} X_i^2}$ be the Euclidean norm of X. Prove that

$$\sqrt{n/3} - 1 \le \mathbf{E}Z \le \sqrt{n/3}.$$

Prove also that

$$\mathbf{P}\left\{Z \ge \mathbf{E}Z + t\sqrt{n}\right\} \le e^{-nt^2/8}.$$

Hint: represent Z as the supremum of an empirical process. Use Theorem 11.1 to check that $\mathrm{Var}(Z) \le 4/3$. Use Theorem 6.10 to establish the tail bound. This provides an example where the weak and wimpy variance estimates (Σ^2 and σ^2) are of the same order of magnitude and where they are both significantly smaller than $\mathbf{E}Z$. The bound provided by Theorem 11.8, $\Sigma^2 \le \sigma^2 + 8\mathbf{E}Z$ is not sharp in this case.

12

Suprema of Empirical Processes: Exponential Inequalities

In this chapter we continue the study of suprema of empirical processes started in Chapter 11. We use the same notation introduced there. Recall that $\mathcal{T}$ denotes a finite or countable index set, and $X_i = (X_{i,s})_{s\in\mathcal{T}}$ for $i = 1, \ldots, n$ are independent (not necessarily identically distributed) real vector-valued random variables. The empirical process indexed by the index set $\mathcal{T}$ is the vector-valued random variable $\sum_{i=1}^n X_i$. Its supremum is defined as

$$Z = \sup_{s\in\mathcal{T}} \sum_{i=1}^{n} X_{i,s}.$$

While Chapter 11 focuses on upper bounds for the variance of Z, in this chapter we proved exponential concentration inequalities. Our main tool is the entropy method introduced in Chapter 6.

The concentration inequalities derived in Section 5.5 for the suprema of Gaussian and Rademacher processes rely on specific tools such as the Bernoulli and the Gaussian logarithmic Sobolev inequalities. To establish analogous bounds for more general distributions, we may start with the modified logarithmic Sobolev inequality of Theorem 6.6. In this chapter we derive extensions of Hoeffding's, Bernstein's, and Bennett's inequalities for suprema of empirical processes. To this end, we tailor the modified logarithmic Sobolev inequality to our needs, in an increasingly sophisticated way, in Sections 12.2 and 12.4. The argument of Section 12.3 combines symmetrization techniques from Section 11.3 with the convex distance inequality to obtain an exponential inequality for suprema of self-normalized empirical processes.

The main result in this chapter is Bousquet's inequality (Theorem 12.5), a Bennett-type inequality for suprema of centered empirical processes, proved in Section 12.4. In Section 12.5 we survey a variety of related results such as tail bounds for sums of possibly non-identically distributed terms and left-tail inequalities.

In Section 12.6 we describe an application to Pearson's chi-square statistics.

12.1 An Extension of Hoeffding's Inequality

We start with the following extensions of Hoeffding's inequality (Theorem 2.2) to empirical processes.

Theorem 12.1 *Assume that the sequences of vectors $(b_{i,s})_{s\in\mathcal{T}}$ and $(a_{i,s})_{s\in\mathcal{T}}$, $i = 1,\dots,n$ are such that $a_{i,s} \le X_{i,s} \le b_{i,s}$ holds for all $i = 1,\dots,n$ and $s \in \mathcal{T}$ with probability 1. Denote*

$$v = \sup_{s\in\mathcal{T}} \sum_{i=1}^{n} (b_{i,s} - a_{i,s})^2 \quad \textit{and} \quad V = \sum_{i=1}^{n} \sup_{s\in\mathcal{T}} (b_{i,s} - a_{i,s})^2 .$$

Then for all $\lambda \in \mathbb{R}$,

$$\log \mathbf{E} e^{\lambda(Z-\mathbf{E}Z)} \le \frac{v\lambda^2}{2} \quad \textit{and} \quad \log \mathbf{E} e^{\lambda(Z-\mathbf{E}Z)} \le \frac{V\lambda^2}{8}.$$

The first inequality is a consequence of Theorem 6.5, while the second follows from the bounded-differences inequality (Theorem 6.2). Clearly, $v \le V$ but the second inequality may be better due to the better constant factor.

12.2 A Bernstein-Type Inequality for Bounded Processes

In this section we describe an improvement of the Hoeffding-type inequalities of the previous section in the same spirit that Bernstein's inequality improves Hoeffding's for sums of independent random variables.

Such an inequality may be proved for suprema of uniformly bounded empirical processes as a simple application of the "exponential Efron–Stein inequality" of Theorem 6.16 combined with concentration of self-bounding functions (Theorem 6.12). Recall that

$$\Sigma^2 = \mathbf{E} \sup_{s\in\mathcal{T}} \sum_{i=1}^{n} X_{i,s}^2 \quad \text{and} \quad \sigma^2 = \sup_{s\in\mathcal{T}} \sum_{i=1}^{n} \mathbf{E}X_{i,s}^2$$

denote the weak variance and the wimpy variance associated with the empirical process.

Theorem 12.2 *Assume that $\mathbf{E}X_{i,s} = 0$, and $|X_{i,s}| \le 1$ for all $s \in \mathcal{T}$ and $i = 1,\dots,n$. Then for all $0 \le \lambda < 1/2$,*

$$\log \mathbf{E} e^{\lambda(Z-\mathbf{E}Z)} \le \frac{2(\Sigma^2 + \sigma^2)\lambda^2}{2(1-2\lambda)}$$

and for $t \ge 0$,

$$\mathbf{P}\{Z \ge \mathbf{E}Z + t\} \le \exp\left(-\frac{t^2}{2\,(2(\Sigma^2+\sigma^2)+t)}\right).$$

Proof For each $i = 1, \ldots, n$, let $Z'_i = \sup_{s \in \mathcal{T}}(X'_{i,s} + \sum_{j \neq i} X_{j,s})$ where $X'_1, \ldots, X'_n$ are independent of each other and of $X_1, \ldots, X_n$, and X'_i has the same distribution as X_i.

Introduce $W = \sup_{s \in \mathcal{T}} \sum_{i=1}^n X_{i,s}^2$ and denote by $\hat{s} \in \mathcal{T}$ the index for which $\sum_{i=1}^n X_{i,s}$ is largest, that is, $Z = \sum_{i=1}^n X_{i,\hat{s}}$. Then clearly, $(Z - Z'_i)_+^2 \leq (X_{i,\hat{s}} - X'_{i,\hat{s}})^2$ for each $i \leq n$. Then, since $\mathbf{E}[X'_{i,\hat{s}} \mid X_1, \ldots, X_n] = 0$,

$$\begin{aligned}\sum_{i=1}^n \mathbf{E}\left[(Z - Z'_i)_+^2 \mid X_1, \ldots, X_n\right] &\leq \sum_{i=1}^n \mathbf{E}\left[(X_{i,\hat{s}} - X'_{i,\hat{s}})^2 \mid X_1, \ldots, X_n\right] \\ &\leq W + \sup_{s \in \mathcal{T}} \sum_{i=1}^n \mathbf{E}\left[(X'_{i,s})^2\right] \\ &= W + \sigma^2.\end{aligned}$$

Now, by the exponential Efron–Stein inequality (Theorem 6.16),

$$\log \mathbf{E} e^{\lambda(Z - \mathbf{E}Z)} \leq \frac{\lambda}{1 - \lambda} \log \mathbf{E} e^{\lambda(W + \sigma^2)}$$

for $\lambda \in [0, 1)$.

As W is a self-bounding function of $X_1, \ldots, X_n$ (see Section 6.7), Theorem 6.12 implies that

$$\log \mathbf{E} e^{\lambda W} \leq \Sigma^2 \left(e^\lambda - 1\right).$$

Combining the last two inequalities, we obtain

$$\log \mathbf{E} e^{\lambda(Z - \mathbf{E}Z)} \leq \frac{\lambda}{1 - \lambda} \left(\Sigma^2 \left(e^\lambda - 1\right) + \lambda \sigma^2\right).$$

Using the fact that $(e^\lambda - 1)(1 - \lambda) \leq (e^\lambda - 1)e^{-\lambda} = 1 - e^{-\lambda} \leq \lambda$ for $\lambda \in [0, 1)$, we have

$$\log \mathbf{E} e^{\lambda(Z - \mathbf{E}Z)} \leq \frac{\lambda^2(\Sigma^2 + \sigma^2)}{(1 - \lambda)^2}.$$

For $\lambda \in [0, 1/2)$ the right-hand side may be upper bounded by

$$\frac{2(\Sigma^2 + \sigma^2)\lambda^2}{2(1 - 2\lambda)},$$

which concludes the proof of the first inequality. To determine the inequalities for the tail probabilities, observe that this bound has the same form as the upper bound for the logarithmic moment-generating function of sub-gamma random variables discussed in

Section 2.4 (with a variance factor $2(\Sigma^2 + \sigma^2)$ and scale parameter 2). The proof of the tail bound follows by the calculations of Section 2.4. □

When the random variables $(X_i)_{1 \le i \le n}$ are identically distributed, Theorem 12.5 represents an improvement on Theorem 12.2. A comparable improvement for non-identically distributed variables is described in Section 12.5.

12.3 A Symmetrization Argument

Next we describe a variant of the Bernstein-type inequality of the previous section. Here the argument is based on symmetrization of tail probabilities and concentration of convex Lipschitz functions of bounded independent random variables (Theorem 6.10). The important difference with respect to Theorem 12.2 is that here we do not assume any boundedness of the random vectors. Instead, the tail inequalities involve a random quantity that may be more difficult to control.

Theorem 12.3 *Let $X'_1, \ldots, X'_n$ be i.i.d. random vectors, independent of $X_1, \ldots, X_n$. Let*

$$W = \mathbf{E}\left[\sup_{s \in \mathcal{T}} \sum_{i=1}^{n} (X_{i,s} - X'_{i,s})^2 \middle| X_1, \ldots, X_n\right].$$

Then for all $t \ge 0$,

$$\mathbf{P}\left\{Z \ge \mathbf{E}Z + 2\sqrt{tW}\right\} \le 4e^{1-t/4}$$

and

$$\mathbf{P}\left\{Z \le \mathbf{E}Z - 2\sqrt{tW}\right\} \le 4e^{1-t/4}.$$

Note that W is a random variable. Its expected value satisfies $\mathrm{Var}(Z) \le \mathbf{E}W \le 4\Sigma^2$. One may interpret the result as sub-Gaussian inequalities for the "self-normalized" variables $(Z - \mathbf{E}Z)/\sqrt{W}$.

The proof relies on the following technical lemma.

Lemma 12.4 *Let $f_1, f_2, f_3 : \mathcal{X}^{2n} \to \mathbb{R}$ be functions of $2n$ variables and define $Z_i = f_i(X_1, \ldots, X_n, X'_1, \ldots, X'_n)$ for $i \in \{1, 2, 3\}$ where $X_1, \ldots, X_n, X'_1, \ldots, X'_n$ are independent random variables taking values in $\mathcal{X}$. Define*

$$Z'_i = \mathbf{E}[Z_i \mid X_1, \ldots, X_n]$$

for $i \in \{1, 2, 3\}$. Assume that $Z_3 \ge 0$ and that there exists $\kappa > 0$ such that for all $t > 0$,

$$\mathbf{P}\{Z_1 \ge Z_2 + (Z_3 t)^{1/2}\} \le \kappa e^{-\gamma t}.$$

Then, for all $t \geq 0$,

$$P\{Z'_1 \geq Z'_2 + (Z'_3 t)^{1/2}\} \leq \kappa e^{1-\gamma t}.$$

Proof As $\sqrt{xy} = \inf_{\theta>0}(\theta x + y/(4\theta))$,

$$Z_1 \geq Z_2 + (Z_3 t)^{1/2} \quad \text{if and only if} \quad \sup_{\theta>0} 4\theta(Z_1 - Z_2 - \theta Z_3) \geq t$$

and similarly,

$$Z'_1 \geq Z'_2 + (Z'_3 t)^{1/2} \quad \text{if and only if} \quad \sup_{\theta>0} 4\theta(Z'_1 - Z'_2 - \theta Z'_3) \geq t.$$

If we define $U = \sup_{\theta>0} 4\theta(Z_1 - Z_2 - \theta Z_3)$ and $U' = \sup_{\theta>0} 4\theta(Z'_1 - Z'_2 - \theta Z'_3)$ then, by Jensen's inequality,

$$U' = \sup_{\theta>0} 4\theta E[Z_1 - Z_2 - \theta Z_3 \mid X_1, \ldots, X_n] \leq E[U \mid X_1, \ldots, X_n]$$

But, by another application of Jensen's inequality, for any nondecreasing convex function φ, we have

$$E\varphi(U') \leq E[\varphi(E[U \mid X_1, \ldots, X_n])] \leq E\varphi(U).$$

We may conclude using the tail comparison inequality of Exercise 2.24 □

Proof The proof uses Lemma 12.4 with $Z_1 = Z$ and $Z_2 = \sup_{s\in\mathcal{T}} \sum_{i=1}^n X'_{i,s}$. Note that $E[Z_2 \mid X_1, \ldots, X_n] = EZ$. By the lemma, it suffices to prove that

$$P\left\{Z_1 \geq Z_2 + 2\sqrt{tW'}\right\} \leq 4e^{-t/4},$$

where $W' = \sup_{s\in\mathcal{T}} \sum_{i=1}^{n}(X_{i,s} - X'_{i,s})^2$.

For each $i = 1, \ldots, n$, introduce $Y_i = (X_i + X'_i)/2$, $Y'_i = (X_i - X'_i)/2$ and also let $\varepsilon_1, \ldots, \varepsilon_n$ be independent Rademacher variables. By exchangeability of X_i and X'_i, the joint distribution of $(Y_i, Y'_i, W')_{i\leq n}$ is the same as that of $(Y_i, \varepsilon_i Y'_i, W')_{i\leq n}$. Note that $W' = 4\sup_{s\in\mathcal{T}} \sum_{i=1}^n Y'^2_{i,s}$, while $Z_1 = \sup_{s\in\mathcal{T}} \sum_{i=1}^n (Y_{i,s} + Y'_{i,s})$ and $Z_2 = \sup_{s\in\mathcal{T}} \sum_{i=1}^n (Y_{i,s} - Y'_{i,s})$. Thus, we have

$$\begin{aligned}
&P\left\{Z_1 \geq Z_2 + 2\sqrt{tW'}\right\} \\
&= P\left\{\sup_{s\in\mathcal{T}} \sum_{i=1}^{n}(Y_{i,s} + \varepsilon_i Y'_{i,s}) \geq \sup_{s\in\mathcal{T}} \sum_{i=1}^{n}(Y_{i,s} - \varepsilon_i Y'_{i,s}) + 2\sqrt{tW'}\right\}.
\end{aligned}$$

We bound the probability above conditionally, by fixing the values of $X_1, \ldots, X_n$ and $X'_1, \ldots, X'_n$. Then

$$\phi_1(\varepsilon_1, \ldots, \varepsilon_n) = \sup_{s \in \mathcal{T}} \sum_{i=1}^{n} (Y_{i,s} + \varepsilon_i Y'_{i,s})$$

and

$$\phi_2(\varepsilon_1, \ldots, \varepsilon_n) = \sup_{s \in \mathcal{T}} \sum_{i=1}^{n} (Y_{i,s} - \varepsilon_i Y'_{i,s})$$

are convex functions on $[-1, 1]^n$ that are Lipschitz with constant $\sqrt{W'}$. If M denotes the common median of $\phi_1 = \phi_1(\varepsilon_1, \ldots, \varepsilon_n)$ and $\phi_2 = \phi_2(\varepsilon_1, \ldots, \varepsilon_n)$, then, denoting by $\mathbf{P}_\varepsilon$ conditional probability with the values of the X_i, X'_i fixed,

$$\begin{aligned}
&\mathbf{P}\left\{Z_1 \geq Z_2 + 2\sqrt{tW'} \mid X_1, \ldots, X_n, X'_1, \ldots, X'_n\right\} \\
&= \mathbf{P}_\varepsilon\left\{\phi_1 \geq \phi_2 + 2\sqrt{tW'}\right\} \\
&\leq \mathbf{P}_\varepsilon\left\{\phi_1 \geq M + \sqrt{tW'}\right\} + \mathbf{P}_\varepsilon\left\{\phi_2 \leq M - \sqrt{tW'}\right\} \\
&\leq 4e^{-t/4} \quad \text{(by Theorem 7.12).}
\end{aligned}$$

The lower tail inequality is proved similarly. □

For uniformly bounded empirical processes, one may recover a version of Theorem 12.2 from Theorem 12.3. Indeed, if $|X_{i,s}| \leq 1$ for all $i = 1, \ldots, n$ and $s \in \mathcal{T}$, then $W/4$ is a self-bounding function. This may be seen by introducing

$$W_i = \mathbf{E}\left[\sup_{s \in \mathcal{T}} \sum_{j \neq i}^{n} (X_{j,s} - X'_{j,s})^2 \middle| X^{(i)}\right]$$

and noting that $0 \leq W - W_i \leq 4$ for all $i = 1, \ldots, n$ and $\sum_{i=1}^{n}(W - W_i) \leq W$. Hence, by Theorem 6.12,

$$\mathbf{P}\left\{W \geq \mathbf{E}W + \sqrt{8t\mathbf{E}W} + 4t/3\right\} \leq e^{-t/4}.$$

The last inequality may be combined with Theorem 12.3 to obtain

$$\mathbf{P}\left\{Z \geq \mathbf{E}Z + 7\sqrt{t\Sigma^2} + 3t\right\} \leq 5e^{-t/4}.$$

12.4 Bousquet's Inequality for Suprema of Empirical Processes

Theorem 12.2 is a useful tool for bounding deviations of the supremum of an empirical process from its mean, but it is not completely satisfactory. Indeed, if the index set $\mathcal{T}$ is reduced to a single element, Theorem 12.2 implies Bernstein's inequality with a sub-optimal constant but one does not recover Bennett's inequality (Theorem 2.9). In this section we prove a Bennett-style concentration inequality for the supremum of an empirical process. As before, let $Z = \sup_{s\in\mathcal{T}} \sum_{i=1}^n X_{i,s}$. The proof is more involved than that of Theorem 12.2 as it does not follow directly from Theorem 6.16.

Theorem 12.5 (BOUSQUET'S INEQUALITY) *Let $X_1, \ldots, X_n$ be independent identically distributed random vectors. Assume that $\mathbf{E}X_{i,s} = 0$, and that $X_{i,s} \leq 1$ for all $s \in \mathcal{T}$. Let $v = 2\mathbf{E}Z + \sigma^2$ (where $\sigma^2 = \sup_{s\in\mathcal{T}} \sum_{i=1}^n \mathbf{E}X_{i,s}^2$ is the wimpy variance). Let $\phi(u) = e^u - u - 1$ and $h(u) = (1+u)\log(1+u) - u$, for $u \geq -1$. Then for all $\lambda \geq 0$,*

$$\log \mathbf{E}e^{\lambda(Z-\mathbf{E}Z)} \leq v\phi(\lambda).$$

Also, for all $t \geq 0$,

$$\mathbf{P}\{Z \geq \mathbf{E}Z + t\} \leq e^{-vh(t/v)}.$$

Recall that by Theorem 11.10, $\mathrm{Var}(Z) \leq v$, which makes the appearance of v natural in the statement of the theorem. By bounding $h(u)$ as in Section 2.7, the theorem implies

$$\mathbf{P}\{Z \geq \mathbf{E}Z + t\} \leq \exp\left(-\frac{t^2}{2(v + t/3)}\right).$$

We introduce some notation used in the proof. For all $i = 1, \ldots, n$, let $Z_i = \sup_{s\in\mathcal{T}} \sum_{j:j\neq i} X_{j,s}$ and let $\hat{s}_i \in \mathcal{T}$ be an index such that $\sum_{j:j\neq i} X_{j,\hat{s}_i} = Z_i$. As $Z \geq \sum_{j=1}^n X_{j,\hat{s}_i}$, we have $X_{i,\hat{s}_i} \leq Z - Z_i \leq X_{i,\hat{s}}$ and

$$\sum_{i=1}^n (Z - Z_i) \leq \sum_{i=1}^n X_{i,\hat{s}} = Z.$$

Denoting $Y_i = Z - Z_i$, we have $\mathbf{E}^{(i)}Y_i \geq \mathbf{E}^{(i)}X_{i,\hat{s}_i} = 0$ (recall that $\mathbf{E}^{(i)}$ denotes conditional expectation conditioned on $X^{(i)} = (X_1, \ldots, X_{i-1}, X_{i+1}, \ldots, X_n)$). We also have $Y_i \leq 1$ with probability one.

As in the proof of Theorem 6.12, the proof of Theorem 12.5 starts from the modified logarithmic Sobolev inequality of Theorem 6.6. The next step is to find an appropriate upper bound for $e^{\lambda Z}\phi(-\lambda(Z - Z_i))$. In the proof of Theorem 6.12 this was achieved by using the elementary inequality $\phi(-\lambda x)/\phi(-\lambda) \leq x$ for $0 \leq x \leq 1$. However, in the proof of Theorem 12.5 we also need to handle negative values of x and therefore we need the following lemma.

Lemma 12.6 *If $\beta \geq 0$, then, for all $\lambda \geq 0$ and $x \leq 1$,*

$$\frac{\phi(-\lambda x)}{\phi(-\lambda)} \leq \frac{x + (\beta x^2 - x)\, e^{-\lambda x}}{1 + (\beta - 1)\, e^{-\lambda}}.$$

The following lemma provides a tool that we may use for proving Lemma 12.6.

Lemma 12.7 *Let I be an interval containing 0. Let $f, g : I \to \mathbb{R}$ be twice differentiable functions such that $f(0) = g(0) = f'(0) = g'(0) = 0$, $g''(0) > 0$, and $xg'(x) > 0$ for every $x \neq 0$. The function ρ defined by $\rho(0) = f''(0)/g''(0)$ and $\rho(x) = f(x)/g(x)$ if $x \in I \setminus \{0\}$ is continuous and nondecreasing on I whenever $f''g' - f'g'' \geq 0$ on I.*

Proof Note first that $g(0) = 0$ and $xg'(x) > 0$ for every $x \neq 0$ implies that $g(x) > 0$ whenever $x \neq 0$. Hence ρ is well defined and twice differentiable on $I \setminus \{0\}$. The continuity of ρ at 0 follows from l'Hôpital's rule. For every $x \neq 0$, $\rho'(x)$ has the same sign as

$$g'(x)\left(\frac{f'(x)}{g'(x)} - \frac{f(x)}{g(x)}\right), \text{ or, equivalently, as } \Delta(x) \stackrel{\text{def}}{=} x\left(\frac{f'(x)}{g'(x)} - \frac{f(x)}{g(x)}\right).$$

Now the extended mean value theorem ensures that for some number c between 0 and x,

$$\frac{f'(c)}{g'(c)} = \frac{f(x)}{g(x)}.$$

Moreover, the function f'/g' (taking value $f''(0)/g''(0)$ at 0) is continuous on I and the assumption $f''g' - f'g'' \geq 0$ ensures that f'/g' is nondecreasing. Hence, for every $x \neq 0$,

$$x\frac{f'(c)}{g'(c)} \leq x\frac{f'(x)}{g'(x)},$$

and therefore $\Delta(x) \geq 0$. This proves that $\rho'(x) \geq 0$ from which the monotonicity of ρ follows. □

Proof of Lemma 12.6. To prove the lemma, it suffices to show that for all $\lambda \geq 0$ and $\beta \geq 0$, the function $\rho(x) = \phi(-\lambda x)/(x + (\beta x^2 - x)\, e^{-\lambda x})$ (with $\rho(0) = \lambda^2/(2(\beta + \lambda)))$ is nondecreasing for $x \in (-\infty, 1]$.

The lemma obviously holds for $\lambda = 0$. Fix $\lambda > 0$ and $\beta \geq 0$. We may rewrite ρ as $\rho = f/g$ with

$$f(x) = e^{\lambda x}\phi(-\lambda x) = \lambda x e^{\lambda x} - e^{\lambda x} + 1 \quad \text{and} \quad g(x) = xe^{\lambda x} + \beta x^2 - x.$$

Then, for every $x \neq 0$,

$$xg'(x) = x^2\left(\lambda e^{\lambda x} + 2\beta\right) + x\left(e^{\lambda x} - 1\right) > 0$$

and

$$f''(x)g'(x) - f'(x)g''(x) = \lambda^2 e^{\lambda x}\left(\phi(\lambda x) + 2\beta\lambda x^2\right) > 0,$$

and therefore Lemma 12.7 implies the monotonicity of ρ. □

The first step of the proof of Bousquet's inequality is the following lemma. It is based on the modified logarithmic Sobolev inequality of Theorem 6.6 and uses Lemma 12.6 to upper bound the expectation of $\exp(\lambda Z)\phi(-\lambda(Z - Z_i))$ conditionally on $X^{(i)} = (X_1, \ldots, X_{i-1}, X_{i+1}, \ldots, X_n)$. Here we use $\beta = 1/2$.

Lemma 12.8 *Let Z, ϕ, and v be defined as in Theorem 12.5. Let $f(\lambda) = \phi(\lambda) + \lambda/2$. If $G(\lambda) = \log \mathbf{E}e^{\lambda(Z-\mathbf{E}Z)}$, then, for $\lambda \geq 0$,*

$$f(\lambda)G'(\lambda) - f'(\lambda)G(\lambda) \leq (v/2)\left(\lambda f'(\lambda) - f(\lambda)\right).$$

Proof Recall the modified logarithmic Sobolev inequality (Theorem 6.6):

$$\mathrm{Ent}(e^{\lambda Z}) = \lambda \mathbf{E}\left[Ze^{\lambda Z}\right] - \mathbf{E}\left[e^{\lambda Z}\right]\log \mathbf{E}\left[e^{\lambda Z}\right] \leq \sum_{i=1}^{n} \mathbf{E}\left[e^{\lambda Z}\phi\left(-\lambda(Z - Z_i)\right)\right].$$

Since $Z - Z_i \leq 1$, Lemma 12.6 with $\beta = 1/2$ implies that for all $i = 1, \ldots, n$,

$$\begin{aligned}
&\phi\left(-\lambda(Z - Z_i)\right) e^{\lambda Z} \\
&\leq \theta(\lambda)\left((Z - Z_i)e^{\lambda Z} + \left(1/2(Z - Z_i)^2 - (Z - Z_i)\right) e^{\lambda Z_i}\right),
\end{aligned}$$

where $\theta(\lambda) = \phi(-\lambda)/(1 - (1/2)\exp(-\lambda))$. Taking conditional expectations, from Lemma 11.11,

$$\mathbf{E}^{(i)}\left[(Z - Z_i)^2/2 - (Z - Z_i)\right] \leq \frac{1}{2}\mathbf{E}^{(i)}[X_{i,\hat{s}_i}^2],$$

and therefore

$$\begin{aligned}
&\mathbf{E}^{(i)}\left[\phi\left(-\lambda(Z - Z_i)\right) e^{\lambda Z}\right] \\
&\leq \theta(\lambda)\left(\mathbf{E}^{(i)}\left[(Z - Z_i)e^{\lambda Z}\right] + \frac{1}{2}\mathbf{E}^{(i)}[X_{i,\hat{s}_i}^2]e^{\lambda Z_i}\right) \\
&\leq \theta(\lambda)\left(\mathbf{E}^{(i)}\left[(Z - Z_i)e^{\lambda Z}\right] + \frac{1}{2}\sup_{s\in\mathcal{T}} \mathbf{E}[X_{i,s}^2]e^{\lambda Z_i}\right).
\end{aligned}$$

Using the fact that $\mathbf{E}^{(i)}[Z - Z_i] \geq 0$ and applying Jensen's inequality,

$$e^{\lambda Z_i} \leq e^{\lambda \mathbf{E}^{(i)} Z} \leq \mathbf{E}^{(i)} e^{\lambda Z}.$$

Thus, for every $i = 1, \ldots, n$, we have

$$\mathbf{E}^{(i)}\left[\phi\left(-\lambda(Z - Z_i)\right) e^{\lambda Z}\right] \leq \theta(\lambda)\mathbf{E}^{(i)}\left[\left(Z - Z_i + \frac{1}{2}\sup_{s\in\mathcal{T}} \mathbf{E}[X_{i,s}^2]\right) e^{\lambda Z}\right].$$

Plugging this last inequality into the modified logarithmic Sobolev inequality and using the fact that $\sum_{i=1}^n (Z - Z_i) \leq Z$,

$$\begin{aligned}\mathrm{Ent}(e^{\lambda Z}) &\leq \theta(\lambda)\left(\mathbf{E}\left[e^{\lambda Z}\left(Z + \frac{1}{2}\sum_{i=1}^{n}\sup_{s\in\mathcal{T}} \mathbf{E}[X_{i,s}^2]\right)\right]\right)\\ &\leq \theta(\lambda)\left(\mathbf{E}\left[e^{\lambda Z}\left(Z - \mathbf{E}Z + \mathbf{E}Z + \frac{\sigma^2}{2}\right)\right]\right),\end{aligned}$$

by recalling the notation $\sigma^2 = \sum_{i=1}^n \sup_{s\in\mathcal{T}} \mathbf{E}[X_{i,s}^2] = \sup_{s\in\mathcal{T}} \sum_{i=1}^n \mathbf{E}[X_{i,s}^2]$. This last inequality may be rewritten as

$$\mathrm{Ent}(e^{\lambda(Z-\mathbf{E}Z)}) \leq \theta(\lambda)\left(\mathbf{E}\left[(Z - \mathbf{E}Z)e^{\lambda(Z-\mathbf{E}Z)}\right] + \frac{v}{2}\mathbf{E}e^{\lambda(Z-\mathbf{E}Z)}\right),$$

or, dividing both sides by $\mathbf{E}\exp(\lambda(Z - \mathbf{E}Z))$, as

$$\lambda G'(\lambda) - G(\lambda) \leq \theta(\lambda)\left(G'(\lambda) + \frac{v}{2}\right).$$

The lemma follows by observing that $f'(\lambda) > 0$ for $\lambda \geq 0$ and rearranging the last inequality. □

Bousquet's inequality now follows easily from Lemma 6.25.

Proof of Theorem 12.5. Let

$$g(\lambda) = \frac{v}{2} \cdot \frac{\lambda f'(\lambda) - f(\lambda)}{f^2(\lambda)}$$

for $\lambda > 0$ and $g(0) = v$, where $f(\lambda) = \phi(\lambda) + \lambda/2$.

It is easy to check that g is continuous on $[0, +\infty)$. By Lemma 12.8, $G(\lambda) = \log \mathbf{E}e^{\lambda(Z-\mathbf{E}Z)}$ satisfies

$$f(\lambda)G'(\lambda) - f'(\lambda)G(\lambda) \leq f^2(\lambda)g(\lambda)$$

for $\lambda \geq 0$. From Lemma 6.25, it follows that for all $\lambda \geq 0$,

$$G(\lambda) \leq \frac{v}{2} f(\lambda) \int_0^\lambda \frac{xf'(x) - f(x)}{f^2(x)} dx.$$

The observation that $(xf'(x) - f(x))/f^2(x) = (-x/f(x))'$ and that $\lim_{x \downarrow 0} x/f(x) = 2$ finally leads to the desired result $G(\lambda) \leq v\phi(\lambda)$.

The tail bound follows using the computation shown in Chapter 2. □

12.5 Non-Identically Distributed Summands and Left-Tail Inequalities

In this section we present, without proof, two inequalities related to the concentration bounds of the previous sections. First note that, unlike Bennett's inequality that holds not only for sums of i.i.d. random variables but also for sums of independent, non-identically distributed bounded and centered random variables, Bousquet's inequality requires that the vectors $X_1, \ldots, X_n$ are identically distributed. The following inequality, though not quite as sharp as Bousquet's, is a step in this direction.

Theorem 12.9 *Let $X_1, \ldots, X_n$ be independent vector-valued random variables and let $Z = \sup_{s \in \mathcal{T}} \sum_{i=1}^n X_{i,s}$. Assume that for all $i \leq n$ and $s \in \mathcal{T}$, $\mathbf{E}X_{i,s} = 0$, and $|X_{i,s}| \leq 1$. Let $v = 2\mathbf{E}Z + \sigma^2$ where $\sigma^2 = \sup_{t \in \mathcal{T}} \sum_{i=1}^n \mathbf{E}X_{i,s}^2$. Then $\mathrm{Var}(Z) \leq v$ and for all $\lambda > 0$,*

$$\log \mathbf{E}e^{\lambda(Z-\mathbf{E}Z)} \leq \frac{v\lambda}{2}\left(\exp((e^{2\lambda} - 1)/2) - 1\right).$$

In particular, for all $t > 0$,

$$\mathbf{P}\{Z \geq \mathbf{E}Z + t\} \leq \exp\left(-\frac{t}{4} \log\left(1 + 2\log(1 + t/v)\right)\right).$$

The proof of this theorem is quite technical and we do not include it here. Note that the variance bound is the same as in the case of identically distributed summands (Theorem 11.10).

Bennett's inequality holds not only for right tails but also for left tails (i.e. for bounding $\mathbf{P}\{Z \leq \mathbf{E}Z - t\}$). Whether such an inequality is true for suprema of centered bounded empirical processes is a natural question. However, the proofs of Theorems 12.5 and 12.9 are tailored to handle deviations above the mean. In view of Theorem 12.5, one may wonder whether $\log \mathbf{E}e^{\lambda(Z-\mathbf{E}Z)} \leq v\phi(\lambda)$ also holds for $\lambda \leq 0$ (where $\phi(\lambda) = \exp(\lambda) - \lambda - 1$). This is still unknown but we do have the following results. Once again, the proofs are omitted.

The next theorem can be proved using variants of the entropy method.

Theorem 12.10 (KLEIN-RIO BOUND) *Using the notation and assumptions of Theorem 12.9, and assuming that the X_i are identically distributed for all $\lambda \leq 0$,*

$$\log \mathbf{E}e^{\lambda(Z-\mathbf{E}Z)} \leq \frac{v}{9}\phi(-3\lambda),$$

and for all $t \geq 0$,

$$\mathbf{P}\{Z \leq \mathbf{E}Z - t\} \leq \exp\left(-\frac{v}{9}h\left(\frac{3t}{v}\right)\right)$$

where $h(t) = (t+1)\log(t+1) - t$.

The next theorem can be proved using a variant of the transportation method.

Theorem 12.11 (SAMSON'S BOUND) *Recall the notation and assumptions of Theorem 12.9 and let*

$$S^2 = \mathbf{E} \sup_{s \in \mathcal{T}} \sum_{i=1}^{n} \mathbf{E}\left[(X_{i,s} - X'_{i,s})_+^2 \mid X_{i,s}\right]$$

where $X'_1, \ldots, X'_n$ are independent copies of $X_1, \ldots, X_n$. For all $\lambda \leq 0$,

$$\log \mathbf{E}e^{\lambda(Z-\mathbf{E}Z)} \leq \frac{S^2}{4}\phi(-2\lambda),$$

and for all $t \geq 0$,

$$\mathbf{P}\{Z \leq \mathbf{E}Z - t\} \leq \exp\left(-\frac{t^2}{2(S^2 + 2t/3)}\right).$$

Note that by Theorem 11.1, S^2 is an upper bound for the variance of Z and $S^2 \leq \Sigma^2 + \sigma^2$, where $\Sigma^2 = \mathbf{E}\sup_{s\in\mathcal{T}} \sum_{i=1}^n X_{i,s}^2$.

Theorems 12.5 and 12.9 are often used through the next corollary which follows from bounds on the inverse of $h : t \to (1+t)\log(1+t) - t$ over $[0, \infty)$. In particular, one may prove that

$$h^{-1}(x) \leq \begin{cases} \sqrt{2x} + 3x & \text{for } x \geq 0 \\ 2x/\log(x) & \text{for } x \geq 3 \\ 2\sqrt{x} & \text{for } 0 \leq x \leq 2/9. \end{cases}$$

Corollary 12.12 *Consider the setup of Theorem 12.5. Let $\phi(u) = e^u - u - 1$, and $h(u) = (1+u)\log(1+u) - u$, for $u \geq -1$. Then for all $\lambda > 0$,*

$$\log \mathbf{E}e^{\lambda(Z-\mathbf{E}Z)} \leq v\phi(\lambda),$$

and

$$\log \mathbf{E}e^{-\lambda(Z-\mathbf{E}Z)} \leq \frac{v}{9}\phi(3\lambda).$$

Also, for all $t \geq 0$,

$$P\left\{Z \geq EZ + \sqrt{2vt} + \frac{t}{3}\right\} \leq e^{-t},$$

$$P\left\{Z \leq EZ - \sqrt{2vt} - \frac{t}{8}\right\} \leq e^{-t},$$

and for $t \geq 3v$,

$$P\left\{Z \geq EZ + 2t/\log(t/v)\right\} \leq e^{-t}.$$

12.6 Chi-Square Statistics and Quadratic Forms

As an illustration of the power of Bousquet's inequality, we present an application to *Pearson's chi-square statistic*, a random variable well known in statistical theory.

Let $p_1, \ldots, p_m > 0$ such that $\sum_{j=1}^{m} p_j = 1$ and suppose that the random vector $(N_1, \ldots, N_m)$ has a multinomial distribution with parameters $n, p_1, \ldots, p_m$. Pearson's chi-square statistic is defined by

$$Z^2 = \sum_{j=1}^{m} \frac{(N_j - np_j)^2}{np_j}.$$

As is well known from classical statistics (and follows easily from a multivariate central limit theorem), if m is fixed and $n \to \infty$, Z^2 converges in distribution to the square of the norm of a standard Gaussian vector. Here we derive a non-asymptotic concentration inequality. To this end, introduce the random variables $W_{i,j}$ for $i = 1, \ldots, n$ and $j = 1, \ldots, m$, defined by

$$W_{i,j} = \begin{cases} 1/\sqrt{p_j} & \text{if } Y_i = j \\ 0 & \text{otherwise} \end{cases}$$

where $Y_1, \ldots, Y_n$ are independent random variables with distribution $P\{Y_i = j\} = p_j$ for $j = 1 \ldots, m$. Then we may write $N_j = \sqrt{p_j} \sum_{i=1}^{n} W_{i,j}$ and Z^2 can be written as

$$Z^2 = \frac{1}{n} \sum_{j=1}^{m} \left(\sum_{i=1}^{n} (W_{i,j} - EW_{i,j}) \right)^2.$$

The key idea is to represent the (nonnegative) random variable Z as the supremum of an empirical process. Indeed, if $\mathcal{T}$ is a dense countable subset of the unit Euclidean ball in $\mathbb{R}^m$, then, by the Cauchy–Schwarz inequality,

$$Z = \sup_{s \in \mathcal{T}} \frac{1}{\sqrt{n}} \sum_{i=1}^{n} \left(\sum_{j=1}^{m} s_j (W_{i,j} - EW_{i,j}) \right),$$

which is of the form $\sup_{s \in \mathcal{T}} \sum_{i=1}^{n} X_{i,s}$ with $X_{i,s} = \sum_{j=1}^{m} s_j (W_{i,j} - EW_{i,j})/\sqrt{n}$.

Note first that by the Cauchy–Schwarz inequality,

$$EZ \leq \sqrt{EZ^2} = \sqrt{m-1}.$$

The wimpy variance may be bounded by straighforward calculation as

$$\sigma^2 = \sup_{s \in \mathcal{T}} \operatorname{Var}\left(\sum_{j=1}^{m} s_j \frac{N_j - np_j}{\sqrt{np_j}}\right) \leq 1.$$

Denoting $p_{\min} = \min_{j=1,\dots,m} p_j$, we may bound the variance of Z by Theorem 11.10 to obtain

$$\operatorname{Var}(Z) \leq \sigma^2 + 2\sqrt{\frac{1}{np_{\min}}}EZ \leq 1 + 2\sqrt{\frac{m-1}{np_{\min}}}.$$

On the other hand, Theorem 12.5 may be used directly to obtain the following exponential tail inequality.

Theorem 12.13 *Let Z^2 be Pearson's chi-square statistic defined above. Then for all $\varepsilon, t > 0$,*

$$P\left\{Z \geq (1+\varepsilon)\sqrt{m-1} + \sqrt{2t} + \kappa(\varepsilon)\sqrt{\frac{1}{np_{min}}}t\right\} \leq e^{-t},$$

where $\kappa(\varepsilon) = 2\left(\frac{1}{3} + \varepsilon^{-1}\right)$.

12.7 Bibliographical Remarks

Theorem 12.1 was noted by Massart (1998). Hoeffding-type inequalities for sums of independent random vectors may also be derived using Marton's transportation method (see Chapter 8).

The material in Section 12.3 is based on Panchenko (2003), though the proof shown here gives slightly worse constants. Panchenko's theorem can be regarded as the latest in a long series of papers beginning with Vapnik and Chervonenkis (1971, 1974, 1981). Some of the results stated in these papers are given below in Exercises 12.1, 12.3, and 12.4. Panchenko was the first to blend the symmetrization and conditioning arguments, which form the heart of the original arguments of Vapnik and Chervonenkis, with Talagrand's convex distance inequality.

Concentration inequalities for suprema of self-normalized empirical processes have been derived using the entropy method by Bercu, Gassiat, and Rio (2002). Adopting the notation of this chapter, they considered

$$Z = \sup_{s \in \mathcal{T}} \frac{\sum_{i=1}^{n} X_{i,s}}{\sqrt{\sum_{i=1}^{n} X_{i,s}^2}}$$

where each $X_{i,s}$ is assumed to be centered with unit variance. Assuming that $\sup_{n=1,2,\ldots} E\left[\sup_{s\in\mathcal{T}} \sum_{i=1}^{n} X_{i,s}\right]/\sqrt{n}$ is finite, Bercu, Gassiat, and Rio (2002) proved exponential tail bounds for Z using a variant of the entropy method starting from Theorem 6.6.

Theorem 12.5 is due to Bousquet (2002*b*). This is a refinement of a series of related results pioneered by Talagrand (1996*b*) (see also Talagrand 1994*b*). Ledoux (1997) showed that the Bennett-type inequality for suprema of bounded empirical processes described in Talagrand (1996*b*) can be derived by the entropy method in a transparent way. Massart (2000*a*) proved that the constants in Talagrand's inequality can be kept reasonable. Panchenko (2001) investigated the potential and limits of Talagrand's approach. Rio (2001, 2002) derived Bennett- and Bernstein-type inequalities with the same variance factor as in Theorem 12.5 but with a sub-optimal scale factor (see Exercise 12.10). The proof in Rio (2002) starts with the modified logarithmic Sobolev inequality of Theorem 6.6. Bousquet (2002*b*) derived a version of Lemma 12.6.

Rio (2012) refined Theorem 12.5: using the notation of this theorem, letting $E = \mathbf{E}Z/n$, he established

$$\log \mathbf{E}e^{\lambda(Z-\mathbf{E}Z)} \leq \frac{(v - E^2)}{(1-E)^2}\phi((1-E)\lambda)$$

(see Exercises 12.7 and 12.8).

Most results of Sections 12.5 come from Klein and Rio (2002, 2005). Theorem 12.11 and many related results come from Samson (2007), who used the infimum-convolution approach to concentration pioneered by Maurey (1991).

The tail bounds for Pearson's chi-square statistic are taken from Castellan (2003) and Massart (2006) who also refine Theorem 12.13 to render it suitable for certain statistical applications. Early tail bounds for Pearson's chi-square statistic were proved by Mason and van Zwet (1987).

This chapter focuses on tail bounds for suprema of centered empirical processes. However, certain non-centered empirical processes appear in some applications in statistics and they also occur in the derivation of left tail bounds in Klein and Rio (2005). Variance and tail bounds for such quantities are described in Boucheron and Massart (2010).

12.8 EXERCISES

12.1. (VAPNIK–CHERVONENKIS INEQUALITIES) Let $\mathcal{C}$ denote a class of subsets of a measurable space $\mathcal{X}$ and let P be a probability distribution over $\mathcal{X}$. Let P_n and P'_n denote the empirical distributions defined by two independent samples of n random variables drawn from the distribution P and let $\nu_n = P_n - P$. Let $h(X_1, \ldots, X_{2n})$ denote the VC-entropy of $\mathcal{C}$ in a $2n$-sample $X_1, \ldots, X_{2n}$. (Recall the definition of VC-entropy from Section 4.5.) prove that for all $\varepsilon > 0$,

$$\mathbf{P}\left\{\sup_{A\in\mathcal{C}} |\nu_n(A)| \geq 2\varepsilon\right\} \leq 4\mathbf{E}\left[2^{h(X_1,\ldots,X_{2n})}\right] e^{-n\varepsilon^2/2}.$$

Hint: prove first that

$$P\left\{\sup_{A\in\mathcal{C}} |\nu_n(A)| \geq 2\varepsilon\right\} \leq 2P\left\{\sup_{A\in\mathcal{C}} |P_n(A) - P'_n(A)| \geq \varepsilon\right\}.$$

Second, use the fact that the distribution of $2n$-samples is invariant under permutation and prove that

$$P\left\{\sup_{A\in\mathcal{C}} |P_n(A) - P'_n(A)| \geq \varepsilon\right\} \leq 2E\left[2^{h(X_1,\ldots,X_{2n})}\right] e^{-2n\varepsilon^2}.$$

Now the improved symmetrization inequality given in Exercise 11.5 can be used to prove in a few lines that

$$P\left\{\sup_{A\in\mathcal{C}} |\nu_n(A)| \geq 2\varepsilon\right\} \leq 2E\left[2^{h(X_1,\ldots,X_n)}\right] e^{-n\varepsilon^2/2}$$

(see Vapnik and Chervonenkis (1971, 1981)). Recall that Theorem 6.14 implies that the so-called annealed VC-entropy $\log_2 E\left[2^{h(X_1,\ldots,X_{2n})}\right]$ and $Eh(X_1,\ldots,X_{2n})$ are within a constant factor of each other.

12.2. (SELF-NORMALIZATION) Let $X_1,\ldots,X_n$ be independent symmetric real random variables. Prove that for all $t > 0$,

$$P\left\{\frac{\sum_{i=1}^n X_i}{\sqrt{\sum_{i=1}^n X_i^2}} \geq t\right\} \leq e^{-t^2/2}.$$

See Bercu, Gassiat, and Rio (2002), Giné, Koltchinskii, and Wellner (2003), Giné and Koltchinskii (2006) and Maurer and Pontil (2009) for more material on concentration for self-normalized empirical processes.

12.3. (VAPNIK–CHERVONENKIS INEQUALITY FOR RELATIVE DEVIATION) Consider the notation introduced in Exercise 12.1. Prove that for all $\varepsilon > 0$,

$$P\left\{\sup_{A\in\mathcal{C}} \frac{P(A) - P_n(A)}{\sqrt{P(A)}} \geq 2\varepsilon\right\} \leq 2E\left[2^{h(X_1,\ldots,X_{2n})}\right] e^{-n\varepsilon^2/2}.$$

Hint: use a symmetrization of the tail probabilities, as in Exercise 12.1, to show that

$$P\left\{\sup_{A\in\mathcal{C}} \frac{P(A) - P_n(A)}{\sqrt{P(A)}} \geq 2\varepsilon\right\} \leq 2P\left\{\sup_{A\in\mathcal{C}} \frac{P'_n(A) - P_n(A)}{\sqrt{(P_n(A) + P'_n(A))/2}} \geq \varepsilon\right\}.$$

(See Vapnik and Chervonenkis (1974) and also Anthony and Shawe-Taylor (1993), Haussler (1992), and Bartlett and Lugosi (1999).)

12.4. (VAPNIK–CHERVONENKIS INEQUALITY FOR RELATIVE DEVIATION, CONTINUED) Prove that for all $\varepsilon > 0$,

$$P\left\{\sup_{A\in\mathcal{C}} \frac{P_n(A) - P(A)}{\sqrt{P_n(A)}} \geq 2\varepsilon\right\} \leq 2E\left[2^{h(X_1,\ldots,X_{2n})}\right] e^{-n\varepsilon^2/2}.$$

Hint: see Exercise 12.3. See Vapnik and Chervonenkis (1974), Anthony and Shawe-Taylor (1993), Haussler (1992), and Bartlett and Lugosi (1999).

12.5. (SUB-GAMMA SUMMANDS) Let $(\alpha_{i,s})$ be a collection of real numbers indexed by $i = 1,\ldots,n$ and let $s \in \mathcal{T}$ be such that $|\alpha_{i,s}| \leq 1$ for all i and s. Let $X_1,\ldots,X_n$ be independent centered random variables such that for all integers $q \geq 2$,

$$E|X_i|^q \leq q!\frac{c^{q-2}\sigma^2/n}{2}$$

for some constants c and σ (note that this implies that X_i is sub-gamma with variance factor σ^2/n and scale factor c). Let $Z = \sup_{s\in\mathcal{T}} \sum_{i\leq n} \alpha_{i,s}X_i$. Let

$$v = \sigma^2 + 2EZ.$$

Prove that for $\lambda \geq 0$,

$$\log Ee^{\lambda(Z-EZ)} \leq \frac{v\lambda^2}{2(1-c\lambda)}$$

(see Bousquet 2002*b*, Theorem 2.12).

12.6. Let Z and $(Z_i)_{i\leq n}$ be defined as in Theorem 12.5. Prove that

$$\sum_{i=1}^{n} E\left[e^{\lambda Z} - e^{\lambda Z_i}\right] \leq Ee^{\lambda Z}\log\left(Ee^{\lambda Z}\right).$$

Hint: use Theorem 6.6 and the self-bounding property $\sum_{i=1}^{n}(Z - Z_i) \leq Z$, Rio (2002).

12.7. Let Z and $(Z_i)_{i\leq n}$ be defined as inTheorem 12.5, that is, $Z = \sup_{s\in\mathcal{T}} \sum_{j=1}^{n} X_{j,s}$ and $Z_i = \sup_{s\in\mathcal{T}} \sum_{j\neq i} X_{j,s}$. Prove that

$$\frac{\log Ee^{\lambda Z}}{n} \leq \frac{\log Ee^{\lambda Z_n}}{n-1}.$$

Hint: use Theorem 6.6 as in Exercise 12.6, but replace Z_i by $\tilde{Z}_i = Z_i + \ln E\exp(\lambda Z) - \ln E\exp(\lambda Z_i)$ (Rio, 2012). Note that it was observed a long time ago that $EZ/n \leq EZ_n/(n-1)$ (Pollard, 1984).

12.8. Letting $\theta(y) = \exp(y)\phi(-y)/(y\phi(y) + y^2/2)$, for $y \geq 0$, prove that for $y \geq 0$ and $x \leq y$,

$$e^x\phi(-x) \leq \theta(y)\left(y\phi(x) + \frac{x^2}{2}\right).$$

Rio (2012) uses this inequality and Exercise 12.7 to refine Theorem 12.5 (see Exercise 12.9).

12.9. Using the notation of Theorem 12.5, letting $E = \mathbf{E}Z/n$ and $\tilde{v} = n\sigma^2 + (2 - E)\mathbf{E}Z$, the aim of this exercise is to prove that, for $\lambda > 0$,

$$\log \mathbf{E}e^{\lambda(Z-\mathbf{E}Z)} \leq \frac{\tilde{v}}{(1-E)^2}\phi((1-E)\lambda).$$

Note first that for $\lambda \geq 0$ satisfying $\lambda\tilde{v} \geq 2n(1-E)$, the inequality follows from the fact that $Z \leq n$. Proceed as in the proof of Theorem 12.5, but replace Z_i by $\tilde{Z}_i = Z_i + E$ for each $1 \leq i \leq n$. Let $\theta(y) = \exp(y)\phi(-y)/(y\phi(y) + y^2/2)$ and use Exercise 12.8 to establish that for $\lambda \geq 0$,

$$\phi\left(-\lambda(Z-\tilde{Z}_i)\right)e^{\lambda Z}$$

$$\leq \lambda\theta(\lambda(1-E))\left((1-E)e^{\lambda Z} + e^{\lambda\tilde{Z}_i}\left[\lambda\left(\frac{(Z-\tilde{Z}_i)^2}{2} - (1-E)(Z-\tilde{Z}_i)\right) - (1-E)\right]\right).$$

Follow the pattern of analysis of Theorem 11.10 and Exercise 11.17 to establish that for $\lambda \geq 0$,

$$\mathbf{E}\left[\phi\left(-\lambda(Z-\tilde{Z}_i)\right)e^{\lambda Z}\right] \leq \lambda\theta(\lambda(1-E))\mathbf{E}\left[(1-E)e^{\lambda Z} + e^{\lambda\tilde{Z}_i}\left(\frac{\lambda\tilde{v}}{2n} - (1-E)\right)\right].$$

Henceforth, assume $0 \leq \lambda\tilde{v}/2n \leq (1-E)$. Use Exercise 12.7 to establish

$$\mathbf{E}e^{\lambda\tilde{Z}_i}\left(\frac{\lambda\tilde{v}}{2n} - (1-E)\right) \leq \left(\frac{\lambda\tilde{v}}{2n} - (1-E)\right)\mathbf{E}\left[e^{\lambda Z}\right]\left(1 - \frac{1}{n}\log\mathbf{E}e^{\lambda(Z-\mathbf{E}Z)}\right),$$

and that

$$\frac{\mathrm{Ent}(e^{\lambda Z})}{\mathbf{E}e^{\lambda Z}} \leq n\lambda\theta(\lambda(1-E))\left((1-E) + \left(\frac{\lambda\tilde{v}}{2n} - (1-E)\right)\left(1 - \frac{1}{n}\log\mathbf{E}e^{\lambda(Z-\mathbf{E}Z)}\right)\right).$$

Letting $G(\lambda) = \log\mathbf{E}e^{\lambda(Z-\mathbf{E}Z)}$, prove that for $0 \leq \lambda \leq 2n(1-E)/\tilde{v}$, G satisfies the differential inequality

$$\lambda G'(\lambda) - G(\lambda) \leq \lambda\theta(\lambda(1-E))\left((1-E)G(\lambda) + \frac{\lambda\tilde{v}}{2}\right).$$

Solve the differential inequality. *Hint:* $\theta(x)$ is the derivative of $\log((\phi(t) + t/2)/t)$.

12.10. (A SUB-OPTIMAL BENNETT-TYPE INEQUALITY) Recall the notation of Theorem 12.5. Prove that for all $\lambda > 0$,

$$\log \boldsymbol{E} e^{\lambda(Z-\boldsymbol{E}Z)} \leq v\lambda\left(e^{\lambda}-1\right)/2.$$

Hint: define $Z_i, \hat{s}_i$ as in the proof of Theorem 12.5. Starting from Theorem 6.6, use Exercise 12.6 to establish

$$\begin{aligned}&\mathrm{Ent}(e^{\lambda Z})\\&\leq \lambda \boldsymbol{E}e^{\lambda Z}\log \boldsymbol{E}e^{\lambda Z}+\sum_{i=1}^{n}\boldsymbol{E}\left[e^{\lambda Z_i}\boldsymbol{E}^{(i)}\left[e^{\lambda X_{i,\hat{s}_i}}\phi(-\lambda X_{i,\hat{s}_i})-\phi(\lambda X_{i,\hat{s}_i})(\lambda X_{i,\hat{s}_i})_+\right]\right].\end{aligned}$$

Use the fact that $e^x\phi(-x)-x_+\phi(x)\leq x^2/2$ for $x\in\mathbb{R}$ in order to conclude. (See Rio (2001, 2002).)

12.11. (SUMS OF INDEPENDENT POSITIVE SEMI-DEFINITE MATRICES) Let $X_1,\ldots,X_n$ be independent random positive semi-definite $d\times d$ matrices. Assume that for all $i\leq n$, the operator norm of X_i satisfies $\|X_i\|\leq a$ almost surely. Let $Z=\|\sum_{i=1}^n X_i\|$. Prove that

$$\mathrm{Var}\left(Z\right)\leq a\boldsymbol{E}Z,$$

and that for all $\lambda\in\mathbb{R}$,

$$\log \boldsymbol{E}e^{\lambda(Z-\boldsymbol{E}Z)}\leq(\boldsymbol{E}Z/a)\phi(a\lambda).$$

Hint: use the fact that $\|\sum_{i=1}^n X_i\|$ is self-bounding. (See Tropp 2010*a*.)

12.12. (SPECTRUM OF A GRAM MATRIX) Let $X_1,\ldots,X_n$ be independent identically distributed random vectors taking values in $\mathbb{R}^d$. The associated Gram matrix G is an $n\times n$ matrix with entries $G_{i,j}=\langle X_i,X_j\rangle$. Let $\lambda_1\geq\cdots\geq\lambda_n$ be a nonincreasing rearrangement of the eigenvalues of G. Assume that the X_i are almost surely bounded. Prove Bennett-like concentration inequalities for $Z=\sum_{j=1}^k\lambda_j$ where $1\leq k\leq d$. *Hint*: the nonzero eigenvalues of G are the same as the nonzero eigenvalues of $\sum_{i=1}^n X_iX_i^T$. Use the Courant–Fisher variational characterization of eigenvalues to check that Z is a self-bounding function. Then use the results from Exercise 12.11. (See Shawe-Taylor and Cristianini 2004, and Zwald and Blanchard 2006.)

12.13. Let $X_1,\ldots,X_n$ be independent sub-gamma random variables with expectation not larger than μ, variance factor smaller than v and scale factor smaller than c (thus, for $t\geq 0$, $\boldsymbol{P}\{X_i\geq\mu+\sqrt{2vt}+ct\}\leq e^{-t}$). Let $M=\max(X_1,\ldots,X_n)$.

Prove that $\boldsymbol{E}M^2$ is not essentially larger than the square of the upper bound on $\boldsymbol{E}M$ derived in Chapter 2 (Theorem 2.6), namely, for all $\lambda>0$, letting $H_n=\sum_{i=1}^n 1/i$, prove that $\boldsymbol{E}M\leq\mu+\sqrt{2vH_n}+cH_n$,

$$\boldsymbol{E}M^2\leq\left(\mu+\sqrt{2vH_n}+cH_n\right)^2+\frac{10v}{\log n}+\frac{c^2\pi^2}{3}$$

and

$$\log Ee^{\lambda M} \leq \lambda\left(\mu + 2\sqrt{vH_n} + cH_n\right) + \frac{\lambda^2(c+\sqrt{v/H_n})^2}{2(1-\lambda(c+\sqrt{v/H_n}))}.$$

Hint: you may assume that there exist independent random variables $Y_1, \ldots, Y_n$ such that $X_i \leq Y_i$ and $P\{Y_i \geq \mu + \sqrt{2vt} + ct\} = e^{-t}$ for all $1 \leq i \leq n$. In order to bound the higher moments of $\max(Y_1, \ldots, Y_n)$, combine Rényi's representation of order statistics (see de Haan and Ferreira (2006, Chapter 2)), the Efron–Stein inequality and Theorem 6.6. Check the tightness of the bounds by assuming that $X_1, \ldots, X_n$ are indeed gamma-distributed. This exercise shows that the right tail of the maximum of n independent sub-gamma random variables with scale factor c is not substantially heavier than the tail of a Gumbel distribution with scale c.

12.14. Let $X_1, \ldots, X_n$ be centered, independent sub-gamma random variables, with variance factor v and scale factor c (i.e. for $t \geq 0$, $P\{X_i \geq \sqrt{2vt} + ct\} \leq e^{-t}$). Let $\tau_n = \sqrt{2v\log n} + c\log n$. Let $Z = \sum_{i=1}^n X_i \mathbb{1}_{\{|X_i| \geq \tau_n\}}$. Prove that $EZ \leq \tau_n + c + \sqrt{v/(2\log n)}$ and that for $0 \leq \lambda \leq \tau_n/2$,

$$\log Ee^{\lambda Z} \leq \frac{e^{\lambda\tau_n} + \lambda\tau_n - 1}{1 - \lambda\tau_n}$$

and $P\{Z \geq t\} \leq 2\exp(-t/(4\tau_n))$ for $t \geq 0$. *Hint*: use a quantile coupling argument: there exists a probability space with random variables $X_1, \ldots, X_n$ as above and independent exponentially distributed random variables $Y_1, \ldots, Y_n$ with $|X_i| \leq \sqrt{2vY_i} + cY_i$ for all $1 \leq i \leq n$, almost surely. Bound $\log(E\exp(\lambda Z))$ by the quantity $n\log E\left[\exp\left(\lambda(\sqrt{2vY_1} + cY_1)\right)\mathbb{1}_{\{Y_1 \geq \log n\}}\right]$. This bound should be compared with those from Exercise 12.13. For small values of λ, they are both of order $\lambda\tau_n$. This should not come as a surprise. With overwhelming probability, Z coincides with $\max(X_1, \ldots, X_n)$, as there is at most one index $1 \leq i \leq n$ such that $|X_i| \geq \tau_n$. This is a special case of the setting of Lemma 11.16.

12.15. (A COROLLARY OF BOUSQUET'S INEQUALITY) Using the notation of Theorem 12.5, prove that for all $0 < \eta \leq 1, \delta > 0$, and for all $t \geq 0$,

$$P\{Z \geq (1+\eta)EZ + t\} \leq \exp\left(-\frac{t^2}{2(1+\delta)\sigma^2}\right) + \exp\left(-\frac{\delta t}{2(1+\delta)(2/\eta + 1/3)}\right).$$

Hint: prove $\exp(-(1/u + v)) \leq \max(\exp(-(\lambda/u)), \exp(-(1-\lambda)/v))$ for $u, v > 0$ and $\lambda \in [0,1]$. (See Lemma 1 in Adamczak (2008).)

12.16. (SUPREMA OF EMPIRICAL PROCESSES WITH UNBOUNDED SUMMANDS) Let $X_1, \ldots, X_n$ be independent identically distributed random vectors. Assume that *(i)* The empirical process is symmetric: for each $s \in \mathcal{T}$, $X_{i,s}$ and $-X_{i,s}$ have the same distribution; *(ii)* There exist independent random variables $(Y_i)_{i \leq n}$ such that $Y_i \geq \max_{s \in \mathcal{T}} |X_{i,s}|$ and Y_i is sub-gamma with variance factor σ^2 and scale factor c (by Theorem 2.3, this entails $\mathrm{Var}(X_{i,s}) \leq 8\sigma^2 + 32c^2$).

Let $Z = \sup_{s\in\mathcal{T}} \sum_{i=1}^n X_{i,s}$ and let $\tau_n = \sqrt{2\sigma^2 \log n} + c\log n$. For each $i \leq n$ and $s \in \mathcal{T}$, let $V_{i,s} = X_{i,s}\mathbb{1}_{\{|X_{i,s}|<\tau_n\}}$ and $W_{i,s} = X_{i,s} - V_{i,s}$. Let $Z_1 = \sup_{s\in\mathcal{T}} \sum_{i=1}^n V_{i,s}$ and $Z_2 = \sup_{s\in\mathcal{T}} \sum_{i=1}^n W_{i,s}$. Check that $\mathbf{E}Z \geq \mathbf{E}Z_1 - \mathbf{E}Z_2$, $\mathbf{E}Z_2 \leq \tau_n(1 + 1/\log n)$, and $\mathbf{P}\{Z_2 \geq t\} \leq 2\exp(-t/(8\tau_n))$. Check also that for $\eta > 0, \varepsilon > 0, t > 0$,

$$\mathbf{P}\{Z \geq (1+\eta)\mathbf{E}Z + t\} \leq \mathbf{P}\Big\{Z_1 \geq (1+\eta)\mathbf{E}Z_1 - (1+\eta)\mathbf{E}Z_2 + (1-\varepsilon)t\Big\} + \mathbf{P}\{Z_2 \geq \varepsilon t\}.$$

Prove that for all $0 < \delta < 1$, for all $t \geq 0$,

$$\mathbf{P}\left\{Z \geq (1+\eta)\mathbf{E}Z + t\right\} \leq \exp\left(-\frac{(1-2\varepsilon)^2 t^2}{16(1+\delta)n(\sigma^2+4c^2)}\right) + \exp\left(-\frac{\delta(1-2\varepsilon)t}{2(1+\delta)\kappa\tau_n}\right) + 2\exp\left(-\frac{\varepsilon t}{8\tau_n}\right).$$

Hint: use the results of Exercises 12.14 and 12.15. Note that the last bound is trivial if $\varepsilon t \leq 4\tau_n$. This truncation-and-separation approach was popularized by Hoffmann-Jørgensen (1974). Chapter 6 of Ledoux and Talagrand (1991) describes the interplay between this approach and concentration of measure. de la Peña and Giné (1999), Giné, Latała, and Zinn (2000), Adamczak (2008), and Mendelson (2010) describe further advances in this direction. This exercise is inspired by Adamczak (2008) who describes more general results and applications.

13

The Expected Value of Suprema of Empirical Processes

In Chapters 11 and 12 we studied deviations of suprema of empirical processes around their expected values and obtained useful, often tight concentration inequalities. A remarkable feature of these inequalities is that a lot can be said about concentration properties without knowing what the expected values are. Bounding the expected value of the supremum of an empirical process is a central object of the study of empirical processes and the purpose of this chapter is to present elements of this rich theory. Interestingly, concentration inequalities provide an important tool in deriving tight upper bounds for such expectations, as pointed out below.

We have already faced simple situations when concentration inequalities help derive upper bounds for suprema of random variables; recall the maximal inequalities of Section 2.5 that, in fact, serve as the basis of some of the arguments to follow.

In Section 13.1 we discuss the perhaps most important basic technique for obtaining sharp upper bounds for suprema of empirical processes, the so-called *chaining* argument (Lemma 13.1). Chaining bounds relate the expected value of the supremum of an empirical process with metric properties of the set indexing the empirical process. Such inequalities proved successful in many areas ranging from the general theory of stochastic processes to statistics. The chaining arguments we present here are not always the sharpest possible and more sophisticated arguments, such as the so-called "generic chaining" approach, sometimes give more accurate results. However, chaining still provides simple and useful answers in many applications. Perhaps the best known bound obtained using classical chaining is Dudley's entropy integral bound for the expectation of the supremum of Gaussian processes (see Corollary 13.2 below). In Section 13.2 we present Sudakov's lower bound for the expected value of the supremum of Gaussian processes which may be regarded as a partial converse to Dudley's entropy integral upper bound.

The rest of this chapter describes examples in which chaining and concentration inequalities interact.

Section 13.3 deals with empirical processes indexed by VC-classes. This application does not fit exactly into the framework of Lemma 13.1. Nevertheless, supplementing chaining with symmetrization (Lemma 11.4) paves the way to sharp bounds.

Section 13.4 and Section 13.5 revisit Nemirovski's inequality already investigated in Section 11.2. While in Section 11.2 we consider $\boldsymbol{E}\|S_n\|^2$ where $S_n = \sum_{i=1}^n X_i$ with X_i independent random vectors with bounded components, in Section 13.4, we are interested in sums of random matrices endowed with the operator norm. The main result of Section 13.4 is Rudelson's inequality that establishes an upper bound for the expected operator norm of Rademacher and Gaussian sums of symmetric matrices.

Section 13.6 takes one step further in the analysis of the Johnson–Lindenstrauss lemma discussed in Sections 2.9 and 5.6. The Klartag–Mendelson theorem presented here describes sufficient conditions on the metric properties of a general set which guarantee that a random projection of the set to a low-dimensional subspace is an approximate isometry, with high probability.

In Section 13.7 Bousquet's inequality (Theorem 12.5) is used in an essential way with techniques known as "peeling (or slicing) and re-weighting" to obtain bounds for normalized empirical processes.

In Section 13.8, Theorem 13.19 is put to work. It allows us to derive an approximate isometry property of the random mapping $L_2(P) \to L_2(P_n)$.

Finally, in Section 13.9 we present an application in which sharp risk bounds are obtained for a classification problem in statistical learning theory.

13.1 Classical Chaining

In this section we describe the basic chaining argument. In the simplest version of chaining, one discretizes the set $\mathcal{T}$ indexing the stochastic process $\{X_t : t \in \mathcal{T}\}$ and the maximal value $\sup_{t\in\mathcal{T}} X_t$ is approximated by maxima over successively refining discretizations. To make this formal, we introduce the notion of δ-nets.

Let $(\mathcal{T}, d)$ be a totally bounded pseudo-metric space and let $\delta > 0$. A *δ-net* is a finite set $\mathcal{T}_\delta \subset \mathcal{T}$ with maximal cardinality such that for all $s, t \in \mathcal{T}_\delta$ with $s \neq t$, one has $d(s,t) > \delta$ (i.e. every pair of distinct elements of $\mathcal{T}$ is δ-separated).

Let $B(t,\delta)$ denote the closed ball of radius δ centered at t. Since $\mathcal{T}_\delta$ has maximal cardinality, the collection of closed balls with radius δ centered at the points of $\mathcal{T}_\delta$ covers $\mathcal{T}$, that is,

$$\mathcal{T} \subseteq \bigcup_{t\in\mathcal{T}_\delta} B(t,\delta).$$

Note that the cardinality $N(\delta,\mathcal{T})$ of a δ-net $\mathcal{T}_\delta$ coincides with the maximal number of disjoint closed balls of radius $\delta/2$ that can be packed into $\mathcal{T}$. $N(\delta,\mathcal{T})$ is called the *δ-packing number* of $\mathcal{T}$.

A *proper δ-covering* of $\mathcal{T}$ is a finite set $\mathcal{T}_\delta \subset \mathcal{T}$ such that

$$\mathcal{T} \subseteq \bigcup_{x\in\mathcal{T}_\delta} B(x,\delta).$$

The minimal cardinality of any δ-covering is denoted by $N'(\delta,\mathcal{T})$. It is called the *δ-covering number* of $\mathcal{T}$.

Packing and covering numbers are closely related as one always has

$$N(2\delta, \mathcal{T}) \le N'(\delta, \mathcal{T}) \le N(\delta, \mathcal{T}).$$

The second inequality follows by the argument above and the first may also be seen easily. These quantities reflect the "size" or "massiveness" of the totally bounded set $\mathcal{T}$.

The *δ-entropy number* $H(\delta, \mathcal{T})$ is defined as the logarithm of the δ-packing number:

$$H(\delta, \mathcal{T}) = \log N(\delta, \mathcal{T}).$$

The function $H(\cdot, \mathcal{T})$ is called the *metric entropy* of $\mathcal{T}$.

The following lemma is at the core of the chaining argument. To avoid worrying about measurability issues, we assume that $\mathcal{T}$ is a finite set. One may extend all results of this chapter to processes indexed by separable metric spaces by standard arguments that we do not detail here.

Lemma 13.1 *Let $\mathcal{T}$ be a finite pseudometric space and let $(X_t)_{t\in\mathcal{T}}$ be a collection of random variables such that for some constants $a, v, c > 0$,*

$$\log \mathbf{E}e^{\lambda(X_t - X_{t'})} \le a\lambda d(t,t') + \frac{v\lambda^2 d^2(t,t')}{2\left(1 - c\lambda d(t,t')\right)}$$

for all $t, t' \in \mathcal{T}$ and all $0 < \lambda < \left(cd(t,t')\right)^{-1}$. Then, for any $t_0 \in \mathcal{T}$,

$$\mathbf{E}\left[\sup_{t\in\mathcal{T}} X_t - X_{t_0}\right] \le 3a\delta + 12\sqrt{v}\int_0^{\delta/2}\sqrt{H(u,\mathcal{T})}du + 12c\int_0^{\delta/2} H(u,\mathcal{T})\,du$$

where $\delta = \sup_{t\in\mathcal{T}} d(t, t_0)$.

Proof For any integer j, let $\delta_j = \delta 2^{-j}$ and let $\mathcal{T}_j$ be a δ_j-net of $\mathcal{T}$. By the definition of the metric entropy, for any integer j we can define a mapping $\Pi_j : \mathcal{T} \to \mathcal{T}_j$ such that

$$d\left(t, \Pi_j(t)\right) \le \delta_j \text{ for all } t \in \mathcal{T}.$$

Since $\mathcal{T}$ is finite, there exists a positive integer J such that for all $t \in \mathcal{T}$,

$$X_t = X_{\Pi_0(t)} + \sum_{j=0}^{J}\left(X_{\Pi_{j+1}(t)} - X_{\Pi_j(t)}\right).$$

Moreover, by the definition of δ, we may assume that $\mathcal{T}_0 = \{t_0\}$, so $\Pi_0(t) = t_0$ and therefore

$$\mathbf{E}\left[\sup_{t\in\mathcal{T}} X_t - X_{t_0}\right] \le \sum_{j=0}^{J}\mathbf{E}\left[\sup_{t\in\mathcal{T}} X_{\Pi_{j+1}(t)} - X_{\Pi_j(t)}\right].$$

Now observe that for every integer j,

$$\left|\left\{\left(\Pi_j(t), \Pi_{j+1}(t)\right) : t \in \mathcal{T}\right\}\right| \le e^{2H(\delta_{j+1},\mathcal{T})}$$

and that by the triangle inequality, for any $t \in \mathcal{T}$,

$$d\left(\Pi_j(t), \Pi_{j+1}(t)\right) \leq 3\delta_{j+1}.$$

Hence, by the maximal inequality of Corollary 2.6,

$$\begin{aligned} &\mathbf{E}\left[\sup_{t\in\mathcal{T}} X_{\Pi_{j+1}(t)} - X_{\Pi_j(t)}\right] \\ &\leq 3\left(a\delta_{j+1} + 2\delta_{j+1}\sqrt{vH(\delta_{j+1},\mathcal{T})} + 2c\delta_{j+1}H(\delta_{j+1},\mathcal{T})\right). \end{aligned}$$

Hence, summing over j,

$$\begin{aligned} \mathbf{E}\left[\sup_{t\in\mathcal{T}} X_t - X_{t_0}\right] &\leq 3a\sum_{j=1}^{J+1}\delta_j + 6\sum_{j=1}^{J+1}\delta_j\left(\sqrt{vH(\delta_j,\mathcal{T})} + cH(\delta_j,\mathcal{T})\right) \\ &\leq 3a\delta + 12\sqrt{v}\int_0^{\delta/2}\sqrt{H(u,\mathcal{T})}du + 12c\int_0^{\delta/2} H(u,\mathcal{T})du \end{aligned}$$

where at the last step we used the fact that metric entropy $H(u,\mathcal{T})$ is nonincreasing as a function of u. □

Letting $a = c = 0$, Lemma 13.1 allows us to recover Dudley's classical bound for suprema of centered processes with sub-Gaussian increments.

Corollary 13.2 (DUDLEY'S ENTROPY INTEGRAL) *Let $\mathcal{T}$ be a finite pseudometric space and let $(X_t)_{t\in\mathcal{T}}$ be a collection of random variables such that*

$$\log \mathbf{E}e^{\lambda(X_t - X_{t'})} \leq \frac{\lambda^2 d^2(t,t')}{2}$$

for all $t, t' \in \mathcal{T}$ and all $\lambda > 0$. Then for any $t_0 \in \mathcal{T}$,

$$\mathbf{E}\left[\sup_{t\in\mathcal{T}} X_t - X_{t_0}\right] \leq 12\int_0^{\delta/2}\sqrt{H(u,\mathcal{T})}du,$$

where $\delta = \sup_{t\in\mathcal{T}} d(t,t_0)$.

This entropic bound is often tight, though in some situations it fails to give sharp bounds (see Exercises 13.4 and 13.5). Note that it also provides an upper bound for suprema of Rademacher processes: if $Z = \sup_{t\in\mathcal{T}}\sum_{i=1}^n \alpha_{i,t}\varepsilon_i$ where the ε_i are independent Rademacher variables, then the condition of Corollary 13.2 is satisfied with $d^2(t,t') = \sum_{i=1}^n(\alpha_{i,t} - \alpha_{i,t'})^2$.

Chaining is by no means the only possible technique for obtaining upper bounds for the expected supremum of empirical processes. In fact, chaining does not always lead to sharp bounds and may often be by-passed by exploiting the special structure of the problem at hand. In the exercise section several such cases are described.

For an example in which Lemma 13.1 fails to provide the best bounds, consider a Gaussian chaos of order 2 defined as follows. Let $X = (X_1, \ldots, X_n)$ be a vector of independent standard normal random variables and let $\mathcal{T}$ be a finite collection of symmetric matrices $A = (a_{i,j})_{n\times n}$ with $a_{i,i} = 0$ for $i = 1, \ldots, n$. Then

$$Z = \sup_{A\in\mathcal{T}} X^T AX$$

is the supremum of a Gaussian chaos process indexed by $\mathcal{T}$. From the analysis of Example 2.12, we find that for any $A, B \in \mathcal{T}$,

$$\log \mathbb{E}e^{\lambda(X^TAX - X^TBX)} \leq \frac{\lambda^2 \|A - B\|_{\mathrm{HS}}^2}{1 - 2\lambda\|A - B\|}$$

where $\|A\|_{\mathrm{HS}} = \left(\sum_{i=1}^n \mu_i^2\right)^{1/2}$ is the Hilbert–Schmidt norm of the matrix A (with eigenvalues $\mu_1, \ldots, \mu_n$) and $\|A\| = \max_i |\mu_i|$ is the operator norm.

Let $H_{\mathrm{HS}}(\delta, \mathcal{T})$ denote the δ-entropy of $\mathcal{T}$ with respect to the Hilbert–Schmidt norm, and let $H_{\mathrm{op}}(\delta, \mathcal{T})$ denote the δ-entropy of $\mathcal{T}$ with respect to the operator norm. Let δ_{HS} and δ_{op} be the diameters of $\mathcal{T}$ under the Hilbert–Schmidt and the operator norms. Since $\|A\| \leq \|A\|_{\mathrm{HS}}$, we may invoke Lemma 13.1 to establish the bound

$$\mathbb{E}Z \leq 12 \int_0^{\delta_{\mathrm{HS}}/2} \left(\sqrt{2H_{\mathrm{HS}}(u, \mathcal{T})} + 2H_{\mathrm{HS}}(u, \mathcal{T})\right) du.$$

However, this upper bound may be improved by a technique known as "generic chaining" to

$$\mathbb{E}Z \leq \kappa \left(\int_0^{\delta_{\mathrm{HS}}/2} \sqrt{2H_{\mathrm{HS}}(u, \mathcal{T})}du + \int_0^{\delta_{\mathrm{op}}/2} 2H_{\mathrm{op}}(u, \mathcal{T})du \right)$$

where $\kappa > 0$ is a universal constant. The proof of this bound is left as a guided exercise (Exercise 13.10). In Section 13.3 we discuss another example in which raw chaining gives suboptimal bounds, though with a simple additional trick one may obtain much tighter bounds.

Note that even in the case of linear Gaussian processes, chaining may not give optimal results. An example is when the process is indexed by an ellipsoid (see Exercises 13.5, 13.19, and 13.20 for some details).

13.2 Lower Bounds for Gaussian Processes

In this section we describe lower bounds for the expected value of the supremum of a Gaussian process. We start with Slepian's lemma, a classical result that relates the maxima of two Gaussian vectors. This result is at the basis of Sudakov's lower bound, a counterpart of Corollary 13.2 for Gaussian processes.

Theorem 13.3 (SLEPIAN'S LEMMA) *Let $X = (X_1, \ldots, X_n)$ and $Y = (Y_1, \ldots, Y_n)$ be Gaussian random vectors with $\mathbf{E}X_i = \mathbf{E}Y_i$ for all $i = 1, \ldots, n$. Let $\delta^X_{i,j} = \mathbf{E}[(X_i - X_j)^2]$ and $\delta^Y_{i,j} = \mathbf{E}[(Y_i - Y_j)^2]$ for all $i, j \in \{1, \ldots, n\}$. If $\delta^X_{i,j} \le \delta^Y_{i,j}$ for all $i, j \in \{1, \ldots, n\}$, then*

$$\mathbf{E} \max_{i=1,\ldots,n} X_i \le \mathbf{E} \max_{i=1,\ldots,n} Y_i.$$

If $|\delta^X_{i,j} - \delta^Y_{i,j}| \le \varepsilon$ for all $i, j \in \{1, \ldots, n\}$, then

$$\left| \mathbf{E} \max_{i=1,\ldots,n} X_i - \mathbf{E} \max_{i=1,\ldots,n} Y_i \right| \le \sqrt{\varepsilon \log n}.$$

Note that by taking $Y = 0$, we recover the inequality for the expected maximum of n Gaussian random variables derived in Section 2.5. While the maximal inequality holds for sub-Gaussian variables, Slepian's lemma uses the Gaussian property in an essential way.

The proof crucially uses the following property of Gaussian vectors: If $F : \mathbb{R}^n \to \mathbb{R}$ is continuously differentiable with moderate growth in the sense that for any $a > 0$, $\lim_{\|x\|\to\infty} F(x)e^{-a\|x\|^2} = 0$ and $X = (X_1, \ldots, X_n)$ is a centered Gaussian vector, then for any $i = 1, \ldots, n$,

$$\mathbf{E}\left[X_i F(X)\right] = \sum_{j=1}^{n} \mathbf{E}[X_i X_j] \mathbf{E} \frac{\partial F}{\partial x_j}(X),$$

see Exercise 13.3. If $F = \partial h / \partial x_i$, this integration-by-parts formula can be rewritten as

$$\begin{aligned}
\mathbf{E}\left[\nabla h(X)^T X\right] &= \sum_{i=1}^{n} \mathbf{E}\left[X_i \frac{\partial h}{\partial x_i}(X)\right] \\
&= \sum_{i=1}^{n} \sum_{j=1}^{n} \mathbf{E}[X_i X_j] \mathbf{E} \frac{\partial^2 h}{\partial x_i \partial x_j}(X) \\
&= \operatorname{trace}\left(\mathbf{E}[XX^T] \mathbf{E} \nabla^2 h(X)\right).
\end{aligned}$$

Proof Without loss of generality, we may assume that X and Y are independent. Let $\lambda > 0$ and let $f : \mathbb{R}^n \to \mathbb{R}$ be defined as

$$f(x_1, \ldots, x_n) = \frac{1}{\lambda} \log \left(\sum_{i=1}^{n} e^{\lambda x_i} \right).$$

Let $\tilde{X}_i = X_i - \mathbf{E}X_i$ and $\tilde{Y}_i = Y_i - \mathbf{E}Y_i$ for $i = 1, \ldots, n$. Introduce the covariance matrices σ^X and σ^Y ($\sigma^X_{i,j} = \mathbf{E}[\tilde{X}_i \tilde{X}_j]$ and $\sigma^Y_{i,j} = \mathbf{E}[\tilde{Y}_i \tilde{Y}_j]$ for $1 \le i, j \le n$). For $0 \le t \le 1$, define $Z_t = (Z_{t,1}, \ldots, Z_{t,n})^T$ as a random vector with components

$$Z_{t,i} = \tilde{X}_i \sqrt{1-t} + \tilde{Y}_i \sqrt{t} + \mathbf{E}X_i.$$

The function $h(t) = \boldsymbol{E}f(Z_t)$ defined for $t \in [0, 1]$ is differentiable with derivative

$$h'(t) = \boldsymbol{E}\left[\nabla f(Z_t)^T \left(\frac{\tilde{Y}}{2\sqrt{t}} - \frac{\tilde{X}}{2\sqrt{1-t}}\right)\right].$$

As $\tilde{X}$ and $\tilde{Y}$ are independent, working conditionally on $\tilde{X} = \tilde{x}$, using the integration-by-parts formula with respect to the Gaussian vector $\tilde{Y}$ leads to

$$\begin{aligned}
&\boldsymbol{E}\left[\nabla f\left(\tilde{x}\sqrt{1-t} + \tilde{Y}\sqrt{t} + \boldsymbol{E}X\right)^T \frac{\tilde{Y}}{2\sqrt{t}}\right] \\
&= \frac{1}{2\sqrt{t}}\text{trace}\left(\sigma^Y \boldsymbol{E}\left[\sqrt{t}\nabla^2 f\left(\tilde{x}\sqrt{1-t} + \tilde{Y}\sqrt{t} + \boldsymbol{E}X\right)\right]\right) \\
&= \frac{1}{2}\text{trace}\left(\sigma^Y \boldsymbol{E}\left[\nabla^2 f\left(\tilde{x}\sqrt{1-t} + \tilde{Y}\sqrt{t} + \boldsymbol{E}X\right)\right]\right).
\end{aligned}$$

Taking expectation with respect to $\tilde{X}$ and proceeding in a similar way to transform $\boldsymbol{E}\left[\nabla f(Z_t)^T \frac{\tilde{X}}{2\sqrt{1-t}}\right]$, we get

$$h'(t) = \frac{1}{2}\boldsymbol{E}\left[\text{trace}\left(\nabla^2 f(Z_t)\left(\sigma^Y - \sigma^X\right)\right)\right].$$

Let $p(z) = \nabla f(z)$. Then straightforward calculation shows that the Hessian of f may be written as

$$\nabla^2 f(z) = \lambda\, \text{diag}\,(p(z)) - \lambda p(z)p(z)^T.$$

As $\sum_{i=1}^n p_i(z) = 1$,

$$\begin{aligned}
\text{trace}\left(\text{diag}\,(p(z))\left(\sigma^Y - \sigma^X\right)\right) &= \sum_{i=1}^n p_i(z)\left(\sigma_{i,i}^Y - \sigma_{i,i}^X\right) \\
&= \frac{1}{2}\sum_{1\le i,j\le d} p_i(z)p_j(z)\left(\sigma_{i,i}^Y - \sigma_{i,i}^X + \sigma_{jj}^Y - \sigma_{j,j}^X\right).
\end{aligned}$$

Substituting the right-hand side into the expansion of $\text{trace}\left(\nabla^2 f(z)\left(\sigma^Y - \sigma^X\right)\right)$ leads to

$$\begin{aligned}
&\text{trace}\left(\nabla^2 f(z)\left(\sigma^Y - \sigma^X\right)\right) \\
&= \frac{\lambda}{2}\sum_{1\le i,j\le n} p_i(z)p_j(z)\left(\sigma_{i,i}^Y - \sigma_{i,i}^X + \sigma_{jj}^Y - \sigma_{j,j}^X - 2\sigma_{i,j}^Y + 2\sigma_{i,j}^X\right) \\
&= \frac{\lambda}{2}\sum_{1\le i,j\le n} p_i(z)p_j(z)\left(\delta_{i,j}^Y - \delta_{i,j}^X\right) \\
&= \frac{\lambda}{2}\text{trace}\left(p(z)p(z)^T\left(\delta^Y - \delta^X\right)\right).
\end{aligned}$$

If $\delta^Y - \delta^X \geq 0$, then $h'(t) \geq 0$ and $h(1) = \mathbf{E}f(Y) \geq \mathbf{E}f(X) = h(0)$. This holds for all choices of $\lambda \geq 0$.

Since

$$\max_{i=1,\dots,n} x_i \leq \frac{1}{\lambda} \log \left(\sum_{i=1}^{n} e^{\lambda x_i} \right) \leq \frac{1}{\lambda} \log n + \max_{i=1,\dots,n} x_i,$$

the first statement of the theorem follows by taking $\lambda \to \infty$.

On the other hand, if $0 \leq \left| \delta^Y_{i,j} - \delta^X_{i,j} \right| \leq \varepsilon$ for all $1 \leq i, j \leq n$, then

$$\begin{aligned}
\left| \mathbf{E}f(Y) - \mathbf{E}f(X) \right| &= \left| \int_0^1 h'(s) ds \right| \\
&\leq \frac{\lambda}{4} \int_0^1 \mathbf{E} \left| \operatorname{trace} \left(p(Z_s) p(Z_s)^T \left(\delta^Y - \delta^X \right) \right) \right| ds \\
&\leq \frac{\lambda}{4} \varepsilon.
\end{aligned}$$

Combining this with the inequalities linking $f(x)$ and $\max_{i=1,\dots,n} x_i$, we have

$$\left| \mathbf{E} \max_{i=1,\dots,n} Y_i - \mathbf{E} \max_{i=1,\dots,n} X_i \right| \leq \frac{\lambda \varepsilon}{4} + \frac{\log n}{\lambda}.$$

Optimizing over λ, we obtain

$$\left| \mathbf{E} \max_{i=1,\dots,n} Y_i - \mathbf{E} \max_{i=1,\dots,n} X_i \right| \leq \sqrt{\varepsilon \log n}. \qquad \square$$

We are now prepared to prove a lower bound that complements Dudley's bound (Corollary 13.2).

Theorem 13.4 (SUDAKOV'S LOWER BOUND) *Let $\mathcal{T}$ be a finite set and let $(X_t)_{t \in \mathcal{T}}$ be a Gaussian vector with $\mathbf{E}X_t = 0$. Then*

$$\mathbf{E} \sup_{t \in \mathcal{T}} X_t \geq \frac{1}{2} \min_{t \neq t' \in \mathcal{T}} \sqrt{\mathbf{E} \left[(X_t - X_{t'})^2 \right] \log |\mathcal{T}|}.$$

Proof Let $(Z_t)_{t \in \mathcal{T}}$ be independent standard Gaussian random variables. Let

$$\delta = \min_{t \neq t'} \left(\mathbf{E} \left[(X_t - X_{t'})^2 \right] \right)^{1/2}$$

and

$$Y_t = \frac{\delta}{\sqrt{2}} Z_t, \text{ for every } t \in \mathcal{T}.$$

As for every $t \neq t' \in \mathcal{T}$, $\mathbf{E}\left[(Y_t - Y_{t'})^2\right] = \delta^2 \leq \mathbf{E}\left[(X_t - X_{t'})^2\right]$, by Theorem 13.3,

$$\delta \mathbf{E} \sup_{t \in \mathcal{T}} Z_t \leq \sqrt{2} \mathbf{E} \sup_{t \in \mathcal{T}} X_t .$$

On the other hand,

$$\mathbf{E} \sup_{t \in \mathcal{T}} Z_t \geq \frac{1}{\sqrt{2}} \sqrt{\log |\mathcal{T}|} .$$

The proof of this last statement is left as an exercise (Exercise 13.6). □

Sudakov's lower bound may be rewritten in terms of metric entropy as follows. Let $(X_t)_{t \in \mathcal{T}}$ be centered Gaussian random variables indexed by the finite set $\mathcal{T}$. Let d be the pseudo-metric on $\mathcal{T}$ defined by $d(t, t')^2 = \mathbf{E}[(X_t - X_{t'})^2]$. Then by Theorem 13.4, for all $\varepsilon > 0$ smaller than the diameter of $\mathcal{T}$,

$$\mathbf{E} \sup_{t \in \mathcal{T}} X_t \geq \frac{1}{2} \varepsilon \sqrt{H(\varepsilon, \mathcal{T})} .$$

Exercise 13.4 provides an example where Sudakov's lower bound is tight while Dudley's entropy integral upper bound is not. Note that Slepian's lemma may also be used to derive upper bounds for the expected value of the supremum of some Gaussian processes such as the largest eigenvalue of some random matrices (see Exercises 13.7 and 13.8) (Fig. 13.1).

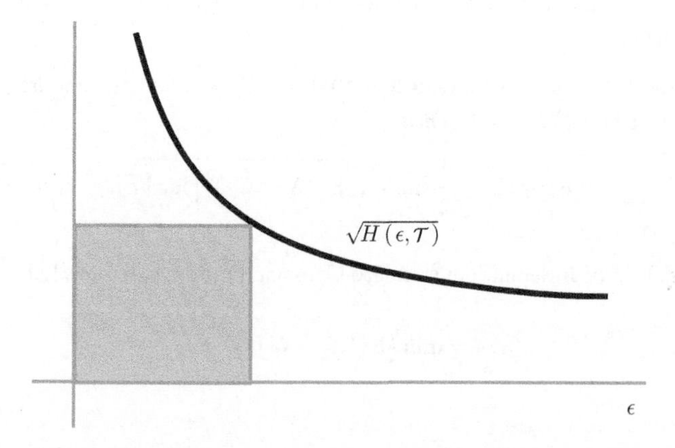

Figure 13.1 Dudley's entropy bound is proportional to the area under the curve $\sqrt{H(\varepsilon, \mathcal{T})}$, while Sudakov's lower bound is the area of the largest rectangle that can be fitted under the same curve

13.3 Chaining and VC-Classes

In this section we are concerned with uniform deviations of relative frequencies from the corresponding probabilities. To be more precise, let $\mathcal{X}$ be some set endowed with a probability measure P and let $X_1, \ldots, X_n$ be independent random variables taking values in $\mathcal{X}$, distributed according to P.

Let $\mathcal{A} = \{A_t : t \in \mathcal{T}\}$ denote a collection of (measurable) subsets of $\mathcal{X}$ indexed by a (finite) set $\mathcal{T}$. We are interested in uniform deviations of empirical averages, that is, in the behavior of the random variable

$$\sup_{t \in \mathcal{T}} \frac{1}{n} \sum_{i=1}^{n} (\mathbb{1}_{\{X_i \in A_t\}} - P(A_t)).$$

For $t \in \mathcal{T}$, denote $Z_t = n^{-1/2} \sum_{i=1}^{n} (\mathbb{1}_{\{X_i \in A_t\}} - P(A_t))$ and

$$Z = \sup_{t \in \mathcal{T}} Z_t.$$

We may introduce a pseudo-metric d on $\mathcal{T}$ defined by

$$d(t, t') = \sqrt{\boldsymbol{P}\{\mathbb{1}_{\{X_i \in A_t\}} \neq \mathbb{1}_{\{X_i \in A_{t'}\}}\}}.$$

Then for all $t, t' \in \mathcal{T}$, $\mathrm{Var}\,(Z_t - Z_{t'}) \leq d^2(t, t')$ and it follows immediately from Bernstein's inequality that

$$\log \boldsymbol{E} e^{\lambda(Z_t - Z_{t'})} \leq \frac{\lambda^2 d^2(t, t')}{2(1 - \lambda/\sqrt{n})}.$$

This bound may be used to apply Lemma 13.1. However, this simple argument may be improved by a simple symmetrization argument that we outline next.

A key ingredient of this approach is the notion of universal entropy. For $\delta > 0$, and a probability measure Q on $\mathcal{X}$, let $N(\delta, \mathcal{A}, Q)$ denote the maximal cardinality N of a subset $\{t, \ldots, t_N\}$ of the index set $\mathcal{T}$ such that $Q(A_{t_i} \Delta A_{t_j}) > \delta^2$ for every $i \neq j$ (here $A \Delta B$ is the symmetric difference of A and B). The *universal δ-metric entropy* (also called *Koltchinskii–Pollard entropy*) of $\mathcal{A}$ is defined by

$$H(\delta, \mathcal{A}) = \sup_{Q} \log N(\delta, \mathcal{A}, Q)$$

where the supremum is taken over the set of all probability measures Q concentrated on some finite subset of $\mathcal{X}$.

Lemma 13.5 *Let $\mathcal{A} = \{A_t : t \in \mathcal{T}\}$ be a countable class of measurable subsets of $\mathcal{X}$ and let $X_1, \ldots, X_n$ be independent random variables taking values in $\mathcal{X}$, with common distribution P. Assume that for some $\sigma > 0$,*

$$P(A_t) \leq \sigma^2 \textit{ for every } t \in \mathcal{T}.$$

Let

$$Z = n^{-1/2} \sup_{t \in \mathcal{T}} \sum_{i=1}^{n} (\mathbb{1}_{\{X_i \in A_t\}} - P(A_t))$$

and denote $D_\sigma = 6\sum_{j=0}^{\infty} 2^{-j}\sqrt{H\left(2^{-(j+1)}\sigma, \mathcal{A}\right)}$. *If* $\sigma^2 \geq D_\sigma^2/(5n)$, *then*

$$EZ \leq 3\sigma D_\sigma.$$

The same upper bound is valid for $Z^- = n^{-1/2} \sup_{t\in\mathcal{T}} \sum_{i=1}^{n} (P(A_t) - \mathbb{1}_{\{X_i \in A_t\}})$.

Proof By the symmetrization inequalities of Lemma 11.4,

$$\begin{aligned} &\mathbf{E} \sup_{t\in\mathcal{T}} \sum_{i=1}^{n} (\mathbb{1}_{\{X_i \in A_t\}} - P(A_t)) \\ &\leq 2\mathbf{E}\left[\mathbf{E}\left[\sup_{t\in\mathcal{T}} \sum_{i=1}^{n} \varepsilon_i \mathbb{1}_{\{X_i\in A_t\}} \middle| X_1, \ldots, X_n\right]\right], \end{aligned}$$

where $\varepsilon_1, \ldots, \varepsilon_n$ are independent Rademacher variables. Define the random variable

$$\delta_n^2 = \max\left(\sup_{t\in\mathcal{T}} \frac{1}{n}\sum_{i=1}^{n} \mathbb{1}_{\{X_i\in A_t\}}, \sigma^2\right).$$

Clearly, $\delta_n^2 \leq \sigma^2 + Z/\sqrt{n}$. As a Rademacher sum is sub-Gaussian, we may use Lemma 13.1 to obtain

$$\mathbf{E}Z \leq 6\mathbf{E}\left[\sqrt{\delta_n^2}\sum_{j=0}^{\infty} 2^{-j}\sqrt{H\left(2^{-j-1}\delta_n, \mathcal{A}\right)}\right] \leq 6\sqrt{\mathbf{E}\delta_n^2}\sum_{j=0}^{\infty} 2^{-j}\sqrt{H\left(2^{-(j+1)}\sigma, \mathcal{A}\right)},$$

where we use the fact that $H(\delta, \mathcal{A})$ is a nonincreasing function of δ. Thus, we have

$$\mathbf{E}Z \leq D_\sigma\sqrt{\sigma^2 + \mathbf{E}Z/\sqrt{n}}.$$

Solving this quadratic inequality for $\mathbf{E}Z$, we get

$$\mathbf{E}Z \leq \frac{D_\sigma^2}{2\sqrt{n}}\left(1 + \sqrt{1 + \frac{4\sigma^2 n}{D_\sigma^2}}\right).$$

When $\sigma^2 \geq D_\sigma^2/(5n)$, the right-hand side may be bounded further by $3\sigma D_\sigma$, as announced.

To bound EZ^-, we may use the inequality just obtained: by the same argument as above,

$$EZ^- \leq D_\sigma \sqrt{\sigma^2 + EZ/\sqrt{n}}.$$

Under the condition $\sigma^2 \geq D_\sigma^2/(5n)$, $EZ/\sqrt{n} \leq 3\sigma D_\sigma/\sqrt{n} \leq 3\sqrt{5}\sigma^2$, and therefore

$$EZ^- \leq \sigma D_\sigma \sqrt{1 + 3\sqrt{5}} \leq 3\sigma D_\sigma. \qquad \square$$

The universal entropy appearing in the bound of Lemma 13.5 may be estimated in an elegant way in terms of the combinatorial notion of the VC dimension of the class $\mathcal{A}$ that we already introduced in Section 3.3. Recall the definition: for an vector $x = (x_1, \ldots, x_n)$ of n points of $\mathcal{X}$, the *trace* of $\mathcal{A}$ on x is defined by

$$\text{tr}(x) = \{A \cap \{x_1, \ldots, x_n\} : A \in \mathcal{A}\}.$$

The VC dimension $D(x)$ of $\mathcal{A}$ (with respect to x) is the cardinality k of the largest subset $\{x_{i_1}, \ldots, x_{i_k}\}$ of $\{x_1, \ldots, x_n\}$ for which $2^k = |\text{tr}(x_{i_1}, \ldots, x_{i_k})|$. $\mathcal{A}$ is called a VC *class* if $V \overset{\text{def}}{=} \sup_{n \geq 1} \sup_{x \in \mathcal{X}^n} D(x) < \infty$. V is called the VC dimension of $\mathcal{A}$.

The next lemma shows how the VC dimension controls the universal entropy.

Lemma 13.6 (HAUSSLER'S VC BOUND FOR UNIVERSAL ENTROPY) *Let $\mathcal{A}$ denote a* VC *class of subsets of $\mathcal{X}$ with* VC *dimension V. For every positive $\delta > 0$,*

$$H(\delta, \mathcal{A}) \leq 2V \log(e/\delta) + \log(e(V+1)) \leq 2V \log(e^2/\delta).$$

The proof of this lemma, which we do not reproduce here, relies on delicate combinatorial properties of the trace of a VC class on a finite sample. A slightly weaker but much easier version is left to the reader as a guided exercise (Exercise 13.11).

In Exercises 13.16 and 13.15, other ways of bounding the universal entropy are shown. Combining Lemma 13.5 and Haussler's bound, we immediately obtain the following.

Theorem 13.7 *Recall the notation of Lemma 13.5. Assume that $\mathcal{A}$ is a* VC *class with* VC *dimension V. Suppose $\sup_{t \in \mathcal{T}} P(A_t) \leq \sigma^2$. Then*

$$\max(EZ, EZ^-) \leq 72\sigma \sqrt{V \log \frac{4e^2}{\sigma}}$$

provided that $\sigma \geq 24\sqrt{V \log(4e^2/\sigma)/(5n)}$.

Proof By Haussler's bound, the quantity D_σ introduced in Lemma 13.5 may be bounded by

$$D_\sigma \leq 24\sqrt{V \log \frac{4e^2}{\sigma}}. \qquad \square$$

Note that the upper bound depends on the sampling distribution P only through the condition $\sigma^2 \geq \sup_{t\in\mathcal{T}} P(A_t)$. If we are ready to upper bound σ by 1, we obtain a distribution-free bound.

Exercise 13.18 provides an example in which the factor $\sqrt{\log(e/\sigma)}$ in the upper bound can be dropped. Exercises 13.15 and 13.16 describe possible generalizations and refinements of Theorem 13.7.

13.4 Gaussian and Rademacher Averages of Symmetric Matrices

In this section we study norms of certain random matrices. More precisely, we consider Gaussian and Rademacher sums of symmetric matrices and investigate the behavior of their operator norm. Since such an operator norm may be considered as the supremum of a stochastic process, one may be tempted to use chaining. However, it is possible to obtain much sharper bounds using the specific features of the process. Recall that the operator norm of a symmetric $d \times d$ matrix M is defined by $\|M\| = \sup_{u\in\mathbb{R}^d:\|u\|_2\leq 1} |u^T Mu|$.

Theorem 13.8 (RUDELSON'S INEQUALITY) *Let $A_1, \ldots, A_n$ be symmetric $d \times d$ matrices. Let $X_1, \ldots, X_n$ be independent standard Gaussian random variables. Let $Z = \| \sum_{i=1}^n X_i A_i \|$. If $\sigma^2 = \left\| \sum_{i=1}^n A_i^2 \right\|$, then*

$$\mathrm{Var}(Z) \leq \sigma^2 \quad \textit{and} \quad \mathbf{E}Z \leq \sqrt{2\log(2d)}\sigma.$$

The theorem also remains valid if the Gaussian coefficients $X_1, \ldots, X_n$ are replaced by independent Rademacher random variables. The details of the easy modification are left to the reader.

Recall that any symmetric matrix A can be diagonalized in an orthogonal basis, that is, there exists an orthogonal matrix O and diagonal matrix D with real diagonal coefficients such that $A = ODO^T$. Then if f is a real-valued function defined on an interval that contains all eigenvalues of A, we may define the matrix $f(A)$ as $f(A) = Of(D)O^T$, where $f(D)$ is the diagonal matrix computed by applying f to each diagonal coefficient of D.

The proof of the bound for the variance is an easy corollary of the Gaussian Poincaré inequality. The non-trivial part is the bound for the expectation. A key ingredient of the proof of this second bound is the Golden–Thompson inequality (see Exercise 13.29) which implies that for symmetric matrices A and B,

$$\mathrm{trace}\left(\exp(A + B)\right) \leq \mathrm{trace}\left(\exp(A)\exp(B)\right).$$

This inequality allows us to bound the moment-generating function of the norm of $\sum_{i=1}^n X_i A_i$ and proceed with an argument similar to the one used in Section 2.5 to bound the expected maximum of sub-Gaussian random variables.

Proof The bound for the variance follows from the Gaussian Poincaré inequality which implies that the variance of the maximum of centered Gaussian variables is always bounded by the maximum of the variances (see Exercise 3.24 and also Theorem 5.8). But Z may be represented as the supremum of a Gaussian process, since

$$Z = \left\| \sum_{i=1}^{n} X_i A_i \right\| = \sup_{u \in \mathbb{R}^d : \|u\|_2 \le 1} \left| u^T \sum_{i=1}^{n} X_i A_i u \right|.$$

This implies

$$\mathrm{Var}(Z) \le \sup_{u \in \mathbb{R}^d : \|u\|_2 \le 1} \mathrm{Var}\left(\sum_{i=1}^{n} X_i u^T A_i u \right) = \sup_{u \in \mathbb{R}^d : \|u\|_2 \le 1} \sum_{i=1}^{n} \left(u^T A_i u\right)^2.$$

Now, let $A_i = \sum_{j=1}^{d} \lambda_{i,j} g_{i,j} g_{i,j}^T$ where $(g_{i,j})_{j=1,\ldots,d}$ is an orthonormal family of eigenvectors of A_i and $(\lambda_{i,j})_{j=1,\ldots,d}$ is the sequence of corresponding eigenvalues. For any $u \in \mathbb{R}^d$ with $\|u\|_2 \le 1$,

$$\begin{aligned} \sum_{i=1}^{n} \left(u^T A_i u\right)^2 &= \sum_{i=1}^{n} \left(\sum_{j=1}^{d} \lambda_{i,j} (u^T g_{i,j})^2 \right)^2 \\ &\le \sum_{i=1}^{n} \left(\sum_{j=1}^{d} \lambda_{i,j}^2 (u^T g_{i,j})^2 \right) \\ &= \sum_{i=1}^{n} \left(u^T A_i^2 u\right) \\ &\le \left\| \sum_{i=1}^{n} A_i^2 \right\|, \end{aligned}$$

where the first inequality follows from Jensen's inequality and the fact that $\sum_{j=1}^{d} (u^T g_{i,j})^2 \le 1$ as $\|u\|_2 \le 1$ and the vectors $(g_{i,j})_{j=1,\ldots,d}$ form an orthonormal basis of $\mathbb{R}^d$. This proves the first inequality.

To prove the bound for the expected value, denote $M = \sum_{i=1}^{n} X_i A_i$. The basic idea is similar to the maximal inequality of Theorem 2.5. To obtain an upper bound for $\boldsymbol{E}\|M\|$, we bound the exponential of $\boldsymbol{E}\|M\|$, via Jensen's inequality, by the moment-generating function of $\|M\|$. Let $s > 0$ be a parameter to be optimized later. Then

$$\begin{aligned} e^{s\boldsymbol{E}\|M\|} &\le \boldsymbol{E} e^{s\|M\|} \quad \text{(by Jensen's inequality)} \\ &= \boldsymbol{E} e^{s \max_{i=1,\ldots,d} |\lambda_i(M)|} \\ &= \boldsymbol{E} \max_{i=1,\ldots,d} \max\left(e^{s\lambda_i(M)}, e^{-s\lambda_i(M)} \right) \\ &\le 2\boldsymbol{E}\,\mathrm{trace}\left(\exp(sM)\right). \end{aligned}$$

We now use the Golden–Thompson inequality to bound $\boldsymbol{E}\,\mathrm{trace}\left(\exp(sM)\right)$. To this end, introduce the matrix $D_0 = s^2 \sum_{i=1}^{n} A_i^2/2$ and, recursively, $D_{j+1} = D_j + sX_{j+1}A_{j+1} - s^2/2A_{j+1}^2$ for $j = 1, \ldots, n$. Thus, $D_n = sM$. Note that, using the fact that the X_i are standard Gaussian random variables,

$$\boldsymbol{E} \exp\left(sX_jA_j - s^2A_j^2/2\right)$$

is the identity matrix. For every $j = 1, \ldots, n$,

$$\begin{aligned}
\boldsymbol{E} \operatorname{trace}\left(\exp(D_{j+1})\right) &= \boldsymbol{E} \operatorname{trace}\left(\exp\left(D_j + sX_{j+1}A_{j+1} - s^2A_{j+1}^2/2\right)\right) \\
&\leq \boldsymbol{E}\left[\operatorname{trace}\left(\exp(D_j)\right) \exp\left(sX_{j+1}A_{j+1} - s^2A_{j+1}^2/2\right)\right] \\
&\quad \text{(by the Golden–Thompson inequality)} \\
&= \operatorname{trace}\left(\boldsymbol{E}\left[\exp(D_j) \exp\left(sX_{j+1}A_{j+1} - s^2A_{j+1}^2/2\right)\right]\right) \\
&\quad \text{(by linearity of the trace)} \\
&= \operatorname{trace}\left(\boldsymbol{E} \exp(D_j) \boldsymbol{E} \exp\left(sX_{j+1}A_{j+1} - s^2A_{j+1}^2/2\right)\right) \\
&\quad \text{(by independence)} \\
&= \operatorname{trace}\left(\boldsymbol{E} \exp(D_j)\right) \\
&= \boldsymbol{E} \operatorname{trace}\left(\exp(D_j)\right).
\end{aligned}$$

Combining these inequalities,

$$\begin{aligned}
\boldsymbol{E} \operatorname{trace}\left(\exp(sM)\right) &= \boldsymbol{E} \operatorname{trace}\left(\exp(D_n)\right) \\
&\leq \boldsymbol{E} \operatorname{trace}\left(\exp(D_0)\right) \\
&= \operatorname{trace}\left(\exp\left(s^2 \sum_{i=1}^n A_i^2/2\right)\right) \\
&\leq d \exp\left(s^2 \left\| \sum_{i=1}^n A_i^2 \right\| /2\right).
\end{aligned}$$

Putting everything together, we have

$$\exp\left(s\boldsymbol{E} \|M\|\right) \leq 2d \left(\exp\left(s^2 \left\| \sum_{i=1}^n A_i^2 \right\| /2\right)\right) = 2de^{s^2\sigma^2/2}.$$

Taking logarithms, dividing both sides by s, and optimizing over s leads to the desired result. □

We emphasize that once one has a good bound for the expected value of $\left\| \sum_{i=1}^n X_iA_i \right\|$, it is easy to obtain bounds that hold with high probability. Indeed, from Theorem 5.6 we get, without further work,

$$\boldsymbol{P}\left\{Z \geq t + \sqrt{2\log(2d)}\sigma\right\} \leq e^{-t^2/(2\sigma^2)}$$

for all $t > 0$. Also, apart from the variance bound given in the theorem, we have $(\mathrm{Var}(Z))^{1/2} \le EZ$ by Exercise 5.17.

Equipped with Rudelson's inequality it is easy to obtain a random-matrix version of Theorem 11.2 where the ℓ_∞^d norm is replaced by the operator norm for $d \times d$ matrices.

Corollary 13.9 *Let $X_1, \ldots, X_n$ be independent random variables taking their values in the space of $d \times d$ symmetric matrices such that they are symmetric (i.e. $-X_i$ has the same distribution as X_i). Let $S_n = \sum_{i=1} X_i$ and $\Sigma^2 = \boldsymbol{E}\| \sum_{i=1}^n X_i^2 \|$. Then*

$$\boldsymbol{E}\left[\|S_n\|^2\right] \le 2(1 + \log(2d))\Sigma^2.$$

The proof parallels the proof of Theorem 11.2, but Rudelson's inequality replaces Corollary 2.6 when there is a need to bound $\boldsymbol{E}\left[\| \sum_{i=1}^n \varepsilon_i X_i \| \mid X_1, \ldots, X_n\right]$.

13.5 Variations of Nemirovski's Inequality

In Section 11.2, we consider norms of sums of $\mathbb{R}^d$-valued independent random variables. However, while Nemirovski's inequality (Theorem 11.2) concerns the ℓ_∞-norm, we use here chaining arguments to derive bounds for the ℓ_p norm for $p \ge 1$.

The setup is as follows: let $X_1, \ldots, X_n$ be independent random vectors in $\mathbb{R}^d$ with $\boldsymbol{E}X_i = 0$ and let

$$S_n = \sum_{i=1}^n X_i.$$

In Section 11.2 we obtained upper bounds for $\boldsymbol{E}\|S_n\|_\infty^2$. Here we deal with ℓ_p norms and derive bounds for $\boldsymbol{E}\|S_n\|_p^2$ for $p \ge 1$.

The key to our approach is to represent the norm as the supremum of an empirical process. Indeed, if $q = p/(p-1)$ (with $q = \infty$ for $p = 1$) and $B_q = \{x \in \mathbb{R} : \|x\|_q \le 1\}$ is the unit ball under the ℓ_q norm, then we may write

$$\|S_n\|_p = \sup_{t \in B_q} \sum_{i=1}^n \langle t, X_i \rangle$$

where $\langle x, y \rangle$ denotes the inner product in $\mathbb{R}^d$.

Based on this representation, it is natural to introduce the weak variance

$$\Sigma_p^2 = \boldsymbol{E} \sup_{t \in B_q} \sum_{i=1}^n \langle t, X_i \rangle^2.$$

Then by Theorem 11.1, $\mathrm{Var}(\|S_n\|_p) \le 2\Sigma_p^2$. The next theorem shows that $\boldsymbol{E}\|S_n\|_p^2$ may also be bounded in terms of the weak variance.

Theorem 13.10 *Let $X_1, \ldots, X_n$ be independent zero-mean random vectors in $\mathbb{R}^d$ and let $S_n = \sum_{i=1}^n X_i$. Then for all $p \ge 1$,*

$$\boldsymbol{E}\,\|S_n\|_p^2 \le 578\, d\, \Sigma_p^2.$$

We need the following estimate of the metric entropy of unit balls whose proof is left as an exercise (Exercise 13.22).

Lemma 13.11 *For all $q \geq 1$ and for all $u \in (0, 1]$, the metric entropy $H(u, B_q)$ of the unit ball B_q under the ℓ_q metric satisfies*

$$H(u, B_q) \leq d \log\left(1 + \frac{2}{u}\right).$$

Proof of Theorem 13.10. Writing

$$\mathbf{E}\|S_n\|_p^2 = \mathrm{Var}\left(\|S_n\|_p\right) + \left(\mathbf{E}\|S_n\|_p\right)^2$$

and recalling that $\mathrm{Var}\left(\|S_n\|_p\right) \leq 2\Sigma_p^2$, it suffices to bound $\mathbf{E}\|S_n\|_p$. We do this by symmetrization, followed by chaining.

By the symmetrization inequalities of Lemma 11.4,

$$\mathbf{E}\|S_n\| \leq 2\mathbf{E} \sup_{t \in B_q} \sum_{i=1}^{n} \varepsilon_i \langle t, X_i \rangle,$$

where $\varepsilon_1, \ldots, \varepsilon_n$ are independent Rademacher variables.

Working conditionally on the X_i, $\sup_{t\in B_q} \sum_{i=1}^n \varepsilon_i \langle t, X_i\rangle$ is the supremum of a process indexed by B_q. This process has sub-Gaussian increments as the following argument shows: by Hoeffding's inequality, for all $\lambda > 0$ and $t, t' \in B_q$,

$$\begin{aligned}
&\mathbf{E}\left[\exp\left(\lambda \sum_{i=1}^{n} \varepsilon_i \langle t - t', X_i \rangle\right) \middle| X_1, \ldots, X_n\right] \\
&\leq \exp\left(\frac{\lambda^2}{2} \sum_{i=1}^{n} \langle t - t', X_i \rangle^2\right) \\
&= \exp\left(\frac{\lambda^2}{2} \|t - t'\|_q^2 \sum_{i=1}^{n} \left\langle \frac{t - t'}{\|t - t'\|_q}, X_i \right\rangle^2\right) \\
&\leq \exp\left(\frac{\lambda^2}{2} \|t - t'\|_q^2 \sup_{t \in B_q} \sum_{i=1}^{n} \langle t, X_i \rangle^2\right).
\end{aligned}$$

Now we may use Dudley's bound (Corollary 13.2) conditionally on $X_1, \ldots, X_n$, to conclude that

$$\mathbf{E}\left[\sup_{t \in B_q} \sum_{i=1}^{n} \varepsilon_i \langle t, X_i \rangle \middle| X_1, \ldots, X_n\right] \leq \left(\sup_{s \in B_q} \sum_{i=1}^{n} \langle s, X_i \rangle^2\right)^{1/2} 12 \int_0^1 \sqrt{H(u, B_q)}du.$$

By Jensen's inequality,

$$(\boldsymbol{E}\left[\|S_n\|\right])^2 \leq 4\left(\boldsymbol{E}\left[\left(\sup_{s\in B_q}\sum_{i=1}^{n}\langle s,X_i\rangle^2\right)^{1/2} 12\int_0^1\sqrt{H\left(u,B_q\right)}du\right]\right)^2$$

$$\leq 576\,\Sigma_p^2\left(\int_0^1\sqrt{H(u,B_q)}du\right)^2.$$

Combining this with the fact that $\mathrm{Var}\left(\|S_n\|_p^2\right) \leq 2\Sigma_p^2$,

$$\boldsymbol{E}\|S_n\|^2 \leq \left(2+576\left(\int_0^1\sqrt{H\left(u,B_q\right)}du\right)^2\right)\Sigma_p^2$$

$$\leq \Sigma_p^2\left(2+576\int_0^1 H\left(u,B_q\right)du\right).$$

By Lemma 13.11, the entropy integral can be upper bounded by $d\log(3^3/2^2)$. □

Recall that in the bound of Theorem 11.2 for the ℓ_∞ norm, the dependence of the upper bound in d is only logarithmic. This suggests that the bound of Theorem 13.10 may not be tight at least for large values of p. Further variations on Nemirovski's inequality for other finite dimensional normed spaces are described in Exercises 13.23, 13.24, and 13.25.

13.6 Random Projections of Sparse and Large Sets

In this section we return to the Johnson–Lindenstrauss problem already studied in Sections 2.9 and 5.6. First recall the setup: we consider $A \subset \mathbb{R}^D$ where D is a large positive integer. Suppose $d < D$ (typically $d \ll D$) and define the random map $W : \mathbb{R}^D \to \mathbb{R}^d$ that assigns to each $\alpha = (\alpha_1, \ldots, \alpha_D) \in \mathbb{R}^D$ the vector $W(\alpha) = (1/\sqrt{d})(W_1(\alpha), \ldots, W_d(\alpha)) \in \mathbb{R}^d$ with

$$W_i\left(\alpha\right) = \sum_{j=1}^{D}\alpha_j X_{i,j}$$

where $X_{i,1}, \ldots, X_{i,d}$ are independent copies of a random variable X satisfying $\boldsymbol{E}X = 0$ and $\mathrm{Var}\left(X\right) = 1$. In this section we only consider the case when X is either a standard Gaussian or a Rademacher random variable. The Johnson–Lindenstrauss lemma (Theorem 2.13) states that if A is finite, then after applying the random projection W, the pairwise distances between elements of A are preserved up to a factor $1 \pm \varepsilon$ with high probability, if d is of the order of $\varepsilon^{-2}\log|A|$. As we argued in Section 5.6, this result depends only on the cardinality of A and it does not take the structure of the set A into account. In particular, it is vacuous if A is an infinite set. There we extended the basic Johnson–Lindenstrauss lemma to take the structure of the set A into account. First we briefly recall this result.

The random map W is called an *ε-isometry* on A if

$$\left| \frac{\|W(\alpha) - W(\alpha')\|^2}{\|\alpha - \alpha'\|^2} - 1 \right| \leq \varepsilon \quad \text{for all distinct } \alpha, \alpha' \in A.$$

Here $\|\cdot\|$ denotes the Euclidean norm. The set T of normalized differences of elements of A plays a crucial role in the analysis.

$$T = \left\{ \frac{a - a'}{\|a - a'\|}, (a, a') \in A \times A \text{ with } a \neq a' \right\}.$$

This set indexes the empirical processes whose suprema are in the focus of our attention:

$$V = \sup_{\alpha \in T} \left(\|W(\alpha)\|^2 - 1 \right) \quad \text{and} \quad V' = \sup_{\alpha \in T} \left(1 - \|W(\alpha)\|^2 \right).$$

In particular, Theorem 5.10 shows that when the quantity

$$\Delta = d\left(\max(\boldsymbol{E}V, \boldsymbol{E}V')\right)^2$$

is small, then one may project to low-dimensional spaces without significantly changing the metric structure of the set. More precisely, we showed that if $d \geq 20(\Delta + \log(2/\delta))\varepsilon^{-2}$, then the random map W is an ε-isometry on A with probability at least $1 - \delta$.

When A is finite, by Corollary 2.6, $\Delta \leq 32 \log |A|$ as long as $d \geq \log |A|$. The goal of this section is to obtain sharp bounds for Δ taking the finer structure of A into account. In particular, Δ may be finite even when A is infinite. At first sight, it might appear to be a routine task to relate Δ with some notion of "richness" of the index set T. Indeed, the process $d(\|W(\alpha)\|^2 - 1)$ and its opposite fit within the scope of Lemma 13.1 (see Exercise 13.26). However, one can do much better by taking into account some specific features of the empirical processes under consideration. Indeed, for all $\alpha \in T$, $\|W(\alpha)\|^2 - 1$ is centered and its variance does not depend on α.

Note that Theorem 5.10 is stated for Gaussian random projections but a minor modification of its proof reveals that it remains valid when the $X_{i,j}$ are Rademacher random variables. In the sequel we restrict our attention to these cases.

A remarkable feature of the Johnson–Lindenstrauss lemma and Theorem 5.10 is that the dimension D of the set A does not play any role. Indeed, we could have formulated Theorem 5.10 so as to accommodate separable Hilbert spaces. The same remark goes for most of the rest of the section. However, for the ease of exposition, we present the results in the finite-dimensional context.

In order to obtain sharper bounds for $\boldsymbol{E}V$ and $\boldsymbol{E}V'$ than one could achieve by ordinary chaining, we use the following "splitting" argument. It is a key ingredient in the proof of both Theorems 13.13 and 13.15 below.

Lemma 13.12 *Let $T \subset \mathbb{R}^D$ denote a finite set of vectors of unit Euclidean norm. Let $W : \mathbb{R}^D \to \mathbb{R}^d$ be the random linear map defined above. Let $\delta \in (0, 1)$ and let T_δ be a δ-net of T. Let*

$$\mathcal{D}_\delta T = \left\{ \alpha - \alpha' : \|\alpha - \alpha'\| \leq \delta, \alpha, \alpha' \in T \right\}.$$

Let V and V' be defined as above. Then, for all $\theta > 0$,

$$\mathbf{E}V \leq (1+\theta)\,\mathbf{E}\sup_{\alpha\in T_\delta}\left(\|W(\alpha)\|^2 - 1\right) + \left(1+\frac{1}{\theta}\right)\mathbf{E}\sup_{\alpha\in\mathcal{D}_\delta T}\|W(\alpha)\|^2 + \theta,$$

and for all $\theta \in (0,1)$,

$$\mathbf{E}V' \leq (1-\theta)\,\mathbf{E}\sup_{\alpha\in T_\delta}\left(1 - \|W(\alpha)\|^2\right) + \left(\frac{1}{\theta}-1\right)\mathbf{E}\sup_{\alpha\in\mathcal{D}_\delta T}\|W(\alpha)\|^2 + \theta.$$

Proof First notice that

$$V = \sup_{\alpha\in T}\left(\|W(\alpha)\|^2 - 1\right) = \left(\sup_{\alpha\in T}\|W(\alpha)\|\right)^2 - 1.$$

Let $\Pi : T \to T_\delta$ be such that for every $\alpha \in T$, $\Pi\alpha$ is a nearest neighbor of α in T_δ. Then

$$\sup_{\alpha\in T}\|W(\alpha)\| \leq \sup_{\alpha\in T_\delta}\|W(\alpha)\| + \sup_{\alpha\in T}\|W(\alpha - \Pi\alpha)\|.$$

As $2ab \leq \theta a^2 + b^2/\theta$,

$$\left(\sup_{\alpha\in T}\|W(\alpha)\|\right)^2$$
$$\leq (1+\theta)\left(\sup_{\alpha\in T_\delta}\|W(\alpha)\|\right)^2 + \left(1+\frac{1}{\theta}\right)\left(\sup_{\alpha\in T}\|W(t-\Pi\alpha)\|\right)^2,$$

and therefore

$$V \leq (1+\theta)\sup_{\alpha\in T_\delta}\left(\|W(\alpha)\|^2 - 1\right) + \left(1+\frac{1}{\theta}\right)\sup_{\alpha\in\mathcal{D}_\delta T}\|W(\alpha)\|^2 + \theta.$$

Taking expectations on both sides leads to the desired result. The proof of the second inequality is similar. □

Before stating the main result of the section, we consider the easier but important special case when A is the collection of k-sparse vectors in $\mathbb{R}^D$, that is, vectors with at most k nonzero coordinates. Thus, the set A is the union of $\binom{D}{k}$ k-dimensional subspaces of $\mathbb{R}^D$. It is usually assumed that $k \ll D$.

Lemma 13.13 *Consider the random projection $W : \mathbb{R}^D \to \mathbb{R}^d$ defined above, where the $X_{i,j}$ are either standard Gaussian or Rademacher random variables. Let A be the set of all k-sparse vectors in $\mathbb{R}^D$. If the unique solution $\varepsilon_* > 0$ of the equation*

$$d\varepsilon^2 = 2k\log\left(\frac{2eD}{k\varepsilon}\right)$$

is smaller than $1/2$, then $\max(\mathbf{E}V, \mathbf{E}V') \leq 20\varepsilon_*$.

Proof The index set $T = \{(a - a')/\|a - a'\| : a, a' \in A\}$ may be partitioned into $\binom{D}{2k}$ subsets defined by picking $2k$ coordinates among the D possible ones. An ε_*-net can be constructed for each subset and the union T_{ε_*} of the $\binom{D}{2k}\varepsilon_*$-nets has cardinality at most $\binom{D}{2k}(1 + 2/\varepsilon_*)^{2k}$ by Lemma 13.11.

The set $\mathcal{D}_{\varepsilon_*}T$, defined as in Lemma 13.12, has the useful property that $\mathcal{D}_{\varepsilon_*}T \subseteq \varepsilon_* T$, which implies

$$\boldsymbol{E} \sup_{\alpha \in \mathcal{D}_{\varepsilon_*}(T)} \|W(\alpha)\|^2 \leq \varepsilon_*^2 \boldsymbol{E} \sup_{\alpha \in T} \|W(\alpha)\|^2 \leq \varepsilon_*^2(1 + \boldsymbol{E}V).$$

By Lemma 13.12, for any $\theta > 0$,

$$\begin{aligned} \boldsymbol{E}V &\leq (1+\theta)\, \boldsymbol{E} \sup_{\alpha \in T_{\varepsilon_*}} \left(\|W(\alpha)\|^2 - 1\right) + \left(1 + \frac{1}{\theta}\right) \boldsymbol{E} \sup_{\alpha \in \mathcal{D}_{\varepsilon_*}(T)} \|W(\alpha)\|^2 + \theta \\ &\leq (1+\theta)\, \boldsymbol{E} \sup_{\alpha \in T_{\varepsilon_*}} \left(\|W(\alpha)\|^2 - 1\right) + \left(1 + \frac{1}{\theta}\right) {\varepsilon_*}^2(1 + \boldsymbol{E}V) + \theta. \end{aligned}$$

We choose $\theta = \varepsilon_*$, so that $(1 + 1/\theta)\, \varepsilon_*^2 = \varepsilon_*(1 + \varepsilon_*) < 1$. Rearranging, we have

$$\boldsymbol{E}V \leq \frac{1}{1 - \varepsilon_*(1 + \varepsilon_*)} \left((1 + \varepsilon_*)\boldsymbol{E}\left[\sup_{\alpha \in T_{\varepsilon_*}} \left(\|W(\alpha)\|^2 - 1\right) \right] + \varepsilon_*(2 + \varepsilon_*) \right).$$

Now $\sup_{\alpha \in T_{\varepsilon_*}} \left(\|W(\alpha)\|^2 - 1\right)$ is the maximum of at most $\binom{D}{2k}(1 + 2/\varepsilon_*)^{2k}$ sub-gamma random variables with variance factor $2/d$ and scale factor $2/d$. By Corollary 2.6,

$$\begin{aligned} &\boldsymbol{E} \sup_{\alpha \in T_{\varepsilon_*}} \left(\|W(\alpha)\|^2 - 1\right) \\ &\leq \sqrt{4\frac{2k}{d} \log\left(\frac{eD}{2k}\left(1 + \frac{2}{\varepsilon_*}\right)\right)} + \frac{1}{d} 2k \log\left(\frac{eD}{2k}\left(1 + \frac{2}{\varepsilon_*}\right)\right) \\ &\leq \sqrt{4\frac{2k}{d} \log\left(\frac{2eD}{k\varepsilon_*}\right)} + \frac{2k}{d} \log\left(\frac{2eD}{k\varepsilon_*}\right) = 2\varepsilon_* + \varepsilon_*^2. \end{aligned}$$

Combining the obtained bounds and using the fact that $\varepsilon_* \leq 1/2$ leads to

$$\boldsymbol{E}V \leq 20\varepsilon_*.$$

In order to upper bound $\mathbf{E}V'$, we use the second inequality in Lemma 13.12. Choosing again $\theta = \varepsilon_*$, and proceeding as in the proof of the upper bound for $\mathbf{E}V$,

$$\begin{aligned}\mathbf{E}V' &\leq (1-\varepsilon_*)\mathbf{E}\sup_{\alpha\in T_{\varepsilon_*}}\left(1-\|W(\alpha)\|^2\right)+\varepsilon_*+\varepsilon_*(1+\mathbf{E}V)\\ &\leq (1-\varepsilon_*)(2\varepsilon_*+\varepsilon_*^2)+\varepsilon_*(2+20\varepsilon_*)\leq 15\varepsilon_*.\end{aligned}$$

□

Combining Theorem 5.10 and Lemma 13.13, we obtain the following, so-called *restricted isometry property* of Gaussian and Rademacher random projections. The remarkable feature is that when d is roughly of the order of $k \log D$, then the random projection preserves the metric structure of the set of *all* k-sparse vectors.

Corollary 13.14 (RESTRICTED ISOMETRY PROPERTY) *Consider the random projection $W : \mathbb{R}^D \to \mathbb{R}^d$ defined above, where the $X_{i,j}$ are either standard Gaussian or Rademacher random variables. Let A be the set of all k-sparse vectors in $\mathbb{R}^D$. If the unique solution $\varepsilon_* > 0$ of the equation*

$$d\varepsilon^2 = 2k\log\left(\frac{2eD}{k\varepsilon}\right)$$

is smaller than $1/2$, then there exists a universal constant κ such that for every $\varepsilon, \delta \in (0,1)$, if $d \geq 20(20^2 d\varepsilon_*^2 + \log(2/\delta))\varepsilon^{-2}$, then the random map W is an ε-isometry on A with probability at least $1-\delta$.

The main result of this section is the next theorem, which gives general conditions for a random projection to be an approximate isometry in terms of the metric entropy of the projected set A. It implies Corollary 13.14 but it is significantly more general.

Theorem 13.15 (KLARTAG–MENDELSON THEOREM) *Let $A \subset \mathbb{R}^D$ and consider the random projection $W : \mathbb{R}^D \to \mathbb{R}^d$ defined above, where the $X_{i,j}$ are either standard Gaussian or Rademacher random variables. Let $T = \{(a-a')/\|a-a'\| : a, a' \in A\}$ and define*

$$\gamma(T) = \int_0^1 \sqrt{H(x,T)}dx,$$

where $H(x,T)$ is the x-entropy of T (with respect to the Euclidean distance). There exists an absolute constant κ'', such that for all $\varepsilon, \delta \in (0,1)$ if $d \geq \kappa''\varepsilon^{-2}(\gamma^2(T) + \log(2/\delta))$, then W is an ε-isometry on A, with probability at least $1-\delta$.

This generalization of the Johnson–Lindenstrauss theorem is relevant even when considering finite sets A. It tells us that the sensitivity of the random projection method depends on the "metric size" of the set A rather than on its cardinality. As an example, consider the case of Corollary 13.14, when A is the set of all k-sparse vectors in $\mathbb{R}^D$. It is not difficult to check that in this case, $\gamma^2(T) \leq 2k\log(D/(2k))$, essentially recovering the result of of Corollary 13.14. In this case, as in many others, the cardinality of A does not matter!

In order to prove the Klartag–Mendelson theorem, by Theorem 5.10, it suffices to show that there exists a universal constant κ such that

$$d \max(EV, EV')^2 \leq \kappa \gamma^2(T).$$

This is shown in Proposition 13.17 below.

The rough idea is as follows. Since V and V' are suprema of centered and normalized chi-square random variables, a glimpse at Lemma 13.1 suggests that a reasonable upper bound should involve both $\int \sqrt{H(x,T)}dx$ and $\int H(x,T)dx$. Indeed when the $X_{i,j}$ are standard Gaussian, it is easy to check that $\|W(\alpha)\|^2 - \|W(\alpha')\|^2$ is sub-Gamma with variance factor proportional to $\|\alpha - \alpha'\|^2$ and scale factor proportional to $\|\alpha - \alpha'\|$ (see Exercise 13.26). Surprisingly, it is possible to engineer an upper bound that only involves $\int \sqrt{H(x,T)}dx$, just as if the process increments were purely sub-Gaussian. The proof of Theorem 13.15 relies on the splitting argument of Lemma 13.12. It improves on the proof of Lemma 13.13 in two respects. First, the chaining lemma (Lemma 13.1) is used to upper bound $E \sup_{\alpha \in T_\delta}(\|W(\alpha)\|^2 - 1)$. The cutoff δ is tuned in such a way that the sub-Gaussian term dominates the sub-gamma term. Second, when upper bounding $E \sup_{\alpha \in \mathcal{D}_\delta T} \|W(\alpha)\|^2$, the key observation is that $\{\|W(\alpha)\|, \alpha \in \mathcal{D}_\delta T\}$ has sub-Gaussian increments. This is established in the next lemma. (Recall the definition of the set $\mathcal{G}(v)$ of sub-Gaussian random variables from Section 2.3.)

Lemma 13.16 *Let $W : \mathbb{R}^D \to \mathbb{R}^d$ be the random map defined as above. Let T be a bounded subset of $\mathbb{R}^D$ and let $\delta > 0$ be such that $\|\alpha\| \leq \delta$ for every $\alpha \in T$. Let $Z = \sqrt{d} \sup_{\alpha \in T} \|W(\alpha)\|$. Then*

$$\mathrm{Var}(Z) \leq \delta^2 \quad \text{and} \quad Z - EZ \in \mathcal{G}(\delta^2)$$

when X is Gaussian and

$$\mathrm{Var}(Z) \leq 2\delta^2 \quad \text{and} \quad Z - EZ \in \mathcal{G}(4\delta^2)$$

when X is Rademacher.

Proof Z may be considered as the supremum of a Gaussian (or a Rademacher) process. Indeed, the representation of the norm as a supremum of linear functions implies that

$$Z = \sup_{u \in \mathbb{R}^d : \|u\|=1} \sup_{\alpha \in T} \sum_{i=1}^{d} \sum_{j=1}^{D} u_i \alpha_j X_{i,j}.$$

On the other hand, the wimpy variance equals

$$\sigma^2 = \sup_{u \in \mathbb{R}^d : \|u\|=1} \sup_{\alpha \in T} \sum_{i=1}^{d} \sum_{j=1}^{D} u_i \alpha_j^2 = \sup_{\alpha \in T} \sum_{j=1}^{D} \alpha_j^2 \leq \delta^2.$$

Now we may use the variance bounds proved in Chapter 3 and the exponential concentration inequalities from Chapter 5 to conclude. □

Now we are ready to state and prove the key ingredient needed to complete the proof of Theorem 13.15.

Proposition 13.17 *Consider the setup of Theorem 13.15. There exists an absolute constant κ' such that*

$$d \max(\boldsymbol{E}V, \boldsymbol{E}V')^2 \leq \kappa' \gamma^2(T) \left(1 + \frac{\gamma(T)}{\sqrt{d}}\right)^2.$$

Proof Thanks to standard separability arguments, we may assume that T is a finite set. By the splitting lemma (Lemma 13.12), for any $\theta > 0$

$$\boldsymbol{E}V \leq (1+\theta)\boldsymbol{E} \sup_{\alpha \in T_\delta} \left(\|W(\alpha)\|^2 - 1\right) + (1 + 1/\theta)\boldsymbol{E} \sup_{\alpha \in \mathcal{D}_\delta T} \|W(\alpha)\|^2 + \theta.$$

We choose $\theta = \gamma(T)/\sqrt{d}$ and bound the two expectations on the right-hand side by $\kappa\gamma(T)/\sqrt{d}$ and $\kappa'\gamma(T)^2/d$, where κ, κ' are universal constants.

Let the cutoff δ be chosen as $\delta = 2^{-J+1}$, where

$$J = \sup \left\{j \geq 0, H\left(2^{-j+1}, T\right) \leq d\right\}.$$

Note that δ is well defined. Indeed, if $H(2, T) = 0$ then $\left\{j \geq 0, H\left(2^{-j+1}, T\right) \leq d\right\}$ is a non-empty set, while $J = \infty$ means that $\delta = 0$.

If $\delta = 0, \boldsymbol{E} \sup_{\alpha \in \mathcal{D}_\delta T} \|W(\alpha)\|^2 = 0$, so we may assume that $\delta > 0$. By the definition of δ, $H(\delta/2, T) > d$. Lemma 13.16 implies that

$$\boldsymbol{E} \sup_{\alpha \in \mathcal{D}_\delta T} \|W(\alpha)\|^2 \leq \left(\boldsymbol{E} \sup_{\alpha \in \mathcal{D}_\delta T} \|W(\alpha)\|\right)^2 + \frac{2\delta^2}{d}.$$

On the other hand, the increments $\|W(\alpha)\| - \|W(\alpha')\| \leq \|W(\alpha - \alpha')\|$ satisfy the condition of Lemma 13.1 with $a = 1$, $v = 4/d$ and $c = 0$. It follows that

$$\boldsymbol{E} \sup_{\alpha \in \mathcal{D}_\delta T} \|W(\alpha)\| \leq 3\delta + \frac{12\delta}{\sqrt{d}} \sum_{j=1}^{\infty} 2^{-j} \sqrt{H(\delta 2^{-j}, \mathcal{D}_\delta T)}.$$

Our choice of the value of δ implies that for every $x > 0$, $H(x, \mathcal{D}_\delta T) \leq 2H(x/2, T)$, and therefore

$$\boldsymbol{E} \sup_{s \in \mathcal{D}_\delta T} \|W(\alpha)\| \leq 3\delta + \frac{24\delta\sqrt{2}}{\sqrt{d}} \sum_{j=2}^{\infty} 2^{-j} \sqrt{H(\delta 2^{-j}, T)}.$$

Since $H(\delta/2, T) > d$, if follows that

$$\mathbf{E} \sup_{\alpha \in \mathcal{D}_\delta T} \left\| W(\alpha) \right\| \leq \frac{24\delta\sqrt{2}}{\sqrt{d}} \sum_{j=1}^{\infty} 2^{-j} \sqrt{H(\delta 2^{-j}, T)}.$$

Setting

$$\Gamma = \sum_{j=1}^{\infty} 2^{-j} \sqrt{H(\delta 2^{-j}, T)},$$

the condition $H(\delta/2, T) > d$ entails $\Gamma^2 > d/4 > 1/4$, so

$$\mathbf{E} \sup_{\alpha \in \mathcal{D}_\delta T} \left\| W(\alpha) \right\|^2 \leq 576 \frac{\delta^2 \Gamma^2}{d} + \frac{4\delta^2}{d} \leq 592 \times \frac{\delta^2 \Gamma^2}{d}.$$

Thanks to the monotonicity of H and to the fact that $\delta \leq 2$,

$$\mathbf{E} \sup_{\alpha \in \mathcal{D}_\delta T} \left\| W(\alpha) \right\|^2 \leq c \frac{\gamma^2(T)}{d}$$

for some absolute constant c. Now, Bernstein's inequality and Lemma 13.1 allow us to derive an upper bound for $\mathbf{E} \sup_{\alpha \in T_\delta} (\|W(\alpha)\|^2 - 1)$ as follows. Recall that for every $\alpha \in \mathbb{R}^d$,

$$\left\| W(\alpha) \right\|^2 = \frac{1}{d} \sum_{i=1}^{d} W_i^2(\alpha).$$

By the Cauchy–Schwarz inequality, for all α, α' and every integer $k \geq 2$,

$$\begin{aligned}
&\mathbf{E}\left[\left| W_i(\alpha)^2 - W_i(\alpha')^2 \right|^k \right] \\
&\leq \left(\mathbf{E}\left[\left| W_i(\alpha) - W_i(\alpha') \right|^{2k} \right] \mathbf{E}\left[\left| W_i(\alpha) + W_i(\alpha') \right|^{2k} \right] \right)^{1/2} \\
&\leq \left\| \alpha - \alpha' \right\|^k \left\| \alpha + \alpha' \right\|^k \sup_{\alpha \in \mathbb{R}^D : \|\alpha\| = 1} \mathbf{E}\left[W_i(\alpha)^{2k} \right] \\
&\leq 4^k \left\| \alpha - \alpha' \right\|^k \left\| \alpha + \alpha' \right\|^k \frac{k!}{2},
\end{aligned}$$

where the last inequality comes from the observation that each $W_i(\alpha)$ is sub-Gaussian (see Section 2.9). Let $\alpha_0 \in T_\delta$, and set $X(\alpha) = \left\| W(\alpha) \right\|^2$. As $\mathbf{E}X(\alpha_0) = 1$,

$$\mathbf{E} \sup_{\alpha \in T_\delta} \left(\left\| W(\alpha) \right\|^2 - 1 \right) = \mathbf{E}\left[\sup_{\alpha \in T_\delta} X(\alpha) - X(\alpha_0) \right].$$

As the conditions of Lemma 13.1 are satisfied with $a = 0$, $v = 8/d$ and $c = 8/d$, we get

$$E \sup_{\alpha \in T_\delta} \left(\|W(\alpha)\|^2 - 1 \right) \le \frac{6}{\sqrt{d}} \sum_{j=1}^{\infty} 2^{-j} \left(\sqrt{8H(2^{-j}, T_\delta)} + \frac{8H(2^{-j}, T_\delta)}{\sqrt{d}} \right).$$

Now, since T_δ is a δ-net, by the definition of δ, $\log |T_\delta| \le d$, and therefore $H(., T_\delta) \le d$, which implies that

$$E \sup_{\alpha \in T_\delta} \left(\|W(\alpha)\|^2 - 1 \right) \le \frac{84}{\sqrt{d}} \sum_{j=1}^{\infty} 2^{-j} \sqrt{H(2^{-j}, T_\delta)}$$

$$\le \frac{84}{\sqrt{d}} \sum_{j=1}^{\infty} 2^{-j} \sqrt{H(2^{-j}, T)} \le \frac{168}{\sqrt{d}} \gamma(T).$$

It is now time to collect bounds. Choose $\theta = \gamma(T)/\sqrt{d}$ and invoke Lemma 13.12 to obtain

$$E \sup_{\alpha \in T} \left(\|W(\alpha)\|^2 - 1 \right) \le 952 \frac{\gamma(T)}{\sqrt{d}} \left(1 + \frac{\gamma(T)}{\sqrt{d}} \right),$$

completing the proof of the upper bound for EV. In order to control EV', we use the second inequality from Lemma 13.12,

$$EV' \le (1 - \theta) E \sup_{\alpha \in T_\delta} \left(1 - \|W(\alpha)\|^2 \right) + \left(\frac{1}{\theta} - 1 \right) E \sup_{\alpha \in \mathcal{D}_\delta(T)} \|W(\alpha)\|^2 + \theta,$$

with $\theta = \gamma(T)/\sqrt{d}$. We can indeed prove that

$$EV' \le C \frac{\gamma(T)}{\sqrt{d}}$$

with $C \ge 2$, by assuming without loss of generality that $\gamma(T)/\sqrt{d} \le 1/2$. □

13.7 Normalized Processes: Slicing and Reweighting

Sometimes one is interested in bounding the supremum of an empirical process $\sup_{s \in \mathcal{T}} \sum_{i=1}^{n} X_{i,s}$ that is quite inhomogeneous in the sense that the variance $\mathrm{Var}\left(\sum_{i=1}^{n} X_{i,s}\right)$ varies with s. To illustrate such a situation, consider the Kolmogorov–Smirnov statistic already discussed in Chapter 11. In this example, $Y_1, \ldots, Y_n$ are independent random variables, uniformly distributed over $[0, 1]$. The (one-sided) Kolmogorov–Smirnov statistic is

$$Z = \sup_{s \in [0,1]} \sum_{i=1}^{n} \left(\mathbb{1}_{\{Y_i \le s\}} - s \right).$$

As half-lines form a VC-class with VC-dimension 1, it follows from Theorem 13.7 that $EZ = O(\sqrt{n})$, and this is the correct order of magnitude. (See Exercise 13.18 for an alternative argument.) For small and large values of s, the variance of $\mathbb{1}_{\{Y_i \le s\}}$ is small and the maximal value is unlikely to be achieved for such indices. Indeed, it is not difficult to see that Z is not very different from $\sup_{s\in[1/4,3/4]} \sum_{i=1}^n (\mathbb{1}_{\{Y_i\le s\}} - s)$. The knowledge of EZ and the availability of concentration inequalities tell us little about the fluctuations of $\sum_{i=1}^n (\mathbb{1}_{\{Y_i\le s\}} - s)$ for small and large values of s. By dividing each $\mathbb{1}_{\{Y_i\le s\}} - s$ by its standard deviation, one obtains a re-weighted process that may contain more interesting information.

In this section we discuss the so-called *peeling* (or *stratification*, or *slicing*) techniques which, in combination with re-weighting, allows one to investigate fine properties of empirical processes. Such techniques will be used in the next two sections to obtain sharp bounds for uniform relative deviations of L_2 distances and the risk of empirical risk minimization in classification.

The basic idea is that by decomposing the class such that each component contains random variables with similar variances, one may take full advantage of Bousquet's inequality (Theorem 12.5).

The following lemma illustrates how slicing the index set $\mathcal{T}$ into sub-collections can be used to investigate re-weighted processes.

Call a function $\psi : [0,\infty) \to [0,\infty)$ *sub-linear* if it is nondecreasing, continuous, $\psi(x)/x$ is nonincreasing, and $\psi(1) \ge 1$. Note that if ψ and ρ are sub-linear, then so are $\psi \circ \rho$ and $\psi + \rho$. Moreover, for any $\alpha \ge \psi(1)$, the equation $\alpha r^2 = \psi(r)$ has a unique solution in $(0,1]$. One can easily check that every sub-linear function ψ is sub-additive in the sense that $\psi(u+v) \le \psi(u) + \psi(v)$ (see Exercise 13.41).

Lemma 13.18 *Let $\mathcal{T}$ be a countable index set and let $L : \mathcal{T} \to [0,\infty)$. Assume that there exists $\bar{s} \in \mathcal{T}$ such that $L(\bar{s}) = \inf_{s\in\mathcal{T}} L(s)$. Let $(Z_s)_{s\in\mathcal{T}}$ denote a stochastic process indexed by $\mathcal{T}$. Assume that there exists a sub-linear function ψ and $r_{cr} > 0$ such that for all $r \ge r_{cr}$,*

$$E \sup_{s: s\in\mathcal{T}, L(s)\le r^2} |Z_s - Z_{\bar{s}}| \le \psi(r).$$

Then, for all $r \ge r_{cr}$,

$$E \sup_{s\in\mathcal{T}} \frac{r^2}{r^2 + L(s)} |Z_s - Z_{\bar{s}}| \le 6\psi(r).$$

Proof Let $r \ge r_{cr}$. We decompose the index set $\mathcal{T}$ into *slices* according to the value of the function L as follows. Let $\mathcal{T}_0 = \left\{s : s \in \mathcal{T},\ L(s) \le r^2\right\}$ and for $k \ge 1$, let

$$\mathcal{T}_k = \left\{s : s \in \mathcal{T},\ r^2 2^{2(k-1)} < L(s) \le r^2 2^{2k}\right\}.$$

Let $V_r = \sup_{s\in\mathcal{T}} \frac{r^2}{r^2+L(s)} |Z_s - Z_{\bar{s}}|$. Then

$$\begin{aligned} EV_r &\le \sum_{k=0}^{\infty} E \sup_{s\in\mathcal{T}_k} r^2 \frac{|Z_{\bar{s}} - Z_s|}{r^2 + L(s)} \\ &\le \psi(r) + \sum_{k=1}^{\infty} \frac{r^2}{r^2 + r^2 2^{2(k-1)}} E \sup_{s\in\mathcal{T}_k} |Z_{\bar{s}} - Z_s| \end{aligned}$$

$$\leq \psi(r) + \sum_{k=1}^{\infty} \frac{1}{1 + 2^{2(k-1)}} \psi(2^k r)$$

$$\leq \psi(r) + 2\sum_{k=1}^{\infty} \frac{2^{k-1}}{1 + 2^{2(k-1)}} \psi(r) \quad \text{(since } \psi \text{ is sub-linear)}$$

$$\leq 2\left(1 + \sum_{k=0}^{\infty} 2^{-k}\right) \psi(r). \qquad \square$$

Theorem 13.19 *For $i = 1, \ldots, n$, let $X_i = (X_{i,s})_{s \in \mathcal{T}}$ be a collection of random variables indexed by a countable set $\mathcal{T}$ and suppose that $X_1, \ldots, X_n$ are independent and identically distributed. Assume that for all $i \leq n$ and $s \in \mathcal{T}$, $\mathbf{E}X_{i,s} = 0$ and that $|X_{i,s}| \leq 1$ almost surely. Let $L : \mathcal{T} \to [0, \infty)$ which achieves its minimum at $\bar{s} \in \mathcal{T}$. Assume that $\sup_{s \in \mathcal{T}, i \leq n} |X_{i,s} - X_{i,\bar{s}}| \leq 1$. Let $\sigma : \mathcal{T} \to [0, \infty)$ be such that for every $s \in \mathcal{T}$, $\mathbf{E}(X_{i,s} - X_{i,\bar{s}})^2 \leq \sigma^2(s)$. Assume there exists a sub-linear function ρ such that for all $s \in \mathcal{T}$, $\sigma(s) \leq \rho\left(L(s)^{1/2}\right)$ and there exists a sub-linear ψ such that for all r satisfying $\sqrt{n}r^2 \geq \psi(r)$,*

$$\sqrt{n}\mathbf{E} \sup_{\substack{s \in \mathcal{T} \\ \sigma(s) \leq r}} \left| \sum_{i=1}^{n} \frac{1}{n}(X_{i,s} - \mathbf{E}X_{i,s} - X_{i,\bar{s}} + \mathbf{E}X_{i,\bar{s}}) \right| \leq \psi(r).$$

Let $\varepsilon, \delta \in (0, 1]$ and let $r(\delta) > 0$ be the unique solution of equation

$$\sqrt{n}r^2 = \frac{1}{\varepsilon}\left(8\psi(\rho(r)) + \rho(r)\sqrt{\log 1/\delta} + \frac{2\log 1/\delta}{3\sqrt{n}}\right).$$

Then, with probability at least $1 - 2\delta$, for all $s \in \mathcal{T}$

$$\left| \sum_{i=1}^{n} \frac{1}{n}(X_{i,s} - \mathbf{E}X_{i,s} - X_{i,\bar{s}} + \mathbf{E}X_{i,\bar{s}}) \right| \leq \varepsilon(L(s) + r^2(\delta)).$$

By taking $\varepsilon = 1$ in Theorem 13.19 and defining r_* as the solution of the equation $\sqrt{n}r^2 = \psi(\rho(r))$, we find that with probability of at least $1 - 2\delta$, for all $s \in \mathcal{T}$

$$\left| \sum_{i=1}^{n} \frac{1}{n}(X_{i,s} - \mathbf{E}X_{i,s} - X_{i,\bar{s}} + \mathbf{E}X_{i,\bar{s}}) \right| \leq L(s) + 130r_*^2 + 4\frac{\log 1/\delta}{n}.$$

This follows in a straightforward manner by observing that if $\varepsilon = 1$,

$$r(\delta)^2 \leq 130r_*^2 + 4\frac{\log 1/\delta}{n}.$$

The proof of this is left to the reader (see Exercise 13.42).

Proof Denote $\overline{X}_{i,s} = X_{i,s} - \mathbf{E}X_{i,s}$ for all $i \leq n$ and $s \in \mathcal{T}$. We prove that with probability at least $1 - \delta$, for all $s \in \mathcal{T}$,

$$\sum_{i=1}^{n} \frac{1}{n}(\overline{X}_{i,s} - \overline{X}_{i,\bar{s}}) \leq \varepsilon(L(s) + r^2(\delta)).$$

A similar argument can be applied to prove that with probability at least $1 - \delta$,

$$\sum_{i=1}^{n} \frac{1}{n}(\overline{X}_{i,\bar{s}} - \overline{X}_{i,s}) \leq \varepsilon(L(s) + r^2(\delta)).$$

Let r be such that $\sqrt{n}r^2 > \psi(r)$. Define the random variable

$$V_r = \sup_{s \in \mathcal{T}} r^2 \frac{\sum_{i=1}^{n}(1/n)(\overline{X}_{i,s} - \overline{X}_{i,\bar{s}})}{L(s) + r^2}.$$

Then V_r is the supremum of a centered empirical process indexed by $\mathcal{T}$. Moreover, as $L(s) \leq r^2$, we have $\sigma(s) \leq \rho(r)$, and therefore, by the assumption of the theorem,

$$\mathbf{E} \sup_{\substack{s \in \mathcal{T} \\ L(s) \leq r^2}} \left| \sum_{i=1}^{n} \frac{1}{n}(\overline{X}_{i,s} - \overline{X}_{i,\bar{s}}) \right| \leq \frac{\psi(\rho(r))}{\sqrt{n}}.$$

Since the class of sub-linear functions is closed by composition, by Lemma 13.18,

$$\mathbf{E}V_r \leq 4\frac{\psi(\rho(r))}{\sqrt{n}}.$$

Note that if $L(s) \leq r^2$,

$$\mathrm{Var}\left(\frac{r^2(\overline{X}_{i,s} - \overline{X}_{i,\bar{s}})}{(L(s) + r^2)}\right) \leq \mathbf{E}(X_{i,s} - X_{i,\bar{s}})^2 \leq \rho^2(r),$$

while for each $s \in \mathcal{T}$, as $2\sqrt{L(s)}r \leq L(s) + r^2$,

$$\mathrm{Var}\left(\frac{r^2(X_{i,s} - X_{i,\bar{s}})}{(L(s) + r^2)}\right) \leq \left(\frac{r\rho(\sqrt{L(s)})}{\sqrt{4L(s)}}\right)^2.$$

Thus if $L(s) \geq r^2$, by the sub-linearity of ρ,

$$\mathrm{Var}\left(\frac{r^2(X_{i,s} - X_{i,\bar{s}})}{L(s) + r^2}\right) \leq \frac{\rho^2(r)}{4}.$$

On the other hand, for all $i \le n, s \in \mathcal{T}$, almost surely

$$\left| r^2 \frac{(X_{i,s} - X_{i,\bar{s}})}{L(s) + r^2} \right| \le 1.$$

Now we may use Bousquet's inequality (Theorem 12.5) to conclude that, with probability at least $1 - \delta$,

$$V_r \le EV_r + \sqrt{\frac{2}{n}(2EV_r + \rho^2(r)) \log \frac{1}{\delta}} + \frac{1}{3n} \log \frac{1}{\delta}$$

$$\le 2EV_r + \rho(r)\sqrt{\frac{\log \frac{1}{\delta}}{n}} + \frac{4}{3}\frac{\log \frac{1}{\delta}}{n}$$

$$\le 8\frac{\psi(\rho(r))}{\sqrt{n}} + \rho(r)\sqrt{\frac{\log \frac{1}{\delta}}{n}} + \frac{4}{3}\frac{\log \frac{1}{\delta}}{n}.$$

Using the definition of $r(\delta)$, this implies that, with probability at least $1 - \delta$, for all $s \in \mathcal{T}$,

$$\sum_{i=1}^{n} \frac{1}{n}(\overline{X}_{i,s} - \overline{X}_{i,\bar{s}}) \le \varepsilon \left(L(s) + r^2(\delta)\right).$$ □

13.8 Relative Deviations for L_2 Distances

This section describes an easy application of the peeling/reweighting technique presented in the previous section. Let $\mathcal{T}$ be a countable set and let $X_1, \ldots, X_n$ be independent identically distributed vector-valued random variables where $X_i = (X_{i,s})_{s \in \mathcal{T}}$. We may define a metric d on $\mathcal{T}$ by $d(s, s') = \left(E(X_{1,s} - X_{1,s'})^2\right)^{1/2}$. The basic question we investigate here is how well the random empirical metric $d_n(s, s') = \left(\sum_{i=1}^{n}(1/n)(X_{i,s} - X_{i,s'})^2\right)^{1/2}$ approximates the metric d.

The next theorem reveals that if the subset of $\mathcal{T}$ formed by those $s \in \mathcal{T}$ with small values of $EX_{1,s}^2$ is not too "rich," the empirical metric space $(\mathcal{T}, d_n)$ faithfully approximates $(\mathcal{T}, d)$, at least above a certain scale.

Theorem 13.20 *Let $X_1, \ldots, X_n$ be defined as above and suppose $EX_{i,s} = 0$ and $|X_{i,s}| \le 1$ almost surely for all $i = 1, \ldots, n$ and $s \in \mathcal{T}$. Assume that there exists $\bar{s} \in \mathcal{T}$ for which $X_{i,\bar{s}} = 0$ almost surely. Assume that there exists a sub-linear function ϕ such that for all $r \ge 0$ such that $\sqrt{n}r^2 \ge \phi(r)$,*

$$\sqrt{n}E \sup_{s \in \mathcal{T}, EX_{1,s}^2 \le r^2} \frac{1}{n} \left| \sum_{i=1}^{n} X_{i,s} \right| \le \phi(r).$$

Let $\varepsilon \in (0,1)$ and let $r(\delta) > 0$ be the unique solution of the equation

$$\sqrt{n}r^2 = \frac{1}{\varepsilon}\left(128\phi(r) + r\sqrt{\log 1/\delta} + \frac{4\log(1/\delta)}{3\sqrt{n}}\right).$$

Then, with probability of at least $1-2\delta$, for all $s \in \mathcal{T}$,

$$\left|\frac{\frac{1}{n}\sum_{i=1}^{n} X_{i,s}^2}{\mathbf{E}X_{1,s}^2} - 1\right| \le \varepsilon\left(1 + \frac{r(\delta)^2}{\mathbf{E}X_{1,s}^2}\right).$$

The theorem implies that if r_* is defined as the positive solution of $\sqrt{n}r^2 = 8\phi(r)$, then with probability of at least $1-2\delta$, for all $s \in \mathcal{T}$ such that $\mathbf{E}X_{i,s}^2 \ge 130r_*^2/\varepsilon^2 + (4/n)\log(1/\delta)$, one has

$$\left|\frac{\frac{1}{n}\sum_{i=1}^{n} X_{i,s}^2}{\mathbf{E}X_{1,s}^2} - 1\right| \le 2\varepsilon.$$

Proof The proof is a simple application of Theorem 13.19 to the family of random variables $\left\{X_{i,s}^2 \,:\, s \in \mathcal{T}, i \le n\right\}$. Note first that

$$\sup_{\substack{s\in\mathcal{T}\\ i\le n}} \left|X_{i,s}^2 - \mathbf{E}X_{i,s}^2\right| \le 1.$$

Second, choosing $L(s) = \sigma^2(s) = \mathbf{E}X_{i,s}^2$, L is minimized by $\bar{s}$ while letting $\rho(r) = r$ for all $r \ge 0$, $\mathbf{E}\left[(X_{i,s}^2)^2\right] \le \rho^2(L(s)^{1/2})$. The only point that needs to be checked is that

$$\sqrt{n}\mathbf{E} \sup_{s\in\mathcal{T},\sigma^2(s)\le r^2} \frac{1}{n}\left|\sum_{i=1}^{n} X_{i,s}^2 - \mathbf{E}X_{i,s}^2\right| \le 8\phi(r).$$

However, this follows by symmetrization (Lemma 11.4) and by the contraction principle (Lemma 11.6) as $x \mapsto x^2$ is 2-Lipschitz over $[-1,1]$. We may apply Theorem 13.19 on $\left\{(X_{i,s}^2 \,:\, s \in \mathcal{T}, i \le n\right\}$ with $\psi = 16\phi$. □

13.9 Risk Bounds in Classification

We close this chapter by describing an application of the techniques introduced in Section 13.7 to construct risk bounds for empirical risk minimization in binary classification. The classification problem is at the heart of statistical learning theory and its analysis served as a driving force for the development of empirical process theory. Here we present just a sample from the rich theory of classification. We apply Theorem 13.19 to obtain sharp bounds for the risk of a classifier that minimizes the empirical risk over a VC class of candidate classifiers.

The setup is described as follows. In binary classification the *observation* X is a random variable taking values in some set $\mathcal{X}$ and its binary *label* Y is a $\{0,1\}$-valued random variable. The joint distribution of X and Y is denoted by P. A *classifier* is a measurable function $s : \mathcal{X} \to \{0,1\}$. The *risk* of classifier s is $\mathbf{P}\{Y \neq s(X)\}$. The so-called *Bayes classifier* $s^*(X) = \mathbb{1}_{\{\mathbf{E}[Y|X] \geq 1/2\}}$ minimizes the risk among all possible classifiers.

In statistical learning, the joint distribution P is unknown but a sample $(X_1, Y_1), \ldots, (X_n, Y_n)$ of independent pairs, distributed according to P, is available. Given a collection $\mathcal{T}$ of classifiers, one may choose $\hat{s} \in \mathcal{T}$ by minimizing the *empirical risk* $\sum_{i=1}^n \mathbb{1}_{\{Y_i \neq s(X_i)\}}$ over $s \in \mathcal{T}$. In this section we work with the simplifying, and perhaps unrealistic, assumption that $s^* \in \mathcal{T}$, that is, the Bayes classifier is in the class of candidate classifiers. The performance of the empirical risk minimizer is measured by the excess risk $\ell(s, s^*) = \mathbf{P}\{Y \neq s(X)\} - \mathbf{P}\{Y \neq s^*(X)\}$. Letting $\eta(X) = \mathbf{E}[Y|X]$, it is straightforward to verify that $\ell(s, s^*) = \mathbf{E}\left[|2\eta(X) - 1||s(X) - s^*(X)|\right]$. We also assume that the collection of sets $\{\{x \in \mathcal{X} : s(x) = 1\} : s \in \mathcal{T}\}$ is a VC-class with VC-dimension V. Introducing $Z_{i,s} = \mathbb{1}_{\{Y_i \neq s(X_i)\}}$ for $i \leq n$ and $s \in \mathcal{T}$, the set of classifiers is endowed with the pseudo-metric $d(s,t) = \sqrt{\mathbf{E}[(Z_{1,s} - Z_{1,t})^2]}$.

Bounds on excess loss depend on the richness of $\mathcal{T}$, the sample size n, but also on how "noisy" the observations are. One way to quantify a "low-noise" assumption is by the *Mammen–Tsybakov noise conditions* according to which there exist $h \in [0,1]$ and $\theta \geq 1$ such that

$$\ell(s, s^*) \geq h^\theta d^{2\theta}(s, s^*), \text{ for all } s \in \mathcal{T}.$$

The simplest and strongest condition belongs to the case $\theta = 1$. In this case one has $|2\eta(X) - 1| = h$ almost surely.

Theorem 13.21 (RISK BOUNDS FOR VC CLASSES) *Assume that $\hat{s}$ minimizes the empirical risk on a sample of size n over a VC-class $\mathcal{T}$ of VC-dimension V. Assume that the Bayes classifier belongs to $\mathcal{T}$, and the Mammen–Tsybakov noise condition is satisfied by some $h > 0$ and $\theta \geq 1$. Then*

$$\mathbf{E}\ell(\hat{s}, s^*) \leq \kappa \left(\frac{V(1 + \log(nh^{2\theta}/V))}{nh} \right)^{\theta/(2\theta - 1)},$$

where κ is a universal constant that does not depend on $n, \mathcal{T}, h, \theta$.

Proof The proof is based on an application of Theorems 13.19 and 13.7. Since $\sum_{i=1}^n Z_{i,\hat{s}} \leq \sum_{i=1}^n Z_{i,s^*}$,

$$\ell(\hat{s}, s^*) \leq \frac{1}{n} \sum_{i=1}^n (Z_{i,s^*} - \mathbf{E}Z_{i,s^*} - Z_{i,\hat{s}} + \mathbf{E}Z_{i,\hat{s}}).$$

Thus, the excess risk is bounded by the oscillation of the centered empirical risk process between s^* and $\hat{s}$.

Let $\phi \colon [0,1] \to \mathbb{R}_+$ be defined by

$$\phi(r) = 72r\sqrt{V \log\left(\frac{4e}{r}\right)}.$$

By Theorem 13.7, for $r \in [0,1]$,

$$\mathbf{E} \sup_{s \in \mathcal{T} : d(s,s^*) \leq r} \left| \sum_{i=1}^{n} (Z_{i,s^*} - \mathbf{E}Z_{i,s^*} - Z_{i,s} + \mathbf{E}Z_{i,s}) \right| \leq \phi(r).$$

One may easily verify that ϕ is sub-linear.

We now apply Theorem 13.19. Since $h^{-1/2}(\sqrt{\ell(s,s^*)})^{1/\theta} \geq d(s,s^*)$, we may choose $\rho(r) = h^{-1/2} r^{1/\theta}$. As $0 \leq \theta \leq 1$, ρ is sub-linear.

Let r_* be the nonnegative solution of $\sqrt{n}r^2 = \phi(\rho(r))$. By Theorem 13.19, with probability at least $1 - 2\exp(-x)$, for all $s \in \mathcal{T}$,

$$\left| \sum_{i=1}^{n} (Z_{i,s^*} - \mathbf{E}Z_{i,s^*} - Z_{i,s} + \mathbf{E}Z_{i,s}) \right| \leq \varepsilon \left(\ell(s,s^*) + 130 r_*^2 + 4\frac{x}{n} \right).$$

Thus, with probability at least $1 - 2e^{-x}$,

$$\ell(\hat{s},s^*) \leq \varepsilon \left(\ell(\hat{s},s^*) + 130 r_*^2 + 4\frac{x}{n} \right).$$

Rearranging, we obtain

$$\ell(\hat{s},s^*) \leq \frac{\varepsilon}{1-\varepsilon} \left(130 r_*^2 + 4\frac{x}{n} \right).$$

Integrating with respect to x leads to

$$\mathbf{E}\ell(\hat{s},s^*) \leq \frac{\varepsilon}{1-\varepsilon} \left(130 r_*^2 + \frac{4}{n} \right).$$

It remains to upper bound r_*^2. From

$$\sqrt{n}r_*^2 = 72 h^{-1/2} r_*^{1/\theta} \sqrt{V \log\left(\frac{4e}{h^{-1/2} r_*^{1/\theta}} \right)},$$

as $r_* \leq 1$, we may deduce

$$r_* \geq \left(72\sqrt{\frac{V}{nh} \log\left(\frac{4e}{h^{-1/2}}\right)} \right)^{\theta/(2\theta-1)}.$$

This entails

$$r_*^2 \leq \left(K^2 \frac{V}{nh} \log \left(\frac{4e}{h^{-1/2} \left(K\sqrt{\frac{V}{nh}} \log \left(\frac{4e}{h^{-1/2}} \right) \right)^{1/(2\theta-1)}} \right) \right)^{\theta/(2\theta-1)} . \qquad \square$$

13.10 Bibliographical Remarks

The notion of metric entropy was introduced by Kolmogorov and Tikhomirov (1961) to quantify the performance of non-linear approximation methods in functional analysis. We refer to the books by DeVore and Lorentz (1993) and Lorentz, Golitschek, and Makovoz (1996) for an in-depth exposition of entropic arguments in approximation theory.

The idea of chaining in order to upper bound the supremum of a Brownian motion was initiated by Kolmogorov (see Slutsky (1937) and Čentsov (1956)). In the context of general Gaussian processes, chaining was introduced by Dudley (1967) in order to provide a sufficient condition for the existence of an almost surely continuous version of a Gaussian process. Since Dudley (1967), chaining provided a generic method to derive tail bounds for suprema of processes as suggested in Exercise 13.9. When dealing with suprema of bounded empirical processes, such bounds can be compared with bounds obtained by combining concentration inequalities and upper bounds for the expectation. The proof pattern used in Section 13.1 is due to Pisier (1983). A general approach would consist of using "majorizing measures" (also called "generic chaining") as introduced by Fernique (1975), rather than metric entropy. We refer to Ledoux and Talagrand (1991) and Talagrand (1994*b*, 1996*a*, 2005) for an extensive study of this topic.

Slepian's lemma (Theorem 13.3) first appears in Slepian (1962), but see also Fernique (1975) and Gordon (1985) for improvements and generalizations. The proof presented here is based on an argument presented by Chatterjee (2005*b*) (see also Piterbarg (1982)). Sudakov's inequality (Theorem 13.4) is from Sudakov (1969). Li and Shao (2001) survey many related inequalities for Gaussian processes.

In empirical process theory, arguments based on uniform entropy numbers were pioneered by Koltchinskii (1981) and Pollard (1984). VC classes of sets were introduced by Vapnik and Chervonenkis (1971). Uniform bounds on L_2 covering numbers for VC-classes were first obtained by Pollard (1982, 1984) and Dudley (1987) (see Exercise 13.11). Lemma 13.6 was proved by Haussler (1995) using combinatorial properties of traces of VC-classes that were first established by Haussler, Littlestone, and Warmuth (1994). Conditions on uniform entropy numbers that generalize those satisfied by VC classes of sets play an important role in the analysis of functional central limit theorems (see van der Vaart and Wellner (1996)). The concept of a VC-class of sets can be used to define VC subgraph classes of functions and VC-major classes (see Exercises 13.13, 13.14). An analog of the VC dimension, called the *fat-shattering dimension* for classes of functions was introduced by Kearns and Schapire (1994) (see Anthony and Bartlett (1999) for a survey and Mendelson and Vershynin (2003) where the relevant generalization of Lemma 13.6 is established).

Upper bounds on the expected value of suprema of empirical prcesses indexed by classes of functions with regularly varying uniform entropy numbers can be found in Talagrand (1994*b*), Mendelson (2002*b*), Giné and Koltchinskii (2006), and Koltchinskii, (2008) (see Exercise 13.18).

Lemma 13.8 was first proved by Rudelson (1999) using non-commutative Khinchine inequalities due to Lust-Piquard and Pisier (1991). The approach described in Section 13.4 was pioneered by Ahlswede and Winter (2002). The presentation given here follows Imbuzeiro Oliveira (2010). Alternative proofs using Lieb's concavity theorem may be found in Tropp (2010*a*, 2010*b*). For a general treatment of matrix inequalities, we recommend Bhatia (1997). Exercises 13.31, 13.32, and 13.33 describe how Rudelson's inequality and concentration inequalities for suprema of empirical processes can be combined in order to establish concentration inequalities for operator norms of sums of random symmetric matrices.

Theorem 13.15 is due to Klartag and Mendelson (2005) who actually proved a stronger result since they were able to replace the functional $\gamma(T)$ that comes from classical chaining by Talagrand's $\gamma_2(T)$ which is obtained by generic chaining (Talagrand, 2005) and is known to sharply characterize the expected value of suprema of Gaussian processes. A purely Gaussian version of this result had previously been established by Gordon (1988). The restricted isometry property described in Corollary 13.14 was introduced by Candès, Romberg, and Tao (2006). Its proof via Lemma 13.13 is due to Baraniuk *et al.* (2008).

Mendelson, Pajor, and Tomczak-Jaegermann (2007) go beyond the scope of Theorem 13.15 and attempt to control $\sum_{s\in\mathcal{T}}(1/n)\sum_{i=1}^{n} X_{i,s}^2 - 1$ where $(X_i)_{i\leq n}$ are independent random vectors with $EX_{i,s} = 0$ and $EX_{i,s}^2 = 1$ for all $s \in \mathcal{T}$, by simply assuming that $X_{i,s} - X_{i,s'}$ is sub-Gaussian with variance factor proportional to the squared distance between s and s'. They provide an extension of Lemma 13.16.

Baraniuk and Wakin (2009) use the same device to perform dimensionality reduction of a manifold of smooth data using random linear projections. They establish an upper bound for the rank of random projections needed to guarantee that, with high probability, all pairwise Euclidean and geodesic distances between points on the manifold are approximately preserved.

Theorem 13.15 and Lemma 13.13 provide transparent proofs of upper bounds on the Gelfand numbers of ℓ_p^n balls derived by Kashin (1977).

The fact that random matrices with independent rows are almost isometric embeddings is central to the emerging field of compressed sensing (see Donoho (2006*a*) and Candès and Tao (2006)). This question is closely related to the control of the largest and smallest singular values of the random matrix $(X_{i,j})_{i\leq d, j\leq D}$ and has been further explored by Rudelson and Vershynin (2010). We refer the reader to Vershynin (2012) and references therein for more details on non-asymptotic results in the booming theory of random matrices.

Adamczak *et al.* (2010) consider extensions of the Johnson–Lindenstrauss problem in which the columns of the random projection matrix do not have i.i.d. sub-Gaussian coefficients but are rather sampled from a log-concave distribution (such as the uniform distribution over a convex body).

A host of useful bounds for the expected value of suprema of empirical and Rademacher processes can be found in van der Vaart and Wellner (1996), Giné and Guillou (2001),

Giné and Koltchinskii (2006), Giné, Koltchinskii, and Wellner (2003), and Massart (2006). Tail bounds for the Kolmogorov–Smirnov statistic have attracted considerable attention. Dvoretzky, Kiefer, and Wolfowitz (1956) were the first to obtain sub-Gaussian inequalities. Massart (1990) proved that

$$P\left\{\sup_{s\in\mathbb{R}} |(P_n - P)((-\infty, s])| \geq t\right\} \leq 2e^{-2nt^2}.$$

We refer to Shorack and Wellner (1986) for classical results on the oscillations of the empirical process indexed by half-lines. The analysis of the modulus of oscillation of empirical processes indexed by general classes of sets goes back to the work of Alexander (1987) (see also van de Geer (2000) for applications to M-estimation). The impact of concentration inequalities on this topic is thoroughly investigated in Giné, Koltchinskii, and Wellner (2003), Giné and Koltchinskii (2006), Massart (2000*b*), Massart and Nédélec (2006), Massart (2006), and Bartlett and Mendelson (2006). Giné, Koltchinskii, and Wellner (2003, 2006) consider different re-weighting techniques. Indeed they normalize $(P_n - P)f_s$ by its standard deviation $\sigma(f_s) = (Pf_s^2 - (Pf_s)^2)^{1/2}$ rather than its variance (see also Bartlett and Mendelson (2006)). This approach allowed them to investigate moduli of continuity of empirical processes, that is, quantities like

$$\sup_{\mathcal{T}_{[r,r')}} \frac{|(P - P_n)f_s|}{\omega(\sigma(f_s))}$$

where $\mathcal{T}_{r_n,r'_n} = \{s : \sigma(f_s) \in [r_n, r'_n)\}$ and ω is some positive nondecreasing function.

Giné, Koltchinskii, and Wellner (2003) and Giné and Koltchinskii (2006) describe improvements which take into account the $L_2(P)$ norm of the envelop of the class.

Versions of Theorems 13.19 and 13.20 can be found in Massart (2000*b*, 2006), Koltchinskii (2006), and Bartlett and Mendelson (2006) The version presented here follows Boucheron, Bousquet, and Lugosi (2005*a*).

Theorem 13.21 was defined by Massart and Nédélec (2006). Matching lower bound for risk estimates can be found in this paper. For surveys on the classification problem, we refer the reader to Devroye, Györfi, and Lugosi (1996) and Boucheron, Bousquet, and Lugosi (2005*a*).

13.11 EXERCISES

The chaining idea

13.1. (ENTROPY NUMBERS AND ε-ENTROPY) Considering entropy numbers rather than packing numbers provides an alternative approach to chaining. Sticking to the notation of Section 13.1, define the nth entropy number $e_n(S)$ for $n \in \mathbb{N}$, as

$$e_n(S) = \inf\left\{\varepsilon : N(\varepsilon, S) \leq 2^{2^n}\right\} = \inf\left\{\varepsilon : H(\varepsilon, S) \leq 2^n \log(2)\right\}.$$

Using the notation of Lemma 13.1, prove that

$$\boldsymbol{E}\left[\sup_{s\in S} X_s - X_{s_0}\right] \leq a\delta + \sqrt{v}\sum_{j=1}^{\infty} e_n(S)2^{n/2} + c\sum_{j=1}^{\infty} e_n(S)2^n.$$

Hint: for $n \in \mathbb{N}$, let S_n be a $e_n(S)$-packed subset of S with maximal cardinality. By the definition of $e_n(S)$, $|S_n| \leq 2^{2^{n+1}}$. For $n \geq 1$, let Π_n map each $s \in S$ on a nearest-neighbor in S_{n-1} and let $\Pi_0(s) = s_0$. (See Talagrand (2005).)

13.2. (ANOTHER LOOK AT COROLLARY 2.6) Let $X_1, \ldots, X_n$ be independent random vectors indexed by $\mathcal{T}$. Let $\varepsilon_1, \ldots, \varepsilon_n$ be independent Rademacher variables. Assume $\max_{i\leq n} \sup_{s\in\mathcal{T}} |X_{i,s}| \leq 1$, $\boldsymbol{E}X_{i,s} = 0$ and let $\sigma^2 \geq \sup_{s\in\mathcal{T}} \sum_{i=1}^n \boldsymbol{E}[X_{i,s}^2]/n$. Prove that there exists a universal constant κ such that:

$$\boldsymbol{E}\sup_{s\in\mathcal{T}}\left|\sum_{i=1}^{n} \varepsilon_i X_{i,s}\right| \leq \kappa \max\left(\sigma\sqrt{n\log|\mathcal{T}|}, \log|\mathcal{T}|\right).$$

Hint: use the contraction principle and the maximal inequality for sub-Gaussian random variables (Theorem 2.5). (See Koltchinskii, 2008.)

13.3. (STEIN'S INTEGRATION-BY-PARTS FORMULA) Prove that if $F : \mathbb{R}^n \to \mathbb{R}$ is continuously differentiable such that for any $a > 0$, $\lim_{\|x\|\to\infty} F(x)\exp(-a\|x\|^2) = 0$ and $X = (X_1, \ldots, X_d)$ is a centered Gaussian vector, then for any $1 \leq i \leq d$,

$$\boldsymbol{E}\left[X_i F(X)\right] = \sum_{j=1}^{d} \boldsymbol{E}[X_i X_j]\boldsymbol{E}\left[\frac{\partial F}{\partial x_j}(X)\right].$$

Hint: use integration-by-parts to establish the formula in dimension 1, then proceed by conditioning. That this is a characteristic property of Gaussian vectors is at the core of Stein's approach to prove central limit theorems; see Chatterjee and Dey (2010), Chatterjee (2005*a*) and references therein.

13.4. (CHAINING AND ITS LIMITATIONS) Let $Y_1, \ldots, Y_n, \ldots$ be a countable collection of independent standard Gaussian random variables. Let

$$Z = \sup_{i=1,2,\ldots} Y_i/\sqrt{\log\max(i,2)}.$$

The random variable Z is the supremum of a Gaussian process. The natural distance associated with this Gaussian process is $d(s,s') = (1/\log s + 1/\log s')^{1/2}$ for $s \neq s', s, s' \geq 2$. Check that for $\delta < 1/2$, the δ-entropy of the index set $\mathbb{N}$ satisfies $\kappa'/\delta^2 \geq H(\delta, \mathbb{N}) \geq \kappa/\delta^2$ for some constants κ, κ'. What kind of upper bound on $\boldsymbol{E}Z$ can be deduced from Corollary 13.2? What kind of lower bound can be deduced from Sudakov's lower bound (Theorem 13.4)? Prove that $\boldsymbol{E}Z < \infty$ (10 is a plausible and generous upper bound). *Hint:* use the fact that for $m \geq 2$,

$\max_{i:1\le i/2^m\le 2} Y_i/\sqrt{\log i} \le \max_{i:1\le i/2^m\le 2} |Y_i|/\sqrt{m\log 2}$. Derive tail bounds for the latter quantity. Use the union bound. See Talagrand (1996*a*, 2005). Note that the random variables $Y_i/\sqrt{\log\max(i,2)}$ have very different variances and that deriving sharp upper bounds for the expectation of the supremum relies on slicing the family of random variables into pieces with similar variances, and computing tight bounds for suprema over the slices.

13.5. (GAUSSIAN PROCESSES INDEXED BY ELLIPSOIDS) Let $X_1,\ldots,X_n$ be independent standard Gaussian random variables. Let $(a_1,\ldots,a_n)\in\mathbb{R}^n$, with $a_i>0$ for $1\le i\le n$. Let $\mathcal{T}=\{s=(s_1,\ldots,s_n):\sum_{i=1}^n s_i^2/a_i^2\le 1\}$. Let $Z=\sup_{s\in\mathcal{T}}\sum_{i=1}^n s_iX_i$. Prove that $\boldsymbol{E}Z\le(\sum_{i=1}^n a_i^2)^{1/2}$. Prove also that

$$\boldsymbol{E}\sup_{s,s'\in\mathcal{T},\|s-s'\|\le c}\sum_{i=1}^n (s_i-s_i')X_i\le\sqrt{8\sum_{i=1}^n\min(a_i^2,c^2)}.$$

What upper bound does Theorem 13.2 imply in this case? (See Talagrand (1996*a*) for a discussion.)

13.6. (EXPECTATION OF A MAXIMUM OF INDEPENDENT GAUSSIAN RANDOM VARIABLES) Let $(X_s)_{s\in\mathcal{T}}$ be independent standard Gaussian random variables. Prove that for $|\mathcal{T}|\ge 2$,

$$\boldsymbol{E}\max_{s\in\mathcal{T}} X_s\ge\frac{1}{\sqrt{2}}\sqrt{\log|\mathcal{T}|}.$$

13.7. (OPERATOR NORM OF A GAUSSIAN MATRIX) Let $X=(X_{i,j})_{1\le i,j\le n}$ be a standard Gaussian vector considered as an $n\times n$ matrix. (This is sometimes called the *Ginibre ensemble.*) Let $Y=(Y_i)_{i\le n}$ be a standard Gaussian vector and let Y' be an independent copy of Y. Let $K=S^{n-1}\times S^{n-1}\subset\mathbb{R}^{2n}$ be the set of pairs of unit vectors from $\mathbb{R}^n$. Let $Z=\sup_{(u,v)\in K}u^TXv$ be the operator norm of X and let $U=\sup_{(u,v)\in K}(\langle u,Y\rangle+\langle v,Y'\rangle)$ be the sum of two independent isonormal processes. Prove that

$$\boldsymbol{E}Z\le\boldsymbol{E}U\le 2\sqrt{n}.$$

Hint: the second inequality is a special case of the bound obtained in Exercise 13.5. The first inequality can be obtained using Theorem 13.3. Note that $u^TXv=\text{trace}(Xvu^T)$. Check that for (u,v) and (s,t) in K, $\boldsymbol{E}\left[\left(u^TXv-s^TXt\right)^2\right]=\|vu^T-ts^T\|_{\text{HS}}^2$ and $\left\|vu^T-ts^T\right\|_{\text{HS}}^2\le\|u-s\|^2+\|v-t\|^2$ (see Davidson and Szarek (2001)).

13.8. (THE LARGEST EIGENVALUE OF A RANDOM MATRIX DISTRIBUTED AS THE GUE) Recall the definition of the GUE from Section 5.10. The largest eigenvalue of an $n\times n$ random matrix from the GUE is the supremum of a Gaussian process indexed by the unit sphere $S^{n-1}=\{u\in\mathbb{C}^n:\|u\|^2=1\}$:

$$Z=\sup_{u\in S^{n-1}}u^*Xu=\sup_{u\in S^{n-1}}\text{trace}(Xuu^*),$$

where u^* is the conjugate transpose of the complex column vector u. Use chaining or a comparison argument as in Exercise 13.7 to prove that $\boldsymbol{E}Z \le C$ for some universal constant $C \ge 1$. *Hint:* check that for $u, v \in S^{n-1}$,

$$\boldsymbol{E}\left[(u^*Xu - v^*Xv)^2\right] = \frac{1}{n}\left\|uu^* - vv^*\right\|_{\mathrm{HS}}^2,$$

and that $\|uu^* - vv^*\|_{\mathrm{HS}}^2 \le 4\|u - v\|^2$.

13.9. (TAIL BOUNDS VIA CHAINING) Let $(\mathcal{T}, d)$, $(X_s)_{s\in\mathcal{T}}$, a, v, c, δ, and s_0 be defined as in Theorem 13.1. Let

$$E = 3a\delta + 6\delta\sum_{j=1}^{\infty} 2^{-j}\left(\sqrt{v\left(j + 2H(\delta_j)\right)} + c\left(j + 2H(\delta_j)\right)\right).$$

Prove that there exists a universal constant κ such that for any $u > 1$

$$\boldsymbol{P}\left\{\sup_{s\in S} X_s - X_{s_0} \ge uE\right\} \le \kappa e^{-u}.$$

Hint: the event

$$\{\exists s \in \mathcal{T}; X_s - X_{s_0} \ge uE\}$$

is included in

$$\bigcup_{s\in\mathcal{T}}\bigcup_{j=0}^{J}\left\{X_{\Pi_{j+1}s} - X_{\Pi_j s} \ge \delta_{j+1}\left(3ua + 2\sqrt{uv(j + 2H(\delta_{j+1}))} + cu\left(j + 2H(\delta_{j+1})\right)\right)\right\}.$$

The probability of the events on the right side are upper bounded by

$$\exp\left(2H(\delta_{j+1})\right)\exp\left(-u(j + 2H(\delta_{j+1}))\right).$$

Up to a constant, E is not larger than the upper bound on expectation described in Lemma 13.1.

13.10. (CHAINING, FAMILIES OF DISTANCES) Recall the definition of a Gaussian chaos of order two from Example 2.12. Let $X = (X_1, \ldots, X_n)$ be a standard Gaussian vector. Let $\mathcal{T}$ be a collection of symmetric $n \times n$ real matrices with zeroes in the diagonal entries. Let $Z = \sup_{A\in\mathcal{T}} X^T(A - A_0)X$ for some fixed $A_0 \in \mathcal{T}$. Let $\|A\|_{\mathrm{op}}$ and $\|A\|_{\mathrm{HS}}$ be the operator and Hilbert–Schmidt norms of A. Let $H_{\mathrm{HS}}(u), H_{\mathrm{op}}(u)$ denote the u-entropy of $\mathcal{T}$ under the two norms. Let

$$E = \int_0^{\delta_{\mathrm{HS}}/2}\sqrt{2H_{\mathrm{HS}}(u, \mathcal{T})}du + \int_0^{\delta_{\mathrm{op}}/2} 2H_{\mathrm{op}}(u, \mathcal{T})du.$$

Prove that there exists a universal constant κ such that for any $u > 1$,

$$P\{Z \geq uE\} \leq \kappa e^{-u}.$$

Hint: for each $n = 1, 2, \ldots$, let $\varepsilon_{op,n}$ be defined by $\inf\left\{\varepsilon : H_{\text{op}}(\varepsilon, \mathcal{T}) \leq 2^n\right\}$ and define $\varepsilon_{\text{HS},n}$ similarly. Let $\mathcal{T}_{op,n}$ and $\mathcal{T}_{\text{HS},n}$ be the corresponding $\varepsilon_{op,n}$– and $\varepsilon_{\text{HS},n}$–nets. For each n, $\mathcal{T}_{op,n}$ and $\mathcal{T}_{\text{HS},n}$ define two partitions $\mathcal{B}_n$ and $\mathcal{C}_n$ of $\mathcal{T}$: the cell associated with an element $A \in \mathcal{T}_{op,n}$ is

$$\left\{A' : A' \in \mathcal{T}, \|A' - A\|_{\text{op}} = \min_{M \in \mathcal{T}_{op,n}} \|M - A'\|\right\}.$$

The cells of $\mathcal{C}_n$ are defined in a similar way. For each n, the partition $\mathcal{A}_n$ is obtained by intersecting the cells of $\mathcal{B}_{n-1}$ and $\mathcal{C}_{n-1}$. For each n, let Π_n map each $A \in \mathcal{T}$ to a distinguished element of A's cell in $\mathcal{A}_n$. Let J be the smallest index n such that $\mathcal{A}_n$ is trivial. Check that for $u > 1$, $\{Z \geq uE\}$ is included in

$$\bigcup_{A \in \mathcal{T}} \bigcup_{j=0}^{J-1} \left\{X_{\Pi_{j+1}s} - X_{\Pi_j s} \geq \left(2\delta_{\text{HS},j-1}\sqrt{u\left(j + 2^{j+1}\log 2\right)} + 2u\delta_{op,j-1}\left(j + 2^{j+1}\log 2\right)\right)\right\}.$$

We refer to Talagrand (2005, Theorem 1.2.7, Section 2.5, and Chapter 5) for a thorough discussion of generic chaining. Generic chaining provides tails bounds for suprema of processes whose increments are controlled by a family of distances. Talagrand discusses the possibility and the difficulties of deriving matching lower bounds.

VC classes and uniform entropy bounds

13.11. (A SUBOPTIMAL BOUND ON THE METRIC ENTROPY OF VC-CLASSES) Using the notation of Lemma 13.6 prove that for all $\kappa > 1$,

$$H(\delta, \mathcal{T}) \leq \frac{\kappa}{\kappa - 1} V \log\left(\frac{2\kappa}{V\delta^2}\right).$$

Hint: assume there exist N elements $A_1, \ldots, A_N$ of the VC-class $\mathcal{T}$ that are δ-separated under $L_2(Q)$ where Q is a probability distribution over $\mathcal{X}$. Pick $n = 2\log N^2/\delta^2$ samples independently at random according to Q. Show that the VC-entropy of $\mathcal{T}$ on this sample is at least $\log N$ while, according to Sauer's lemma it is also less than $V\log(en/V)$ if $n \geq V$ where V is the VC-dimension of $\mathcal{T}$ in $\mathcal{X}$. Proofs of Sauer's lemma can be found in Sauer (1972), Frankl (1983), Bollobás (1986), Ledoux and Talagrand (1991). See Haussler (1995) and also Pollard (1990).

13.12. (DENSITY OF 1-INCLUSION GRAPHS) Let $\mathcal{C}$ be a collection of subsets of $\mathcal{X}$. Let $x = (x_1, \ldots, x_n)$ be a sample of n (not necessarily distinct) elements from $\mathcal{X}$. Recall that the trace of $\mathcal{C}$ on x is defined as

$$\text{tr}(x) = \{J : J \subseteq \{1, \ldots, n\}, \exists A \in \mathcal{C}, \forall i \in \{1, \ldots, n\}, i \in J \Leftrightarrow x_i \in A\}.$$

As the trace $\text{tr}(x)$ may be identified as a subset of $\{0,1\}^n$, it induces a subgraph of the n-cube. This subgraph is called the 1-inclusion graph. The density $\text{dens}(x)$ of this subgraph is the ratio between the number of edges and the number of vertices. Prove that $\text{dens}(x) \leq \frac{1}{2}\log_2 |\text{tr}(x)|$. Prove that if $\mathcal{C}$ is a VC-class with VC-dimension V, then $\text{dens}(x) \leq V/2$. Is this bound tight? Let $X_1, \ldots, X_n$ be independent identically distributed $\mathcal{X}$-valued random variables. Let $Z = \text{dens}(X_1, \ldots, X_n)$. Prove that Z is a self-bounding random variable. 1-inclusion graphs were introduced in Haussler, Littlestone, and Warmuth (1994). Their combinatorial properties were used to investigate the behavior of some online classification algorithms. The analysis of 1-inclusion graphs proved also useful when deriving sharp bounds on the universal entropy of VC-classes.

13.13. (VC-SUBGRAPH CLASSES OF FUNCTIONS) The subgraph of a function $f : \mathcal{X} \to \mathbb{R}$ is the set $\{(x,t) : x \in \mathcal{X}, t \in \mathbb{R}, t < f(x)\}$. A class $\mathcal{F}$ of functions $\mathcal{X} \to \mathbb{R}$ is VC-subgraph with VC-dimension V if the collection of all subgraphs defined by choosing $f \in \mathcal{F}$ is a VC-class of subsets of $\mathcal{X} \times \mathbb{R}$ with VC-dimension V. For any probability measure Q on $\mathcal{X}$, let $\|f\|_Q$ be the $L_2(Q)$ norm of f. Prove that for any VC-subgraph class of functions with VC-dimension V and envelope F (i.e. $\sup_{f\in\mathcal{F}} |f(x)| \leq F(x)$ for all $x \in \mathcal{X}$), and any Q, the packing numbers of $\mathcal{F}$ with respect to the $L_2(Q)$-pseudometric satisfy

$$N\left(\delta \|F\|_Q, \mathcal{F}, Q\right) \leq \kappa (V+1)(16e)^{V+1} \left(\frac{1}{\delta}\right)^{2V}.$$

Hint: use $\mathbf{E}[|f(X) - f'(X)|] = Q \otimes \lambda\{(x,t) : f(x) \leq t < f'(x) \vee f'(x) \leq t < f(x)\}$ where X is distributed according to Q and also that $\mathbf{E}[|f(X) - f'(X)|^2] \leq \mathbf{E}[|f(X) - f'(X)| 2F(X)]$. Use Lemma 13.6. See van der Vaart and Wellner (1996, Chapter 2.6).

13.14. (VC-MAJOR CLASSES OF FUNCTIONS) A class $\mathcal{F}$ of real-valued functions defined on a set $\mathcal{X}$ is a VC-*major* class with VC-dimension V if the collection of all subsets $\{x : x \in \mathcal{X} f(x) > t\}$ defined by choosing $f \in \mathcal{F}$ and $t \in \mathbb{R}$ is a VC-class of subsets of $\mathcal{X}$ with VC-dimension V. Prove that a bounded VC-major class of functions is a multiple of the symmetric convex hull of a VC-class of indicator functions. *Hint:* use the fact that if $0 \leq f \leq 1$, $f(x) = \lim_{m\to\infty} \sum_{i=1}^m 1/m \mathbb{1}_{\{f(x)>i/m\}}$ (van der Vaart and Wellner, 1996, Chapter 2.6).

13.15. (REGULARLY VARYING UNIVERSAL ENTROPY NUMBERS AND SUPREMA OF EMPIRICAL PROCESSES) A measurable function $f : \mathbb{R}^+ \to \mathbb{R}$ is said to be *regularly varying* at ∞ with regular variation index $\alpha \in \mathbb{R}$, if $f(x) > 0$ for all sufficiently large x and for all $x \in \mathbb{R}^+$, $\lim_{t\to\infty} f(tx)/f(t) = x^\alpha$. Let $(A_s)_{s\in\mathcal{T}}$ be some countable class of measurable subsets of $\mathcal{X}$. Assume there exists a nonincreasing function $\psi : (0,1] \to \mathbb{R}$ such that $\psi(1/x)$ is regularly varying with regular variation index smaller than 2 such that

$$H(\delta, \mathcal{T}) \leq \psi(\delta)$$

where $H(\delta) = \sup_Q \log N(\delta, \mathcal{T}, Q)$ is the universal δ-metric entropy of $\mathcal{T}$ as defined in Section 13.3. Assume that $\sigma \in (0,1)$ is such that $P(A_s) \leq \sigma^2$, for every $s \in \mathcal{T}$. Let $X_1, \ldots, X_n$ drawn i.i.d. from the distribution P on $\mathcal{X}$ and let

$$Z_{\mathcal{T}}^{+} = \sup_{s \in \mathcal{T}} \frac{1}{\sqrt{n}} \sum_{i=1}^{n} (\mathbb{1}_{\{X_i \in A_s\}} - \boldsymbol{P}\{X_i \in A_s\})$$

$$Z_{\mathcal{T}}^{-} = \sup_{s \in \mathcal{T}} \frac{1}{\sqrt{n}} \sum_{i=1}^{n} (\boldsymbol{P}\{X_i \in A_s\} - \mathbb{1}_{\{X_i \in A_s\}}).$$

Then there exists a universal constant K (that may depend on ψ but not on σ) such that

$$\max\left(\boldsymbol{E}Z_{\mathcal{T}}^{-}, \boldsymbol{E}Z_{\mathcal{T}}^{+}\right) \leq K\sigma\sqrt{\psi(\sigma/2)}.$$

Hint: check first that

$$\boldsymbol{E}Z_{\mathcal{T}}^{+} \leq K\sqrt{\sigma^2 + \boldsymbol{E}Z_{\mathcal{T}}^{+}/\sqrt{n}} \int_0^{\sigma/2} \sqrt{\psi(u)}du.$$

Check that $\sqrt{\psi(1/x)}/x^2$ is regularly varying of index smaller than -1 and use Karamata's theorem (Bingham, Goldie, and Teugels, 1987) to deduce that

$$\lim_{\sigma \to 0^+} \int_0^{\sigma/2} \sqrt{\psi(u)}du = \frac{1}{1-\alpha/2} \frac{\sigma}{2} \sqrt{\psi(\sigma/2)}.$$

See also Giné, Koltchinskii, and Wellner (2003), and Giné and Koltchinskii (2006).

13.16. (EMPIRICAL PROCESSES INDEXED BY CLASSES OF BOUNDED FUNCTIONS) Consider the same setting as in Exercise 13.15, but instead of assuming that the functions indexed by $\mathcal{T}$ are $\{0,1\}$-valued, assume that they take their values in $[0,1]$. Derive comparable upper bounds for suprema of empirical processes. *Hint:* the only difficult part consists of upper bounding $\boldsymbol{E}\delta_n^2 = \boldsymbol{E}\sup_{s \in \mathcal{T}} P_n f_s^2$ as a function of σ^2 and $\boldsymbol{E}Z_{\mathcal{T}}^{+}$. Use symmetrization and the contraction principle to prove that $\boldsymbol{E}\delta_n^2 \leq \sigma^2 + 8\boldsymbol{E}Z_{\mathcal{T}}^{+}/\sqrt{n}$.

13.17. Assume $\{f_s : s \in \mathcal{T}\}$ is a pointwise separable class of functions mapping from $\mathcal{X} \to \mathbb{R}$ with envelope function g (i.e. $\forall x \in \mathcal{X}, \forall s \in \mathcal{T}, |f_s(x)| \leq g(x) \leq b$ where $b \in \mathbb{R}$). Assume there exists $\kappa > 0$ and a nonincreasing function $\psi : \mathbb{R}^+ \to \mathbb{R}$ such that $\psi(1/x)$ is regularly varying with index smaller than 2 and for all probability distribution Q,

$$\log(\delta, \mathcal{T}, Q) \leq \psi\left(\frac{\kappa \|g\|_{L_2(Q)}}{\delta}\right).$$

Let $X_1, \ldots, X_n$ be independently distributed according to P. Assume that $\boldsymbol{E}f_s(X)^2 \leq \sigma^2$ for all $s \in \mathcal{T}$. Let Z and Z^- be defined as in Theorem 13.7. Prove that there exists a universal constant κ' (that may depend on ψ but not on σ, P, or $\mathcal{T}$) such that

$$\max(EZ, EZ^-) \leq \kappa'\sigma \left(\psi \left(\frac{\kappa \|g\|_{L_2(P)}}{\sigma} \right) \right)^{1/2}$$

provided $\sigma \geq \sqrt{\psi(\kappa \|g\|_{L_2(P)}/\sigma)}$. This result generalizes Theorem 13.7 where the trivial constant envelope is used (see Theorem 3.1, Giné and Koltchinskii 2006).

13.18. (THE LOGARITHMIC FACTOR IN THEOREM 13.7) The Kolmogorov–Smirnov statistics is an example where the factor $\sqrt{\log(E/\sigma)}$ can be dropped in Theorem 13.7. Let $\mathcal{T} = \mathbb{Q} \cap [0, \sigma]$ for some $\sigma \in (0, 1]$. For $s \in \mathcal{T}$, let $A_s = [0, s]$. Let $U_1, \ldots, U_n$ be independently and uniformly distributed over $[0, 1]$ and let $X_{i,s} = \mathbb{1}_{\{U_i \in A_s\}} - s$. Check that $(A_s)_{s \in \mathcal{T}}$ is a VC class with VC-dimension 1. Prove that for $\sigma \in (0, 1]$,

$$E\left[\frac{1}{\sqrt{n}} \sup_{s \in [0, \sigma^2]} \sum_{i=1}^{n} X_{i,s} \right] \leq 4\sigma.$$

Hint: use the symmetrization inequalities (Theorem 11.4), Lemma 11.12, and Hoeffding's inequality. Note that Theorem 3.1 from Giné and Koltchinskii (2006) can be used to prove that the logarithmic term is not necessary. This can be done by choosing carefully the envelope function. This result can be generalized to multidimensional cumulative distribution functions.

13.19. (SUPREMA OF EMPIRICAL PROCESSES INDEXED BY BALLS IN HILBERT SPACES OF FUNCTIONS) Let $X, X_1, \ldots, X_n$ be independently distributed according to P. Let $L_2(P)$ be the set of functions such that $Ef(X)^2 < \infty$. Let $\mathcal{T}$ be a d-dimensional subspace of centered functions from $L_2(P)$. Prove that

$$E \sup_{\substack{f \in \mathcal{T} \\ E[f(X)^2] \leq R^2}} \sum_{i=1}^{n} f(X_i) \leq 2R\sqrt{nd}.$$

Hint: use symmetrization. See also Exercise 13.5. Note that if the unit ball of $\mathcal{T}$ has an envelope function F that satisfies $E \max_{i=1,\ldots,n} F(X_i)^2 \leq nd/\kappa^2$ then the expectation can be lower bounded by $R\sqrt{nd}/(2\kappa)$. See Koltchinskii, (2008).

13.20. (SUPREMA OF EMPIRICAL PROCESSES INDEXED BY INTERSECTION OF ELLIPSOIDS) Let $X, X_1, \ldots, X_n$ be independently distributed over $\mathcal{X}$ according to P. Let $L_2(P)$ be the set of functions on $\mathcal{X}$ such that $Ef(X)^2 < \infty$. Assume $(g_j)_{1 \leq j \leq d}$ is an orthonormal system of centered functions in $L_2(P)$, and let $(\lambda_j)_{j \leq d}$ be a sequence of positive integers. Let

$$\mathcal{E}_1(R) = \left\{ f : f = \sum_{j=1}^{d} \alpha_j g_j, \sum_{i=1}^{d} \frac{\alpha_j^2}{\lambda_j} \leq R^2 \right\},$$

$$\mathcal{E}_2(R) = \left\{ f : f = \sum_{j=1}^{d} \alpha_j g_j, \sum_{i=1}^{d} \alpha_j^2 \leq R^2 \right\}.$$

Prove that

$$\mathbf{E} \sup_{f \in \mathcal{E}_1(1) \cap \mathcal{E}_2(R)} \sum_{i=1}^{n} f(X_i) \leq 4 \sqrt{n \sum_{j=1}^{d} \lambda_j \wedge R^2}.$$

Hint: use symmetrization. See also Exercise 13.5. This problem arises in statistical learning theory in the analysis of the so-called kernel machines (see Mendelson (2002*a*), Cucker and Zhou (2007), Steinwart and Christmann (2008) for more material on the role of reproducing kernel Hilbert spaces in statistical learning theory).

13.21. (MAXIMAL INEQUALITIES FOR CONVEX HULLS OF VC-CLASSES) The symmetric convex hull of a collection $\mathcal{F}$ of functions on $\mathcal{X}$ is defined by

$$\overline{\text{sconv}}(\mathcal{F}) = \left\{ \sum_{i=1}^{k} \lambda_i f_i : k \in \mathbb{N}, \sum_{i=1}^{k} |\lambda_i| \leq 1, f_i \in \mathcal{F} \text{ for } 1 \leq i \leq k \right\}.$$

Prove that for some universal constant $\kappa > 0$,

$$\mathbf{E} \sup_{g \in \overline{\text{sconv}}(\mathcal{F})} \left| \sum_{i=1}^{n} \varepsilon_i g(X_i) \right| \leq \kappa \mathbf{E} \sup_{f \in \mathcal{F}} \left| \sum_{i=1}^{n} \varepsilon_i f(X_i) \right|,$$

where the $X_1, \ldots, X_n$ are independently distributed over $\mathcal{X}$ and $\varepsilon_1, \ldots, \varepsilon_n$ are independent Rademacher variables. See van der Vaart and Wellner (1996).

Norms of sums of random vectors

13.22. (METRIC ENTROPY OF UNIT BALLS) For $p \in [1, \infty]$, and positive integer d, let B_p^d be the unit ball of ℓ_p^d. Prove that ε-entropy of B_p^d under the ℓ_p^d metric $H\left(\varepsilon, B_p^d\right)$ satisfies

$$d \log \frac{1}{\varepsilon} \leq H\left(\varepsilon, B_p^d\right) \leq d \log\left(1 + \frac{2}{\varepsilon}\right).$$

Hint: use a volume argument. For any $0 < \varepsilon < 1$, ℓ_p^d balls of radius $\varepsilon/2$ centered on an ε-net for B_p^d are disjoint and included in $(1 + \varepsilon/2)B_p^d$. The volume of εB_p^d is ε^d times the volume of B_p^d. Meanwhile, B_p^d is included in the union of ℓ_p^d balls of radius ε centered on a ε-net for B_p^d.

13.23. (ON THE CONSTANTS IN NEMIROVSKI'S INEQUALITY) Recall the notation and setup of Theorem 13.10 and let $V = \sum_{i=1}^{n} \mathbf{E}\|X_i\|_p^2$ denote the "strong variance." Prove that

$$\mathbf{E}\|S_n\|_p^2 \leq K(p, d)V,$$

where

$$K(p,d) = \begin{cases} d^{2/p-1} & \text{if } 1 \le p \le 2 \\ d^{1-2/p} & \text{if } 2 \le p \le \infty. \end{cases}$$

(Nemirovski, 2000.)

13.24. (AN IMPROVEMENT) Using the notation of the previous exercise, prove that for $p \geq 2$, one may choose

$$K(p,d) \leq \inf_{q \in [2,p] \cup \mathbb{R}} (q-1) d^{2/q-2/p}$$

(see Duembgen *et al.* 2010).

13.25. (NEMIROVSKI'S INEQUALITY AND THE BANACH–MAZUR DISTANCE) Let $\mathbb{B}$ and $\mathbb{B}'$ denote two Banach spaces. The Banach–Mazur distance $d_{\mathrm{bm}}(\mathbb{B}, \mathbb{B}')$ between $\mathbb{B}$ and $\mathbb{B}'$ is defined as

$$\inf\left\{ \|T\| \cdot \|T^{-1}\| \,:\, T \text{ is an isomorphism between } \mathbb{B} \text{ and } \mathbb{B}' \right\},$$

where $\|T\| = \sup\{\|Tx\|_{\mathbb{B}'}/\|x\|_{\mathbb{B}} \,:\, x \in \mathbb{B}\}$. Prove that if X_i are independent random vectors from $\mathbb{B}$, and $S_n = \sum_{i=1}^n X_i$, then

$$\boldsymbol{E}\|S_n\|_{\mathbb{B}}^2 \leq (d_{\mathrm{bm}}(\mathbb{B}, \mathbb{H}))^2\, V$$

where $\mathbb{H}$ is any Hilbert space (Duembgen *et al.* 2010).

13.26. (DISTRIBUTION OF $W_i(\alpha)^2 - W_i(\alpha')^2$) Using the notation of Section 13.6 and assuming that $(X_{i,j})_{i \leq d, j \leq D}$ are independent standard Gaussian, prove that $W_i(\alpha)^2 - W_i(\alpha')^2$ is distributed like $\sin(\theta)Y_1 - \sin(\theta)Y_2$ where θ is the angle between α and α', and Y_1, Y_2 are independent χ_1^2-distributed random variables. Deduce from this exact representation that for $\lambda < d/(8\|\alpha - \alpha'\|)$,

$$\begin{aligned} \log \boldsymbol{E} e^{\lambda(\|W(\alpha)\|^2 - \|W(\alpha')\|^2)} &= -\frac{d}{2} \log\left(1 - \frac{4\lambda^2(1 - \langle \alpha, \alpha' \rangle^2)}{d^2}\right) \\ &\leq \frac{\lambda^2 8\|\alpha - \alpha'\|^2)/d}{2\,(1 - 8\lambda\|\alpha - \alpha'\|/d)}. \end{aligned}$$

13.27. (STAR SHAPING) For $i = 1, \ldots, n$, let $X_i = (X_{i,s})_{s \in \mathcal{T}}$ be independent identically distributed centered random vectors. For $r \geq 0$, let $\mathcal{T}_r = \{s : s \in \mathcal{T},\ \boldsymbol{E}[X_{i,s}^2] \leq r^2\}$. Let L be a function on $\mathcal{T}$ that satisfies $\boldsymbol{E}[X_{i,s}^2]^{1/2} \leq L(s)$. Assume that for all $s \in \mathcal{T}$, $|X_{i,s}| \leq 1$ almost surely. Prove that the function

$$\psi^*(r) = \boldsymbol{E} \sup_{\substack{s \in \mathcal{T}, \alpha \in [0,1] \\ \alpha L(s) \leq r}} \sum_{i=1}^n \alpha X_{i,s}.$$

defined for $r \geq 0$ is sub-linear. Prove that for every $r > 0$,

$$E \sup_{s \in \mathcal{T}} \frac{r}{\max(r, L(s))} \sum_{i=1}^{n} X_{i,s} \leq \psi^*(r).$$

The so-called "star-shaping" technique, originally used in asymptotic geometry, has been successfully used in statistical learning theory (see Bartlett and Mendelson (2006), Mendelson and Philips (2004), Mendelson (2003, 2002*b*, 2002*a*), Bartlett, Bousquet, and Mendelson (2002*b*), and Bartlett and Mendelson (2002)).

13.28. (LIE PRODUCT FORMULA) Let A, B be $n \times n$ matrices (not necessarily symmetric or Hermitian). The exponential of a matrix $\exp(A)$ is defined by the power series expansion $\exp(A) = \sum_{n=0}^{\infty} A^n/n!$. Prove that

$$\exp(A + B) = \lim_{m \to \infty} \left(\exp \frac{A}{m} \exp \frac{B}{m} \right)^m .$$

Hint: first check that for any $n \times n$ matrices X, Y,

$$\|X^m - Y^m\| \leq m(\max(\|X\|, \|Y\|))^{m-1} \|X - Y\|.$$

Then apply this bound to $X_m = \exp\left(\frac{A+B}{m}\right)$ and $Y_m = \exp \frac{A}{m} \exp \frac{B}{m}$ to show that $\|X_m^m - Y_m^m\| = O(1/m)$. See Bhatia (1997, Chapter IX), and the references therein.

13.29. (GOLDEN–THOMPSON INEQUALITY) Let A and B be two $n \times n$ Hermitian matrices. Prove that

$$\operatorname{trace}(\exp(A + B)) \leq \operatorname{trace}(\exp(A) \exp(B)).$$

Hint: check first that for any two Hermitian positive semi-definite matrices X, Y, for all $m = 1, 2, \ldots$,

$$\operatorname{trace}\left((XY)^{2m}\right) \leq \operatorname{trace}\left((X^2Y^2)^m\right).$$

Combine this inequality with the Lie product formula (Exercise 13.28) using $X = \exp(A/m), Y = \exp(B/m)$ with $m = 2^{k+1}$, take k to infinity and use the continuity of the trace to conclude. The Golden–Thompson inequality is a special case of the following more general statement: if f is a complex-valued function over the space of matrices that satisfies $f(XY) = f(YX)$ and $|f(X^{2m})| \leq |f((XX^*)^m)|$, then $0 \leq f(\exp(A + B)) \leq f(\exp A \exp B)$ for any Hermitian matrices A, B. See Bhatia (1997, Chapter IX).

13.30. (RUDELSON'S INEQUALITY FOR RADEMACHER SUMS) Let $A_1, \ldots, A_n$ be symmetric $d \times d$ matrices. Let $X_1, \ldots, X_n$ be independent Rademacher random variables. Let $Z = \| \sum_{i=1}^{n} X_i A_i \|$ and $\sigma^2 = \left\| \sum_{i=1}^{n} A_i^2 \right\|$. Prove that

$$\operatorname{Var}(Z) \leq \sigma^2 \quad \text{and} \quad EZ \leq \sqrt{\pi \log(2d)}\sigma.$$

Hint: the proof of the variance bound parallels the proof of the Gaussian case in Theorem 13.8. Replace Exercise 3.24 by Example 3.6. The proof of the upper bound on expectation follows from Theorem 13.8 by a general comparison argument (see Exercise 11.14).

13.31. (MATRIX HOEFFDING INEQUALITIES) Let $X_1, \ldots, X_n$ be independent symmetric $d \times d$ matrices. Let $A_1, \ldots, A_n$ be deterministic symmetric $d \times d$ matrices. Let $\|\cdot\|$ denote the operator norm. Assume $\mathbf{E}X_i = 0$ and $A_i^2 - X_i^2$ is positive semi-definite almost surely for all $1 \leq i \leq n$. Let $Z = \left\|\sum_{i=1}^n X_i\right\|$ and $\sigma^2 = \left\|\sum_{i=1}^n A_i^2\right\|$. Prove that

$$\mathbf{E}Z \leq 2\sqrt{2\log d}\sigma$$
$$\mathrm{Var}(Z) \leq \sigma^2$$
$$\mathbf{P}\{Z \geq \mathbf{E}Z + t\} \leq e^{-t^2/8}.$$

Hint: use results and methods from Section 13.4, symmetrization inequalities (Lemma 11.4), and the bounded-differences inequality. See Tropp (2010*a*).

13.32. (SUMS OF POSITIVE SEMI-DEFINITE MATRICES) Let $X_1, \ldots, X_n$ be independent symmetric positive semi-definite $d \times d$ matrices. Let $\|\cdot\|$ denote the operator norm. Assume that $\|X_i\| \leq 1$ almost surely for all $1 \leq i \leq n$. Let $Z = \|\sum_{i=1}^n X_i\|$. Prove that $\mathrm{Var}(Z) \leq \mathbf{E}Z$ and that Z satisfies the following Bennett-style inequalities:

$$\mathbf{P}\{Z \geq \mathbf{E}Z + t\} \leq \exp\left(-\mathbf{E}Zh\left(\frac{t}{\mathbf{E}Z}\right)\right)$$

for $t \geq 0$, while

$$\mathbf{P}\{Z \leq \mathbf{E}Z - t\} \leq \exp\left(-\mathbf{E}Zh\left(\frac{-t}{\mathbf{E}Z}\right)\right)$$

for $0 \leq t \leq \mathbf{E}Z$, where $h(t) = (t+1)\log(t+1) - t$. Prove that $\mathbf{E}Z$ satisfies

$$\mathbf{E}Z \leq 2\sqrt{2\log d}\sqrt{\mathbf{E}Z} + \left\|\mathbf{E}\sum_{i=1}^n X_i\right\|.$$

Hint: to prove the first part, verify that $Z = \|\sum_{i=1}^n X_i\|$ is self-bounding and use Theorem 6.12. The last relation entails $\mathbf{E}Z \leq 8\log d + 2\left\|\mathbf{E}\sum_{i=1}^n X_i\right\|$. It is immediately seen that $\mathbf{E}Z \geq \left\|\mathbf{E}\sum_{i=1}^n X_i\right\|$. On the other hand, letting $n = d$, and letting all X_i be uniformly distributed among the orthogonal projections on the lines generated by vectors of the canonical basis, it is not hard to verify that $\mathbf{E}Z$ is not smaller than the maximum number of balls that fall into one bin when throwing d balls into d bins at random. The latter is known to be tightly concentrated around $\log(d)/\log(\log(d))$. Compare with matrix Chernoff bounds in Tropp (2010*a*) and in Ahlswede and Winter (2002). The latter upper bounds the probability that Z is larger than $\mu + t$ by $d\exp(-\mu h(t/\mu))$ where $\mu = \left\| E\sum_{i=1}^n X_i \right\|$. Obviously, $d\exp(-\mu h(t/\mu)) \geq \exp\left(-EZh\left(\frac{t}{EZ}\right)\right)$.

13.33. (BENNETT- AND BERNSTEIN-TYPE INEQUALITIES FOR MATRICES) Let $X_1, \ldots, X_n$ be independent symmetric random $d \times d$ matrices. Let $\| \cdot \|$ denote the operator norm. Assume that $EX_i = 0$ and $\|X_i\| \leq 1$ almost surely for all $1 \leq i \leq n$. Let $Z = \| \sum_{i=1}^n X_i \|$. Prove that $\mathrm{Var}(Z) \leq v = \|E\sum_{i=1}^n X_i^2\| + 2EZ$ and that Z satisfies Bennett- and Bernstein-style inequalities with variance factor v and scale factor 1. *Hint:* use Bousquet's inequality. See Tropp (2010*a*).

Suprema of some classical processes

13.34. (LE CAM'S POISSONIZATION LEMMA) Let $\mathcal{T}$ be a finite index set. Let $X_i = (X_{i,s})_{s\in\mathcal{T}}$, $i = 1, 2, \ldots$ be independently identically distributed and centered. Let N_n be a Poisson random variable with expectation n, independent of the X_i. Prove that

$$\left(1 - \frac{1}{e}\right) E \sup_{s\in\mathcal{T}} \sum_{i=1}^{n} X_{i,s} \leq E \sup_{s\in\mathcal{T}} \sum_{i=1}^{N_n} X_{i,s}.$$

Hint: let $(Y_i)_{i\in\mathbb{N}}$ be Poisson random variables with expectation 1 independent of the X_i. Use the fact that the left-hand side equals $E\sup_{s\in\mathcal{T}} \sum_{i=1}^n (E[Y_i \wedge 1])X_{i,s}$ (van der Vaart and Wellner, 1996).

13.35. (VARIANCE OF THE SUPREMUM OF THE KAC PROCESS) Let $\mathcal{T}$ be a finite set. Let $X_i = (X_{i,s})_{s\in\mathcal{T}}$, $i = 1, 2, \ldots$ be independently identically distributed and centered random vectors with $|X_i| \leq 1$. Let N_n be a Poisson random variable with expectation n, independent of the X_i. Let N_n be Poisson distributed with expectation n and independent of the X_i. Let $Z_k = \sup_{s\in\mathcal{T}} \sum_{i=1}^k X_{i,s}$. The supremum of the Kac process is defined by

$$Z = \sup_{s\in\mathcal{T}} \sum_{i=1}^{N_n} X_{i,s} = Z_{N_n}.$$

Let $\sigma^2 = \sup_{s\in\mathcal{T}} EX_{1,s}^2$. Prove that

$$\begin{aligned}
\mathrm{Var}(Z) &\leq EN_n\sigma^2 + 2EZ \\
&\quad + EN_n E\left[\left(E\left[\sup_{s\in\mathcal{T}} \sum_{i=1}^{N_n+1} X_{i,s} - \sup_{s\in\mathcal{T}} \sum_{i=1}^{N_n} X_{i,s} \middle| N_n\right]\right)^2\right] \\
&\leq EN_n\sigma^2 + 2EZ + EN_n E\left[\left(\frac{E[Z_{\max(N_n,1)}|N_n]}{\max(N_n, 1)}\right)^2\right].
\end{aligned}$$

Hint: use the Poisson Poincaré inequality (Exercise 3.21) and Theorem 11.10.

13.36. (VARIANCE OF THE SUPREMUM OF THE KAC PROCESS, CONTINUED) Using the notation of Exercise 13.35, prove that

$$\mathrm{Var}(Z) \leq \boldsymbol{E} \sup_{s \in \mathcal{T}} \sum_{i=1}^{N_n} X_{i,s}^2 + \boldsymbol{E} N_n \sup_{s \in \mathcal{T}} \boldsymbol{E} X_{1,s}^2,$$

See Reynaud-Bouret (2003, page 109).

13.37. (VARIANCE OF THE SUPREMUM OF THE KAC PROCESS INDEXED BY A VC-CLASS) Use the notation of Theorem 13.7 and Exercise 13.35 to prove that the variance of the Kac process indexed by a VC class with VC dimension V, where each set has probability at most than σ^2, is upper bounded by

$$n\sigma^2 + \kappa\sigma\sqrt{Vn \log \frac{e}{\sigma}} + \kappa V \sigma^2 \log \frac{e}{\sigma},$$

where κ is a universal constant.

13.38. (CRAMÉR–VON MISES STATISTIC). The Cramér–von Mises statistis is defined as

$$Z^2 = n \int_0^1 (P_n([0,x]) - x)^2 dx,$$

where $P_n([0,x]) = \sum_{i=1}^n \mathbb{1}_{\{X_i \leq x\}}$ is the empirical measure defined by $X_1, \ldots, X_n$ that are independently distributed according to the uniform distribution. Show that Z is the supremum of an empirical process. Derive an upper bound for $\boldsymbol{E}Z$. Compute the Efron–Stein upper bounds for Var (Z). *Hint:* use the Riesz–Fischer theorem to represent Z as the supremum of an empirical process. Denoting by $\mathcal{T}$ the class of rational sequences $(s_i)_{i \leq \mathbb{N}}$ with $\sum_{i \in \mathbb{N}} s_i^2 = 1$, prove that

$$Z = \sup_{s \in \mathcal{T}} \sum_{i=1}^{\infty} \sqrt{n} \int_0^1 s_i \sqrt{2} \sin(2\pi i x)(P_n([0,x]) - x) dx$$

(van der Vaart and Wellner, 1996, Chapter 2.13).

13.39. (ANDERSON–DARLING STATISTIC) The Anderson–Darling statistic is defined by

$$Z^2 = n \int_0^1 \frac{(P_n([0,x]) - x)^2}{x(1-x)} dx.$$

Prove that Z is the supremum of an empirical process. Compute the Efron–Stein upper bound on the variance of Z and compare it to $\boldsymbol{E}Z^2 = 1$. *Hint:* proceed as in the previous exercise.

13.40. (HIGHER CRITICISM STATISTIC) Define the *higher-criticism statistic* by

$$Z = \sup_{s \in [1/\sqrt{n}, 1-1/\sqrt{n}]} \sqrt{n} \frac{|P_n([0,s]) - s|}{\sqrt{s(1-s)}}$$

where P_n denotes the empirical distribution defined by a sample $X_1, \ldots, X_n$ drawn independently from the uniform distribution over $[0,1]$, Use the tools of Section 13.7 to show that $\boldsymbol{E}Z \leq \sqrt{2 \log \log n}$. Prove that the variance of Z is bounded by a function of $1/(2 \log \log n)$. *Hint:* it is known that

$$\boldsymbol{P}\left\{\sqrt{2\log\log n}\left(Z - \left(\sqrt{2\log\log n} + \frac{\log\sqrt{\frac{\log\log n}{2\pi}}}{\sqrt{2\log\log n}}\right)\right) < t\right\} \to e^{-e^{-t}},$$

that is, that after centering and rescaling, Z converges in law to a Gumbel distribution. See Donoho and Jin (2004), Jaeschke (1979), and de Haan and Ferreira (2006).

13.41. (PROPERTIES OF SUB-LINEAR FUNCTIONS) Recall that $\psi : [0,\infty) \to [0,\infty)$ is *sub-linear* if it is non-decreasing, continuous, $\psi(x)/x$ is nonincreasing, and $\psi(1) \geq 1$. Prove the following: (a) If ψ and ρ are sub-linear then so are $\psi \circ \rho$ and $\psi + \rho$. (b) For any $\alpha \geq \psi(1)$, the equation $\alpha r^2 = \psi(r)$ has a unique solution in $(0,1]$. (c) If ψ is sub-linear, then $\psi(u+v) \leq \psi(u) + \psi(v)$. (d) If X is a positive random variable, then $\boldsymbol{E}\psi(X) \leq 2\psi(\boldsymbol{E}X)$.

13.42. Let ψ and ρ denote non-trivial sub-linear functions. Let r_* denote the unique positive solution of the equation $\sqrt{n}r^2 = \psi(\rho(r))$. For some $a, b, c \in [0,\infty)$ with $a \geq 1$, let u denote the unique solution of equation $r^2 = \frac{a}{\sqrt{n}}\psi(\rho(r)) + \frac{b}{\sqrt{n}}\rho(r) + c$. Check that

$$u^2 \leq 2\left(a^2 + b^2\right) r_*^2 + 2c.$$

See Koltchinskii (2006), Massart and Nédélec (2006), and Massart (2000*b*, 2006).

13.43. (JOHNSON–LINDENSTRAUSS THEOREM FOR SPARSE VECTORS) Consider the notation of Lemma 13.13. For a subset $A \subseteq \mathbb{R}^D$, let conv(A) denote the convex hull of A and for $\lambda > 0$, let $\lambda A = \{\lambda x : x \in A\}$ and $A - A = \{x - y : x, y \in A\}$. Let A be the subset of unit vectors of $\mathbb{R}^D$ whose components of index larger than k are zero. For $0 < \varepsilon < 1$, let A_ε be an ε-net for A. Prove that for $0 < \varepsilon < 1$, $A \subseteq 2\text{conv}(A_\varepsilon)$. Use this inclusion and the contraction principle (see Lemma 11.5 and Exercise 11.11) to establish that

$$\boldsymbol{E} \sup_{\alpha \in \mathcal{A}} \|W(\alpha)\| \leq 2\boldsymbol{E} \sup_{\alpha \in \mathcal{A}_\varepsilon} \|W(\alpha)\|.$$

Use this result to give another proof of Lemma 13.13. See Mendelson, Pajor, and Tomczak-Jaegermann (2008).

14

Φ-Entropies

In Chapter 3 we introduced a machinery that allows us to derive bounds for the variance of a function of independent random variables. Then, in Chapters 5 and 6, with the help of logarithmic Sobolev inequalities and their modifications, we were able to derive exponential concentration inequalities, somewhat analogous to the Efron–Stein inequality. We call this the entropy method because it is based on a crucial sub-additivity property of the entropy, shown in Chapter 4. A necessary condition for the entropy method to work is the finiteness of the moment-generating function of the random variable of interest.

The purpose of this chapter and the next is to introduce a methodology to bound higher moments of functions of independent random variables. This method, though more technical than the entropy method, is at least as powerful and works for random variables that are not necessarily exponentially integrable.

Our approach is based on a generalization of the entropy method. The basic pillar of the method is the introduction of certain convex functionals of random variables that we call *Φ-entropies*. These functionals may be thought of as a common generalization of the variance and the entropy of a random variable.

In Section 14.1 we start by investigating the sub-additivity properties of Φ-entropies. We establish a duality formula, generalizing the one proved for the "ordinary" entropy in Section 4.9 and characterize Φ-entropies that are sub-additive.

The next step in our program of extending the entropy method consists of deriving inequalities that we name "Φ-Sobolev inequalities," generalizing the modified logarithmic Sobolev inequalities obtained in Sections 6.3 and 6.8. This is done in Section 14.2.

We close this chapter by deriving, in Section 14.3, sharp Φ-Sobolev inequalities for Bernoulli distributions. As a corollary, we obtain the optimal constant of the logarithmic Sobolev inequality for unbalanced Bernoulli distributions.

14.1 Φ-Entropy and its Sub-Additivity

Let $\Phi : [0, \infty) \to \mathbb{R}$ be a convex function and assign, to every nonnegative integrable random variable Z, the number

$$H_\Phi(Z) = \boldsymbol{E}\Phi(Z) - \Phi(\boldsymbol{E}Z).$$

By Jensen's inequality, $H_\Phi(Z)$ is always nonnegative. We call $H_\Phi(Z)$ the Φ-entropy of Z.

Observe that with $\Phi(x) = x^2$, the Φ-entropy is just the variance of Z, while for $\Phi(x) = x \log x$, $H_\Phi(Z)$ reduces to the "ordinary" notion of entropy introduced in Chapter 4. In the next chapter we show that other choices of Φ, in particular, $\Phi(x) = x^a$ for $a \in (1, 2]$, yield interesting variants and an appropriate modification of the entropy method based on such Φ-entropies leads to non-trivial moment inequalities.

As before, we are interested in random variables Z that are functions of independent random variables. In particular, we consider $Z = f(X_1, \ldots, X_n)$ where $X_1, \ldots, X_n$ are independent random variables taking values in a set $\mathcal{X}$ and f is a nonnegative function on $\mathcal{X}^n$.

The key property that we need is the following sub-additivity inequality of Φ-entropies:

$$H_\Phi(Z) \leq \mathbf{E} \sum_{i=1}^{n} H_\Phi^{(i)}(Z)$$

where $H_\Phi^{(i)}(Z) = \mathbf{E}^{(i)}\Phi(Z) - \Phi(\mathbf{E}^{(i)}Z)$ is the conditional entropy and, as before, $\mathbf{E}^{(i)}$ denotes conditional expectation conditioned on the $n-1$-vector $X^{(i)} = (X_1, \ldots, X_{i-1}, X_{i+1}, \ldots, X_n)$.

When $\Phi(x) = x^2$, this sub-additivity property is just the Efron–Stein inequality (Theorem 3.1), while with $\Phi(x) = x \log x$ it becomes the sub-additivity inequality of the "ordinary" entropy (see Theorem 4.22).

Here we show that Φ-entropies are sub-additive for a large class of convex functions Φ. In fact, we characterize the class of functions Φ that give rise to entropy functionals with the sub-additive property.

First we point out that sub-additivity is equivalent to a simple "Jensen-type" inequality. On the one hand, observe that for $n = 2$ and setting $Z = f(X_1, X_2)$, the sub-additivity property reduces to

$$H_\Phi\left(\int f(x, X_2) d\mu_1(x)\right) \leq \int H_\Phi\left(f(x, X_2)\right) d\mu_1(x), \tag{14.1}$$

where μ_1 denotes the distribution of X_1. On the other hand, (14.1) implies the sub-additivity property. Indeed let Y_1 be distributed like X_1, and let Y_2 be distributed like the $n-1$-tuple $X_2, \ldots, X_n$. Let μ_1 and μ_2 denote the corresponding distributions. Then $Z = f(Y_1, Y_2)$ is a measurable function of the two independent random variables Y_1 and Y_2. By the Tonelli–Fubini theorem,

$$\begin{aligned} H_\Phi(Z) = \iint \Bigg(& \Phi(f(y_1, y_2)) - \Phi\left(\int f(y_1', y_2) d\mu_1(y_1')\right) \\ & + \Phi\left(\int f(y_1', y_2) d\mu_1(y_1')\right) \\ & - \Phi\left(\iint f(y_1', y_2') d\mu_1(y_1') d\mu_2(y_2')\right) \Bigg) d\mu_1(y_1) d\mu_2(y_2) \end{aligned}$$

$$= \int \left(\int \left[\Phi(f(y_1, y_2)) - \Phi\left(\int f(y_1', y_2) d\mu_1(y_1') \right) \right] d\mu_1(y_1) \right) d\mu_2(y_2)$$
$$+ \int \left(\Phi\left(\int f(y_1', y_2) d\mu_1(y_1') \right) \right.$$
$$\left. - \Phi\left(\iint f(y_1', y_2') d\mu_1(y_1') d\mu_2(y_2') \right) \right) d\mu_2(y_2)$$
$$= \int H_\Phi(f(Y_1, y_2)) d\mu_2(y_2) + H_\Phi\left(\int f(y_1', Y_2) d\mu_1(y_1') \right)$$
$$\leq \int H_\Phi(f(Y_1, y_2)) d\mu_2(y_2) + \int H_\Phi\left(f(y_1', Y_2)\right) d\mu_1(y_1'),$$

where the last step follows from (14.1). In other words, we get

$$H_\Phi(Z) \leq EH_\Phi^{(1)}(Z) + \int H_\Phi\left(f(x_1, X_2, \ldots, X_n)\right) d\mu_1(x_1).$$

Proceeding by induction, (14.1) leads to the sub-additivity property for every n.

Thus, the sub-additivity property of H_Φ is equivalent to what we could call the *Jensen property*, that is, (14.1). This implies that in order to prove sub-additivity of a Φ-entropy, it suffices to show that it has the Jensen property.

We establish that the functional H_Φ satisfies the Jensen property by following the lines of the proof of the sub-additivity property for the "usual" entropy shown in Section 4.13. The key of this proof is a duality formula that expresses H_Φ as a supremum of affine functions.

Theorem 14.1 (SUB-ADDITIVITY OF Φ-ENTROPY) *Let $\mathcal{C}$ denote the class of functions $\Phi : [0,\infty) \to \mathbb{R}$ that are continuous and convex on $[0,\infty)$, twice differentiable on $(0,\infty)$, and such that either Φ is affine or Φ'' is strictly positive and $1/\Phi''$ is concave. For all $\Phi \in \mathcal{C}$, the entropy functional H_Φ is sub-additive.*

As mentioned above, the main ingredient of the proof of Theorem 14.1 is a duality formula for Φ-entropy of the form

$$H_\Phi(Z) = \sup_{T \in \mathcal{T}} \mathbf{E}\left[\psi_1(T) Z + \psi_2(T)\right],$$

for convenient functions ψ_1 and ψ_2 and a suitable class of nonnegative variables $\mathcal{T}$. Such a formula obviously implies that the functional H_Φ is convex. On the other hand, it also implies the Jensen property and therefore the sub-additivity property for H_Φ by the following simple argument: consider again $Z = f(Y_1, Y_2)$ as a function of $Y_1 = X_1$ and $Y_2 = (X_2, \ldots X_n)$. Then

$$
\begin{aligned}
&H_\Phi\left(\int f(y_1, Y_2)d\mu_1(y_1)\right) \\
&= \sup_{T\in\mathcal{T}} \int \left[\psi_1(T(y_2)) \int f(y_1,y_2)d\mu_1(y_1) + \psi_2(T(y_2))\right] d\mu_2(y_2) \\
&\text{(by Fubini's theorem)} \\
&= \sup_{T\in\mathcal{T}} \int \left(\int \left[\psi_1(T(y_2))f(y_1,y_2) + \psi_2(T(y_2))\right] d\mu_2(y_2)\right) d\mu_1(y_1) \\
&\leq \int \left(\sup_{T\in\mathcal{T}} \int \left[\psi_1(T(y_2))f(y_1,y_2) + \psi_2(T(y_2))\right] d\mu_2(y_2)\right) d\mu_1(y_1) \\
&= \int H_\Phi(f(y_1, Y_2))d\mu_1(y_1).
\end{aligned}
$$

Thus, in order to complete the proof of Theorem 14.1, the following lemma is sufficient. Denote the convex set of nonnegative and integrable random variables Z by $\mathbb{L}_1^+$.

Lemma 14.2 (DUALITY FORMULA FOR Φ-ENTROPIES) *Let $\Phi \in \mathcal{C}$ and $Z \in \mathbb{L}_1^+$. If $\Phi(Z)$ is integrable, then*

$$H_\Phi(Z) = \sup_{T\in\mathbb{L}_1^+, T\neq 0} \left\{\mathbf{E}\left[(\Phi'(T) - \Phi'(\mathbf{E}T))(Z - T) + \Phi(T)\right] - \Phi(\mathbf{E}T)\right\}.$$

Proof The case when Φ is affine is trivial: H_Φ equals zero, and so does the expression defined by the duality formula.

Note that the expression within the brackets on the right-hand side equals $H_\Phi(Z)$ for $T = Z$, so the proof of Lemma 14.2 amounts to checking that

$$H_\Phi(Z) \geq \mathbf{E}\left[(\Phi'(T) - \Phi'(\mathbf{E}T))(Z - T) + \Phi(T)\right] - \Phi(\mathbf{E}T)$$

under the assumption that $\Phi(Z)$ is integrable and $T \in \mathbb{L}_1^+$.

Assume first that Z and T are bounded and bounded away from 0. For any $\lambda \in [0, 1]$, we set $T_\lambda = (1 - \lambda) Z + \lambda T$ and

$$g(\lambda) = \mathbf{E}\left[\left(\Phi'(T_\lambda) - \Phi'(\mathbf{E}T_\lambda)\right)(Z - T_\lambda)\right] + H_\Phi(T_\lambda).$$

Our aim is to show that the function g is nonincreasing on $[0, 1]$. Noticing that $Z - T_\lambda = \lambda(Z - T)$ and using our boundedness assumptions to differentiate under the expectation, we have

$$
\begin{aligned}
g'(\lambda) = &-\lambda\left(\mathbf{E}\left[(Z - T)^2\,\Phi''(T_\lambda)\right] - (\mathbf{E}[Z - T])^2\,\Phi''(\mathbf{E}T_\lambda)\right) \\
&+\mathbf{E}\left[(\Phi'(T_\lambda) - \Phi'(\mathbf{E}T_\lambda))(Z - T)\right] \\
&+\mathbf{E}\left[\Phi'(T_\lambda)(T - Z)\right] - \Phi'(\mathbf{E}T_\lambda)\,\mathbf{E}[T - Z],
\end{aligned}
$$

that is,

$$g'(\lambda) = -\lambda\left(\mathbf{E}\left[(Z-T)^2\,\Phi''(T_\lambda)\right] - (\mathbf{E}[Z-T])^2\,\Phi''(\mathbf{E}T_\lambda)\right).$$

Now, by the Cauchy–Schwarz inequality,

$$\begin{aligned}(\mathbf{E}[Z-T])^2 &= \left(\mathbf{E}\left[(Z-T)\sqrt{\Phi''(T_\lambda)}\frac{1}{\sqrt{\Phi''(T_\lambda)}}\right]\right)^2 \\ &\le \mathbf{E}\left[\frac{1}{\Phi''(T_\lambda)}\right]\mathbf{E}\left[(Z-T)^2\,\Phi''(T_\lambda)\right].\end{aligned}$$

Using the concavity of $1/\Phi''$, Jensen's inequality implies that

$$\mathbf{E}\left[\frac{1}{\Phi''(T_\lambda)}\right] \le \frac{1}{\Phi''(\mathbf{E}T_\lambda)},$$

which leads to

$$(\mathbf{E}[Z-T])^2 \le \frac{1}{\Phi''(\mathbf{E}T_\lambda)}\mathbf{E}\left[(Z-T)^2\,\Phi''(T_\lambda)\right],$$

which is equivalent to $g'(\lambda) \le 0$ and therefore $g(1) \le g(0) = H_\Phi(Z)$. This means that for any T, $\mathbf{E}\left[(\Phi'(T) - \Phi'(\mathbf{E}T))(Z-T)\right] + H_\Phi(T) \le H_\Phi(Z)$.

In the general case we consider the sequences $Z_n = (Z \vee 1/n) \wedge n$ and $T_k = (T \vee 1/k) \wedge k$ and our purpose is to take the limit, as $k, n \to \infty$, in the inequality

$$H_\Phi(Z_n) \ge \mathbf{E}\left[(\Phi'(T_k) - \Phi'(\mathbf{E}T_k))(Z_n - T_k) + \Phi(T_k)\right] - \Phi(\mathbf{E}T_k),$$

which we can also write as

$$\mathbf{E}\left[\psi(Z_n, T_k)\right] \ge -\Phi'(\mathbf{E}T_k)\,\mathbf{E}[Z_n - T_k] - \Phi(\mathbf{E}T_k) + \Phi(\mathbf{E}Z_n), \tag{14.2}$$

where $\psi(z,t) = \Phi(z) - \Phi(t) - (z-t)\Phi'(t)$. Since we have to show that

$$\mathbf{E}[\psi(Z,T)] \ge -\Phi'(\mathbf{E}T)\,\mathbf{E}[Z-T] - \Phi(\mathbf{E}T) + \Phi(\mathbf{E}Z) \tag{14.3}$$

with $\psi \ge 0$, we can always assume $\psi(Z,T)$ to be integrable (since otherwise (14.3) is trivially satisfied). Taking the limit when n and k go to infinity on the right-hand side of (14.2) is easy, while the treatment of the left-hand side requires some care. Note that $\psi(z,t)$, as a function of t, decreases on $(0,z)$ and increases on (z,∞). Similarly, as a function of z, $\psi(z,t)$ decreases on $(0,t)$ and increases on $(t,+\infty)$. Hence, for every t, $\psi(Z_n,t) \le \psi(1,t) + \psi(Z,t)$ while for every z, $\psi(z,T_k) \le \psi(z,1) + \psi(z,T)$. Hence, given k,

$$\psi(Z_n, T_k) \le \psi(1, T_k) + \psi(Z, T_k),$$

as $\psi((z \vee 1/n) \wedge n, T_k) \to \psi(z, T_k)$ for every z, we can apply the dominated convergence theorem to conclude that $\mathbf{E}\psi(Z_n, T_k)$ converges to $\mathbf{E}\psi(Z, T_k)$ as $n \to \infty$. Hence, we have

$$\mathbf{E}\psi(Z, T_k) \geq -\Phi'(\mathbf{E}T_k)\,\mathbf{E}[Z - T_k] - \Phi(\mathbf{E}T_k) + \Phi(\mathbf{E}Z).$$

Now we also have $\psi(Z, T_k) \leq \psi(Z, 1) + \psi(Z, T)$ and we can apply the dominated convergence theorem again to ensure that $\mathbf{E}\psi(Z, T_k)$ converges to $\mathbf{E}\psi(Z, T)$ as $k \to \infty$. Taking the limit as $k \to \infty$ implies that (14.3) holds for every $T, Z \in \mathbb{L}_1^+$ such that $\Phi(Z)$ is integrable and $\mathbf{E}T > 0$. If $Z \neq 0$ a.s., (14.3) is achieved for $T = Z$ while if $Z = 0$ a.s., it is achieved for $T = 1$ and the proof of the lemma is now complete in its full generality. □

Remark 14.4 Note that since the supremum in the duality formula of Lemma 14.2 is achieved for $T = Z$ (or $T = 1$ if $Z = 0$), the duality formula remains true if the supremum is restricted to the class $\mathcal{T}_\Phi$ of variables T such that $\Phi(T)$ is integrable. Hence, we may also write the alternative formula

$$H_\Phi(Z) = \sup_{T \in \mathcal{T}_\Phi} \left\{ \mathbf{E}\left[(\Phi'(T) - \Phi'(\mathbf{E}T))(Z - T) \right] + H_\Phi(T) \right\}.$$

Remark 14.5 Note that Lemma 14.2 generalizes the duality formula of Theorem 4.13 for the "usual" entropy. Indeed, taking $\Phi(x) = x \log x$, we get

$$\mathrm{Ent}(Z) = \sup_T \left\{ \mathbf{E}\left[(\log(T) - \log(\mathbf{E}T)) Z \right] \right\}$$

where the supremum is extended to the set of nonnegative and integrable random variables T with $\mathbf{E}T > 0$. Another case of interest is $\Phi(x) = x^p$, with $p \in (1, 2]$. In this case, one has, by the previous remark,

$$H_\Phi(Z) = \sup_T \left\{ p\mathbf{E}\left[Z\left(T^{p-1} - (\mathbf{E}T)^{p-1} \right) \right] - (p-1) H_\Phi(T) \right\},$$

where the supremum is extended to the set of nonnegative variables in $\mathbb{L}_p$.

Remark 14.6 For the sake of simplicity we have focused on nonnegative variables and convex functions Φ on $[0, \infty)$. This restriction can be suppressed and one may consider Φ that is a convex function on $\mathbb{R}$ and define the Φ-entropy of a real-valued integrable random variable Z by the same formula as in the nonnegative case. Assuming this time that Φ is differentiable on $\mathbb{R}$ and twice differentiable on $\mathbb{R} \setminus \{0\}$, the proof of the duality formula above can be easily adapted to cover this case provided that $1/\Phi''$ can be extended to a concave function on $\mathbb{R}$. In particular, if $\Phi(x) = |x|^p$, where $p \in (1, 2]$, one gets

$$H_\Phi(Z) = \sup_T \left\{ p\mathbf{E}\left[Z\left(\frac{|T|^p}{T} - \frac{|\mathbf{E}T|^p}{\mathbf{E}T} \right) \right] - (p-1) H_\Phi(T) \right\}$$

where the supremum is extended to $\mathbb{L}_p$. Note that for $p = 2$ this formula reduces to the classical one for the variance

$$\mathrm{Var}(Z) = \sup_T \left\{2\,\mathrm{Cov}(Z,T) - \mathrm{Var}(T)\right\},$$

where the supremum is extended to the set of square integrable variables. This means that the sub-additivity inequality for the Φ-entropy also holds for convex functions Φ on $\mathbb{R}$ under the condition that $1/\Phi''$ is the restriction to $\mathbb{R} \setminus \{0\}$ of a concave function on $\mathbb{R}$.

We close this section by pointing out that, provided that Φ'' is strictly positive, the condition $1/\Phi''$ concave is necessary for the sub-additivity property to hold. In fact, even more is true: the concavity of $1/\Phi''$ is necessary for the Φ-entropy H_Φ to be convex on the set of bounded and nonnegative random variables.

Proposition 14.3 *Let $\Phi : [0,\infty) \to \mathbb{R}$ be a strictly convex function which is twice differentiable on $(0,\infty)$. Let the probability space $(\Omega, \mathcal{A}, \mathbf{P})$ be rich enough in the sense that $\mathbf{P}$ maps $\mathcal{A}$ onto $[0,1]$. If H_Φ is convex on the set of of bounded, nonnegative random variables, then $\Phi''(x) > 0$ for every $x > 0$ and $1/\Phi''$ is concave on $(0,\infty)$.*

Proof Let $\theta \in [0,1]$ and let $x, x', y, y' > 0$. By the assumption on the probability space, we may define a pair of random variables (X,Y) by

$$(X,Y) = \begin{cases} (x,y) & \text{with probability } \theta \\ (x',y') & \text{with probability } 1-\theta. \end{cases}$$

Then convexity of H_Φ means that

$$H_\Phi(\lambda X + (1-\lambda)Y) \le \lambda H_\Phi(X) + (1-\lambda) H_\Phi(Y)$$

for every $\lambda \in (0,1)$. Defining, for every $u, v > 0$,

$$F_\lambda(u,v) = -\Phi(\lambda u + (1-\lambda)v) + \lambda\Phi(u) + (1-\lambda)\Phi(v),$$

the inequality is equivalent to

$$F_\lambda(\theta(x,y) + (1-\theta)(x',y')) \le \theta F_\lambda(x,y) + (1-\theta)F_\lambda(x',y').$$

Hence, F_λ is convex on $(0,\infty)^2$. This implies, in particular, that the determinant of the Hessian matrix of F_λ is nonnegative at each point (x,y). Thus, setting $x_\lambda = \lambda x + (1-\lambda)y$,

$$\left[\Phi''(x) - \lambda\Phi''(x_\lambda)\right]\left[\Phi''(y) - (1-\lambda)\Phi''(x_\lambda)\right] \ge \lambda(1-\lambda)\left[\Phi''(x_\lambda)\right]^2,$$

which means that

$$\Phi''(x)\Phi''(y) \geq \lambda\Phi''(y)\Phi''(x_\lambda) + (1-\lambda)\Phi''(x)\Phi''(x_\lambda).$$

If $\Phi''(x) = 0$ for some point x, we see that either $\Phi''(y) = 0$ for every y, which is impossible because Φ is assumed to be strictly convex, or there exists some y such that $\Phi''(y) > 0$ and then Φ'' is identically equal to 0 on the nonempty open interval with endpoints x and y, which also contradicts the assumption of strict convexity of Φ. Hence Φ'' is strictly positive at each point of $(0,\infty)$ and the inequality above becomes

$$\frac{1}{\Phi''(\lambda x + (1-\lambda)y)} \geq \frac{\lambda}{\Phi''(x)} + \frac{(1-\lambda)}{\Phi''(y)}$$

which implies that $1/\Phi''$ is concave. □

14.2 From Φ-Entropies to Φ-Sobolev Inequalities

Now we describe the next step in our program of deriving moment inequalities for functions of independent random variables. The program follows the outline of the entropy method for proving exponential concentration inequalities. Recall that, after establishing the sub-additive property of the entropy, we used symmetrization and variational arguments to derive modified logarithmic Sobolev inequalities (recall Theorems 6.6 and 6.15). The following lemma generalizes these symmetrization and variational arguments.

Lemma 14.4 *Let Φ be a continuous and convex function on $[0,\infty)$. Then, denoting by Φ' the right derivative of Φ, for every nonnegative and integrable random variable Z,*

$$H_\Phi(Z) = \inf_{u\geq 0} \mathbf{E}\left[\Phi(Z) - \Phi(u) - (Z-u)\Phi'(u)\right].$$

Let Z' be an independent copy of Z. Then

$$H_\Phi(Z) \leq \frac{1}{2}\mathbf{E}\left[(Z-Z')\left(\Phi'(Z)-\Phi'(Z')\right)\right] = \mathbf{E}\left[(Z-Z')_+\left(\Phi'(Z)-\Phi'(Z')\right)\right].$$

If, moreover, $\psi(x) = (\Phi(x)-\Phi(0))/x$ is concave on $(0,\infty)$, then

$$H_\Phi(Z) \leq \frac{1}{2}\mathbf{E}\left[(Z-Z')\left(\psi(Z)-\psi(Z')\right)\right] = \mathbf{E}\left[(Z-Z')_+\left(\psi(Z)-\psi(Z')\right)\right].$$

Proof Without loss of generality, we assume that $\Phi(0) = 0$. By the convexity of Φ, for all $u > 0$,

$$-\Phi(\mathbf{E}Z) \leq -\Phi(u) - (\mathbf{E}Z-u)\Phi'(u),$$

and therefore

$$H_\Phi(Z) \le \mathbf{E}\left[\Phi(Z) - \Phi(u) - (Z-u)\Phi'(u)\right].$$

Since the latter inequality becomes an equality when $u = \mathbf{E}Z$, the variational formula is proven. Further, since Z' is an independent copy of Z, we find that

$$\begin{aligned} H_\Phi(Z) &\le \mathbf{E}\left[\Phi(Z) - \Phi(Z') - (Z-Z')\,\Phi'(Z')\right] \\ &\le -\mathbf{E}\left[(Z-Z')\,\Phi'(Z')\right] \end{aligned}$$

and by symmetry,

$$2H_\Phi(Z) \le -\mathbf{E}\left[(Z'-Z)\,\Phi'(Z)\right] - \mathbf{E}\left[(Z-Z')\,\Phi'(Z')\right],$$

which leads to the second inequality of the lemma. To prove the third inequality, we simply note that

$$\frac{1}{2}\mathbf{E}\left[(Z-Z')\,(\psi(Z)-\psi(Z'))\right] - H_\Phi(Z) = -\mathbf{E}Z\mathbf{E}\psi(Z) + \Phi(\mathbf{E}Z).$$

But the concavity of ψ implies that $\mathbf{E}\psi(Z) \le \psi(\mathbf{E}Z) = \Phi(\mathbf{E}Z)/\mathbf{E}Z$ and the result follows. □

The next lemma shows that we can apply the third inequality of the lemma whenever $\Phi \in \mathcal{C}$. In particular, for $\Phi(x) = x^p$, with $p \in (1,2]$, it improves on the second inequality by a factor of p.

Lemma 14.5 *If $\Phi \in \mathcal{C}$, then both Φ' and $\psi(x) = (\Phi(x) - \Phi(0))/x$ are concave on $(0,\infty)$.*

Proof Without loss of generality we may assume that $\Phi(0) = 0$. If Φ is strictly convex,

$$\begin{aligned} \frac{1}{\Phi''((1-\lambda)u+\lambda x)} &\ge \frac{1-\lambda}{\Phi''(u)} + \frac{\lambda}{\Phi''(x)} \quad \text{(by the concavity of } 1/\Phi''\text{)} \\ &\ge \frac{\lambda}{\Phi''(x)} \quad \text{(since by the strict convexity of } \Phi,\ \Phi''(x) > 0\text{)}. \end{aligned}$$

In any case, the concavity of $1/\Phi''$ implies that for every $\lambda \in (0,1)$ and every $x, u > 0$,

$$\lambda\Phi''\left((1-\lambda)u + \lambda x\right) \le \Phi''(x),$$

that is, for all $t > 0$,

$$\lambda\Phi''(t+\lambda x) \le \Phi''(x).$$

Letting $\lambda \to 1$, we see that Φ'' is nonincreasing, that is, Φ' is concave. Setting $\psi(x) = \Phi(x)/x$, one has

$$x^3\psi''(x) = x^2\Phi''(x) - 2x\Phi'(x) + 2\Phi(x).$$

The convexity of Φ and its continuity at 0 imply that $\lim_{x\to 0} x\Phi'(x) = 0$. Also, by concavity of Φ',

$$x^2\Phi''(x) \leq 2x\left(\Phi'(x) - \Phi'(x/2)\right),$$

so $\lim_{x\to 0} x^2\Phi''(x) = 0$ and therefore $\lim_{x\to 0} x^3\psi''(x) = 0$. Denoting (abusively) by $\Phi^{(3)}$ the right derivative of Φ'' (which is well defined since $1/\Phi''$ is concave) and by γ the right derivative of $x^3\psi''(x)$, we have $\gamma(x) = x^2\Phi^{(3)}(x)$. Then $\gamma(x) \leq 0$ since Φ'' is nonincreasing. Thus, $x^3\psi''(x)$ is nonincreasing. Since $x^3\psi''(x)$ tends to 0 at 0, this means that $x^3\psi''(x) \leq 0$ and therefore $\psi''(x) \leq 0$, proving the concavity of ψ. □

Now we are prepared to prove analogs of the "modified logarithmic Sobolev inequalities" of Theorems 6.6 and 6.15. Analogous with this terminology, we may refer to the following two theorems as *modified Φ-Sobolev inequalities.* The purpose of these inequalities is to upper bound the Φ-entropy of a conveniently chosen convex function of the variable of interest Z.

In the following, $X_1, \ldots, X_n$ denote independent random variables, taking values in some space $\mathcal{X}$ and $f : \mathcal{X}^n \to \mathcal{I}$ is a function mapping into a (possibly infinite) interval $\mathcal{I} \subset \mathbb{R}$. Let $Z = f(X_1, \ldots, X_n)$ and let $Z_i' = f(X_1, \ldots, X_i', \ldots, X_n)$ be obtained by replacing the variable X_i by an independent copy X_i'.

As in Section 6.9, we introduce the random variables

$$V^+ = \sum_{i=1}^{n} \mathbf{E}'\left[(Z - Z_i')_+^2\right]$$

and

$$V^- = \sum_{i=1}^{n} \mathbf{E}'\left[(Z - Z_i')_-^2\right],$$

where $\mathbf{E}'$ denotes expectation with respect to the variables $X_1', \ldots, X_n'$ only.

If $f_i : \mathcal{X}^{n-1} \to \mathcal{I}$ are arbitrary measurable functions, we write $Z_i = f_i(X^{(i)}) = f_i(X_1, \ldots, X_{i-1}, X_{i+1}, \ldots, X_n)$ and

$$V = \sum_{i=1}^{n} (Z - Z_i)^2.$$

Then we have the following "Φ-Sobolev" inequalities.

Theorem 14.6 *Let $\Phi \in \mathcal{C}$ and let η be a nondecreasing, nonnegative and differentiable convex function on $\mathcal{I}$. Let $\psi(x) = (\Phi(x) - \Phi(0))/x$. If the function $\psi \circ \eta$ is convex, then*

$$H_\Phi(\eta(Z)) \leq \mathbf{E}\left[V^+\eta'^2(Z)\psi'(\eta(Z))\right].$$

On the other hand, if $\Phi' \circ \eta$ is convex and $Z_i \leq Z$ for all $i = 1, \ldots, n$, then

$$H_\Phi(\eta(Z)) \leq \frac{1}{2}\mathbf{E}\left[V\eta'^2(Z)\Phi''(\eta(Z))\right].$$

Proof First fix $x < y$ and assume that $g = \Phi' \circ \eta$ is convex. Setting

$$h(t) = \Phi(\eta(y)) - \Phi(\eta(t)) - \left(\eta(y) - \eta(t)\right)g(t),$$

we have

$$h'(t) = -g'(t)\left(\eta(y) - \eta(t)\right).$$

But for every $t \leq y$, the monotonicity and convexity assumptions on η and g yield

$$0 \leq g'(t) \leq g'(y) \quad \text{and} \quad 0 \leq \eta(y) - \eta(t) \leq (y-t)\eta'(y),$$

hence

$$-h'(t) \leq (y-t)\,\eta'(y)g'(y).$$

Integrating this inequality with respect to t on $[x, y]$, we obtain

$$\Phi(\eta(y)) - \Phi(\eta(x)) - (\eta(y) - \eta(x))\,\Phi'(\eta(x)) \leq \frac{1}{2}(y-x)^2\,\eta'^2(y)\Phi''(\eta(y)).$$

Now sub-additivity of the Φ-entropy (Theorem 14.1), combined with the variational inequality of Lemma 14.4 and the inequality above lead to

$$H_\Phi(\eta(Z)) \leq \frac{1}{2}\sum_{i=1}^{n}\mathbf{E}\left[(Z-Z_i)^2\,\eta'^2(Z)\Phi''(\eta(Z))\right]$$

and therefore to the second inequality of the theorem.

Under the assumption that $\psi \circ \eta$ is convex, we have

$$0 \leq \eta(y) - \eta(x) \leq (y-x)\eta'(y)$$

and

$$0 \leq \psi(\eta(y)) - \psi(\eta(x)) \leq (y-x)\eta'(y)\psi'(\eta(y)),$$

which implies

$$(\eta(y) - \eta(x)) (\psi(\eta(y)) - \psi(\eta(x))) \leq (x - y)^2 \eta'^2(y)\psi'(\eta(y)).$$

The first inequality of the theorem follows from here in a similar way, but using the last inequality of Lemma 14.4. □

The case when η is nonincreasing is handled by the following theorem.

Theorem 14.7 *Let $\Phi \in \mathcal{C}$ and let η be a nonnegative, nonincreasing, and differentiable convex function on $\mathcal{I}$. Let $\psi(x) = (\Phi(x) - \Phi(0))/x$. If $\tilde{Z}$ is a random variable satisfying $\tilde{Z} \leq \min_{1 \leq i \leq n} Z_i$ and if $\Phi' \circ \eta$ is convex, then*

$$H_\Phi(\eta(Z)) \leq \frac{1}{2}\mathbf{E}\left[V\eta'^2\left(\tilde{Z}\right)\Phi''\left(\eta\left(\tilde{Z}\right)\right)\right],$$

while if $\psi \circ \eta$ is convex, we have

$$H_\Phi(\eta(Z)) \leq \mathbf{E}\left[V^+\eta'^2\left(\tilde{Z}\right)\psi'\left(\eta\left(\tilde{Z}\right)\right)\right]$$

and

$$H_\Phi(\eta(Z)) \leq \mathbf{E}\left[V^-\eta'^2(Z)\psi'(\eta(Z))\right].$$

The proof of Theorem 14.7 parallels the proof of Theorem 14.6 and it is left to the reader as an exercise (see Exercise 14.1).

Observe that by taking $\eta(z) = \exp(\lambda z)$ and $\Phi(x) = x \log x$ in Theorems 14.6 and 14.7, we obtain

$$\mathrm{Ent}(e^{\lambda Z}) \leq \lambda^2 \mathbf{E}\left[V^+ e^{\lambda Z}\right]$$

for $\lambda \geq 0$, while if $\lambda \leq 0$, one has

$$\mathrm{Ent}(e^{\lambda Z}) \leq \lambda^2 \mathbf{E}\left[V^- e^{\lambda Z}\right].$$

We have already derived these inequalities as consequences of the modified logarithmic Sobolev inequalities of Theorem 6.15.

14.3 Φ-Sobolev Inequalities for Bernoulli Random Variables

In this section we present Φ-Sobolev inequalities for functions of Bernoulli random variables. In the first part of the section we consider symmetric Bernoulli random variables and prove a Φ-Sobolev inequality that contains the Efron–Stein (or Poincaré) inequality and the logarithmic Sobolev inequality (Theorem 5.1) as special cases for such distributions. Also, we obtain a family of inequalities that "interpolate" between these extremes.

In the second half of the section we extend these results to unbalanced Bernoulli distributions. As a special case, we obtain the logarithmic Sobolev inequality of Theorem 5.2 with the optimal constants.

Suppose first that the random vector X is uniformly distributed over $\{-1,1\}^n$, and let $f:\{-1,1\}^n \to [0,\infty)$ be defined on the n-dimensional binary hypercube.

In Chapter 5 we introduced the functional

$$\mathcal{E}(f) = \frac{1}{4}\mathbf{E}\sum_{i=1}^{n}\left(f(X) - f(\overline{X}^{(i)})\right)^2 = \frac{1}{2}\mathbf{E}\sum_{i=1}^{n}\left(f(X) - f(\overline{X}^{(i)})\right)_+^2,$$

where the random binary vector $\overline{X}^{(i)} = (X_1,\dots,X_{i-1},-X_i,X_{i+1},\dots,X_n)$ is obtained by flipping the i-th component of X while leaving the others unchanged. In Chapter 5 we proved that $\mathrm{Var}(f) \le \mathcal{E}(f)$ and $\mathrm{Ent}(f^2) \le 2\mathcal{E}(f)$. Both of these results may be regarded as Φ-Sobolev inequalities with $\Phi(x) = x^2$ and $\Phi(x) = x\log x$, respectively.

The second inequality – the logarithmic Sobolev inequality for symmetric Bernoulli distributions – allowed us to establish the Bonami–Beckner inequality (Theorem 5.18). Here we show that the Bonami–Beckner inequality may, in turn, be used to deduce sharp Φ-Sobolev inequalities for $\Phi(z) = z^{2/r}$ for all $r \in [1,2)$. This collection of Φ-Sobolev inequalities "interpolate" between the two cases mentioned above, in a sense that we explain below.

Theorem 14.8 (Φ-SOBOLEV INEQUALITIES FOR BALANCED BERNOULLI RANDOM VARIABLES) *Let $f:\{-1,1\}^n \to [0,\infty)$ and assume that X is uniformly distributed over $\{-1,1\}^n$. Then for all $r \in [1,2)$, letting $\Phi(z) = z^{2/r}$,*

$$H_\Phi(f^r) \le (2-r)\mathcal{E}(f).$$

Proof If $r = 1$, the result follows from the Efron–Stein inequality, so we may assume $1 < r < 2$. Recall the notation of Section 5.8: for any $S \subseteq \{1,\dots,n\}$, $u_S(x) = \prod_{i\in S} x_i$ where $x \in \{-1,1\}^n$. For $\gamma > 0$, the operator T_γ maps a function $f = \sum_{S\subset\{1,\dots,n\}} \alpha_S u_S$ to

$$T_\gamma f = \sum_{S\subset\{1,\dots,n\}} \gamma^{|S|}\alpha_S u_S.$$

If $\gamma = \sqrt{r-1}$, then by Theorem 5.18,

$$\|T_\gamma f\|_2 \le \|f\|_r.$$

By the definition of T_γ and the orthogonality of $(u_S)_{S\subseteq\{0,1\}^n}$,

$$\|T_\gamma f\|_2^2 = \mathbf{E}\left[((T_\gamma f)(X))^2\right] = \mathbf{E}\left[f(X)(T_{\gamma^2} f)(X))\right].$$

Denoting by Id the identity operator (i.e. $\mathrm{Id}f = f$), the statement of Theorem 5.18 may be rewritten as

$$\mathbf{E}\left[f(X)^2\right] - \mathbf{E}\left[f(X)^r\right]^{2/r} \le \mathbf{E}\left[f(X)(\mathrm{Id} - T_{\gamma^2})f(X)\right] = \sum_{S\subseteq\{1\dots,n\}} \alpha_S^2 (1 - (r-1)^{|S|}).$$

We may further bound the right-hand side by noticing that $1 - (r-1)^{|S|} \le (2-r)|S|$ for all $S \subset \{1, \dots, n\}$. Indeed, it holds trivially for $|S| = 0$ and for $|S| \ge 1$ it follows by the fact that $(2-r)|S| + (r-1)^{|S|} - 1$ is decreasing over $[1, 2]$ (it is convex over $[1, 2]$ and has zero derivative at $r = 2$) and equals 0 for $r = 2$.

Thus,

$$\mathbf{E}\left[f(X)^2\right] - \mathbf{E}\left[f(X)^r\right]^{2/r} \le (2-r) \sum_{S\subseteq\{1\dots,n\}} \alpha_S^2 |S| = (2-r)\mathcal{E}(f)$$

where we use the fact that $\sum_{S\subseteq\{1\dots,n\}} \alpha_S^2 |S| = \mathcal{E}(f)$, as established in Section 9.4. □

Observe that one may recover the logarithmic Sobolev inequality of Theorem 5.1 from Theorem 14.8. Indeed, letting $r \to 2$,

$$\lim_{r\to 2^-} \frac{\mathbf{E}[f(X)^2] - (\mathbf{E}[f(X)^r]^{2/r})}{2-r} = \frac{\mathrm{Ent}(f^2)}{2}.$$

Next we address the analog question for unbalanced Bernoulli distributions. We derive directly a family of optimal Φ-Sobolev inequalities for this case. Thus, let $p \in (0, 1)$, $p \neq 1/2$. $X = (X_1, \dots, X_n)$ is a vector of independent random variables with $\mathbf{P}\{X_i = 1\} = p = 1 - \mathbf{P}\{X_i = -1\}$. The functional $\mathcal{E}$ is defined accordingly:

$$\mathcal{E}(f) = p(1-p)\mathbf{E}\sum_{i=1}^{n} \left(f(X) - f(\overline{X}^{(i)})\right)^2.$$

Theorem 14.9 (Φ-SOBOLEV INEQUALITIES FOR UNBALANCED BERNOULLI RANDOM VARIABLES) *Let* $f : \{-1, 1\}^n \to [0, \infty)$ *and let* $\Phi(z) = z^{2/r}$. *Then for all* $r \in [1, 2)$,

$$H_\Phi(f^r) \le C_{p,r}\mathcal{E}(f),$$

where

$$C_{p,r} = \frac{p^{1-2/r} - (1-p)^{1-2/r}}{(1-p)p^{1-2/r} - p(1-p)^{1-2/r}}.$$

The constant $C_{p,r}$ *is optimal.*

Proof Thanks to the sub-additivity of Φ-entropies and to the definition of $\mathcal{E}$, it suffices to prove the inequality for $n = 1$.

First observe that for any $\kappa > 0$ and $f : \{-1, 1\} \to [0, \infty)$, if $H_\Phi(f^r) \leq \kappa\mathcal{E}(f)$, then $H_\Phi((\lambda f)^r) \leq \kappa\mathcal{E}(\lambda f)$ for all $\lambda > 0$. Thus, without loss of generality, we may rescale f so that $(p^{1/r}f(1) + (1-p)^{1/r}f(-1))/2 = 1$. Then f is entirely determined by the number $y = ((1-p)^{1/r}f(-1) - p^{1/r}f(1))/2 \in (-1, 1)$.

Now consider the function $f_0 : \{-1, 1\} \to [0, \infty)$ determined by

$$y_0 = \frac{(1-p)^{1-1/r} - p^{-1/r}}{(1-p)^{1/r} + p^{1/r}}.$$

Then $f_0(1) = p^{-1/r}(1 - y_0) = (1-p)^{-1/r}(1 + y_0) = f_0(-1)$ and therefore f_0 is constant, implying $H_\Phi(f_0^r) = 0 = \mathcal{E}(f_0)$.

Both $\mathcal{E}(f)$ and $H(f^r)$ may be written as functions of y. As $\mathcal{E}(f)$ is a quadratic polynomial of y, the first step is to bound $H(f^r)$ by a polynomial. Observe that

$$\mathbf{E}\left[f(X)^r\right]^{2/r} = ((1-y)^r + (1+y)^r)^{2/r}.$$

Let $\rho(z) = \left((1-\sqrt{z})^r + (1+\sqrt{z})^r\right)^{2/r}$ for $z \in [0, 1]$. The function ρ is convex and differentiable over $(0, 1)$ (see Exercise 14.4). Hence $\rho(y^2) \geq \rho(y_0^2) + \rho'(y_0^2)(y^2 - y_0^2)$ for all $y \in (-1, 1)$. Noting that

$$\rho'(y_0^2) = 2\frac{(1-p)^{1-1/r} - p^{1-1/r}}{(1-p)^{1/r} - p^{1/r}},$$

we have

$$\begin{aligned}
H(f^r) &= H(f^r) - H(f_0^r) \\
&\leq p^{1-2/r}((1-y)^2 - (1-y_0)^2) + (1-p)^{1-2/r}((1+y)^2 - (1+y_0)^2) \\
&\quad -\rho'(y_0^2)(y^2 - y_0^2) \\
&= \left(p^{1-2/r} + (1-p)^{1-2/r}\right)(y^2 - y_0^2) \\
&\quad -2\left(p^{1-2/r} - (1-p)^{1-2/r}\right)(y - y_0) \\
&\quad -2\frac{(1-p)^{1-1/r} - p^{-1/r}}{(1-p)^{1/r} - p^{1/r}}(y^2 - y_0^2) \\
&= \left(p^{1-2/r} - (1-p)^{1-2/r}\right)\left(\frac{(1-p)^{1-1/r} + p^{-1/r}}{(1-p)^{1/r} - p^{1/r}}(y^2 - y_0^2) - 2(y - y_0)\right) \\
&= \left(p^{1-2/r} - (1-p)^{1-2/r}\right)\frac{(y - y_0)^2}{y_0}.
\end{aligned}$$

On the other hand,

$$\begin{aligned}(f(1)-f(-1))^2 &= (f(1)-f_0(1)-(f(-1)-f_0(-1)))^2\\ &= (y-y_0)^2(p^{-1/r}+(1-p)^{-1/r})^2.\end{aligned}$$

We choose κ to equate the two quadratic functions above, that is,

$$\begin{aligned}&\left(p^{1-2/r}-(1-p)^{1-2/r}\right)\frac{(y-y_0)^2}{y_0}\\ &= \kappa\mathcal{E}(f) = \kappa p(1-p)(y-y_0)^2(p^{-1/r}+(1-p)^{-1/r})^2.\end{aligned}$$

This yields $\kappa = C_{p,r}$.

The optimality of $C_{p,r}$ can be verified by choosing $f(-1) = p^{2/r}$ and $f(1) = (1-p)^{2/r}$. □

As in the case of balanced Bernoulli random variables, Theorem 14.9 may be used to derive the optimal logarithmic Sobolev inequalities for unbalanced Bernoulli random variables as announced in Theorem 5.2.

Corollary 14.10 (LOGARITHMIC SOBOLEV INEQUALITIES FOR UNBALANCED BERNOULLI RANDOM VARIABLES) *For any function* $f : \{-1,1\}^n \to \mathbb{R}$,

$$\mathrm{Ent}(f^2) \leq c(p)\mathcal{E}(f)$$

with

$$c(p) = \frac{1}{1-2p}\log\frac{1-p}{p}.$$

Proof By the remark following the proof of Theorem 14.8, the proof reduces to noting that

$$\lim_{r\to 2^-}\frac{C_{p,r}}{2-r} = \frac{c(p)}{2}.$$ □

14.4 Bibliographical Remarks

Early results on Φ-entropies derive from (among others) Csiszár (1967, 1972) who defined the related notion of ϕ-divergence, see Brègman (1967), Hu (2000), and Arnold *et al.* (2001).

The Φ-Sobolev inequalities explored in this chapter, when used with $\Phi(x) = x^a$ with $a \in (1,2]$, may be thought of as interpolation between Poincaré (when $\Phi(x) = x^2$) and logarithmic Sobolev (with $\Phi(x) = x\log x$) inequalities. Such interpolations go back to Beckner (1989). The duality formula of Lemma 14.2 is due to Bobkov (see Ledoux 1997, Latała and Oleszkiewicz 2000, Chafaï 2002, and Boucheron *et al.* 2005*b*).The treatment given in Section 14.2 follows Boucheron *et al.* (2005*b*). Chafaï (2002) developed a related framework for Φ-entropies and Φ-Sobolev inequalities.

It is shown by Latała and Oleszkiewicz (2000) (see also Ledoux 1997) that there is a tight connection between the convexity of H_Φ and the sub-additivity property. Latała and Oleszkiewicz (2000) show that $\Phi \in \mathcal{C}$ implies the convexity of H_Φ. The Φ-Sobolev inequalities of Theorems 14.6 and 14.7 are from Boucheron *et al.* (2005*b*). Some methods used to derive inequalities for Φ-entropies rely on auxiliary assumptions on concerning Φ (see Chafaï (2006); see also Exercise 14.2).

Section 14.3 is based on Latała and Oleszkiewicz (2000). Theorem 14.8 is a special case of Theorem 2 in Latała and Oleszkiewicz (2000), but see also Kwapień, Latała, and Oleszkiewicz (1996). Theorem 14.9 comes from Remark 2 in Latała and Oleszkiewicz (2000). Chafaï (2006) describes Φ-Sobolev inequalities for binomial and Poisson distributions.

14.5 EXERCISES

14.1. Prove Theorem 14.7. (Boucheron *et al.* 2005*b*).

14.2. Some inequalities for Φ-entropies rely on assumptions on Φ. Prove that the following statements are equivalent:
i) convexity of $(u,v) \mapsto A^\Phi(u,v) = \Phi(u+v) - \Phi(u) - \Phi'(u)v$ (the Brègman divergence defined by Φ);
ii) convexity of $(u,v) \mapsto B^\Phi(u,v) = (\Phi'(u+v) - \Phi'(u))v$;
iii) convexity of $(u,v) \mapsto C^\Phi(u,v) = \Phi''(u)v^2$;
iv) Φ is affine or $\Phi'' > 0$ and $-1/\Phi''$ is convex (the condition stated by Latała and Oleszkiewicz (2000) and used in the statement of 14.1);
v) Φ is affine or $\Phi'' > 0$ and $\Phi''''\Phi'' \geq 2\Phi'''^2$;
vi) $(a,b) \mapsto t\Phi(a) + (1-t)\Phi(b) - \Phi(ta + (1-t)b)$ is convex for any $0 \leq t \leq 1$;
See Chafaï (2006) for a discussion.

14.3. (A FAMILY OF Φ-ENTROPIES) Let X be an $\mathcal{X}$-valued random variable and f a nonnegative measurable function on $\mathcal{X}$. Prove that

$$\theta(r) = 2r\frac{\mathbf{E}[f^2(X)] - (\mathbf{E}[f(X)^r])^{2/r}}{2-r}$$

is nondecreasing in $r \in [1,2)$ (Latała and Oleszkiewicz, 2000).

14.4. (CALCULUS) Let $r \in [1,2]$, and for $z \in [0,1]$, define $\rho(z) = ((1+\sqrt{z})^r + (1-\sqrt{z})^r)^{2/r}$. Prove that ρ is differentiable and convex over $[0,1]$.

14.5. (Φ-SOBOLEV INEQUALITIES FOR GAUSSIAN DISTRIBUTIONS) Let $f : \mathbb{R}^n \to \mathbb{R}$ be a nonnegative differentiable function. Assume that X is a standard Gaussian vector. Then for all $r \in [1,2)$, letting $\Phi(z) = z^{2/r}$,

$$H_\Phi(f^r) \le (2-r)\boldsymbol{E}[\|\nabla f\|^2].$$

Hint: start from Theorem 14.8 and proceed as in the proof of Theorem 5.5.

14.6. (Φ-SOBOLEV INEQUALITIES FOR POISSON DISTRIBUTION) Let $\Phi(x) = x^{2/r}$ for some $r \in (1,2)$. Let X be distributed according to a Poisson distribution. Prove that for $f : \mathbb{N} \to [0,\infty)$,

$$H_\Phi(f(X)) \le (\boldsymbol{E}X)\boldsymbol{E}\left[D(\Phi \circ f)(X) - 2/rf(X)^{2/r-1}Df(X)\right]$$

where $Df(x) = f(x+1) - f(x)$. See Chafaï (2003, 2006). See also Exercises 3.21 and 6.12.

14.7. (KHINCHINE'S INEQUALITIES FOR UNBALANCED BERNOULLI RANDOM VARIABLES) Let $p \in (0,1/2)$ and assume that $X_1,\ldots,X_n$ are independent random variables with $\boldsymbol{P}\{X_i = -p\} = 1-p$ and $\boldsymbol{P}\{X_i = 1-p\} = p$. For $r > 2$, let

$$C_{r,p} = \begin{cases} (1/p)^{1/2-1/r} & \text{if } r \le \log(1/p) \\ \sqrt{\frac{1/p}{\log(1/p)}}\sqrt{r} & \text{if } r \ge \log(1/p). \end{cases}$$

Prove that there exists a universal constant κ such that for all $r > 2$, for all $\alpha_1,\ldots,\alpha_n \in \mathbb{R}^n$, letting $Z = \sum_{i=1}^n \alpha_i X_i$, we have

$$\boldsymbol{E}\left[|Z|^r\right]^{1/r} \le \kappa C_{r,p}\boldsymbol{E}[Z^2]^{1/2}.$$

Hint: use Theorem 14.9. (Oleszkiewicz 2003.)

15

Moment Inequalities

This chapter is dedicated to upper bounds for higher centered moments of functions of independent random variables. The bounds derived here may be regarded as generalizations of the Efron–Stein inequality.

As before, $X = (X_1, \ldots, X_n)$ denotes a vector of independent random variables taking values in a set $\mathcal{X}$ and $f : \mathcal{X}^n \to \mathbb{R}$ is a measurable function. We are interested in bounds for the moments of the random variable $Z = f(X)$.

Recall that in Section 6.9 we introduced the random variables V^+ and V^- as

$$V^+ = \sum_{i=1}^{n} \boldsymbol{E}' \left[(Z - Z_i')_+^2 \right]$$

and

$$V^- = \sum_{i=1}^{n} \boldsymbol{E}' \left[(Z - Z_i')_-^2 \right],$$

where $X_1', \ldots, X_n'$ are independent copies of $X_1, \ldots, X_n$, and the random variable Z_i' is obtained by replacing the variable X_i by an independent copy X_i', that is, $Z_i' = f(X_1, \ldots, X_i', \ldots, X_n)$. (Here $\boldsymbol{E}'[\cdot] = \boldsymbol{E}[\cdot|X]$ denotes expectation with respect to the variables X_i' only.)

Recall also that if $f_i : \mathcal{X}^{n-1} \to \mathbb{R}$ are measurable functions, we define $Z_i = f_i(X^{(i)}) = f_i(X_1, \ldots, X_{i-1}, X_{i+1}, \ldots, X_n)$ and

$$V = \sum_{i=1}^{n} (Z - Z_i)^2.$$

According to the Efron–Stein inequality, the variance of Z may be bounded by $\boldsymbol{E}V^+ = \boldsymbol{E}V^-$, and by $\boldsymbol{E}V$. At the same time, by Theorem 6.16, the moment-generating function of Z may be bounded in terms of the moment-generating function of V^+ and V^-. In this chapter we show that, even when the moment-generating function of V^+ (or Z) does not exist, we may bound the moments of Z in terms of moments of V^+, V^-, or V).

Our approach is reminiscent of the entropy method that led us to the "exponential Efron–Stein" inequalities of Theorem 6.16. However, instead of using modified logarithmic Sobolev inequalities to obtain differential inequalities for the moment-generating function of Z, here we use the Φ-Sobolev inequalities of Section 14.2 to obtain recursive inequalities for the moments of Z. Solving these recursions leads us to the main results of this chapter.

In Section 15.1 we start by deriving inequalities that relate higher moments of Z to the sum of a lower moment and another term that involves V (or V^+). These bounds are then used, by inductive arguments, in Sections 15.2 and 15.3, to establish the main results of this chapter. The use of results are then illustrated in Sections 15.4, 15.5, and 15.6 by describing moment inequalities for sums of independent random variables, empirical processes, and conditional Rademacher averages.

15.1 Generalized Efron–Stein Inequalities

We start with simple corollaries of the ϕ-Sobolev inequalities of Theorems 14.6 and 14.7. In a sense, these bounds may be regarded as generalized versions of the Efron–Stein inequality as they bound moments of Z by moments of lower order and functions of V^+, V^-, and V.

Lemma 15.1 *Let $q > 2$ and let α satisfy $q/2 \le \alpha \le q - 1$. Then*

$$\mathbf{E}\left[(Z - \mathbf{E}Z)_+^q\right] \le \mathbf{E}\left[(Z - \mathbf{E}Z)_+^\alpha\right]^{q/\alpha} + \frac{q(q-\alpha)}{2}\mathbf{E}\left[V(Z - \mathbf{E}Z)_+^{q-2}\right],$$

$$\mathbf{E}\left[(Z - \mathbf{E}Z)_+^q\right] \le \mathbf{E}\left[(Z - \mathbf{E}Z)_+^\alpha\right]^{q/\alpha} + \alpha\,(q-\alpha)\,\mathbf{E}\left[V^+(Z - \mathbf{E}Z)_+^{q-2}\right],$$

and

$$\mathbf{E}\left[(Z - \mathbf{E}Z)_-^q\right] \le \mathbf{E}\left[(Z - \mathbf{E}Z)_-^\alpha\right]^{q/\alpha} + \alpha\,(q-\alpha)\,\mathbf{E}\left[V^-(Z - \mathbf{E}Z)_-^{q-2}\right].$$

Proof Let q and α be such that $1 \le q/2 \le \alpha \le q - 1$. Let $\phi(x) = x^{q/\alpha}$. Applying Theorem 14.6 with $\eta(z) = (z - \mathbf{E}Z)_+^\alpha$ leads to the first two inequalities. Finally, we may apply the third inequality of Theorem 14.7 with $\eta(z) = (z - \mathbf{E}Z)_-^\alpha$ to obtain the third inequality of the lemma. □

The next lemma is a variant of Lemma 15.1 that works for nonnegative random variables.

Lemma 15.2 *Let $q \ge 2$ and $q/2 \le \alpha \le q - 1$. If for all $i = 1, \ldots, n$*

$$0 \le Z_i \le Z \text{ almost surely,}$$

then

$$E[Z^q] \le E[Z^\alpha]^{q/\alpha} + \frac{q(q-\alpha)}{2} E\left[VZ^{q-2}\right].$$

Proof The lemma follows by taking $\Phi(x) = x^{q/\alpha}$ and applying Theorem 14.6 with $\eta(z) = z^\alpha$. □

The third lemma bounds "left" moments in terms of V and V^+ and requires an additional "bounded differences" condition.

Lemma 15.3 *If the increments $Z - Z_i$ are bounded by a random variable $M \ge 0$ for all $i = 1, \ldots, n$, then*

$$E\left[(Z - EZ)_-^q\right] \le E\left[(Z - EZ)_-^\alpha\right]^{q/\alpha} + \frac{q(q-\alpha)}{2} E\left[V(Z - EZ - M)_-^{q-2}\right]$$

If $Z - Z_i' \le M$ for $i = 1, \ldots, n$ for a random variable $M \ge 0$, then

$$E\left[(Z - EZ)_-^q\right] \le E\left[(Z - EZ)_-^\alpha\right]^{q/\alpha} + \alpha(q-\alpha) E\left[V^+(Z - EZ - M)_-^{q-2}\right]$$

Proof The proof follows from Theorem 14.7. □

The inequalities of the lemmas above may now be used by induction to obtain the moment inequalities that are the principal results of this chapter.

15.2 Moments of Functions of Independent Random Variables

We present the main moment inequalities in this section. For a random variable Y and $q > 0$, introduce the notation

$$\|Y\|_q = \left(E|Y|^q\right)^{1/q}.$$

In Section 3.6, we used Efron–Stein inequality, the simplest Φ-Sobolev inequality in order to show that if the Efron–Stein estimate V^+ of the variance of a function of many independent random variables $Z = f(X_1, \ldots, X_n)$ is upper bounded by a constant c, then Z has sub-exponential tails. As a warm-up illustration of how our inductive arguments work, we re-prove this simple result, starting, once again, from the Efron–Stein inequality.

Recall from Theorem 2.1 the fact that the q-th moment is bounded by a constant multiple of q for all $q \ge 1$ is equivalent to sub-exponential tails.

We verify, by induction, that for all integers $k \ge 1$ and for all $q \in [k, k+1)$, $\left\|(Z - EZ)_+\right\|_q \le \sqrt{cq}$.

For $q \in [1, 2]$, by Hölder's inequality, $\|(Z - EZ)_+\|_q \leq \|(Z - EZ)_+\|_2$ while $\|(Z - EZ)_+\|_2 \leq \sqrt{c}$ by the Efron–Stein's inequality. For $q = 3$, from the second inequality of Lemma 15.1 with $\alpha = q/2$, we obtain

$$\begin{aligned} \|(Z - EZ)_+\|_q^q &\leq \|(Z - EZ)_+\|_{q/2}^q + \frac{cq^2}{4} \|(Z - EZ)_+\|_{q-2}^{q-2} \\ &\leq \frac{9c^{3/2}}{4} \leq (3/2)^2 \sqrt{c}^3. \end{aligned}$$

By Hölder's inequality, for all $q \in [2, 3]$, $\|(Z - EZ)_+\|_q \leq \|(Z - EZ)_+\|_3 \leq q\sqrt{c}$.

Assume now that the moment bound holds for all integers smaller than some $k \geq 3$. Then for $q \in [k, k + 1)$, from the second inequality of Lemma 15.1 with $\alpha = q/2$, and the induction hypothesis, we obtain

$$\begin{aligned} \|(Z - EZ)_+\|_q^q &\leq \|(Z - EZ)_+\|_{q/2}^q + \frac{cq^2}{4} \|(Z - EZ)_+\|_{q-2}^{q-2} \\ &\leq (q/2\sqrt{c})^q + \frac{cq^2}{4} ((q - 2)\sqrt{c})^{q-2} \\ &\leq q^q \sqrt{c}^q \left(\left(\frac{1}{2}\right)^q + \left(1 - \frac{2}{q}\right)^{q-2} \right) \\ &\leq q^q \sqrt{c}^q. \end{aligned}$$

Even though this is our third and simplest proof of a sub-optimal result, it illustrates the pattern of several proofs in this section. However, in order to obtain improved, sometimes tight, bounds, we choose values of α close to q in Lemma 15.1, rather than $\alpha = q/2$.

Before stating the most general results, we start with the following simple sub-Gaussian bound.

Theorem 15.4 *If $V^+ \leq c$ for some constant $c \geq 0$, then for all integers $q \geq 2$,*

$$\|(Z - EZ)_+\|_q \leq \sqrt{Kqc},$$

where $K = 1/\left(e - \sqrt{e}\right) < 0.935$. If furthermore $V^- \leq c$ then for all integers $q \geq 2$,

$$\|Z\|_q \leq EZ + 2^{1/q}\sqrt{Kqc}.$$

Recall from Theorem 2.1 that the fact that the q-th moment is bounded by a constant multiple of $\sqrt{q}$ for all q is equivalent to sub-Gaussian tails and therefore Theorem 15.4 is essentially equivalent to Theorem 6.7. However, the proof is quite different and it illustrates the essence of the techniques of the more general results below in a transparent way.

Proof Define

$$m_q = \left\| (Z - \mathbf{E}Z)_+ \right\|_q .$$

From the second inequality of Lemma 15.1 with $\alpha = q - 1$, we obtain, for $q \geq 3$,

$$m_q^q \leq m_{q-1}^q + c\,(q-1)\, m_{q-2}^{q-2}. \tag{15.1}$$

We use this inequality to show by induction that, for all $q \geq 2$,

$$m_q^q \leq (Kqc)^{q/2} .$$

For $q = 2$ this holds since by the Efron–Stein inequality, $m_2^2 \leq \mathbf{E}V^+ \leq c$. The case $q = 3$ follows from (15.1), since using $m_1 \leq m_2 \leq \sqrt{c}$, we have

$$m_3^3 \leq 3c^{3/2}.$$

Consider now $q \geq 4$ and assume that

$$m_j \leq \sqrt{Kjc}$$

for every $j \leq q - 1$. Then, it follows from (15.1) and two applications of the induction hypothesis that

$$\begin{aligned} m_q^q &\leq K^{q/2}c^{q/2}\sqrt{q-1}\left(\sqrt{q-1}\right)^{q-1} + \frac{K^{q/2}}{K}c^{q/2}\,(q-1)\left(\sqrt{q-2}\right)^{q-2} \\ &= (Kqc)^{q/2}\left(\left(\frac{q-1}{q}\right)^{q/2} + \frac{q-1}{Kq}\left(\frac{q-2}{q}\right)^{(q-2)/2}\right) \\ &= (Kqc)^{q/2}\left(\frac{q-1}{q}\right)^{q/2}\left(1 + \frac{1}{K}\left(\frac{q-2}{q-1}\right)^{(q-2)/2}\right). \end{aligned}$$

The first part of the theorem then follows from the fact that the factor multiplying $(Kqc)^{q/2}$ on the right-hand side is bounded by 1 for all $q \geq 4$ (Exercise 15.1).

To prove the second part, observe that if, in addition, $V^- \leq c$, then we may apply the first inequality to $-Z$ to obtain

$$\left\| (Z - \mathbf{E}Z)_- \right\|_q \leq K\sqrt{qc}.$$

The statement follows since

$$\mathbf{E}|Z - \mathbf{E}Z|^q = \mathbf{E}(Z - \mathbf{E}Z)_+^q + \mathbf{E}(Z - \mathbf{E}Z)_-^q \leq 2\left(K\sqrt{qc}\right)^q . \qquad \square$$

Now we are ready for the main results of this chapter. The next theorem shows that the q-th moment of Z may be bounded in terms of the $q/2$-th moment of V^+, V^-, and V, thus generalizing the Efron–Stein inequality which only treats the case $q = 2$.

Let $\kappa_1 = 1$ and for any integer $q \geq 2$, define

$$\kappa_q = \frac{1}{2}\left(1 - \left(1 - \frac{1}{q}\right)^{q/2}\right)^{-1}.$$

Then $\kappa_q \nearrow \kappa$ as $q \to \infty$, where

$$\kappa = \frac{\sqrt{e}}{2\left(\sqrt{e} - 1\right)} < 1.271.$$

Theorem 15.5 *For any real $q \geq 2$*

$$\left\|(Z - \mathbf{E}Z)_+\right\|_q \leq \sqrt{\left(1 - \frac{1}{q}\right) 2\kappa_q q \left\|V^+\right\|_{q/2}}$$
$$\leq \sqrt{2\kappa q \left\|V^+\right\|_{q/2}} = \sqrt{2\kappa q}\left\|\sqrt{V^+}\right\|_q,$$

and

$$\left\|(Z - \mathbf{E}Z)_-\right\|_q \leq \sqrt{\left(1 - \frac{1}{q}\right) 2\kappa_q q \left\|V^-\right\|_{q/2}}$$
$$\leq \sqrt{2\kappa q \left\|V^-\right\|_{q/2}} = \sqrt{2\kappa q}\left\|\sqrt{V^-}\right\|_q.$$

Proof It suffices to prove the first inequality, as the second follows from the first by replacing Z by $-Z$.

We prove by induction on k that for all integers $k \geq 1$, and all $q \in (k, k+1]$,

$$\left\|(Z - \mathbf{E}Z)_+\right\|_q \leq \sqrt{q\kappa_q c_q},$$

where $c_q = 2\left\|V^+\right\|_{q/2 \vee 1}(1 - 1/q)$.

For $k = 1$ it follows from Hölder's inequality and the Efron–Stein inequality that

$$\left\|(Z - \mathbf{E}Z)_+\right\|_q \leq \sqrt{\left\|V^+\right\|_1} \leq \sqrt{2\kappa_q \left\|V^+\right\|_{1 \vee q/2}(1 - 1/q)}.$$

Assume now that the property holds for all integers smaller than some $k > 1$, and consider $q \in (k, k+1]$. By Hölder's inequality,

$$\mathbf{E}\left[V^+(Z - \mathbf{E}Z)_+^{q-2}\right] \leq \left\|V^+\right\|_{q/2}\left\|(Z - \mathbf{E}Z)_+\right\|_q^{q-2},$$

so using Lemma 15.1 with $\alpha = q - 1$, we get

$$\left\|(Z - EZ)_+\right\|_q^q \le \left\|(Z - EZ)_+\right\|_{q-1}^q + \frac{q}{2}c_q \left\|(Z - EZ)_+\right\|_q^{q-2}.$$

Defining

$$x_q = \left\|(Z - EZ)_+\right\|_q^q \left(q\kappa_q c_q\right)^{-q/2},$$

it suffices to prove that $x_q \le 1$. With this notation, the previous inequality becomes

$$x_q q^{q/2} c_q^{q/2} \kappa_q^{q/2} \le x_{q-1}^{q/(q-1)} (q-1)^{q/2} c_{q-1}^{q/2} \kappa_{q-1}^{q/2} + \frac{1}{2} x_q^{1-2/q} q^{q/2} c_q^{q/2} \kappa_q^{q/2-1},$$

from which, using $c_{q-1} \le c_q$ and $\kappa_{q-1} \le \kappa_q$, we have

$$x_q \le x_{q-1}^{q/q-1} \left(1 - \frac{1}{q}\right)^{q/2} + \frac{1}{2\kappa_q} x_q^{1-2/q}.$$

Assuming, by induction, that $x_{q-1} \le 1$, this implies that

$$x_q \le \left(1 - \frac{1}{q}\right)^{q/2} + \frac{1}{2\kappa_q} x_q^{1-2/q}.$$

Since the function

$$f_q(x) = \left(1 - \frac{1}{q}\right)^{q/2} + \frac{1}{2\kappa_q} x^{1-2/q} - x$$

is strictly concave on $[0, \infty)$ and positive at $x = 0, f_q(1) = 0$ and $f_q(x_q) \ge 0$ imply that $x_q \le 1$ as desired. □

15.3 Some Variants and Corollaries

Next we present some variants of Theorem 15.5. The first result may be proved by an argument essentially identical to the proof of Theorem 15.5. The details are left to the reader.

Theorem 15.6 *Assume that $Z_i \le Z$ for all $1 \le i \le n$. Then for any real $q \ge 2$,*

$$\left\|(Z - EZ)_+\right\|_q \le \sqrt{\kappa_q q \left\|V\right\|_{q/2}} \le \sqrt{\kappa q \left\|V\right\|_{q/2}}.$$

Even though Theorem 15.5 provides some information concerning the growth of moments of $(Z - \boldsymbol{E}[Z])_-$, this information may be difficult to extract in concrete cases. The following result relates the moments of $(Z - \mathbf{E}[Z])_-$ with $\|V^+\|_q$ rather than with $\|V^-\|_q$. This requires certain boundedness assumptions on the increments of Z.

Theorem 15.7 *Suppose that for every $i = 1, \ldots, n$,*

$$(Z - Z_i')_+ \leq M$$

for a random variable M. Then for every real number $q \geq 2$,

$$\left\|(Z - \mathbf{E}Z)_-\right\|_q \leq \sqrt{Cq\left(\|V^+\|_{q/2} \vee q\,\|M\|_q^2\right)},$$

where $C < 4.16$.

Proof We use the notation $m_q = \left\|(Z - \mathbf{E}Z)_-\right\|_q$. Note that the continuous function

$$e^{-1/2} + \frac{1}{x}e^{1/\sqrt{x}} - 1$$

decreases from ∞ to $e^{-1/2} - 1 < 0$ on $(0, \infty)$. Define C as the unique zero of this function.

Since $C > 1/2$, it follows from Hölder's inequality and the Efron–Stein inequality that for $q \in [1, 2]$,

$$\left\|(Z - \mathbf{E}Z)_-\right\|_q \leq \sqrt{2\,\|V^+\|_1} \leq \sqrt{2\kappa_q\,\|V^+\|_{1 \vee q/2}}.$$

Define

$$c_q = \left\|V^+\right\|_{1 \vee q/2} \vee q\,\|M\|_q^2.$$

For $q \geq 2$, Lemma 15.3 (with $\alpha = q - 1$) implies

$$m_q^q \leq m_{q-1}^q + q\mathbf{E}\left[V^+\left((Z - \mathbf{E}Z)_- + M\right)^{q-2}\right]. \tag{15.2}$$

We first deal with the case $q \in [2, 3)$. By the subadditivity of x^{q-2} for $q \in [2, 3]$, we have

$$\left((Z - \mathbf{E}Z)_- + M\right)^{q-2} \leq M^{q-2} + (Z - \mathbf{E}[Z])_-^{q-2}.$$

Using Hölder's inequality we obtain from (15.2) that

$$m_q^q \leq m_{q-1}^q + q\,\|M\|_q^{q-2}\left\|V^+\right\|_{q/2} + q\left\|V^+\right\|_{q/2} m_q^{q-2}.$$

Using the fact that $m_{q-1} \leq \sqrt{c_{q-1}} \leq \sqrt{c_q}$, this implies

$$m_q^q \leq c_q^{q/2} + q^{2-q/2}c_q^{q/2} + qc_q m_q^{q-2}.$$

Let $x_q = \left(\frac{m_q}{\sqrt{Cqc_q}} \right)^q$. Then the preceding inequality becomes

$$x_q \le \left(\frac{1}{Cc_q} \right)^{q/2} + \frac{1}{C} \left(\left(\sqrt{Cq} \right)^{-q+2} + x_q^{1-2/q} \right)$$

which in turn implies

$$x_q \le \frac{1}{2C} + \frac{1}{C} \left(1 + x_q^{1-2/q} \right)$$

since $q \ge 2$ and $C \ge 1$. The function

$$g_q(x) = \frac{1}{2C} + \frac{1}{C} \left(1 + x^{1-2/q} \right) - x$$

is strictly concave on $[0, \infty)$ and positive at 0. Furthermore, $g_q(1) = 5/(2c) - 1 < 0$, since $C > 5/2$. Hence $g_q(x_q) \ge 0$ only if $x_q \le 1$, which settles the case $q \in [2, 3]$.

We now turn to the case $q \ge 3$. We prove by induction on $k \ge 2$, that for all $q \in [k, k+1)$, $m_q \le \sqrt{qC\kappa_q c_q}$. By the convexity of x^{q-2} we have, for every $\theta \in (0, 1)$,

$$\begin{aligned} ((Z - \mathbf{E}Z)_- + M)^{q-2} &= \left(\theta \frac{(Z - \mathbf{E}Z)_-}{\theta} + (1-\theta) \frac{M}{1-\theta} \right)^{q-2} \\ &\le \theta^{-q+3} M^{q-2} + (1-\theta)^{-q+3} (Z - \mathbf{E}[Z])_-^{q-2}. \end{aligned}$$

Using Hölder's inequality we obtain from (15.2) that

$$m_q^q \le m_{q-1}^q + q\theta^{-q+3} \|M\|_q^{q-2} \left\| V^+ \right\|_{q/2} + q (1-\theta)^{-q+3} \left\| V^+ \right\|_{q/2} m_q^{q-2}.$$

Now assume by induction that $m_{q-1} \le \sqrt{C(q-1) c_{q-1}}$. Since $c_{q-1} \le c_q$, we have

$$m_q^q \le C^{q/2} (q-1)^{q/2} c_q^{q/2} + q^{-q+2} \theta^{-q+3} q^{q/2} c_q^{q/2} + q (1-\theta)^{-q+3} c_q m_q^{q-2}.$$

Let $x_q = C^{-q/2} m_q^q (qc_q)^{-q/2}$. Then it suffices to show that $x_q \le 1$ for all $q > 2$. Observe that

$$x_q \le \left(1 - \frac{1}{q} \right)^{q/2} + \frac{1}{C} \left(\theta^{-q+3} \left(\sqrt{Cq} \right)^{-q+2} + (1-\theta)^{-q+3} x_q^{1-2/q} \right).$$

We choose θ minimizing

$$g(\theta) = \theta^{-q+3} \left(\sqrt{Cq} \right)^{-q+2} + (1-\theta)^{-q+3},$$

that is, $\theta = 1/\left(\sqrt{Cq}+1\right)$. Since for this value of θ,

$$g(\theta) = \left(1 + \frac{1}{\sqrt{Cq}}\right)^{q-2},$$

the bound on x_q becomes

$$x_q \le \left(1 - \frac{1}{q}\right)^{q/2} + \frac{1}{C}\left(1 + \frac{1}{\sqrt{Cq}}\right)^{q-2}\left(1 + \left(\frac{\sqrt{Cq}}{1+\sqrt{Cq}}\right)\left(x_q^{1-2/q} - 1\right)\right).$$

Hence, using the elementary inequalities

$$\left(1 - \frac{1}{q}\right)^{q/2} \le e^{-1/2} \quad \text{and} \quad \left(1 + \frac{1}{\sqrt{Cq}}\right)^{q-2} \le e^{1/\sqrt{C}},$$

we get

$$x_q \le e^{-1/2} + \frac{e^{1/\sqrt{C}}}{C}\left(\frac{\sqrt{Cq}}{1+\sqrt{Cq}}\right)\left(x_q^{1-2/q} - 1\right).$$

Since the function

$$f_q(x) = e^{-1/2} + \frac{e^{1/\sqrt{C}}}{C}\left(1 + \left(\frac{\sqrt{Cq}}{1+\sqrt{Cq}}\right)\left(x^{1-2/q} - 1\right)\right) - x$$

is strictly concave on $[0,\infty)$ and positive at 0 and C is defined in such a way that $f_q(1) = 0$, f_q can be nonnegative at x_q only if $x_q \le 1$ which proves the theorem by induction. □

The next corollary allows us to deal with "generalized" self-bounding functions.

Corollary 15.8 *Assume that $Z_i \le Z$ for all $i = 1, \ldots, n$ and $V \le WZ$ for a random variable $W \ge 0$. Then for all real numbers $q \ge 2$,*

$$\|Z\|_q \le 2\mathbf{E}Z + \kappa q \|W\|_q.$$

Also,

$$\left\|(Z - \mathbf{E}Z)_+\right\|_q \le \sqrt{2\kappa q \|W\|_q \mathbf{E}Z} + \kappa q \|W\|_q.$$

Proof Let $q \geq 2$. Then

$$\begin{aligned}
&\left\|(Z - EZ)_+\right\|_q \\
&\leq \sqrt{\kappa q \left\|WZ\right\|_{q/2}} \quad \text{(by Theorem 15.6)} \\
&\leq \sqrt{\kappa q \left\|Z\right\|_q \left\|W\right\|_q} \quad \text{(by Hölder's inequality)} \\
&\leq \frac{1}{2}\left\|Z\right\|_q + \frac{\kappa q}{2}\left\|W\right\|_q \quad \text{since } \sqrt{ab} \leq (a+b)/2 \text{ for } a, b \geq 0.
\end{aligned}$$

Now $Z \geq 0$ implies that $\left\|(Z - EZ)_-\right\|_q \leq EZ$ and we have $\left\|Z\right\|_q \leq EZ + \left\|(Z - EZ)_+\right\|_q$. Hence,

$$\left\|Z\right\|_q \leq 2EZ + \kappa q \left\|W\right\|_q,$$

concluding the proof of the first statement. To prove the second inequality, note that

$$\begin{aligned}
&\left\|(Z - EZ)_+\right\|_q \\
&\leq \sqrt{\kappa q \left\|WZ\right\|_{q/2}} \quad \text{(by Theorem 15.6)} \\
&\leq \sqrt{\kappa q \left\|W\right\|_q \left\|Z\right\|_q} \quad \text{(by Hölder's inequality)} \\
&\leq \sqrt{\kappa q \left\|W\right\|_q \left(2EZ + \kappa q \left\|W\right\|_q\right)} \quad \text{(by the first inequality)} \\
&\leq \sqrt{2\kappa q \left\|W\right\|_q EZ} + \kappa q \left\|W\right\|_q
\end{aligned}$$

as desired. □

15.4 Sums of Random Variables

In this section we apply the results stated in Sections 15.2 and 15.3 for sums of independent random variables. As a result, we recover versions of some classical moment inequalities such as the Khinchine–Kahane, Marcinkiewicz, and Rosenthal inequalities. We emphasize that rather than offering an exhaustive account of moment inequalities for sums of independent random variables, we illustrate how the machinery developed in the previous sections may be used to obtain such inequalities. In all cases, the proof does not require much further work. Also, we obtain explicit constants which only depend on q. These constants are not optimal, though in some cases their dependence on q is of the right order.

The simplest example is the case of the Khinchine's inequality which states that for all $q \geq 2$, there exists a B_q such that for all $a_1, \ldots, a_n > 0$,

$$\sqrt{\sum_{i=1}^{n} a_i^2} \leq \left(\mathbf{E} \left| \sum_{i=1}^{n} a_i X_i \right|^q \right)^{1/q} \leq B_q \sqrt{\sum_{i=1}^{n} a_i^2},$$

where $X_1, \ldots, X_n$ are independent Rademacher variables. The inequality on the left-hand side is a simple application of Jensen's inequality, while the upper bound follows from Theorem 15.4 as follows.

Theorem 15.9 (KHINCHINE'S INEQUALITY) *Let $a_1, \ldots, a_n > 0$ be constants and let $X_1, \ldots, X_n$ be independent Rademacher variables (i.e. with $\mathbf{P}\{X_i = -1\} = \mathbf{P}\{X_i = 1\} = 1/2$). If $Z = \sum_{i=1}^{n} a_i X_i$ then for any integer $q \geq 2$,*

$$\left\| (Z)_+ \right\|_q = \left\| (Z)_- \right\|_q \leq \sqrt{2Kq} \sqrt{\sum_{i=1}^{n} a_i^2}$$

and

$$\left\| Z \right\|_q \leq 2^{1/q} \sqrt{2Kq} \sqrt{\sum_{i=1}^{n} a_i^2}$$

where $K = 1/\left(e - \sqrt{e}\right) < 0.935$.

Proof We may use Theorem 15.4. Since

$$V^+ = \sum_{i=1}^{n} \mathbf{E}\left[(a_i(X_i - X_i'))_+^2 \mid X_i \right] = 2 \sum_{i=1}^{n} a_i^2 \mathbb{1}_{a_i X_i > 0} \leq 2 \sum_{i=1}^{n} a_i^2,$$

the result follows. □

Note also that using a symmetrization argument (see Exercise 15.5), Khinchine's inequality above implies Marcinkiewicz' inequality: if $X_1, \ldots, X_n$ are independent centered random variables then for any $q \geq 2$,

$$\left\| \sum_{i=1}^{n} X_i \right\|_q \leq 2^{1+1/q} \sqrt{2Kq} \sqrt{\left\| \sum_{i=1}^{n} X_i^2 \right\|_{q/2}} .$$

Another classical moment inequality for sums of independent random variables is Rosenthal's inequality that bounds the q-th moment of the sum in terms of the q-th moment of the individual variables. The case of nonnegative and centered summands are usually dealt with separately. Next we prove two such results that we obtain from our general moment inequalities.

Theorem 15.10 *Define*

$$Z = \sum_{i=1}^{n} X_i,$$

where $X_1, \ldots, X_n$ are independent and nonnegative random variables. Then for all integers $q \geq 1$,

$$\left\| (Z - EZ)_+ \right\|_q \leq \sqrt{2\kappa q \left\| \max_{i=1,\ldots,n} X_i \right\|_q EZ} + \kappa q \left\| \max_{i=1,\ldots,n} X_i \right\|_q,$$

$$\left\| (Z - EZ)_- \right\|_q \leq \sqrt{Kq \sum_{i=1}^{n} EX_i^2}.$$

Also,

$$\|Z\|_q \leq 2EZ + \kappa q \left\| \max_{i=1,\ldots,n} X_i \right\|_q.$$

Proof To prove the first and the third inequalities, we may use Corollary 15.8. Simply note that

$$V = \sum_{i=1}^{n} X_i^2 \leq WZ,$$

where

$$W = \max_{i=1,\ldots,n} X_i.$$

In order to obtain the second inequality, just observe that

$$V^- \leq \sum_{i=1}^{n} E\left[X_i'^2\right],$$

and apply Theorem 15.4 to $-Z$. □

Note that Rosenthal's inequality – and its variants – typically bound the moments of $\sum_{i=1}^n X_i$ in terms of $\sum_{i=1}^n E|X_i|^q$ and not in terms of $\| \max_{i=1,\ldots,n} X_i \|_q$ as in the theorem above. However, by bounding $E| \max_{i=1,\ldots,n} X_i|^q \leq \sum_{i=1}^n E|X_i|^q$ we recover inequalities of the usual form.

We may use the previous result to derive a Rosenthal-type inequality for sums of centered variables.

Theorem 15.11 *Let $X_1, \ldots, X_n$ be independent real-valued random variables with $EX_i = 0$. Define*

$$Z = \sum_{i=1}^{n} X_i, \quad \sigma^2 = \sum_{i=1}^{n} EX_i^2, \quad Y = \max_{i=1,\ldots,n} |X_i|.$$

Then for any integer $q \geq 2$,

$$\|Z_+\|_q \leq \sigma\sqrt{6\kappa q} + q\kappa\sqrt{2}\,\|Y\|_q .$$

Proof We use Theorem 15.5. Note that V^+ (defined at the beginning of the chapter) equals

$$V^+ = \sum_{i=1}^n X_i^2 + \sum_{i=1}^n EX_i^2.$$

Thus,

$$\begin{aligned}
\left\|(Z)_+\right\|_q &\leq \sqrt{2\kappa q\,\|V^+\|_{q/2}} \quad \text{(by Theorem 15.5)},\\
&\leq \sqrt{2\kappa q}\sqrt{\left(\sum_{i=1}^n EX_i^2\right) + \left\|\sum_{i=1}^n X_i^2\right\|_{q/2}}\\
&\leq \sqrt{2\kappa q}\sqrt{\sum_{i=1}^n EX_i^2 + 2\sum_{i=1}^n EX_i^2 + 2\kappa q\,\|Y^2\|_{q/2}}\\
&\quad \text{(by Theorem 15.10)}\\
&= \sqrt{2\kappa q}\sqrt{3\sum_{i=1}^n EX_i^2 + \kappa q\,\|Y^2\|_{q/2}}\\
&\leq \sigma\sqrt{6\kappa q} + q\kappa\sqrt{2}\,\|Y\|_q .
\end{aligned}$$

□

15.5 Suprema of Empirical Processes

Next we apply our general moment inequalities to derive bounds for the moments of suprema of empirical processes. The arguments are no more difficult than those of the previous section for sums of independent random variables. As a first illustration, we point out that the proof of Khinchine's inequality in the previous section extends, in a straightforward way, to an analogous supremum. The basic notation and conventions for empirical processes are introduced in Chapter 11.

Theorem 15.12 *Let $\mathcal{T} \subset \mathbb{R}^n$ be a (countable) set of vectors $t = (t_1, \ldots, t_n)$ and let $X_1, \ldots, X_n$ be independent Rademacher variables. If $Z = \sup_{t\in\mathcal{T}} \sum_{i=1}^n t_i X_i$ then for any integer $q \geq 2$,*

$$\left\|(Z - EZ)_+\right\|_q \leq \sqrt{2Kq}\,\sup_{t\in\mathcal{T}}\sqrt{\sum_{i=1}^n t_i^2}$$

where $K = 1/\left(e - \sqrt{e}\right)$, *and*

$$\left\|(Z - EZ)_-\right\|_q \leq \sqrt{2Cq}\sup_{t\in\mathcal{T}}\sqrt{\sum_{i=1}^n t_i^2} \vee 2\sqrt{Cq}\sup_{i,t}|t_i|.$$

where C is defined as in Theorem 15.7.

Before stating the main result of the section, we mention the following consequence of Corollary 15.8.

Theorem 15.13 *Let* $X_1, \ldots, X_n$ *denote a collection of independent random vectors with nonnegative coordinates indexed by the countable set* $\mathcal{T}$. *Let* $Z = \sup_{s\in\mathcal{T}}\sum_{i=1}^n X_{i,s}$ *and let*

$$M = \max_{i=1,\ldots,n}\sup_{s\in\mathcal{T}} X_{i,s}.$$

Then, for all $q \geq 2$,

$$\|Z\|_q \leq 2EZ + \kappa q\,\|M\|_q\,.$$

Next we turn to the case of centered processes. Let $\mathcal{T}$ denote a countable index set. Let $X_1, \ldots, X_n$ denote independent random vectors indexed by $\mathcal{T}$ such that for all $s \in \mathcal{T}$ and $i = 1, \ldots, n$, $EX_{i,s} = 0$. Let

$$Z = \sup_{s\in\mathcal{T}}\left|\sum_{i=1}^n X_{i,s}\right|.$$

Recall from Chapter 11 the definition of the weak variance Σ^2 and the wimpy variance σ^2:

$$\Sigma^2 = E\sup_{s\in\mathcal{T}}\sum_{i=1}^n X_{i,s}^2, \quad \sigma^2 = \sup_{s\in\mathcal{T}} E\sum_{i=1}^n X_{i,s}^2.$$

A third quantity appearing in the moment and tail bounds is

$$M = \max_{i=1,\ldots,n} Y_i$$

where $Y_i = \sup_{s\in\mathcal{T}}|X_{i,s}|$. The random variable Y_i is often called the *envelope* of the collection of coordinates.

Before stating the main theorem, we recall the connection between the wimpy and the weak variances established by Theorem 11.17:

$$\Sigma^2 \leq \sigma^2 + 32\sqrt{EM^2}EZ + 8EM^2.$$

The next theorem offers two upper bounds for the moments of suprema of centered empirical processes.

Theorem 15.14 *Let $X_1, \ldots, X_n$ denote independent random vectors indexed by $\mathcal{T}$ such that for all $s \in \mathcal{T}$ and $i = 1, \ldots, n$, $\mathbf{E}X_{i,s} = 0$. Let*

$$Z = \sup_{s \in \mathcal{T}} \left| \sum_{i=1}^{n} X_{i,s} \right|.$$

Then for all $q \geq 2$,

$$\left\| (Z - \mathbf{E}Z)_+ \right\|_q \leq \sqrt{2\kappa q}\,(\Sigma + \sigma) + 2\kappa q\Big(\|M\|_q + \sup_{\substack{s \in \mathcal{T} \\ i=1,\ldots,n}} \|X_{i,s}\|_2 \Big),$$

and furthermore

$$\|Z\|_q \leq 2\mathbf{E}Z + 2\sigma\sqrt{2\kappa q} + 20\kappa q \, \|M\|_q + 4\sqrt{\kappa q}\, \|M\|_2 .$$

Proof The proof is based on the Theorem 15.5 which states that

$$\left\| (Z - \mathbf{E}Z)_+ \right\|_q \leq \sqrt{2\kappa q \, \|V^+\|_{q/2}}.$$

We may bound V^+ as follows.

$$\begin{aligned} V^+ &\leq \sup_{s \in \mathcal{T}} \sum_{i=1}^{n} \mathbf{E}\left[(X_{i,s} - X'_{i,s})^2 \mid X_1^n \right] \\ &\leq \sup_{s \in \mathcal{T}} \sum_{i=1}^{n} \left(\mathbf{E}X_{i,s}^2 + X_{i,s}^2 \right) \\ &\leq \sup_{s \in \mathcal{T}} \sum_{i=1}^{n} \mathbf{E}X_{i,s}^2 + \sup_{s \in \mathcal{T}} \sum_{i=1}^{n} X_{i,s}^2. \end{aligned}$$

Thus, by Minkowski's inequality and the Cauchy–Schwarz inequality,

$$\begin{aligned} &\sqrt{\|V^+\|_{q/2}} \\ &\leq \sqrt{\sup_{s \in \mathcal{T}} \sum_{i=1}^{n} \mathbf{E}X_{i,s}^2 + \left\| \sup_{s \in \mathcal{T}} \sum_{i=1}^{n} X_{i,s}^2 \right\|_{q/2}} \\ &\leq \sigma + \left\| \sup_{s \in \mathcal{T}} \sqrt{\sum_{i=1}^{n} X_{i,s}^2} \right\|_q \\ &= \sigma + \left\| \sup_{s \in \mathcal{T}} \sup_{\alpha : \|\alpha\|_2 \leq 1} \sum_{i=1}^{n} \alpha_i X_{i,s} \right\|_q \\ &\leq \sigma + \Sigma + \left\| \left(\sup_{s \in \mathcal{T}, \alpha : \|\alpha\|_2 \leq 1} \sum_{i=1}^{n} \alpha_i X_{i,s} - \mathbf{E} \sup_{s \in \mathcal{T}, \alpha : \|\alpha\|_2 \leq 1} \sum_{i=1}^{n} \alpha_i X_{i,s} \right)_+ \right\|_q . \end{aligned}$$

The last term may be upper bounded again by Theorem 15.5. Indeed, the corresponding V^+ is not more than

$$\max_{i=1,\ldots,n} \sup_{s\in\mathcal{T}} X_{i,s}^2 + \max_{i=1,\ldots,n} \sup_{s\in\mathcal{T}} \boldsymbol{E}[X_{i,s}^2],$$

and thus

$$\left\| \left(\sup_{s\in\mathcal{T},\alpha:\|\alpha\|_2\le 1} \sum_{i=1}^n \alpha_i X_{i,s} - \boldsymbol{E} \sup_{s\in\mathcal{T},\alpha:\|\alpha\|_2\le 1} \sum_{i=1}^n \alpha_i X_{i,s} \right)_+ \right\|_q$$

$$\le \sqrt{2\kappa q}\left(\|M\|_q + \max_{i=1,\ldots,n} \sup_{s\in\mathcal{T}} \left\|X_{i,s}\right\|_2 \right).$$

This completes the proof of the first inequality of the theorem. The second inequality follows because by nonnegativity of Z, $\left\|(Z-\boldsymbol{E}Z)_-\right\|_q \le \boldsymbol{E}Z$ and therefore $\|Z\|_q \le \boldsymbol{E}Z + \left\|(Z-\boldsymbol{E}Z)_+\right\|_q$ and because by the first inequality, combined with Lemma 11.17, we have

$$\left\|(Z-\boldsymbol{E}Z)_+\right\|_q \le \sqrt{2\kappa q}\left(\sigma + \sqrt{32\sqrt{\boldsymbol{E}M^2}\boldsymbol{E}Z} + \sqrt{8\boldsymbol{E}M^2} + \sigma \right)$$

$$+2\kappa q\left(\|M\|_q + \sup_{i,s\in\mathcal{T}} \left\|X_{i,s}\right\|_2 \right)$$

$$\le \boldsymbol{E}Z + 2\sigma\sqrt{2\kappa q} + 16\kappa\sqrt{\boldsymbol{E}M^2} + \sqrt{16\kappa q\boldsymbol{E}M^2}$$

$$+2\kappa q\left(\|M\|_q + \sup_{i,s\in\mathcal{T}} \left\|X_{i,s}\right\|_2 \right)$$

(using the inequality $\sqrt{ab} \le a + b/4$).

Using $\|M\|_2 \le \|M\|_q$ and $\sup_{s\in\mathcal{T},i=1,\ldots,n} \left\|X_{i,s}\right\|_2 \le \|M\|_2$, we obtain the desired result. □

15.6 Conditional Rademacher Averages

As another easy application of the general moment bounds, we now study conditional Rademacher averages. We have already met these functions in Section 3.3 but there we assumed that the class $\mathcal{F}$ only contains bounded functions. When this is not the case, the result below may be useful.

Let $\mathcal{F}$ be a countable class of measurable real-valued functions. The conditional Rademacher average is defined by

$$Z = \boldsymbol{E}\left[\sup_{f\in\mathcal{F}} \left| \sum_{i=1}^n \varepsilon_i f(X_i) \right| \mid X_1^n \right]$$

where the ε_i are independent Rademacher random variables.

Theorem 15.15 *Let Z denote a conditional Rademacher average and let* $M = \sup_{i,f} f(X_i)$. *Then*

$$\left\| (Z - EZ)_+ \right\|_q \le \sqrt{2\kappa q \left\| M \right\|_q EZ} + \kappa q \left\| M \right\|_q ,$$

and

$$\left\| (Z - EZ)_- \right\|_q \le \sqrt{2C_2} \left(\sqrt{q \left\| M \right\|_q EZ} + 2q \left\| M \right\|_q \right)$$

where C_2 *is the constant of Exercise 15.3.*

Proof Define

$$Z_i = \mathbf{E}\left[\sup_{f \in \mathcal{F}} \left| \sum_{j \neq i} \varepsilon_j f(X_j) \right| \mid X_1^n \right].$$

Recall from Section 3.3 the self-bounding property of conditional Rademacher averages. In particular, even without the boundedness assumption, we still have that for all i, $Z - Z_i \ge 0$ and

$$\sum_{i=1}^{n} (Z - Z_i) \le Z.$$

Thus, we have

$$V \le ZM, \text{ and } Z - Z_i \le M.$$

The result now follows by Corollary 15.8, noting that $M = W$. □

15.7 Bibliographical Remarks

The material of this chapter is mostly based on Boucheron *et al.* (2005*b*).

Recall Burkholder's inequalities from martingale theory. Burkholder's inequalities may be regarded as extensions of Marcinkiewicz's inequalities to sums of martingale increments. They are natural candidates for deriving moment inequalities for a function $Z = f(X_1, \ldots, X_n)$ of many independent random variables. The approach mimics the method of bounded differences (see Section 6.1) classically used to derive Bernstein- or Hoeffding-like inequalities under similar circumstances. The method works as follows: let $\mathcal{F}_i$ denote the σ-algebra generated by the sequence (X_1^i). Then the sequence $M_i = \mathbf{E}[Z|\mathcal{F}_i]$ is an $\mathcal{F}_i$-adapted martingale (the Doob martingale associated with Z). Let $\langle Z \rangle$ denote the associated *quadratic variation*

$$\langle Z \rangle = \sum_{i=1}^{n} (M_i - M_{i-1})^2,$$

let $[Z]$ denote the *predictable quadratic variation*

$$[Z] = \sum_{i=1}^{n} \boldsymbol{E}\left[(M_i - M_{i-1})^2 \mid \mathcal{F}_{i-1}\right],$$

and let D be defined as $D = \max_{i=1,\ldots,n} |M_i - M_{i-1}|$. Burkholder's inequalities imply that for $q \geq 2$,

$$\|Z - \boldsymbol{E}Z\|_q \leq (q-1)\sqrt{\|\langle Z\rangle\|_{q/2}} = (q-1)\left\|\sqrt{\langle Z\rangle}\right\|_q.$$

Note that the dependence on q in this inequality differs from the dependence in Theorem 15.5. The Burkholder–Rosenthal–Pinelis inequality (Pinelis, 1994, Theorem 4.1) implies that there exists a universal constant C such that

$$\|Z - \boldsymbol{E}Z\|_q \leq C\left(\sqrt{q\left\|[Z]\right\|_{q/2}} + q\,\|D\|_q\right).$$

With some extra information on the sensitivity of Z with respect to its arguments, such inequalities may be used to develop a strict analog of the method of bounded differences for moment inequalities. In principle such an approach should provide tight results, but finding good bounds on the moments of the quadratic variation process often proves quite difficult.

The inequalities introduced in this chapter have a form similar to those obtained by Doob's martingale representation and Burkholder's inequality. But, instead of relying on the quadratic variation process, they rely on a more tractable quantity. Indeed, in many cases V^+ and V^- are easier to deal with than $[Z]$ or $\langle Z\rangle$.

For more information on moment inequalities for sums of independent random variables, we refer to the de la Peña and Giné (1999).

For some historical notes on Khinchine's inequality, see the bibliographical remarks of Chapter 5. For Marcinkiewicz' inequalities see, for example, de la Peña and Giné (1999, page 34).

There are numerous versions of Rosenthal's inequality, the first dating back to Rosenthal (1970).

Burkolder's inequalities are described and surveyed in Burkholder (1988, 1989), but see also Chow and Teicher (1978, page 384). It is known that for general martingales, Burkholder's inequality is essentially unimprovable (see Burkholder 1989, Theorem 3.3). However, for the special case of Doob martingale associated with Z this bound is perhaps improvable.

Theorem 15.14, may be regarded as an analog of Talagrand's inequality (Talagrand, 1996*b*) for moments. Indeed, Talagrand's exponential inequality may be easily deduced from Theorem 15.14 by bounding the moment-generating function by moments.

Theorem 15.10 is similar to inequality (H_r) in Giné, Latała, and Zinn (2000), which follows from an improved Hoffmann–Jørgensen inequality by Kwapień and Woyczyńsky (1992).

The first inequality in Theorem 15.14 improves inequality (3) of Pinelis (1995). The second inequality is a version of Proposition 3.1 of Giné, Latała and Zinn (2000).

Pinelis (1995) extends Theorem 15.11 for martingales.

The paper by Boucheron *et al.* (2005*b*) contains applications to Rademacher chaos and Boolean polynomials. Clémençon, Lugosi, and Vayatis (2008) apply these inequalities to obtain moment inequalities for U-statistics.

Conditional Rademacher averages appeared at the core of the early concentration inequalities used in the theory of probability in Banach spaces (see Ledoux and Talagrand (1991)).

15.8 EXERCISES

15.1. Prove that for all integers $q \geq 4$,

$$x_q = \left(\frac{q-1}{q}\right)^{q/2}\left(1 + \frac{1}{K}\left(\frac{q-2}{q-1}\right)^{(q-2)/2}\right) \leq 1.$$

Also, $\lim_{q\to\infty} x_q = 1$.

15.2. Mimic the argument of Theorem 15.5 to prove Theorem 15.6.

15.3. Prove the following variant of Theorem 15.7. Suppose that for every $i = 1\dots,n$, $0 \leq Z - Z_i \leq M$. Then

$$\left\|(Z - \mathbf{E}Z)_-\right\|_q \leq \sqrt{C_2 q\left(\|V\|_{q/2} \vee q\,\|M\|_q^2\right)},$$

where $C_2 < 2.42$.

15.4. Combine the previous exercise with the proof of Corollary 15.8 to show the following. Assume that $Z_i \leq Z$ for all $i = 1,\dots,n$ and $V \leq WZ$ for a random variable $W \geq 0$. Suppose also that for every $1 \leq i \leq n$,

$$0 \leq Z - Z_i \leq M$$

for some random variable M. Then for all $q \geq 2$,

$$\left\|(Z - \mathbf{E}Z)_-\right\|_q \leq \sqrt{C_2 q\left(\|M\|_q\left(2\mathbf{E}Z + 2q\,\|W\|_q\right) \vee q\,\|M\|_q^2\right)}$$

where $C_2 < 2.42$ is as in Exercise 15.3.

15.5. Use symmetrization and Theorem 15.9 to derive the following version of Marcinkiewicz' inequality: if $X_1,\dots,X_n$ are independent centered random variables then for any $q \geq 2$,

$$\left\|\sum_{i=1}^n X_i\right\|_q \leq 2^{1+1/q}\sqrt{2Kq}\sqrt{\left\|\sum_{i=1}^n X_i^2\right\|_{q/2}}.$$

15.6. Let $X_1,\ldots,X_n$ be independent standard Gaussian random variables. Let $\mu = \mathbf{E}X_i^4 = 3$. Let $a_1,\ldots,a_n$ be real numbers. Let $Z = \sum_{i=1}^n a_i(X_i^4 - \mu)$. Find upper bounds for the variance of Z and $\|Z_+\|_q$ for $q \geq 2$.

15.7. Let $X_1,\ldots,X_n$ be symmetric exponentially distributed independent random variables such that $\mathbf{P}\{|X_i| > x\} = e^{-x}$ for $x \geq 0$. Let $s \in [0,\infty)^n$ have nonincreasing coordinates $s_1 \geq s_2 \geq \cdots \geq s_n$. Prove that there exists $\kappa > 0$ such that for all $p \geq 2$,

$$\mathbf{E}\left[\left|\sum_{i=1}^n s_i X_i\right|^p\right]^{1/p} \leq \kappa\left(p\sum_{i=1}^p s_i + \sqrt{p\sum_{i=p+1}^n s_i^2}\right).$$

15.8. Let $X_1,\ldots,X_n$ be independent standard Gaussian random variables. Let $\mathcal{T}$ be a countable index set. Let $a_1,\ldots,a_n$ be vectors indexed by $\mathcal{T}$. Let $Z = \sup_{s\in\mathcal{T}} \sum_{i=1}^n a_{i,s}(X_i^4 - \mu)$. Find upper bounds for the variance of Z and $\|Z_+\|_q$ for $q \geq 2$.

15.9. Let Z satisfy Bernstein's inequality with variance factor $\sigma^2 + 2\mathbf{E}Z$ and scale factor $1/3$. Prove that for $\theta > 0$ and $\lambda \in [0,1]$,

$$\mathbf{P}\{Z - \mathbf{E}Z \geq \theta\mathbf{E}Z + t\} \leq \exp\left(-\frac{\lambda t^2}{2\sigma^2}\right) \vee \exp\left(-\frac{(1-\lambda)t}{2(1/3 + 2/\theta)}\right).$$

Hint: verify that for all $u,v > 0$ and all $0 \leq \lambda \leq 1$, $\exp(-1/(u+v)) \leq \exp(-\lambda/u) \vee \exp(-(1-\lambda)/v)$. See Adamczak (2008).

15.10. Let $X_1,\ldots,X_n$ be independent identically distributed random vectors indexed by the countable set $\mathcal{T}$. Assume that for all $i = 1\ldots,n$, $s \in \mathcal{T}$, $\mathbf{E}X_{i,s} = 0$. For $i = 1,\ldots,n$, let $Y_i = \sup_{s\in\mathcal{T}} |X_{i,s}|$. Assume that for some $b > 0$, $\mathbf{E}e^{Y_i/b} \leq 2$. Let $Z = \sup_{s\in\mathcal{T}} \left|\sum_{i=1}^n X_{i,s}\right|$ and $\sigma^2 = \sup_{s\in\mathcal{T}} \sum_{i=1}^n \mathbf{E}X_{i,s}^2$. Prove that for all $0 < \varepsilon < 1$ and $\delta > 0$ there exists $\kappa = \kappa(\varepsilon,\delta)$ such that for all $t > 0$,

$$\mathbf{P}\left\{Z \geq (1+\varepsilon)\mathbf{E}Z + t\right\} \leq \exp\left(-\frac{t^2}{2(1+\delta)\sigma^2}\right) \vee 3\exp\left(-\frac{t}{\kappa b\log n}\right).$$

See Adamczak (2008).

REFERENCES

Achlioptas, D. (2003). Data base friendly random projections: Johnson–Lindenstrauss with binary coins. *Journal of Computer and System Sciences*, **66**, 671–687.

Adamczak, R. (2006). Moment inequalities for *u*-statistics. *The Annals of Probability*, **34**, 2288–2314.

Adamczak, R. (2008). A tail inequality for suprema of unbounded empirical processes with applications to Markov chains. *Electronic Journal of Probability*, **13**, 1000–1034.

Adamczak, R., Litvak, E., Pajor, A., and Tomczak-Jaegermann, N. (2010). Quantitative estimates of the convergence of the empirical covariance matrix in log-concave ensembles. *Journal of the American Mathematical Society*, **23**, 535–561.

Ahlswede, R. and Winter, A. (2002). Strong converse for identification via quantum channels. *IEEE Transactions on Information Theory*, **48**, 569–579.

Ahlswede, R., Gács, P., and Körner, J. (1976). Bounds on conditional probabilities with applications in multi-user communication. *Zeitschrift für Wahrscheinlichkeitstheorie und verwandte Gebiete*, **34**, 157–177. (Correction in **39**: 353–354, (1977)).

Aida, S. and Stroock, D. (1994). Moment estimates derived from Poincaré and logarithmic Sobolev inequalities. *Mathematical Research Letters*, **1**, 75–86.

Aida, S., Masuda, T., and Shigekawa, I. (1994). Logarithmic Sobolev inequalities and exponential integrability. *Journal of Functional Analysis*, **126**, 83–101.

Aldous, D.J. (2001). The $\zeta(2)$ limit in the random assignment problem. *Random Structures & Algorithms*, **18**, 381–418.

Aldous, D.J. and Diaconis, P. (1995). Hammersley's interacting particle process and longest increasing subsequences. *Probability Theory and Related Fields*, **103**, 199–213.

Alexander, K. (1987). Rates of growth and sample moduli for weighted empirical processes indexed by sets. *Probability Theory and Related Fields*, **75**, 379–423.

Alon, N. and Milman, V.D. (1985). λ_1, isoperimetric inequalities for graphs, and superconcentrators. *Journal of Combinatorial Theory. Series B*, **38**, 73–88.

Alon, N. and Spencer, J.H. (1992). *The Probabilistic Method*. Wiley, New York.

Alon, N., Krivelevich, M., and Vu, V.H. (2002). On the concentration of eigenvalues of random symmetric matrices. *Israel Journal of Mathematics*, **131**, 259–267.

Amsalu, S., Houdré, C., and Matzinger, H. (2012). Sparse long blocks and the variance of the LCS. Arxiv preprint arXiv:1204.1009.

Anderson, G.W., Guionnet, A., and Zeitouni, O. (2010). *An Introduction to Random Matrices*, Volume 118 of *Cambridge Studies in Advanced Mathematics*. Cambridge University Press, Cambridge.

Ané, C. and Ledoux, C. (2000). On logarithmic Sobolev inequalities for continuous time random walks on graphs. *Probability Theory and Related Fields*, **116**, 573–602.

Ané, C., Blachère, S., Chafaï, D., Fougères, P., Gentil, I., Malrieu, F., Roberto, C., and Scheffer, G. (2000). *Sur les inégalités de Sobolev logarithmiques*, Volume 10 of *Panoramas et Synthèses*. Société Mathématique de France, Paris.

Angluin, D. and Valiant, L.G. (1979). Fast probabilistic algorithms for Hamiltonian circuits and matchings. *Journal of Computing System Science*, **18**, 155–193.

Anthony, M. and Bartlett, P.L. (1999). *Neural Network Learning: Theoretical Foundations*. Cambridge University Press, Cambridge.

Anthony, M. and Shawe-Taylor, J. (1993). A result of Vapnik with applications. *Discrete Applied Mathematics*, **47**, 207–217.

Apostol, T.M. (1969). *Calculus II*, Volume II. Math. Assoc. Amer, Washington DC.

Arcones, M.A. and Giné, E. (1993). On decoupling, series expansions, and tail behavior of chaos processes. *Journal of Theoretical Probability*, **6**, 101–122.

Arlot, S. (2007). Rééchantillonnage et Sélection de modèles. Ph. D. thesis, Université Paris-Sud.

Arnold, A., Markowich, P., Toscani, G., and Unterreiter, A. (2001). On convex Sobolev inequalities and the rate of convergence to equilibrium for Fokker–Planck type equations. *Communications in Partial Differential Equations*, **26**, 43–100.

Aubrun, G. (2005). A sharp small deviation inequality for the largest eigenvalue of a random matrix. In *Séminaire de Probabilités XXXVIII*, Volume 1857 of *Lecture Notes in Math.*, pp. 320–337. Springer, Berlin.

Azuma, K. (1967). Weighted sums of certain dependent random variables. *Tohoku Mathematical Journal*, **68**, 357–367.

Bai, Z. and Silverstein, J.W. (2010). *Spectral Analysis of Large Dimensional Random Matrices*. Springer Verlag New York.

Baik, J., Deift, P., and Johansson, K. (1999). On the distribution of the length of the longest increasing subsequence of random permutations. *Journal of the American Mathematical Society*, **12**, 1119–1178.

Baik, J., Deift, P., and Johansson, K. (2000). On the distribution of the length of the second row of a Young diagram under Plancherel measure. *Geometric and Functional Analysis*, **10**, 702–731.

Bakry, D. and Ledoux, M. (1996). Lévy-Gromov's isoperimetric inequality for an infinite dimensional diffusion generator. *Inventiones Mathematicae*, **123**, 259–281.

Ball, K. (1997). An elementary introduction to modern convex geometry. In *Flavors of Geometry* (ed. S. Levy). Cambridge University Press, Cambridge.

Baraniuk, R.G., Davenport, M., DeVore, R.A., and Wakin, M.B. (2008). A simple proof of the restricted isoperimetry property for random matrices. *Constructive Approximation*, **28**, 253–263.

Baraniuk, R.G. and Wakin, M.B. (2009). Random projections of smooth manifolds. *Foundation of Computational Mathematics*, **9**, 51–77.

Barbour, A.D., Holst, L., and Janson, S. (1992). *Poisson Approximation*. Oxford Studies in Probability. 2. Oxford: Clarendon Press.

Barron, A.R., Cohen, A., Dahmen, W., and DeVore, R.A. (2008). Approximation and learning by greedy algorithms. *The Annals of Statistics*, **36**, 64–94.

Barthe, F. (2003). Autour de l'inégalité de Brunn–Minkowski. *Annales de la Faculté des Sciences de Toulouse*, **xii**, 127–178.

Barthe, F. and Maurey, B. (2000). Some remarks on isoperimetry of Gaussian type. *Annales de l'institut Henri Poincaré (B) Probabilités et Statistiques*, **36**, 419–434.

Bartlett, P.L. and Lugosi, G. (1999). An inequality for uniform deviations of sample averages from their means. *Statistics and Probability Letters*, **44**, 55–62.

Bartlett, P.L. and Mendelson, S. (2002). Rademacher and Gaussian complexities: risk bounds and structural results. *Journal of Machine Learning Research*, **3**, 463–482.

Bartlett, P.L. and Mendelson, S. (2006). Empirical minimization. *Probability Theory and Related Fields*, **135**, 311–334.

Bartlett, P.L., Boucheron, S., and Lugosi, G. (2002*a*). Model selection and error estimation. *Machine Learning*, **48**, 85–113.

Bartlett, P.L., Bousquet, O., and Mendelson, S. (2002*b*). Localized Rademacher complexities. In *Proceedings of the 15th Annual Conference on Computational Learning Theory*, pp. 44–48. Springer, New York.

Bartlett, P.L., Mendelson, S., and Neeman, J. (2012). ℓ_1-regularized linear regression: Persistence and oracle inequalities. *Probability Theory and Related Fields*, **June** 1–32.

Beckner, W. (1975). Inequalities in Fourier analysis. *Annals of Mathematics*, **102**, 159–182.

Beckner, W. (1989). A generalized Poincaré inequality for Gaussian measures. *Proceedings of the American Mathematical Society*, **105**, 397–400.

Ben-Or, M. and Linial, N. (1990). Collective coin flipping. In *Randomness and Computation* (ed. S. Micali), pp. 91–115. Academic Press, New York.

Benaïm, M. and Rossignol, R. (2006). A Poincaré-type inequality and its application to first passage percolation. Technical report, Université de Neuchatel.

Benjamini, I., Kalai, G., and Schramm, O. (2003). First passage percolation has sublinear distance variance. *The Annals of Probability*, **31**, 1970–1978.

Bennett, G. (1962). Probability inequalities for the sum of independent random variables. *Journal of the American Statistical Association*, **57**, 33–45.

Bercu, B., Gassiat, E., and Rio, E. (2002). Concentration inequalities, large and moderate deviations for self-normalized empirical processes. *The Annals of Probability*, **30**(4), 1576–1604.

Bernstein, S.N. (1946). *The Theory of Probabilities*. Gastehizdat Publishing House, Moscow.

Bezrukov, S.L. (1994). Isoperimetric problems in discrete spaces. In *Extremal problems for finite sets (Visegrád, 1991)*, Volume 3 of *Bolyai Soc. Math. Stud.*, pp. 59–91. János Bolyai Math. Soc., Budapest.

Bezrukov, S.L. and Serra, O. (2002). A local-global principle for vertex-isoperimetric problems. *Discrete Mathematics*, **257**, 285–309.

Bhatia, R. (1997). *Matrix Analysis*. Springer-Verlag, New York.

Bickel, P.J., Ritov, Y., and Tsybakov, A.B. (2009). Simultaneous analysis of lasso and Dantzig selector. *The Annals of Statistics*, **37**, 1705–1732.

Bingham, N.H., Goldie, C.M., and Teugels, J.L. (1987). *Regular Variation*. Cambridge University Press, Cambridge.

Birgé, L. (2005). A new lower bound for multiple hypothesis testing. *IEEE Transactions on Information Theory*, **51**, 1611–1615.

Birgé, L. and Massart, P. (1998). Minimum contrast estimators on sieves: exponential bounds and rates of convergence. *Bernoulli*, **4**, 329–375.

Birgé, L. and Massart, P. (2001). Gaussian model selection. *Journal of the European Mathematical Society (JEMS)*, **3**, 203–268.

Birnbaum, Z.W. (1942). An inequality for Mill's ratio. *Annals of Mathematical Statistics*, **13**, 245–246.

Bobkov, S. (1996). Some extremal properties of the Bernoulli distribution. *Teoriya Veroyatnosteui i ee Primeneniya*, **41**, 877–884.

Bobkov, S. (1997). An isoperimetric inequality on the discrete cube, and an elementary proof of the isoperimetric inequality in Gauss space. *The Annals of Probability*, **25**, 206–214.

Bobkov, S. and Götze, F. (1999*a*). Discrete isoperimetric and Poincaré-type inequalities. *Probability Theory and Related Fields*, **114**, 245–277.

Bobkov, S. and Götze, F. (1999*b*). Exponential integrability and transportation cost under logarithmic Sobolev inequalities. *Journal of Functional Analysis*, **163**, 1–28.

Bobkov, S. and Houdré, C. (1996). Variance of Lipschitz functions and an isoperimetric problem for a class of product measures. *Bernoulli*, **2**, 249–255.

Bobkov, S. and Houdré, C. (1997). Isoperimetric constants for product probability measures. *The Annals of Probability*, **25**, 184–205.

Bobkov, S. and Ledoux, M. (1997). Poincaré's inequalities and Talagrands's concentration phenomenon for the exponential distribution. *Probability Theory and Related Fields*, **107**, 383–400.

Bobkov, S. and Ledoux, M. (1998). On modified logarithmic Sobolev inequalities for Bernoulli and Poisson measures. *Journal of Functional Analysis*, **156**(2), 347–365.

Bobkov, S. and Tetali, P. (2006). Modified logarithmic Sobolev inequalities in discrete settings. *Journal of Theoretical Probability*, **19**(2), 289–336.

Bollobás, B. (1986). *Combinatorics*. Cambridge University Press, Cambridge.

Bollobás, B. (2001). *Random Graphs*. Cambridge University Press, Cambridge.

Bollobás, B. and Brightwell, G. (1992). The height of a random partial order: Concentration of measure. *The Annals of Applied Probability*, **2**, 1009–1018.

Bollobás, B. and Leader, I. (1991*a*). Compressions and isoperimetric inequalities. *Journal of Combinatorial Theory. Series A*, **56**, 47–62.

Bollobás, B. and Leader, I. (1991*b*). Isoperimetric inequalities and fractional set systems. *Journal of Combinatorial Theory. Series A*, **56**, 63–74.

Bollobás, B. and Riordan, O. (2006*a*). The critical probability for random Voronoi percolation in the plane is 1/2. *Probability Theory and Related Fields*, **136**(3), 417–468.

Bollobás, B. and Riordan, O. (2006*b*). Sharp thresholds and percolation in the plane. *Random Structures & Algorithms*, **29**(4), 524–548.

Bollobás, B. and Thomason, A.G. (1987). Threshold functions. *Combinatorica*, **7**, 35–58.

Bolthausen, E., Comets, F., and Dembo, A. (2009). Large deviations for random matrices and random graphs. Preprint.

Bonami, A. (1970). Étude des coefficients de Fourier des fonctions de $L^p(G)$. *Annales de l'Institut Fourier (Grenoble)*, **20**, 335–402.

Bordenave, C., Caputo, P., and Chafaï, D. (2011). Spectrum of non-Hermitian heavy tailed random matrices. *Communications in Mathematical Physics*, **307**, 513–560.

Borell, C. (1975). The Brunn–Minkowski inequality in Gauss space. *Inventiones Mathematicae*, **30**, 207–216.

Borell, C. (1986). A brief survey of Antoine Ehrhard's scientific work. In *Geometrical and Statistical Aspects of Probability in Banach Spaces (Strasbourg, 1985)*, Volume 1193 of *Lecture Notes in Math.*, pp. 1–3. Springer, Berlin.

Boucheron, S. and Massart, P. (2010). A high-dimensional Wilks phenomenon. *Probability Theory and Related Fields*, **148**, 1–29.

Boucheron, S., Bousquet, O., and Lugosi, G. (2005*a*). Theory of classification: a survey of some recent advances. *ESAIM Probability and Statistics*, **9**, 323–375.

Boucheron, S., Bousquet, O., Lugosi, G., and Massart, P. (2005*b*). Moment inequalities for functions of independent random variables. *The Annals of Probability*, **33**, 514–560.

Boucheron, S., Lugosi, G., and Massart, P. (2000). A sharp concentration inequality with applications. *Random Structures & Algorithms*, **16**, 277–292.

Boucheron, S., Lugosi, G., and Massart, P. (2003). Concentration inequalities using the entropy method. *The Annals of Probability*, **31**, 1583–1614.

Boucheron, S., Lugosi, G., and Massart, P. (2009). On concentration of self-bounding functions. *Electronic Journal of Probability*, **14**, 1884–1899.

Bourgain, J. and Kalai, G. (1997). Influences of variables and threshold intervals under group symmetries. *Geometric and Functional Analysis*, **7**, 438–461.

Bourgain, J., Kahn, J., Kalai, G., Katznelson, Y., and Linial, N. (1992). The influence of variables in product spaces. *Israel Journal of Mathematics*, **77**, 55–64.

Bousquet, O. (2002*a*). A Bennett concentration inequality and its application to suprema of empirical processes. *Comptes Rendus de l'Académie des Sciences de Paris Série I Mathématiques*, **334**, 495–500.

Bousquet, O. (2002*b*). Concentration inequalities and empirical processes theory applied to the analysis of learning algorithms. Ph. D. thesis, Ecole Polytechnique.

Bramson, M. and Durrett, R. (ed.) (1999). *Perplexing Problems in Probability. Festschrift in Honor of Harry Kesten.*, Volume 44 of *Progress in Probability*. Birkhäuser, Boston, Inc.

Brascamp, H.J. and Lieb, E.H. (1976). On extensions of the Brunn–Minkowski and Prékopa–Leindler theorems, including inequalities for log concave functions, and with an application to the diffusion equation. *Journal of Functional Analysis*, **22**, 366–389.

Brègman, L.M. (1967). A relaxation method of finding a common point of convex sets and its application to the solution of problems in convex programming. *Akademija Nauk SSSR. Žurnal Vyčislitelc' noĭ Matematiki i MatematičeskoĭFiziki*, **7**, 620–631.

Brieden, A., Gritzmann, P., Kannan, R., Klee, V., Lovász, L., and Simonovits, M. (2001). Deterministic and randomized polynomial-time approximation of radii. *Mathematika. A Journal of Pure and Applied Mathematics*, **48**(1-2), 63–105.

Bunea, F., Tsybakov, A., and Wegkamp, M. (2007). Sparsity oracle inequalities for the Lasso. *Electronic Journal of Statistics*, **1**, 169–194.

Burkholder, D.L. (1988). Sharp inequalities for martingales and stochastic integrals. *Astérisque*, **157–158**, 75–94.

Burkholder, D.L. (1989). Explorations in martingale theory and its applications. In *Ecole d'Eté de Probabilités de Saint-Flour* XIX, Lecture Notes in Mathematics #1464. pp. 1–66. Springer, New York.

Candès, E. and Tao, T. (2005). Decoding by linear programming. *IEEE Transactions on Information Theory*, **51**, 4203–4215.

Candès, E. and Tao, T. (2006). Near optimal system recovery from random projections: universal encoding strategies. *IEEE Transactions on Information Theory*, **52**, 5406–5425.

Candès, E. and Tao, T. (2007). The Dantzig selector: statistical estimation when p is much larger than n. *The Annals of Statistics*, **35**, 2313–2351.

Candès, E., Romberg, J., and Tao, T. (2006). Robust uncertainty principles: exact signal reconstruction from highly incomplete frequency information. *IEEE Transactions on Information Theory*, **52**, 489–509.

Capitaine, M., Hsu, E.P., and Ledoux, M. (1997). Martingale representation and a simple proof of logarithmic Sobolev inequalities on path spaces. *Electronic Communication in Probability*, **2**, 71–81.

Castellan, G. (2003). Density estimation via exponential model selection. *IEEE Transactions on Information Theory*, **49**, 2052–2060.

Catoni, O. (2003). Laplace transform estimates and deviation inequalities. *Annales de l'Institut Henri Poincaré (B) Probabilités et Statistiques*, **39**, 1–26.

Čentsov, N. (1956). La convergence faible des processus stochastiques à trajectoires sans discontinuités de seconde espèce et l'approche dite "heuristique" au tests du type de Kolmogorov-Smirnov. *Teoriya Veroyatnosteui i ee Primeneniya*, **1**, 155–161.

Chafaï, D. (2002). On ϕ-entropies and ϕ-Sobolev inequalities. Technical report, arXiv.math.PR/0211103.

Chafaï, D. (2003). Entropies, convexity, and functional inequalities. arXiv.math.PR0211103.

Chafaï, D. (2006). Binomial-Poisson entropic inequalities and the M/M/∞ queue. *ESAIM Probability and Statistics*, **10**, 317–339.

Chatterjee, S. (2005*a*). Concentration inequalities for exchangeable pairs. Ph. D. thesis, Stanford University.

Chatterjee, S. (2005*b*). An error bound in the Sudakov–Fernique inequality. Arxiv preprint math/0510424.

Chatterjee, S. (2007). Stein's method for concentration inequalities. *Probability Theory and Related Fields*, **138**(1), 305–321.

Chatterjee, S. (2010). The missing log in large deviations for subgraph counts. Preprint.

Chatterjee, S. and Dey, P.S. (2010). Applications of Stein's method for concentration inequalities. *The Annals of Probability*, **38**, 2443–2485.

Chazottes, J.R., Collet, P., Külske, C., and Redig, F. (2007). Concentration inequalities for random fields via coupling. *Probability Theory and Related Fields*, **137**, 201–225.

Chernoff, H. (1952). A measure of asymptotic efficiency of tests of a hypothesis based on the sum of observations. *Annals of Mathematical Statistics*, **23**, 493–507.

Chow, Y.S. and Teicher, H. (1978). *Probability Theory, Independence, Interchangeability, Martingales.* Springer-Verlag, New York.

Chung, F.R.K. and Lu, L. (2006*a*). *Complex graphs and networks*, Volume 107 of *CBMS Regional Conference Series in Mathematics.* Published for the Conference Board of the Mathematical Sciences, Washington, DC.

Chung, F.R.K. and Lu, L. (2006*b*). Concentration inequalities and martingale inequalities: a survey. *Internet Mathematics*, **3**, 79–127.

Chvátal, V. and Sankoff, D. (1975). Longest common subsequences of two random sequences. *Journal of Applied Probability*, **12**, 306–315.

Cirel'son, B.S. and Sudakov, V.N. (1974). Extremal properties of half spaces for spherically invariant measures. *Zapiski Naučnyh Seminarov Leningradskogo Otdelenija Matematičeskogo Instituta im. V. A. Steklova Akademii Nauk SSSR (LOMI)*, **41**, 14–24.

Clémençon, S., Lugosi, G., and Vayatis, N. (2008). Ranking and empirical minimization of *U*-statistics. *The Annals of Statistics*, **36**, 844–874.

Collet, P. (2006). Variance and exponential estimates via coupling. *Bulletin of the Brazilian Mathematical Society*, **37**, 461–475.

Cordero-Erausquin, D. and Ledoux, M. (2011). Hypercontractive measures, Talagrand's inequality, and influences. Arxiv preprint arXiv:1105.4533.

Cover, T.M. and Thomas, J.A. (1991). *Elements of Information Theory*. John Wiley, New York.

Craig, C. (1933). On the Tchebychef inequality of Bernstein. *Annals of Mathematical Statistics*, **4**, 94–102.

Cramér, H. (1938). Sur un nouveau théorème-limite de la théorie des probabilités. *Actualités Scientifiques et Industrielles*, **736**, 5–23.

Csiszár, I. (1967). Information-type measures of difference of probability distributions and indirect observations. *Studia Scientiarum Mathematicarum Hungarica*, **2**, 299–318.

Csiszár, I. (1972). A class of measures of informativity of observation channels. Collection of articles dedicated to the memory of Alfréd Rényi. *Periodica Mathematica Hungarica*, **2**, 191–213.

Csiszár, I. and Körner, J. (1981). *Information Theory: Coding Theorems for Discrete Memoryless Systems*. Academic Press, New York.

Cucker, F. and Zhou, D.X. (2007). *Learning Theory: An Approximation Theory Viewpoint*. Cambridge University Press, Cambridge.

Dančík, V. and Paterson, M. (1994). Upper bound for the expected. In *Proceedings of STACS'94. Lecture notes in Computer Science*, 775, pp. 669–678. Springer, New York.

Davidson, K.R. and Szarek, S.J. (2001). Local operator theory, random matrices and Banach spaces. In *Handbook of the Geometry of Banach Spaces*, pp. 317–366. Elsevier, Amsterdam.

Davies, E.B. and Simon, B. (1984). Ultracontractivity and the heat kernel for Schrödinger operators and Dirichlet Laplacians. *Journal of Functional Analysis*, **59**, 335–395.

de Haan, L. and Ferreira, A. (2006). *Extreme Value Theory*. Springer-Verlag, New York.

de la Peña, V.H. and Giné, E. (1999). *Decoupling: from Dependence to Independence*. Springer, New York.

Deken, J.P. (1979). Some limit results for longest common subsequences. *Discrete Mathematics*, **26**, 17–31.

DeMarco, B. and Kahn, J. (2010). Upper tails for triangles. *Random Structures & Algorithms*. arXiv:1005.4471.

Dembo, A. (1997). Information inequalities and concentration of measure. *The Annals of Probability*, **25**, 927–939.

Dembo, A. and Zeitouni, O. (1998). *Large Deviation Techniques and Applications*. Springer, New York.

Deuschel, J.-D. and Stroock, D. (1989). *Large Deviations*. American Mathematical Society, Washington DC.

DeVore, R.A. and Lorentz, G.G. (1993). *Constructive approximation*, Volume 303 of *Grundlehren der Mathematischen Wissenschaften [Fundamental Principles of Mathematical Sciences]*. Springer-Verlag, Berlin.

Devroye, L. (1988). The kernel estimate is relatively stable. *Probability Theory and Related Fields*, **77**, 521–536.

Devroye, L. (1991). Exponential inequalities in nonparametric estimation. In *Nonparametric Functional Estimation and Related Topics* (ed. G. Roussas), pp. 31–44. NATO ASI Series: Kluwer Academic Publishers, Dordrecht.

Devroye, L. (2002). Laws of large numbers and tail inequalities for random tries and Patricia trees. *Journal of Computational and Applied Mathematics*, **142**, 27–37.

Devroye, L. and Györfi, L. (1985). *Nonparametric Density Estimation: The L_1 View*. John Wiley, New York.

Devroye, L. and Lugosi, G. (2000). *Combinatorial Methods in Density Estimation*. Springer-Verlag, New York.

Devroye, L. and Lugosi, G. (2008). Local tail bounds for functions of independent random variables. *The Annals of Probability*, **36**, 143–159.

Devroye, L., Györfi, L., and Lugosi, G. (1996). *A Probabilistic Theory of Pattern Recognition*. Springer-Verlag, New York.

Diaconis, P. and Saloff-Coste, L. (1996). Logarithmic Sobolev inequalities for finite Markov chains. *The Annals of Applied Probability*, **6**, 695–750.

Diaconis, P. and Saloff-Coste, L. (1998). What do we know about the Metropolis algorithm? *Journal of Computer and System Sciences*, **57**, 20–36.

Dobrushin, R.L. (1970). Prescribing a system of random variables by conditional distributions. *Theory of Probability and its Applications*, **15**, 458–486.

Donoho, D.L. (2006*a*). Compressed sensing. *IEEE Transactions on Information Theory*, **52**, 1289–1306.

Donoho, D.L. (2006*b*). For most large underdetermined systems of equations, the minimal l_1-norm near-solution approximates the sparsest near-solution. *Communications on Pure and Applied Mathematics*, **59**, 907–934.

Donoho, D.L. (2006*c*). For most large underdetermined systems of linear equations the minimal l_1-norm solution is also the sparsest solution. *Communications on Pure and Applied Mathematics*, **59**, 797–829.

Donoho, D.L. and Jin, J. (2004). Higher criticism for detecting sparse heterogeneous mixtures. *The Annals of Statistics*, **32**, 952–994.

Döring, H. and Eichelsbacher, P. (2009). Moderate deviations in a random graph and for the spectrum of Bernoulli random matrices. *Electronic Journal of Probability*, **14**, 2636–2656. Available at http://arxiv.org/pdf/0901.3246v1.

Dubhashi, D. and Panconesi, A. (2009). *Concentration of Measure for the Analysis of Randomized Algorithms*. Cambridge University Press, New York.

Dubhashi, D. and Ranjan, D. (1998). Balls and bins: a study in negative dependence. *Random Structures & Algorithms*, **13**, 99–124.

Dudley, R.M. (1967). The sizes of compact subsets of Hilbert space and continuity of Gaussian processes. *Journal of Functional Analysis*, **1**, 290–330.

Dudley, R.M. (1987). Universal Donsker classes and metric entropy. *The Annals of Probability*, **15**, 1306–1326.

Dudley, R.M. (1999). *Uniform Central Limit Theorems*. Cambridge University Press, Cambridge.

Dudley, R.M. (2002). *Real Analysis and Probability*, Volume 74 of *Cambridge Studies in Advanced Mathematics*. Cambridge University Press, Cambridge.

Duembgen, L., van de Geer, S.A., Veraar, M.C., and Wellner, J.A. (2010). Nemirovski's inequalities revisited. *American Mathematical Monthly*, **117**, 138–160.

Dupuis, P. and Ellis, R.S. (1997). *A Weak Convergence Approach to the Theory of Large Deviations*. John Wiley, New York.

Dvoretzky, A., Kiefer, J., and Wolfowitz, J. (1956). Asymptotic minimax character of a sample distribution function and of the classical multinomial estimator. *Annals of Mathematical Statistics*, **33**, 642–669.

Efron, B. and Stein, C. (1981). The jackknife estimate of variance. *The Annals of Statistics*, **9**, 586–596.

Efron, B. and Tibshirani, R.J. (1994). *An Introduction to the Bootstrap*. Chapman and Hall, New York.

Ehrhard, A. (1982). Un principe de symétrisation dans les espaces de Gauss: interprétation géométrique de l'inégalité de Brunn–Minkowski de Borell. *Comptes Rendus de l'Académie des Sciences de Paris Série I Mathématiques*, **294**, 589–591.

Ehrhard, A. (1983*a*). Symétrisation dans l'espace de Gauss. *Mathematica Scandinavica*, **53**(2), 281–301.

Ehrhard, A. (1983*b*). Un principe de symétrisation dans les espaces de Gauss. In *Probability in Banach spaces, IV (Oberwolfach, 1982)*, Volume 990 of *Lecture Notes in Math.*, pp. 92–101. Springer-Verlag, Berlin.

Ehrhard, A. (1984). Inégalités isopérimétriques et intégrales de Dirichlet gaussiennes. *Annales Scientifiques de l'École Normale Supérieure*, **17**, 317–332.

Ehrhard, A. (1986). Éléments extrémaux pour les inégalités de Brunn-Minkowski gaussiennes. *Annales de l'institut Henri Poincaré (B) Probabilités et Statistiques*, **22**, 149–168.

Erdős, P. and Rényi, A. (1960). On the evolution of random graphs. *Publications of the Mathematical Institute of the Hungarian Academy of Sciences, Ser. A*, **5**, 17–61.

Erdős, L. and Yau, H.T. (2012). Universality of local spectral statistics of random matrices. *Bulletin of the American Mathematical Society*, **49**, 377–414.

Falik, D. and Samorodnitsky, A. (2007). Edge-isoperimetric inequalities and influences. *Combinatorics, Probability, and Computing*, **16**, 693–712.

Fernique, X. (1975). Régularité des trajectoires des fonctions aléatoires gaussiennes, *Ecole d'Eté de Probabilités de Saint-Flour IV—1974*, Volume 480 of *Lecture Notes in Mathematics*, pp. 1–96. Springer-Verlag, Berlin.

Fortuin, C.M., Kasteleyn, P.W., and Ginibre, J. (1971). Correlation inequalities on some partially ordered sets. *Communications in Mathematical Physics*, **22**, 89–103.

Frankl, P. (1983). On the trace of finite sets. *Journal of Combinatorial Theory, Series A*, **34**, 41–45.

Frankl, P. and Füredi, Z. (1981). A short proof for a theorem of Harper about Hamming-spheres. *Discrete Mathematics*, **34**, 311–313.

Frankl, P. and Maehara, H. (1988). The Johnson–Lindenstrauss lemma and the sphericity of some graphs. *Journal of Combinatorial Theory, Series B*, **44**, 355–362.

Frankl, P. and Maehara, H. (1990). Some geometric applications of the beta distribution. *Annals of the Institute of Statistical Mathematics*, **42**, 463–474.

Fréchet, M. (1957). Sur la distance de deux lois de probabilité. *Comptes Rendus de l'Académie des Sciences de Paris*, **244**, 689–692.

Friedgut, E. (1998). Boolean functions with low average sensitivity depend on few coordinates. *Combinatorica*, **18**, 27–36.

Friedgut, E. (1999). Sharp threshold of graph properties and the k-sat problem (with an appendix by Jean Bourgain). *Journal of the American Mathematical Society*, **12**, 1017–1054.

Friedgut, E. (2005). Hunting for sharp thresholds. *Random Structures & Algorithms*, **26**, 37–51.

Friedgut, E. and Kalai, G. (1996). Every monotone graph property has a sharp threshold. *Proceedings of the American Mathematical Society*, **124**, 2993–3002.

Frieze, A.M. (1985). On the value of a random minimum spanning tree problem. *Discrete Applied Mathematics*, **10**, 47–56.

Frieze, A.M. (1991). On the length of the longest monotone subsequence in a random permutation. *The Annals of Applied Probability*, **1**, 301–305.

Füredi, Z. and Komlós, J. (1981). The eigenvalues of random symmetric matrices. *Combinatorica*, **1**, 233–241.

Galambos, J. (1987). *The Asymptotic Theory of Extreme Order Statistics*. R.E. Kreiger Florida.

Gallager, R.G. (1968). *Information Theory and Reliable Communication*. John Wiley, New York.

Gardner, R.J. (2002). The Brunn–Minkowski inequality. *Bulletin of the American Mathematical Society*, **39**, 355–405.

Garling, D.J.H. (2007). *Inequalities: A Journey into Linear Analysis*. Cambridge University Press, Cambridge.

Giannopoulos, A.A. and Milman, V.D. (2001). Euclidean structure in finite dimensional normed spaces. In *Handbook of the Geometry of Banach Spaces*, pp. 707–779. Elsevier, Amsterdam.

Giné, E. and Guillou, A. (2001). On consistency of kernel density estimators for randomly censored data: rates holding uniformly over adaptive intervals. *Annales de l'institut Henri Poincaré (B) Probabilités et Statistiques*, **37**, 503–522.

Giné, E. and Koltchinskii, V. (2006). Concentration inequalities and asymptotic results for ratio type empirical processes. *The Annals of Probability*, **34**, 1143–1216.

Giné, E. and Zinn, J. (1984). Some limit theorems for empirical processes. *The Annals of Probability*, **12**, 929–989.

Giné, E., Koltchinskii, V., and Wellner, J.A. (2003). Ratio limit theorems for empirical processes. In *Stochastic Inequalities and Applications*, pp. 249–278. Birkhaüser, Basel.

Giné, E., Latała, R., and Zinn, J. (2000). Exponential and moment inequalities for U-statistics. In *High Dimensional Probability II—Progress in Probability*, pp. 13–38. Birkhäuser, Basel.

Gordon, R.D. (1941). Values of Mills' ratio of area to bounding ordinate and of the normal probability integral for large values of the argument. *Annals of Mathematical Statistics*, **12**, 364–366.

Gordon, Y. (1985). Some inequalities for Gaussian processes and applications. *Israel Journal of Mathematics*, **50**(4), 265–289.

Gordon, Y. (1988). On Milman's inequality and random subspaces which escape through a mesh in $\mathbb{R}^n$, *Geometric aspects of functional analysis (1986/87)*, Volume 1317 of *Lecture Notes in Mathematics*, pp. 84–106. Springer-Verlag, Berlin.

Götze, F. and Tikhomirov, A. (2003). Rate of convergence to the semi-circular law. *Probability Theory and Related Fields*, **127**, 228–276.

Götze, F. and Tikhomirov, F. (2005). The rate of convergence for spectra of GUE and LUE matrix ensembles. *Central European Journal of Mathematics*, **3**, 666–704.

Greenshtein, E. and Ritov, Y. (2004). Persistence in high-dimensional linear predictor selection and the virtue of overparametrization. *Bernoulli*, **10**(6), 971–988.

Grimmett, G. (1989). *Percolation*. Springer-Verlag, New York.

Groeneboom, P. (2002). Hydrodynamical methods for analyzing longest increasing subsequences. probabilistic methods in combinatorics and combinatorial optimization. *Journal of Computational and Applied Mathematics*, **142**, 83–105.

Gromov, M. (1980). Paul Lévy's isoperimetric inequality. *Inst. Hautes Etudes Sci., Publ. Math.* Preprint.

Gross, L. (1975). Logarithmic Sobolev inequalities. *American Journal of Mathematics*, **97**, 1061–1083.

Guntuboyina, A. and Leeb, H. (2009). Concentration of the spectral measure of large Wishart matrices with dependent entries. *Electronic Communications of Probability*, **14**, 334–342.

Gupta, S.D.A. and Dasgupta, S. (2002). An elementary proof of the Johnson–Lindenstrauss lemma. *Random Structures & Algorithms*, **22**, 60–65.

Haagerup, U. (1981). The best constants in the Khintchine inequality. *Studia Mathematica*, **70**, 231–283.

Hagerup, T. and Rüb, C. (1990). A guided tour of Chernoff bounds. *Information Processing Letters*, **33**, 305–308.

Hammersley, J.M. and Welsh, D.J.A. (1965). First-passage percolation, subadditive processes, stochastic networks, and generalized renewal theory. In *Bernoulli–Bayes–Laplace Anniversary Volume*, pp. 61–110. Springer-Verlag, Berlin.

Hammersley, J.M. (1972). A few seedlings of research. In *Proceedings of the Sixth Berkeley Symposium on Mathematical Statistics and Probability (Univ. California, Berkeley, Calif., 1970/1971), Vol. I: Theory of statistics*, pp. 345–394. Berkeley, Calif., Univ. California Press.

Han, T.S. (1978). Nonnegative entropy measures of multivariate symmetric correlations. *Information and Control*, **36**, 133–156.

Hanson, D.L. and Wright, F.T. (1971). A bound on tail probabilities for quadratic forms in independent random variables. *Annals of Mathematical Statistics*, **42**, 1079–1083.

Hardy, G.H., Littlewood, J.E., and Pólya, G. (1952). *Inequalities*. Cambridge University Press, London.

Harper, L.H. (1966). Optimal numberings and isoperimetric problems on graphs. *Journal of Combinatorial Theory*, **1**, 385–393.

Harris, T.E. (1960). A lower bound for the critical probability in a certain percolation process. *Proceedings of the Cambridge Philosophical Society*, **56**, 13–20.

Hart, S. (1976). A note on the edges of the n-cube. *Discrete Mathematics*, **14**, 157–163.

Hatami, H. (2012). A structure theorem for Boolean functions with small total influences. *Annals of Mathematics*, **176**, 509–533.

Haussler, D. (1992). Decision theoretic generalizations of the PAC model for neural net and other learning applications. *Information and Computation*, **100**, 78–150.

Haussler, D. (1995). Sphere packing numbers for subsets of the Boolean n-cube with bounded Vapnik–Chervonenkis dimension. *Journal of Combinatorial Theory, Series A*, **69**, 217–232.

Haussler, D., Littlestone, N., and Warmuth, M. (1994). Predicting $\{0, 1\}$-functions on randomly drawn points. *Information and Computation*, **115**, 248–292.

Higuchi, Y. and Yoshida, N. (1995). Analytic conditions and phase transitions in ising models. Unpublished notes (in Japanese).

Hoeffding, W. (1948). A class of statistics with asymptotically normal distribution. *Annals of Mathematical Statistics*, **19**, 293–325.

Hoeffding, W. (1963). Probability inequalities for sums of bounded random variables. *Journal of the American Statistical Association*, **58**, 13–30.

Hoffmann-Jørgensen, J. (1974). Sums of independent Banach space valued random variables. *Studia Mathematica*, **52**, 159–186.

Horn, R.A. and Johnson, C.R. (1990). *Matrix Analysis*. Cambridge University Press, Cambridge.

Houdré, C. and Litherland, T.L. (2009). On the longest increasing subsequence for finite and countable alphabets. *High Dimensional Probability V: The Luminy Volume IMS Collections*, **5**, 185–212.

Houdré, C. and Reynaud-Bouret, P. (2003). Exponential inequalities, with constants, for U-statistics of order two. In *Stochastic inequalities and applications*, Volume 56 of *Progr. Probab.*, pp. 55–69. Birkhäuser, Basel.

Houdré, C., Lember, J., and Matzinger, H. (2006). On the longest common increasing binary subsequence. *Comptes Rendus Mathematique*, **343**, 589–594.

Hu, Y. (2000). A unified approach to several inequalities for Gaussian and diffusion measures. In *Séminaire de Probabilités, XXXIV*, Volume 1729 of *Lecture Notes in Math.*, pp. 329–335. Springer-Verlag, Berlin.

Huang, C., Cheang, G.L.H., and Barron, A.R. (2010). Risk of penalized least squares, greedy selection and ℓ_1 penalization for flexible function libraries. preprint.

Ibragimov, I.A. and Khasminskii, R.Z. (1981). *Statistical Estimation: Asymptotic Theory*. Springer-Verlag, New York.

Imbuzeiro Oliveira, R. (2010). Sums of random Hermitian matrices and an inequality by Rudelson. *Electronic Communication in Probability*, **15**, 203–212.

Indyk, P. and Motwani, R. (1998). Approximate nearest neighbors: Towards removing the curse of dimensionality. In *Proceedings of the 30th Symposium on Theory of Computing*, pp. 604–613. ACM, New York.

Its, A.R., Tracy, C.A., and Widom, H. (2001). Random words, Toeplitz determinants and integrable systems I. In *Random Matrix Models and Their Applications* (ed. P. Bleher and A.R. Its), Volume 40 of *Math. Sci. Res. Inst. Publ.*, pp. 245–258. Cambridge University Press, Cambridge.

Jaeschke, D. (1979). The asymptotic distribution of the supremum of the standardized empirical distribution function on subintervals. *The Annals of Statistics*, 7(1), 108–115.

Janson, S. (1990). Poisson approximation for large deviations. *Random Structures & Algorithms*, **1**, 221–230.

Janson, S. (1995). The minimal spanning tree in a complete graph and a functional limit theorem for trees in a random graph. *Random Structures & Algorithms*, 7, 337–355.

Janson, S. and Ruciński, A. (2002). The infamous upper tail. *Random Structures & Algorithms*, **20**, 317–342.

Janson, S. and Ruciński, A. (2004). The deletion method for upper tail estimates. *Combinatorica*, **24**, 615–640.

Janson, S., Łuczak, T., and Ruciński, A. (2000). *Random Graphs*. John Wiley, New York.

Janson, S., Oleszkiewicz, K., and Ruciński, A. (2004). Upper tails for subgraph counts in random graphs. *Israel Journal of Mathematics*, **142**, 61–92.

Johansson, K. (2001). Discrete orthogonal polynomial ensembles and the Plancherel measure. *Annals of Mathematics*, **153**, 259–296.

Johnson, W. and Lindenstrauss, J. (1984). Extensions of Lipschitz maps into a Hilbert space. *Contemporary Mathematics*, **26**, 189–206.

Kahane, J.P. (1964). Sur les sommes vectorielles $\sum \pm u_n$. *Comptes Rendus de l'Académie des Sciences de Paris Série I Mathématiques*, **259**, 2577–2580.

Kahn, J., Kalai, G., and Linial, N. (1988). The influence of variables on Boolean functions. In *Proceedings of the 29th Annual Symposium on the Foundations of Computer Science*, pp. 68–80. White Plains, New York.

Kalai, G. (2004). Social indeterminacy. *Econometrica*, 72(5), 1565–1581.

Kalai, G. and Safra, S. (2006). Threshold phenomena and influence: perspectives from mathematics, computer science, and economics. In *Computational Complexity and Statistical Physics*, pp. 25–60. Oxford University Press, New York.

Karp, R.M. (1988). *Probabilistic Analysis of Algorithms*. Class Notes, University of California, Berkeley.

Kashin, B. (1977). The widths of certain finite dimensional sets and classes of smooth functions. *Izv. Akad. Nauk SSSR Ser. Mat*, **41**(2), 334–351.

Katona, G.O.H. (1975). The Hamming-sphere has minimum boundary. *Studia Scientiarum Mathematicarum Hungarica*, **10**, 131–140.

Kearns, M. and Schapire, R.E. (1994). Efficient distribution-free learning of probabilistic concepts. *Journal of Computer Systems Sciences*, **48**, 464–497.

Keller, N., Mossel, E., and Sen, A. (2012*a*). Geometric influences. *The Annals of Probability*, to appear.

Keller, N., Mossel, E., and Sen, A. (2012*b*). Geometric influences II: Correlation inequalities and noise sensitivity. Technical report, arXiv:1206.1210v1 [math.PR].

Kesten, H. (1993). On the speed of convergence in first passage percolation. *The Annals of Applied Probability*, **3**, 296–338.

Khintchine, A. (1923). Über dyadische Brüche. *Mathematische Zeitschriften*, **18**, 109–116.

Kim, J.H. and Vu, V. (2000). Concentration of multivariate polynomials and applications. *Combinatorica*, **20**, 417–434.

Kim, J.H. and Vu, V. (2004). Divide and conquer martingales and the number of triangles in a random graph. *Random Structures & Algorithms*, **24**, 166–174.

Klaassen, C.A.J. (1985). On an inequality of Chernoff. *The Annals of Probability*, **13**(3), 966–974.

Klartag, B. and Mendelson, S. (2005). Empirical processes and random projections. *Journal of Functional Analysis*, **225**, 229–245.

Klein, T. (2002). Une inégalité de concentration à gauche pour les processus empiriques. *Comptes Rendus de l'Académie des Sciences de Paris Série I Mathématiques*, **334**(6), 501–504.

Klein, T. and Rio, E. (2005). Concentration around the mean for maxima of empirical processes. *The Annals of Probability*, **33**, 1060–1077.

Kleinberg, J.M. (1997). Two algorithms for nearest-neighbor search in high dimensions. In *Proceedings of the 29th Annual ACM Symposium on Theory of Computing*, pp. 599–608. ACM, New York.

Kleitman, D.J. (1979). Extremal hypergraph problems. In *Surveys in combinatorics (Proc. Seventh British Combinatorial Conf., Cambridge, 1979)*, Volume 38, pp. 44–65. Cambridge University Press, Cambridge.

Kolmogorov, A.N. and Tikhomirov, V.M. (1961). ε-entropy and ε-capacity of sets in function spaces. *Translations of the American Mathematical Society*, **17**, 277–364.

Koltchinskii, V. (1981). On the central limit theorem for empirical measures. *Theory of Probability and Mathematical Statistics*, **24**, 71–82.

Koltchinskii, V. (2001). Rademacher penalties and structural risk minimization. *IEEE Transactions on Information Theory*, **47**, 1902–1914.

Koltchinskii, V. (2006). Local Rademacher complexities and oracle inequalities in risk minimization. *The Annals of Statistics*, **36**, 00–00.

Koltchinskii, V. (2008). Oracle inequalities in empirical risk minimization and sparse recovery problems. Lectures from the 38th Probability Summer School, Saint-Flour, Volume 2033 of *Lecture Notes in Mathematics*. Springer-Verlag, Berlin.

Koltchinskii, V. (2009*a*). The Dantzig selector and sparsity oracle inequalities. *Bernoulli*, **15**, 799–828.

Koltchinskii, V. (2009*b*). Sparsity in penalized empirical risk minimization. *Annales de l'Institut Henri Poincaré (B) Probabilités et Statistiques*, **45**, 7–57.

Koltchinskii, V. and Panchenko, D. (2000). Rademacher processes and bounding the risk of function learning. In *High Dimensional Probability II* (ed. E. Giné, D. Mason, and J. Wellner), pp. 443–459. Birkhäuser, Basel.

Kontorovich, L.A. and Ramanan, K. (2008). Concentration inequalities for dependent random variables via the martingale method. *The Annals of Probability*, **36**, 2126–2158.

Kontoyiannis, I., Harremoës, P., and Johnson, O. (2005). Entropy and the law of small numbers. *IEEE Transactions on Information Theory*, **51**(2), 466–472.

Külske, C. (2003). Concentration inequalities for functions of Gibbs fields with application to diffraction and random Gibbs measures. *Communications in Mathematical Physics*, **239**, 29–51.

Kwapień, S. and Woyczyński, W. (1992). *Random Series and Stochastic Integrals: Single and Multiple.* Birkhäuser, Boston.

Kwapień, S., Latała, R., and Oleszkiewicz, K. (1996). Comparison of moments of sums of independent random variables and differential inequalities. *Journal of Functional Analysis*, **136**, 258–268.

Lang, S. (1965). *Algebra.* Addison-Wesley Reading, Mass.

Latała, R. (1997). Estimation of moments of sums of independent real random variables. *The Annals of Probability*, **25**, 1502–1513.

Latała, R. (1999). Tail and moment estimates for some types of chaos. *Studia Mathematica*, **135**, 39–53.

Latała, R. (2006). Estimates of moments and tails of Gaussian chaoses. *The Annals of Probability*, **34**, 2315–2331.

Latala, R. and Oleszkiewicz, K. (1994). On the best constant in the Khinchin–Kahane inequality. *Studia Mathematica*, **109**, 101–104.

Latała, R. and Oleszkiewicz, K. (1999). Gaussian measures of dilatations of convex symmetric sets. *The Annals of Probability*, **27**, 1922–1938.

Latała, R. and Oleszkiewicz, K. (2000). Between Sobolev and Poincaré. In *Geometric Aspects of Functional Analysis, Israel Seminar (GAFA), 1996-2000*, pp. 147–168. Lecture Notes in Mathematics, Volume 1745. Springer-Verlag, Berlin.

Leader, I. (1991). Discrete isoperimetric inequalities. In *Probabilistic combinatorics and its applications (San Francisco, CA, 1991)*, Volume 44, pp. 57–80. American Mathematical Society, Providence, RI.

Ledoux, M. (1996). Isoperimetry and Gaussian analysis. In *Lectures on Probability Theory and Statistics* (ed. P. Bernard), pp. 165–294. Springer-Verlag, Berlin.

Ledoux, M. (1997). On Talagrand's deviation inequalities for product measures. *ESAIM Probability and Statistics*, **1**, 63–87.

Ledoux, M. (1999). Concentration of measure and logarithmic Sobolev inequalities. In *Séminaire de Probabilités XXXIII. Lecture Notes in Mathematics* Volume 1709, pp. 120–216. Springer-Verlag, Berlin.

Ledoux, M. (2000). The geometry of Markov diffusion generators. *Annales de la Faculté des Sciences de Toulouse*, IX, 305–366.

Ledoux, M. (2001). *The Concentration of Measure Phenomenon.* American Mathematical Society, Providence, RI.

Ledoux, M. (2003). A remark on hypercontractivity and tail inequalities for the largest eigenvalues of random matrices, *Séminaire de Probabilités XXXVII*, Volume 1832 of *Lecture Notes in Mathematics*, pp. 360–369. Springer Verlag, Berlin.

Ledoux, M. (2005). Distributions of invariant ensembles from the classical orthogonal polynomials: the discrete case. *Electronic Journal of Probability*, **10**, 1116–1146.

Ledoux, M. (2007). Deviation inequalities on largest eigenvalues. In *Geometric aspects of functional analysis*, Volume 1910 of *Lecture Notes in Mathematics*, pp. 167–219. Springer-Verlag, Berlin.

Ledoux, M. and Talagrand, M. (1991). *Probability in Banach Space*. Springer-Verlag, Berlin.

Leindler, L. (1972). On a certain converse of Hölder's inequality II. *Acta Universitatis Szegediensis. Acta Scientiarum Mathematicarum*, **33**, 217–223.

Lember, J. and Matzinger, H. (2009). Standard deviation of the longest common subsequence. *The Annals of Probability*, **37**, 1192–1235.

Lévy, P. (1951). *Problèmes concrets d'analyse fonctionelle*. Gauthier-Villars, Paris.

Li, W.V. and Shao, Q.M. (2001). Gaussian processes: inequalities, small ball probabilities and applications. *Stochastic Processes: Theory and Methods*, **19**, 533–597.

Lidskii, V.B. (1950). The proper values of the sum and product of symmetric matrices. *Doklady Akademiya Nauk SSSR*, **75**, 769–772.

Linial, N. and Rozenman, E. (2002). An extremal problem on degree sequences of graphs. *Graphs and Combinatorics*, **18**, 573–582.

Linial, N., London, E., and Rabinovich, Y. (1995). The geometry of graphs and some of its algorithmic applications. *Combinatorica*, **15**, 215–245.

Linusson, S. and Wästlund, J. (2004). A proof of Parisi's conjecture on the random assignment problem. *Probability Theory and Related Fields*, **128**, 419–440.

Littlewood, J.E. (1930). On bounded bilinear forms in an infinite number of variables. *Quarterly Journal of Mathematics, Oxford Ser. 1*, 164–174.

Logan, B.F. and Shepp, L.A. (1977). A variational problem for Young tableaux. *Advances in Mathematics*, **26**, 206–222.

Loomis, L. and Whitney, H. (1949). An inequality related to the isoperimetric inequality. *Bulletin of the Americal Mathematical Society*, **55**, 961–962.

Lorentz, G.G., Golitschek, M.V., and Makovoz, Y. (1996). *Constructive approximation*, Volume 304 of *Grundlehren der Mathematischen Wissenschaften*. Springer-Verlag, Berlin.

Lust-Piquard, F. and Pisier, G. (1991). Non commutative Khintchine and Paley inequalities. *Arkiv für Matematik*, **29**(1), 241–260.

MacKay, D.J.C. (2003). *Information Theory, Inference, and Learning Algorithms*. Cambridge University Press, Cambridge.

Major, P. (2006). A multivariate generalization of Hoeffding's inequality. *Electronic Communications in Probability*, **2**, 220–229.

Major, P. (2007). On a multivariate version of Bernstein's inequality. *Electronic Journal of Probability*, **12**, 966–988.

Major, P. (2005). Tail behaviour of multiple random integrals and *U*-statistics. *Probability Surveys*, **2**, 448–505.

Margulis, G. (1974). Probabilistic characteristics of graphs with large connectivity. *Akademiya Nauk SSSR. Institut Problem Peredachi Informatsii*, **10**, 101–108.

Marshall, A.W. and Olkin, I. (1979). *Inequalities: Theory of Majorization and its Applications.* Academic Press, New York.

Martinelli, F. (1997). Lectures on Glauber dynamics for discrete spin models, *Ecole d'été de Probabilités de Saint-Flour,* Volume 1717 of *Lecture Notes in Mathematics,* pp. 93–191. Springer-Verlag, Berlin.

Marton, K. (1986). A simple proof of the blowing-up lemma. *IEEE Transactions on Information Theory,* **32,** 445–446.

Marton, K. (1996*a*). Bounding $\bar{d}$-distance by informational divergence: a way to prove measure concentration. *The Annals of Probability,* **24,** 857–866.

Marton, K. (1996*b*). A measure concentration inequality for contracting Markov chains. *Geometric and Functional Analysis,* **6,** 556–571. Erratum: 7:609–613, 1997.

Marton, K. (2003). Measure concentration and strong mixing. *Studia Scientiarum Mathematicarum Hungarica,* **1,** 95–113.

Marton, K. (2004). Measure concentration for Euclidean distance in the case of dependent random variables. *The Annals of Probability,* **32,** 2526–2544.

Mason, D.M. and van Zwet, W.R. (1987). A refinement of the KMT inequality for the uniform empirical process. *The Annals of Probability,* **15**(3), 871–884.

Massart, P. (1990). The tight constant in the Dvoretzky–Kiefer–Wolfowitz inequality. *The Annals of Probability,* **18,** 1269–1283.

Massart, P. (1998). Optimal constants for Hoeffding type inequalities. Technical report, Mathematiques, Université de Paris-Sud, Report 98.86.

Massart, P. (2000*a*). About the constants in Talagrand's concentration inequalities for empirical processes. *The Annals of Probability,* **28,** 863–884.

Massart, P. (2000*b*). Some applications of concentration inequalities to statistics. *Annales de la Faculté des Sciences de Toulouse,* **IX,** 245–303.

Massart, P. (2006). In *Ecole d'été de Probabilités de Saint-Flour 2003.* Volume 1717 *Lecture Notes in Mathematics.* Concentration inequalities and model selection. Springer-Verlag, Berlin.

Massart, P. and Meynet, C. (2010). An ℓ_1 oracle inequality for the Lasso. *Electronic Journal of Statistics,* **5,** 669–687.

Massart, P. and Nédélec, E. (2006). Risk bounds for statistical learning. *The Annals of Statistics,* **34,** 2326–2366.

Maurer, A. (2006). Concentration inequalities for functions of independent variables. *Random Structures & Algorithms,* **29,** 121–138.

Maurer, A. and Pontil, M. (2009). Empirical Bernstein bounds and sample variance penalization. In *Proceedings of the 22nd Annual Conference on Learning Theory,* 73–82.

Maurey, B. (1979). Construction de suites symétriques. *Comptes Rendus de l'Académie des Sciences de Paris Série I Mathématiques,* **288,** 679–681.

Maurey, B. (1991). Some deviation inequalities. *Geometric and Functional Analysis,* **1**(2), 188–197.

McDiarmid, C. (1989). On the method of bounded differences. In *Surveys in Combinatorics 1989* (ed. J. Siemons), pp. 148–188. Cambridge University Press, Cambridge.

McDiarmid, C. (1998). Concentration. In *Probabilistic Methods for Algorithmic Discrete Mathematics* (ed. M. Habib, C. McDiarmid, J. Ramirez-Alfonsin, and B. Reed), pp. 195–248. Springer, New York.

McDiarmid, C. and Reed, B. (2006). Concentration for self-bounding functions and an inequality of Talagrand. *Random Structures & Algorithms*, **29**, 549–557.

Meckes, E. and Meckes, M. (2012). Concentration and convergence rates for spectral measures of random matrices. *Probability Theory and Related Fields*, arXiv:1109.5997.

Mehta, M.L. (2004). *Random Matrices* (3rd edn). Elsevier, Amsterdam.

Mendelson, S. (2002*a*). Geometric parameters of kernel machines. In *Proceedings of COLT 2002*. Springer-Verlag, New York.

Mendelson, S. (2002*b*). Improving the sample complexity using global data. *IEEE Transactions on Information Theory*, **48**, 1977–1991.

Mendelson, S. (2003). A few notes on statistical learning theory. In *Advanced Lectures in Machine Learning* (ed. S. Mendelson and A. Smola), LNCS 2600, pp. 1–40. Springer-Verlag, Berlin.

Mendelson, S. (2010). Empirical processes with a bounded ψ_1 diameter. *Geometric and Functional Analysis*, **20**, 988–1027.

Mendelson, S. and Philips, P. (2004). On the importance of "small" coordinate projections. *Journal of Machine Learning Research*, **5**, 219–238.

Mendelson, S. and Vershynin, R. (2003). Entropy and the combinatorial dimension. *Inventiones Mathematicae*, **152**, 37–55.

Mendelson, S., Pajor, A., and Tomczak-Jaegermann, N. (2007). Reconstruction and subgaussian operators in asymptotic geometric analysis. *Geometric and Functional Analysis*, **17**, 1248–1282.

Mendelson, S., Pajor, A., and Tomczak-Jaegermann, N. (2008). Uniform uncertainty principle for Bernoulli and subgaussian ensembles. *Constructive Approximation*, **28**, 277–289.

Milman, V. and Schechtman, G. (1986). *Asymptotic Theory of Finite-Dimensional Normed Spaces*. Springer-Verlag, New York.

Molloy, M. and Reed, B. (2002). *Graph Colouring and the Probabilistic Method*, Volume 23 of *Algorithms and Combinatorics*. Springer-Verlag, Berlin.

Mossel, E., O'Donnell, R., and Oleszkiewicz, K. (2010). Noise stability of functions with low influences: invariance and optimality. *Annals of Mathematics*, **171**, 295–341.

Nair, C., Prabhakar, B., and Sharma, M. (2005). Proofs of the Parisi and Coppersmith–Sorkin random assignment conjectures. *Random Structures & Algorithms*, **27**, 413–444.

Nelson, E. (1973). The free Markoff field. *Journal of Functional Analysis*, **12**, 211–227.

Nemirovski, A. (2000). Topics in non-parametric statistics. In *Lectures on Probability Theory and Statistics St-Flour, 1998*, Volume 1738 of *Lecture Notes in Mathematics*, pp. 85–277. Springer-Verlag, Berlin.

O'Donnell, R. and Wimmer, K. (2009). KKL, Kruskal-Katona, and monotone nets. In *50th Annual IEEE Symposium on Foundations of Computer Science, 2009. FOCS'09.*, pp. 725–734. IEEE.

Okamoto, M. (1958). Some inequalities relating to the partial sum of binomial probabilities. *Annals of the Institute of Statistical Mathematics*, **10**, 29–35.

Oleszkiewicz, K. (2003). On a nonsymmetric version of the Khinchine–Kahane inequality. In *Stochastic Inequalities and Applications*, Volume 56 of *Progress in Probability*, pp. 157–168. Birkhäuser, Basel.

Ordentlich, E. and Weinberger, M.J. (2005). A distribution dependent refinement of Pinsker's inequality. *IEEE Transactions on Information Theory*, **51**, 1836–1840.

Paley, R.E. and Zygmund, A. (1930). On some series of functions. *Proceedings of the Cambridge Philosophical Society*, **26**, 337–357.

Palmer, E.M. (1985). *Graphical Evolution*. John Wiley & Sons, New York.

Panchenko, D. (2001). A note on Talagrand's concentration inequality. *Electronic Communications in Probability*, **6**, 55–65.

Panchenko, D. (2003). Symmetrization approach to concentration inequalities for empirical processes. *The Annals of Probability*, **31**, 2068–2081.

Philips, T.K. and Nelson, R. (1995). The moment bound is tighter than Chernoff's bound for positive tail probabilities. *The American Statistician*, **49**, 175–178.

Pinelis, I. (1994). Optimum bounds for the distributions of martingales in Banach spaces. *The Annals of Probability*, **22**, 1679–1706.

Pinelis, I. (1995). Optimum bounds on moments of sums of independent random vectors. *Siberian Advances in Mathematics*, **5**, 141–150.

Pinsker, M.S. (1964). *Information and Information Stability of Random Variables and Processes*. Holden-Day, New York.

Pisier, G. (1983). Some applications of the metric entropy condition to harmonic analysis. In *Banach Spaces, Harmonic Analysis and Probability*, Volume 995 of *Lecture Notes in Mathematics*, pp. 123–159. University of Connecticut 1980–81. Springer, Berlin, Heidelberg.

Piterbarg, V.I. (1982). Gaussian random processes. In *Probability theory. Mathematical statistics. Theoretical cybernetics*, Volume 19, pp. 155–199. Akad. Nauk SSSR.

Politis, D.N., Romano, J.P., and Wolf, M. (1999). *Subsampling*. Springer, New York.

Pollard, D. (1982). A central limit theorem for k-means clustering. *The Annals of Probability*, **10**, 919–926.

Pollard, D. (1984). *Convergence of Stochastic Processes*. Springer-Verlag, New York.

Pollard, D. (1990). *Empirical Processes: Theory and Applications*. NSF-CBMS Regional Conference Series in Probability and Statistics, Institute of Mathematical Statistics, Hayward, CA.

Pollard, D. (2007). A note on Talagrand's convex hull concentration inequality. In *Asymptotics: particles, processes and inverse problems*, Volume 55 of *IMS Lecture Notes Monograph Series*, pp. 196–203. IMS, Beachwood, OH.

Prékopa, A. (1971). Logarithmic concave measures with application to stochastic programming. *Acta Universitatis Szegediensis. Acta Scientiarum Mathematicarum*, **32**, 1.

Prékopa, A. (1973). On logarithmic concave measures and functions. *Acta Universitatis Szegediensis. Acta Scientiarum Mathematicarum*, **34**, 335–343.

Quenouille, M. (1949). Approximate test of correlation in time series. *Journal of the Royal Statistical Society, Series B*, **11**, 68–84.

Raab, M. and Steger, A. (1998). "Balls into bins"—a simple and tight analysis. *Randomization and Approximation Techniques in Computer Science*, **1518**, 159–170.

Rachev, S.T. (1991). *Probability Metrics and the Stability of Stochastic Models*. Wiley Series in Probability and Mathematical Statistics. Wiley, New York.

Reynaud-Bouret, P. (2003). Adaptive estimation of the intensity of inhomogeneous Poisson processes via concentration inequalities. *Probability Theory and Related Fields*, **126**, 103–153.

Rhee, W. (1993). A matching problem and subadditive Euclidean functionals. *The Annals of Applied Probability*, **3**, 794–801.

Rhee, W.T. and Talagrand, M. (1986). Martingale inequalities and the jackknife estimate of variance. *Statistics and Probability Letters*, **4**, 5–6.

Rhee, W.T. and Talagrand, M. (1987). Martingales, inequalities, and NP-complete problems. *Mathematics of Operations Research*, **12**, 177–181.

Richardson, T.J. and Urbanke, R.L. (2008). *Modern Coding Theory*. Cambridge University Press, Cambridge.

Rio, E. (1993). Covariance inequalities for strongly mixing processes. *Annales de l'Institut Henri Poincare (B) Probability and Statistics*, **29**, 587–597.

Rio, E. (2000). Inégalités de Hoeffding pour les fonctions lipschitziennes de suites dépendantes. *Comptes Rendus de l'Académie des Sciences-Series I-Mathématiques*, **330**, 905–908.

Rio, E. (2001). Inégalités de concentration pour les processus empiriques de classes de parties. *Probability Theory and Related Fields*, **119**, 163–175.

Rio, E. (2002). Une inégalité de bennett pour les maxima de processus empiriques. *Annales de l'institut Henri Poincaré (B) Probabilités et Statistiques*, **38**, 1053–1057.

Rio, E. (2012). Sur les fonctions de taux dans les inégalités de Talagrand pour les processus empiriques. *Comptes Rendus de l'Académie des Sciences de Paris Série I Mathématiques*, **350**, 303–305.

Rosenthal, H.P. (1970). On the subspace of l_p $(p > 2)$ spanned by the sequences of independent random variables. *Israel Journal of Mathematics*, **8**, 273–303.

Rossignol, R. (2006). Threshold for monotone symmetric properties through a logarithmic Sobolev inequality. *The Annals of Probability*, **34**, 1707–1725.

Rudelson, M. (1999). Random vectors in isotropic position. *Journal of Functional Analysis*, **164**, 60–72.

Rudelson, M. and Vershynin, R. (2010). *Non-Asymptotic Theory of Random Matrices: Extreme Singular Values*, pp. 1576–1602. Hindustan Book Agency, New Delhi.

Russo, L. (1982). An approximate zero-one law. *Zeitschrift für Wahrscheinlichkeitstheorie und verwandte Gebiete*, **61**, 129–139.

Saloff-Coste, L. (1997). Lectures on finite Markov chains. In *Lectures on Probability Theory and Statistics*, pp. 301–413. Springer, New York.

Samson, P.-M. (2000). Concentration of measure inequalities for Markov chains and ϕ-mixing processes. *The Annals of Probability*, **28**, 416–461.

Samson, P.-M. (2003). Concentration inequalities for convex functions on product spaces. In *Stochastic Inequalities and Applications* (ed. E. Giné and C. Houdré), Volume 56. Birkhäuser, Basel.

Samson, P.-M. (2007). Infimum-convolution description of concentration properties of product probability measures, with applications. *Annales de l'institut Henri Poincaré (B) Probabilités et Statistiques*, **43**, 321–338.

Sauer, N. (1972). On the density of families of sets. *Journal of Combinatorial Theory, Series A*, **13**, 145–147.

Schechtman, G. (2003). Concentration results and applications. In *Handbook of the Geometry of Banach Spaces*, Volume 2, pp. 1603–1634. Elsevier, Amsterdam.

Scheffé, H. (1947). A useful convergence theorem for probability distributions. *Annals of Mathematical Statistics*, **18**, 434–458.

Schmidt, E. (1948). Die Brunn-Minkowskische Ungleichung und ihr Spiegelbild sowie die isoperimetrische Eigenschaft der Kugel in der euklidischen und nichteuklidischen Geometrie. *Mathematische Nachrichten*, **1**, 81–115.

Schneider, R. (1993). *Convex Bodies: The Brunn–Minkowski Theory*. Cambridge University Press, Cambridge.

Schudy, W. and Sviridenko, M. (2012). Concentration and moment inequalities for polynomials of independent random variables. In *Proceedings of the Twenty-Third Annual ACM-SIAM Symposium on Discrete Algorithms*, pp. 437–446.

Shamir, E. and Spencer, J. (1987). Sharp concentration of the chromatic number on random graphs $G_{n,p}$. *Combinatorica*, 7, 374–384.

Shannon, C.E. (1948). A mathematical theory of communication. *Bell System Technical Journal*, **27**, 379–423.

Shawe-Taylor, J. and Cristianini, N. (2004). *Kernel Methods for Pattern Analysis*. Cambridge University Press, Cambridge.

Shorack, G.R. and Wellner, J.A. (1986). *Empirical Processes with Applications in Statistics*. Wiley, New York.

Sion, M. (1958). On general minimax theorems. *Pacific Journal of Mathematics*, **8**, 171–176.

Slepian, D. (1962). The one-sided barrier problem for Gaussian noise. *Bell System Technical Journal*, **41**(2), 463–501.

Slutsky, E. (1937). Alcuni proposizioni sulla teoria degli funzioni aleatorie. *Giornale dell'Istituto Italiano degli Attuari*, **8**, 183–199.

Soshnikov, A. (1999). Universality at the edge of the spectrum in Wigner random matrices. *Communications in Mathematical Physics*, **207**, 697–733.

Steele, J.M. (1982). Long common subsequences and the proximity of two random strings. *SIAM Journal of Applied Mathematics*, **42**, 731–737.

Steele, J.M. (1986). An Efron–Stein inequality for nonsymmetric statistics. *The Annals of Statistics*, **14**, 753–758.

Steele, J.M. (1996). *Probability Theory and Combinatorial Optimization*. SIAM, CBMS-NSF Regional Conference Series in Applied Mathematics 69, 3600 University City Science Center, Philadelphia, PA 19104.

Steele, J.M. (2004). *The Cauchy–Schwarz Master Class: An Introduction to the Art of Mathematical Inequalities*. Cambridge University Press and the Mathematical Association of America, Cambridge, UK and Washington DC.

Stein, E.M. and Shakarchi, R. (2005). *Real Analysis*. Princeton Lectures in Analysis III. Princeton University Press, New Jersey.

Steinwart, I. and Christmann, A. (2008). *Support Vector Machines*. Information Science and Statistics. Springer, New York.

Strassen, V. (1965). The existence of probability measures with given marginals. *Annals of Mathematical Statistics*, **36**, 423–439.

Sudakov, V.N. (1969). Gaussian measures, Cauchy measures and ε-entropy. In *Soviet Math. Dokl*, **10**, pp. 310–313.

Szarek, S.J. (1976). On the best constants in the Khintchine inequality. *Studia Mathematica*, **63**, 197–208.

Talagrand, M. (1993). Isoperimetry, logarithmic Sobolev inequalities on the discrete cube and Margulis' graph connectivity theorem. *Geometric and Functional Analysis*, **3**, 296–314.

Talagrand, M. (1994*a*). On Russo's approximate zero-one law. *The Annals of Probability*, **22**, 1576–1587.

Talagrand, M. (1994*b*). Sharper bounds for Gaussian and empirical processes. *The Annals of Probability*, **22**, 28–76.

Talagrand, M. (1995). Concentration of measure and isoperimetric inequalities in product spaces. *Publications Mathématiques de l'I.H.E.S.*, **81**, 73–205.

Talagrand, M. (1996*a*). Majorizing measures: the generic chaining. *The Annals of Probability*, **24**, 1049–1103. (Special Invited Paper.)

Talagrand, M. (1996*b*). New concentration inequalities in product spaces. *Inventiones Mathematicae*, **126**, 505–563.

Talagrand, M. (1996*c*). A new look at independence. *The Annals of Probability*, **24**, 1–34. (Special Invited Paper.)

Talagrand, M. (1996*d*). Transportation cost for Gaussian and other product measures. *Geometric and Functional Analysis*, **6**, 587–600.

Talagrand, M. (1997). On boundaries and influences. *Combinatorica*, **17**, 275–285.

Talagrand, M. (1999). On concentration and influences. *Israel Journal of Mathematics*, **111**, 275–284.

Talagrand, M. (2005). *The Generic Chaining*. Springer, New York.

Tao, T. (2012). *Topics in Random Matrix Theory*, Volume 132 of *Graduate Studies in Mathematics*. American Mathematics Society, New York.

Tibshirani, R. (1996). Regression shrinkage and selection via the lasso. *Journal of the Royal Statistical Society, Series B.*, **58**, 267–288.

Tillich, J.-P. and Zémor, G. (2001). Discrete isoperimetric inequalities and the probability of a decoding error. *Combinatorics, Probability and Computing*, **9**(05), 465–479.

Tracy, C.A. and Widom, H. (1994). Level-spacing distributions and the Airy kernel. *Communications in Mathematical Physics*, **159**, 151–174.

Tracy, C.A. and Widom, H. (2001). On the distributions of the lengths of the longest monotone subsequences in random words. *Probability Theory and Related Fields*, **119**, 350–380.

Tropp, J.A. (2010*a*). User-friendly bounds for sums of random matrices. *Foundations of Computational Mathematics*, **11**, 1–46.

Tropp, J.A. (2010*b*). User-friendly tail bounds for matrix martingales. Preprint available at http://arxiv. org/abs/1004.4389.

Tsirelson, B.S., Ibragimov, I.A., and Sudakov, V.N. (1976). Norm of Gaussian sample function. In *Proceedings of the 3rd Japan-U.S.S.R. Symposium on Probability Theory*, Volume 550 of *Lecture Notes in Mathematics*, pp. 20–41. Springer-Verlag, Berlin.

Tukey, J.W. (1958). Bias and confidence in not quite large samples. *Annals of Mathematical Statistics*, **29**, 614.

Uspensky, J.V. (1937). *Introduction to Mathematical Probability*. McGraw-Hill, New York.

Valiant, L.G. (1982). A scheme for fast parallel communication. *SIAM Journal on Computing*, **11**(2), 350–361.

Valiant, L.G. and Brebner, G.J. (1981). Universal schemes for parallel communication. In *Proceedings of the Thirteenth Annual ACM Symposium on Theory of Computation, STOC*, pp. 263–277. ACM, New York.

van de Geer, S. (2000). *Applications of Empirical Process Theory*. Cambridge University Press, Cambridge.

van de Geer, S. (2008). High-dimensional generalized linear models and the LASSO. *The Annals of Statistics*, **36**, 614–645.

van den Berg, J. (2008). Approximate zero-one laws and sharpness of the percolation transition in a class of models including two-dimensional Ising percolation. *The Annals of Probability*, **36**, 1880–1903.

van der Vaart, A.W. (1998). *Asymptotic Statistics*. Cambridge University Press, New York.

van der Vaart, A.W. and Wellner, J.A. (1996). *Weak Convergence and Empirical Processes*. Springer-Verlag, New York.

Vapnik, V.N. (1982). *Estimation of Dependencies Based on Empirical Data*. Springer-Verlag, New York.

Vapnik, V.N. (1995). *The Nature of Statistical Learning Theory*. Springer-Verlag, New York.

Vapnik, V.N. (1998). *Statistical Learning Theory*. John Wiley, New York.

Vapnik, V.N. and Chervonenkis, A.Ya. (1971). On the uniform convergence of relative frequencies of events to their probabilities. *Theory of Probability and its Applications*, **16**, 264–280.

Vapnik, V.N. and Chervonenkis, A.Ya. (1974). *Theory of Pattern Recognition*. Nauka, Moscow (in Russian); German translation: *Theorie der Zeichenerkennung*, Akademie Verlag, Berlin, 1979.

Vapnik, V.N. and Chervonenkis, A.Ya. (1981). Necessary and sufficient conditions for the uniform convergence of means to their expectations. *Theory of Probability and its Applications*, **26**, 821–832.

Varadhan, S.R.S (1984). *Large Deviations and Applications*. Springer, New York.

Vempala, S.S. (2004). *The Random Projection Method*. DIMACS Series in Discrete Mathematics and Theoretical Computer Science, 65. American Mathematical Society, Providence, RI.

Vershynin, R. (2012). Introduction to non-asymptotic analysis of random matrices. In *Compressed Sensing, Theory and Applications*, pp. 210–268. Cambridge University Press, Cambridge.

Verzelen, N. (2010). Adaptive estimation of stationary Gaussian fields. *The Annals of Statistics*, **38**, 1363–1402.

Vu, V. (2000). On the concentration of multivariate polynomials with small expectation. *Random Structures & Algorithms*, **16**, 344–363.

Vu, V. (2001). A large deviation result on the number of small subgraphs of a random graph. *Combinatorics, Probability, and Computing*, **10**, 79–94.

Wästlund, J. (2005). Evaluation of Janson's constant for the variance in the random minimum spanning tree problem. Linköping Studies in Mathematics, No. 7.

Wigner, E.P. (1955). Characteristic vectors of bordered matrices with infinite dimensions. *Annals of Mathematics*, **62**, 548–564.

Wishart, J. (1928). The generalized product moment distribution in samples from a normal multivariate distribution. *Biometrika*, **20A**, 32–52.

Wu, L. (2000). A new modified logarithmic Sobolev inequality for Poisson point processes and several applications. *Probability Theory and Related Fields*, **118**, 427–438.

Yurinskii, V.V. (1976). Exponential inequalities for sums of random vectors. *Journal of Multivariate Analysis*, **6**, 472–499.

Yurinsky, V.V. (1995). *Sums and Gaussian Vectors*. Springer, New York.

Zwald, L. and Blanchard, G. (2006). On the convergence of eigenspaces in kernel principal component analysis. In *Neural Information Processing Systems, NIPS 2005, Canada* (ed. Y. Weiss, L. Bottou, and J. Platt), Volume 18, pp. 1649–1656. MIT Press, Cambridge, MA.

Zygmund, A. (1959). *Trigonometric Series I*. Cambridge University Press, Cambridge.

AUTHOR INDEX

SUBJECT INDEX

Printed and bound by CPI Group (UK) Ltd, Croydon, CR0 4YY